Undergraduate Lecture Notes in Physics

Undergraduate Lecture Notes in Physics (ULNP) publishes authoritative texts covering topics throughout pure and applied physics. Each title in the series is suitable as a basis for undergraduate instruction, typically containing practice problems, worked examples, chapter summaries, and suggestions for further reading.

ULNP titles must provide at least one of the following:

- An exceptionally clear and concise treatment of a standard undergraduate subject.
- A solid undergraduate-level introduction to a graduate, advanced, or non-standard subject.
- A novel perspective or an unusual approach to teaching a subject.

ULNP especially encourages new, original, and idiosyncratic approaches to physics teaching at the undergraduate level.

The purpose of ULNP is to provide intriguing, absorbing books that will continue to be the reader's preferred reference throughout their academic career.

More information about this series at http://www.springer.com/series/8917

Paul J. Nahin

Inside Interesting Integrals

(With an Introduction to Contour Integration)

A Collection of Sneaky Tricks, Sly Substitutions, and Numerous Other Stupendously Clever, Awesomely Wicked, and Devilishly Seductive Maneuvers for Computing Hundreds of Perplexing Definite Integrals From Physics, Engineering, and Mathematics (Plus Numerous Challenge Problems with Complete, Detailed Solutions)

Second Edition

 Springer

Paul J. Nahin
Department of Electrical Engineering
University of New Hampshire
Durham, NH, USA

ISSN 2192-4791   ISSN 2192-4805   (electronic)
Undergraduate Lecture Notes in Physics
ISBN 978-3-030-43787-9   ISBN 978-3-030-43788-6   (eBook)
https://doi.org/10.1007/978-3-030-43788-6

This Springer imprint is published by the registered company Springer Nature Switzerland AG
The registered company address is: Gewerbestrasse 11, 6330 Cham, Switzerland

Bernhard Riemann (1826–1866), the German mathematical genius whose integral is the subject of this book

AIP Emilio Segrè Visual Archives, T. J. J. See Collection

This book is dedicated to all who, when they read the following line from John le Carré's 1989 Cold War spy novel The Russia House, immediately know they have encountered a most interesting character:

"Even when he didn't follow what he was looking at, he could relish a good page of mathematics all day long."

as well as to all who understand how frustrating is the lament in Anthony Zee's book Quantum Field Theory in a Nutshell:

"Ah, if we could only do the integral .... But we can't."

and to all (as they open this book of examples) who smile with anticipation
at the now-famous words of UC/Berkeley applied mathematician Beresford Parlett,

"Only wimps do the general case. True teachers tackle examples."

Finally, for those readers who appreciate the fact that every new encounter with an unknown integral is akin to engaging in mortal combat (but without, it is to be hoped, bloodshed—or, at the worst, not too much of it), consider these words from the world's most highly skilled assassin about what it takes to be really good at the task at hand (who was not speaking of doing integrals, but the sentiment still applies):

"One must be both a mathematician and a poet. As though poetry were a science, or mathematics an art."

—from Trevanian's 1979 novel Shibumi (a Japanese reference to a state of effortless perfection)

Preface to the New Edition

"I want to point out that learning the tricks of calculus is quite important because with a few tricks you can solve almost any solvable problem. *But you will need the help of a number of key integrals from which all the others are derivable* [my emphasis]."

—Caltech professor of physics Richard Feynman, lecturing to scientists and engineers at the Hughes Research Laboratories in Malibu, California[1]

Without (I hope) appearing to be too puffed-up, let me say right off-the-bat that the first printing of this book was generally well-received. I am very grateful to all those readers who took the time to write to me to express their pleasure at opening a math book written to be read with a twinkle in the eye, rather than as a needle in the rear-end.

One of the nightmares of the writer of a math/physics book is the receipt of a communication from a reader pointing out that the book contains incredibly stupid errors that reveal the author to be a raging ignoramus. In a math book, virtually every symbol has a crucial role to play, as opposed to works of fiction where individual typos are almost always either missed or, if spotted, are dismissed with a chuckle. The readers of *math* books, however, are not so easily amused. I became an even bigger fan of the brilliant English mathematician G. H. Hardy than I had already been when, after reading that even he had suffered this unhappy experience, I learned how he had handled it. When a talented student once pointed out to Hardy some

---

[1]Feynman (1918–1988), who received the 1965 Nobel Prize in physics, gave a series of weekly talks at Hughes during the years 1966–1971. (Those were the very years I was a Member of the Technical Staff at Hughes but, alas, not in Malibu. This was my *third* "near miss" at meeting Feynman; I will tell you of the first two in just a bit.) The talks covered a very broad range of topics—all delivered with Feynman's signature style of "tough guy charm"—and the quotation is from a lecture on integration methods. Feynman's lectures were widely attended, and one of those who regularly listened was John Neer who took almost verbatim notes. Neer has posted pdf scans of those notes on the Web, and they make fascinating reading. Less well-known are the notes made by the physicist James Keck (1924–2010) from the course "Mathematical Methods" that Feynman gave at Cornell University during the Fall of 1946 and the Spring of 1947. Keck's notes are also on the Web.

errors in equations appearing in one of his books, the great man had simply slipped his glasses down to the end of his nose, peered briefly at the printing goofs, and then dismissed them with the words "Yes, you're right, those are mistakes." And that was the end of *that!*[2]

With that said, however, I can hardly avoid admitting that the first edition of this book was not perfect. I am sure this new, corrected, and expanded edition is not either, but things have been greatly improved because numerous correspondents have been tremendously generous in sharing with me their discoveries of misprints, typos, and other exasperating screw-ups that I should have caught when reading page proofs (but did not). All the corrections brought to my attention have been addressed, and I am particularly in the debt of Daniel Zwillinger, author of *The Handbook of Integration*, and Editor-in-Chief of the *CRC Standard Mathematical Tables and Formulae* (both published by CRC Press). Also, a detailed, multipage errata from Bruce Roth in Perth, Scotland was extremely helpful. Bruce, who is a professional proofreader of mathematical works, demonstrated just how good he is at his job. Finally, a long, beautifully handwritten letter (in ink!) from Robert Burkel, professor emeritus of mathematics at Kansas State University, and a similarly detailed e-mail from David Short were equally helpful.

Okay, on now to new business. While writing this book, it was never far from my thoughts that I was attempting a borderline audacious task. Would it *really* be possible, I wondered, as I lay sleepless in bed at night, to produce a book of reasonable size on the truly vast topic of definite integrals without overlooking *something* of an elementary yet central nature? The word *elementary* is important because, clearly, I have left out a whole lot of complicated stuff! You will not find much here, for example, at the level of Victor Moll's terrific books on the often fantastic entries in "Gradshteyn and Ryzhik."[3] Moll's two books are herculean efforts at explaining where many of the mind-boggling entries in that incredible collection of integrals (famous among physicists, engineers, and mathematicians), named after its Soviet authors, come from—many have been deep mysteries since their first appearance in 1943. *My* concerns were not nearly so grand, however.

---

[2]Since writing those words I have, after reading the 2019 novel *Game of Snipers* by Pulitzer Prize winner Stephen Hunter, somewhat modified my views on the "benign" nature of typos in works of fiction. In an endnote to the novel, Hunter reveals that in one of his earlier works, he had been energetically dragged over the coals by readers enraged by a seemingly harmless printing goof. As Hunter so entertainingly writes, "For some reason, this mistake seemed to trigger certain . . . reviewers into psychotic episodes. Folks, calm down, have a drink, hug someone soft. It'll be all right." I cannot imagine Hardy reading a Stephen Hunter novel (there is math in them but, as the ballistics of high-velocity rifle bullets fired to extreme distances, it is all *applied* math for which Hardy had a well-known dislike), but I bet he would agree (if he were still among us) with the spirit of Hunter's suggestions.

[3]Moll is Professor of Mathematics at Tulane University; as mentioned in the original Preface, the terrific book he and the late George Boros wrote in 2004 (*Irresistible Integrals*) was my inspiration for this book. Moll's new books are *Special Integrals in Gradshteyn and Ryzhik: The Proofs*, volume 1 (2014) and volume 2 (2015), both published by CRC Press. Every serious student of mathematical physics should read them.

No, my big worry was that I would be buried under an avalanche of e-mails from outraged readers opening with "You *dolt*, how could you possibly have left out [fill-in with your favorite definite integral]?" So, after this book was published in August 2014, I have ever since been cautiously opening texts and reading journal papers with trepidation, always with the dread that I would turn a page and suddenly be faced with a definite integral with these words stamped in big, bold, red words across it: "How *could* you have overlooked *me*?" Even with the appearance of this corrected printing, this could of course still happen.

Now, mere simple variations on an entry in the book would not be in that unhappy category. If I did happen across a new (to me) integral, I would not lose any sleep over it if I could easily derive it from one of the integrals already in the book. Let me give you four examples of this.

Consider first

$$\int_0^1 \frac{\ln(x)}{1+x}\,dx,$$

which is not in the original edition. We start by writing

$$-\int_0^1 \frac{x\,\ln(x)}{1-x^2}\,dx = \frac{1}{2}\int_0^1 \frac{\ln(x)}{1+x}\,dx - \frac{1}{2}\int_0^1 \frac{\ln(x)}{1-x}\,dx,$$

and what makes this interesting is that the integral on the left appears early in the derivation of (5.2.4), which says

$$\int_0^{\pi/2} \cot(x)\ln\{\sec(x)\}\,dx = \frac{\pi^2}{24} - \int_0^1 \frac{x\,\ln(x)}{1-x^2}\,dx.$$

That is,

$$\frac{1}{2}\int_0^1 \frac{\ln(x)}{1+x}\,dx - \frac{1}{2}\int_0^1 \frac{\ln(x)}{1-x}\,dx = \frac{\pi^2}{24}.$$

Integrating the second integral on the left by-parts, that is, in

$$\int_0^1 u\,dv = (uv)\Big|_0^1 - \int_0^1 v\,du$$

let $u = \ln(x)$ and $dv = \frac{dx}{1-x}$, and so $du = \frac{dx}{x}$ and $v = -\ln(1-x)$. This gives us

$$\int_0^1 \frac{\ln(x)}{1-x}\,dx = -\ln(x)\ln(1-x)\Big|_0^1 + \int_0^1 \frac{\ln(1-x)}{x}\,dx$$

or, since

$$\ln(x)\ln(1-x)\big|_0^1 = 0$$

because the expression tends to the same value (whatever it is) as $x \to 0+$ and $x \to 1-$, then

$$\int_0^1 \frac{\ln(x)}{1-x}\,dx = \int_0^1 \frac{\ln(1-x)}{x}\,dx = -\frac{\pi^2}{6},$$

where the right-most integral is evaluated in (5.2.2). So,

$$\frac{1}{2}\int_0^1 \frac{\ln(x)}{1+x}\,dx - \frac{1}{2}\left(-\frac{\pi^2}{6}\right) = \frac{\pi^2}{24}$$

or, at last,

$$\int_0^1 \frac{\ln(x)}{1+x}\,dx = -\frac{\pi^2}{12}.$$

This is equal to -0.8225 and the MATLAB numerical integration command *integral* (see note 10) agrees: *integral(@(x)log(x)./(1+x),0,1) = −0.8225*.

A little aside: I say a lot more, in the original Preface, about this book's use of computers (MATLAB) to "confirm" our theoretical calculations and about how pure mathematicians may object to that. Now, please understand, I am definitely not arguing that there is something "wrong" about invoking theorems, lemmas, and other sorts of highly technical, heavy mathematical weaponry to justify a step. But for what we are doing *here*, I claim that going to that level is often a bit like using a flamethrower to cook a hamburger. Feynman often wrote of *his* feeling about the appearance of overly abstract mathematics in physics, essentially calling it "carrying rigor to the point of rigor mortis' and, if there is anything in mathematics 'like doing physics,' it is doing definite integrals."[4]

A joke I liked to tell my students nicely illustrates this difference in thinking. A practical engineer and an abstract mathematician are observing what they think is an empty house, when they see two people go in and then, soon after, three people exit. The engineer concludes that their initial belief must have been in error. The mathematician, on the other hand, states that if now one person enters the house, it will again be empty.

There are, of course, similar jokes that reverse the roles and, in the interest of symmetry (thereby allowing me to offend just about *everybody* reading this book),

---

[4]See, for example, Feynman's *Lectures on Gravitation,* based on presentations he gave at Caltech during the 1962–1963 academic year, the very year I was there doing my MSEE (presentations, to my everlasting regret, I did not attend—my *second* near miss at meeting Feynman), where you will find some quite pointed comments on his unhappy view of excessive mathematical abstraction.

here is an example of that. Our engineer and mathematician are listening to a theoretical physicist lecture on the super-dimensional string theory of quantum gravity, which the mathematician follows with ease but which leaves the engineer in a daze. "Holy Cow," exclaims the engineer, "How can you visualize a space with eleven dimensions?" His friend replies, with a knowing chuckle, "It's easy—just imagine n-dimensional space and then set n equal to eleven."

If you at least smiled at one (or even better, both) of these little tales, then this *is* a book for you.

Okay, let us continue with my second example. Shortly after this book appeared, I had an occasion to check out of my university's math library a copy of Reinhold Remmert's elegant 1989 Springer book (published in English in 1991), *Theory of Complex Functions.* As I have mentioned, I have made it a habit to scan every math and physics book that comes into my hands for any discussion of definite integrals and to be on the lookout for entries that appear "interesting." Well, on page 408 of Remmert I found *two*. In his chapter on integration in the complex plane (discussed in this book in Chap. 8), he asks readers to use the residue calculus to show

$$\int_0^\infty \frac{\sqrt{x}}{x^2 + a^2} \, dx = \frac{\pi}{\sqrt{2a}}, \quad a > 0 \tag{A}$$

and

$$\int_{-\infty}^\infty \frac{\cos(x)}{(x^2 + a^2)^2} \, dx = \frac{\pi(1 + a)}{2a^3 e^a}, \quad a > 0. \tag{B}$$

The first one, (A), I did not recall doing, but I almost immediately suspected that the change of variable $x = at^2$ could lead to something interesting. Indeed, with that change

$$\frac{dx}{dt} = 2at$$

and so

$$dx = 2at \, dt.$$

This results in

$$\int_0^\infty \frac{\sqrt{x}}{x^2 + a^2} \, dx = \int_0^\infty \frac{\sqrt{at}}{a^2 t^4 + a^2} 2at \, dt = \frac{2a\sqrt{a}}{a^2} \int_0^\infty \frac{t^2}{t^4 + 1} \, dt = \frac{2}{\sqrt{a}} \int_0^\infty \frac{t^2}{t^4 + 1} \, dt.$$

Now, in (2.3.4) it is shown by *elementary means* (no complex integration) that

$$\int_0^\infty \frac{t^2}{t^4 + 1} dt = \frac{\pi\sqrt{2}}{4}$$

and so, just like that, we immediately have

$$\int_0^\infty \frac{\sqrt{x}}{x^2 + a^2} dx = \left(\frac{2}{\sqrt{a}}\right)\frac{\pi\sqrt{2}}{4} = \frac{\pi}{\sqrt{2a}}$$

as was to be shown. So, BIG sigh of relief, and even a smile, because I did not need to use residues at all. Freshman calculus is all that is needed.

What about (B)? Well, as it turns out I have that in this book's Chap. 8, done by complex integration, and in fact you will find a more general form of (B) in Challenge Problem 8.3(d) and its solution. But now, with the success of (A) puffing me up, I wondered if I could do (B), too, with just elementary means? And, sure enough, with (3.1.7) and using what in the book I call "Feynman's favorite trick"—differentiating under the integral sign—I *could*. From (3.1.7), elementary analysis gives us

$$\int_0^\infty \frac{\cos(ax)}{x^2 + b^2} dx = \frac{\pi}{2b} e^{-ab}$$

and so, trivially, we have

$$\int_{-\infty}^\infty \frac{\cos(ax)}{x^2 + b^2} dx = \frac{\pi}{b} e^{-ab}.$$

Then, differentiating with respect to the parameter b,

$$\int_{-\infty}^\infty \frac{-\cos(ax)2b}{(x^2 + b^2)^2} dx = \pi\left[\frac{-abe^{-ab} - e^{-ab}}{b^2}\right]$$

and so

$$\int_{-\infty}^\infty \frac{\cos(ax)}{(x^2 + b^2)^2} dx = \pi\left[\frac{abe^{-ab} + e^{-ab}}{2b^3}\right] = \frac{\pi(ab + 1)e^{-ab}}{2b^3} = \frac{\pi(ab + 1)}{2b^3 e^{ab}}.$$

Or, as Remmert's a is our b, in our last result let us swap a and b to get

$$\int_{-\infty}^\infty \frac{\cos(bx)}{(x^2 + a^2)^2} dx = \frac{\pi(ab + 1)}{2a^3 e^{ab}},$$

which is more general than the original (B) (set $b = 1$ to answer Remmert's problem).

As another example of establishing an interesting definite integral that, while not specifically in this book, is in fact a "close cousin" to one that is, consider the "interesting" claim

$$\int_0^\infty \frac{a\sin(x) - \sin(ax)}{x^2} \, dx = a\ln|a|,$$

which is obviously true for $a = \pm 1$. And for $a = \pi$, for example, this formula says

$$\int_0^\infty \frac{\pi\sin(x) - \sin(\pi x)}{x^2} \, dx = \pi\ln|\pi| = 3.59627\ldots$$

and the MATLAB command *integral* agrees:

$$integral\left(@(x)(pi * \sin(x) - \sin(pi * x))./(x.^2), 0, \inf\right) = 3.59628\ldots.$$

We can derive this integral as follows, using integration-by-parts. Starting with

$$\int_0^\infty u \, dv = (uv)\Big|_0^\infty - \int_0^\infty v \, du$$

write $u = a\sin(x) - \sin(ax)$ and $dv = \frac{dx}{x^2}$, and so $du = [a\cos(x) - a\cos(ax)]dx$ and $v = -\frac{1}{x}$. Thus,

$$\int_0^\infty \frac{a\sin(x) - \sin(ax)}{x^2} \, dx = -\left(\frac{a\sin(x)}{x} - \frac{\sin(ax)}{x}\right)\Big|_0^\infty$$
$$+ \int_0^\infty \frac{a\cos(x) - a\cos(ax)}{x} \, dx$$
$$= a\int_0^\infty \frac{\cos(x) - \cos(ax)}{x} \, dx.$$

Then, recalling (3.4.6), which says

$$\int_0^\infty \frac{\cos(qx) - \cos(px)}{x} \, dx = \ln\left(\frac{p}{q}\right),$$

and setting $q = 1$ and $p = a$, we immediately have our result

$$\int_0^\infty \frac{a\sin(x) - \sin(ax)}{x^2} \, dx = a\ln|a|,$$

where we use |a| for the log argument because the value of the integral

$$\int_0^\infty \frac{\cos(x) - \cos(ax)}{x}\,dx$$

clearly does *not* depend on the sign of a. Indeed, this last observation can be applied to (3.4.6), itself, which means I could have more generally have written it as

$$\int_0^\infty \frac{\cos(qx) - \cos(px)}{x}\,dx = \ln\left|\frac{p}{q}\right|.$$

A pleasant surprise for me was how readers took some of the solutions I give in the book as motivation to find *better* solutions. For example, I mention in Challenge Problem C2.1 that it took me five hours to show that

$$I = \int_0^4 \frac{\ln(x)}{\sqrt{4x - x^2}}\,dx = 0.$$

I do recall *briefly* thinking I was perhaps asking for trouble with my admission of sluggishness but, alas, that very character defect rendered me too lazy to delete the comment. So, sure enough, one reader promptly sent me the following response. Let $x = t^2$ and so $dx = 2t\,dt$. Then,

$$I = \int_0^2 \frac{\ln(t^2)}{\sqrt{4t^2 - t^4}}\,2t\,dt = \int_0^2 \frac{2\ln(t)}{t\sqrt{4 - t^2}}\,2t\,dt = 4\int_0^2 \frac{\ln(t)}{\sqrt{4 - t^2}}\,dt.$$

Next, let $t = 2\sin(u)$ and so $dt = 2\cos(u)du$. Then

$$I = 4\int_0^{\pi/2} \frac{\ln[2\sin(u)]}{\sqrt{4 - 4\sin^2(u)}}\,2\cos(u)du = 4\int_0^{\pi/2} \frac{\ln[2\sin(u)]}{2\cos(u)}\,2\cos(u)du$$

$$= 4\int_0^{\pi/2} \ln[2\sin(u)]du = 0,$$

as we can see if we set $a = 2$ in (2.4.1).

This *is* a nice alternative derivation to the one I give in the back of the book, but here is another reader's solution that I received some time later that is even better. Write the integral as the sum of two parts:

$$I = \int_0^4 \frac{\ln(x)}{\sqrt{4x - x^2}}\,dx = \int_0^2 \frac{\ln(x)}{\sqrt{4x - x^2}}\,dx + \int_2^4 \frac{\ln(x)}{\sqrt{4x - x^2}}\,dx.$$

In the last integral, make the change of variable $y = 4 - x$ (and so $dx = -dy$ and $x = 4 - y$). Therefore,

$$I = \int_0^2 \frac{\ln(x)}{\sqrt{4x - x^2}} dx + \int_2^0 \frac{\ln(4-y)}{\sqrt{16 - 4y - (16 - 8y + y^2)}} (-dy)$$

$$= \int_0^2 \frac{\ln(x)}{\sqrt{4x - x^2}} dx + \int_0^2 \frac{\ln(4-y)}{\sqrt{4y - y^2}} dy$$

$$= \int_0^2 \frac{\ln(x)}{\sqrt{4x - x^2}} dx + \int_0^2 \frac{\ln(4-x)}{\sqrt{4x - x^2}} dx$$

or, as $\ln(x) + \ln(4 - x) = \ln(4x - x^2)$,

$$I = \int_0^2 \frac{\ln(4x - x^2)}{\sqrt{4x - x^2}} dx.$$

Next, change variable to $u = 4x - x^2$ and so $\frac{du}{dx} = 4 - 2x$ or,

$$dx = \frac{du}{4 - 2x} = \frac{du}{2(2 - x)} = -\frac{du}{2(x - 2)}.$$

Now, $x^2 - 4x = -u$ and so $x^2 - 4x + 4 = 4 - u$ and so $(x - 2)^2 = 4 - u$ or, $x - 2 = \sqrt{4 - u}$ which means $dx = -\frac{du}{2\sqrt{4-u}}$. Thus,

$$I = \int_0^4 \frac{\ln(u)}{\sqrt{u}} \left(-\frac{du}{2\sqrt{4 - u}} \right) = -\frac{1}{2} \int_0^4 \frac{\ln(u)}{\sqrt{4u - u^2}} du = -\frac{1}{2} I.$$

This can be so only if $I = 0$, and we are done. This solution requires no use of any other integral, and that is why I think it the superior derivation.[5]

As another example of a reader going me one better, consider Challenge Problem C1.2, where I ask for a proof that $I = \int_1^\infty \frac{dx}{\sqrt{x^3 - 1}}$ exists by showing that 4 is an upper bound on the value of I. (I am clearly bounded from below by zero, as the integrand is never negative over the entire interval of integration.) We start by letting $e^y = x^3 - 1$. So, when $x = 1$ then $e^y = 0$ or, $y = -\infty$, and when $x = \infty$ then $y = +\infty$. Now,

$$3x^2 \frac{dx}{dy} = e^y$$

or,

$$dx = \frac{e^y}{3x^2} dy.$$

Since $x^3 = e^y + 1$, then $x = (e^y + 1)^{1/3}$. Therefore,

---

[5] This analysis is due to Arthemy Lugin, a Massachusetts high school student when he wrote to me in 2014.

$$I = \int_{-\infty}^{\infty} \frac{1}{e^{y/2}} \frac{e^y}{3(e^y+1)^{2/3}} dy = \frac{1}{3} \int_{-\infty}^{\infty} \frac{e^{y/2}}{(e^y+1)^{2/3}} dy$$

$$= \frac{1}{3}\int_{-\infty}^{0} \frac{e^{y/2}}{(e^y+1)^{2/3}} dy + \frac{1}{3}\int_{0}^{\infty} \frac{e^{y/2}}{(e^y+1)^{2/3}} dy < \frac{1}{3}\int_{-\infty}^{0} \frac{e^{y/2}}{1} dy + \frac{1}{3}\int_{0}^{\infty} \frac{e^{y/2}}{e^{(\frac{2}{3})y}} dy$$

where the denominator in each integral has been replaced by a *smaller* quantity (because $e^y + 1 > 1$ for $y < 0$ and $e^y + 1 > e^y$ for $y > 0$). So,

$$I < \frac{1}{3}\int_{-\infty}^{0} e^{y/2} dy + \frac{1}{3}\int_{0}^{\infty} e^{-y/6} dy = \frac{1}{3}\left[\left(2e^{y/2}\right)\Big|_{-\infty}^{0} - \left(6e^{-y/6}\right)\Big|_{0}^{\infty}\right]$$

$$= \frac{1}{3}[2 - (-6)] = \frac{8}{3} = 2.6666\ldots.$$

This is a *very good* upper bound (certainly better than my value of 4) because if we use MATLAB to actually calculate the integral (as is done at the end of Sect. 3.6) we get 2.4286. To simply show the *existence* of the integral, of course, *any* finite upper bound is just as good as any other.

To show you how really inventive one reader was in topping me, consider Challenge Problem C8.7. There I ask you to show a result attributed to the great Cauchy, himself:

$$\int_{0}^{\infty} \frac{e^{\cos(x)} \sin\{\sin(x)\}}{x} dx = \frac{\pi}{2}(e-1),$$

using contour integration, a result I introduce with words asserting this integral would "otherwise be pretty darn tough [to do]." Well, it is actually *not* so tough, even with just "routine" methods, and here is how I learned how to do it without unleashing the power of contour integration.[6]

From Euler's well-known formula, we can write

$$e^{e^{ix}} = e^{\cos(x)+i\sin(x)} = e^{\cos(x)}e^{i\sin(x)}$$

$$= e^{\cos(x)}[\cos\{\sin(x)\} + i\sin\{\sin(x)\}]. \qquad (A')$$

---

[6]The analysis I am about to show you is due to Cornel Ioan Vălean, who wrote to me from his home in Timis, Romania. Cornel's e-mail was brief, and I have filled-in some of the intermediate details he skipped over. Cornel's subsequent communications with me were so impressive that I urged him to get in touch with my editor at Springer (Dr. Sam Harrison), to query him about the possibility of authoring his own book. He did, with the result that you can find an ocean more of interesting integrals in which to swim in his *(almost) Impossible Integrals, Sums, and Series*, Springer 2019.

Now, from the equally well-known power-series expansion of the exponential, we have

$$e^y = \frac{1}{0!} + \frac{y}{1!} + \frac{y^2}{2!} + \ldots + \frac{y^n}{n!} + \ldots$$

and so, with $y = e^{ix}$, we can write

$$e^{e^{ix}} = \frac{1}{0!} + \frac{e^{ix}}{1!} + \frac{e^{i2x}}{2!} + \ldots + \frac{e^{inx}}{n!} + \ldots$$

or, using Euler's formula again,

$$e^{e^{ix}} = 1 + \frac{\cos(x) + i \sin(x)}{1!} + \frac{\cos(2x) + i \sin(2x)}{2!} + \ldots$$
$$+ \frac{\cos(nx) + i \sin(nx)}{n!} + \ldots \tag{B'}$$

If we equate the imaginary parts of (A') and (B'), we arrive at

$$e^{\cos(x)} \sin\{\sin(x)\} = \sum_{n=1}^{\infty} \frac{\sin(nx)}{n!}.$$

Thus,

$$\int_0^\infty \frac{e^{\cos(x)} \sin\{\sin(x)\}}{x} dx = \int_0^\infty \frac{1}{x} \sum_{n=1}^{\infty} \frac{\sin(nx)}{n!} dx - \sum_{n=1}^{\infty} \frac{1}{n!} \int_0^\infty \frac{\sin\{nx\}}{x} dx.$$

Each of the infinity of integrals in the sum is, by (3.2.1) (since n is positive in all of the integrands) equal to $\frac{\pi}{2}$, and so

$$\int_0^\infty \frac{e^{\cos(x)} \sin\{\sin(x)\}}{x} dx = \frac{\pi}{2} \sum_{n=1}^{\infty} \frac{1}{n!}.$$

Since

$$e = \sum_{n=0}^{\infty} \frac{1}{n!} = 1 + \sum_{n=1}^{\infty} \frac{1}{n!}$$

then

$$\sum_{n=1}^{\infty} \frac{1}{n!} = e - 1$$

and so we have our (Cauchy's) result *without* contour integration:

$$\int_0^\infty \frac{e^{\cos(x)} \sin\{\sin(x)\}}{x}\,dx = \frac{\pi}{2}(e-1).$$

As I said, we will do this integral again when we get to the development of contour integration in Chap. 8. Those new calculations will prove to be significantly more involved than what we have just done and so, you might wonder, what is the big deal about learning a *harder* way to do integrals? The big deal is that contour integration is *not* always harder than is doing an integral by "routine" methods. For example, in Chap. 8, you will find a contour integral evaluation of

$$\int_{-\infty}^\infty \frac{e^{ax}}{1-e^x}\,dx = \frac{\pi}{\tan(a\pi)}, \quad 0 < a < 1,$$

and it will not really be all that difficult. And then, new for this edition, I will show you how this result—(8.6.9)—can be found *without* the use of contour integration. In that case, as you will see, it is the "routine" approach that is the (far) more complicated of the two solutions. So, when faced with a tough integral, having experience with both approaches increases the chance for success.

As I prepare to bring this essay to an end, let me give you three specific examples of a central feature of this book: a routine acceptance of the role of a computer in mathematical work. (G. H. Hardy was a superb mathematician but, given his famous mistrust of "modern" gadgets—he used a telephone only on those occasions when a postcard would be too slow—I must admit I fear he would have been absolutely appalled at my enthusiasm for the use of electronic computers in mathematics.) First, you will recall (if you read the original Preface before reading this) the words of Professor Moll on the persuasive power of computational agreement with theory. I recalled those words when, one day recently, I was cruising around on the Web and came across an essay by a university mathematics professor on the topic of differentiating an integral.

In his essay, the professor derived what he called a "curious result," one that he said he had never seen before:

$$\int_0^\infty x^n e^{-tx} \sin(x)\frac{dx}{x} = (n-1)!\,\frac{\sin\left\{n\,\sin^{-1}\left(\frac{1}{\sqrt{1+t^2}}\right)\right\}}{(1+t^2)^{n/2}},$$

which is said to hold for all real $t \geq 0$ and integer $n \geq 1$. The professor provided an outline of how he derived this impressive formula, and I had absolutely no reason to question any part of his analysis. I was, however, simply too lazy to grind through what gave every sign of being a long, grubby exercise in algebra. At one point, he admitted that to get to the final result one needed to do "a bit of tidying up" and, oh boy, we all know what *that* means, don't we? Nothing good, I promise you!

So, what I did, instead, was to fire up my computer and use MATLAB to numerically evaluate both sides of his "curious result" for some randomly selected values of n and t. Here is a sampling of the results, with all calculations truncated to fifteen decimal places. You will notice that the agreement between theory and calculation is astonishingly good, and so the integral formula is either correct or we have witnessed a series of highly improbable (indeed, *incredibly unlikely*) coincidences. When he wrote, the professor apparently did not realize (and I did not remember until after I had done my computer check) that Euler had, in fact, derived this very formula long ago, in 1781. When you read the end of Sect. 7.5, and the derivation there of (7.5.7), you will see how Euler did it in his typically brilliant fashion (Euler's footprints are *everywhere* in mathematics).

n	t	Integral	Formula
2	3	0.060000000000000	0.060000000000000
3	1	0.500000000000002	0.500000000000000
5	6	0.002118483085700	0.002118483085700
7	4	0.035175431782907	0.035175431782907
11	5	0.049416872616243	0.049416872616242

For my second example of computers in mathematics, you will find that later in this book reference will be made, numerous times, to what mathematicians, physicists, and engineers alike call *Euler's identity*: $e^{iz} = \cos(z) + i\sin(z)$, where z is any complex number and $i - \sqrt{-1}$. If $z = \pi$, in particular, this reduces to $e^{i\pi} = -1$ or, equivalently, to the incredibly mysterious $e^{i\pi} + 1 = 0$. This statement is said to be incredibly mysterious because it links five of the most important numbers in mathematics, numbers that at first encounter seem to have little to do with each other (with the exceptions of 0 and 1, of course). In a notebook entry dated April 1933, the then not quite fifteen-year-old Feynman enthusiastically declared this to be "the most remarkable formula in math."[7] I suspect that even after a lifetime in world-class physics, he had encountered at most only a few other formulas that would have caused him to adjust that ranking.[8]

My reason for telling you all this is to motivate showing you a beautiful way, using a computer, to graphically illustrate the truth of $e^{i\pi} = -1$. From the definition of the exponential function, we have

---

[7]An image of this notebook entry is reproduced as the frontispiece photograph in my book *Dr. Euler's Fabulous Formula*, Princeton, 2011.

[8]Here is a runner-up in the "remarkable" contest, which is itself a result of Euler's identity. Since $e^{i\frac{\pi}{2}} = \cos\left(\frac{\pi}{2}\right) + i\sin\left(\frac{\pi}{2}\right) = i$, then $i^i = \left(e^{i\frac{\pi}{2}}\right)^i = e^{-\frac{\pi}{2}} = 0.20978\ldots$. That is, an imaginary number raised to an imaginary power is *real*! Indeed, matters are even *more* wondrous than this. That is because $i = e^{i\left(\frac{\pi}{2}+2\pi k\right)}$ because $e^{i2\pi k} = 1$ for *any* integer k (not just k = 0), and so $i^i = e^{-2\pi\left(k+\frac{1}{4}\right)}$, k *any* integer, and so there are *infinitely many* distinct, real values for i^i. Who would have even *guessed* such a thing?

$$e^z = \lim_{n \to \infty} \left(1 + \frac{z}{n}\right)^n$$

and so, if $z = i\pi$, then

$$e^{i\pi} = \lim_{n \to \infty} \left(1 + \frac{i\pi}{n}\right)^n.$$

It is easy, with modern computer software (like MATLAB), to calculate $\left(1 + \frac{i\pi}{n}\right)^n$ for any given value of n and then to extract the real and imaginary parts of the result. One can do that as a sequence of individual multiplications: for example, if $n = 3$, then we can calculate the three complex numbers

$$w_1 = \left(1 + \frac{i\pi}{3}\right),$$

$$w_2 = \left(1 + \frac{i\pi}{3}\right)^2 = w_1\left(1 + \frac{i\pi}{3}\right),$$

$$w_3 = \left(1 + \frac{i\pi}{3}\right)^3 = w_2\left(1 + \frac{i\pi}{3}\right).$$

Each of these three complex numbers has a real and an imaginary part, so we can write

$$w_1 = R_1 + i\,I_1,$$

$$w_2 = R_2 + i\,I_2,$$

$$w_3 = R_3 + i\,I_3.$$

If we plot the points (R_1, I_1), (R_2, I_2), and (R_3, I_3) and join them with straight lines, we can see the individual w's in the complex plane and, in particular, where $\left(1 + \frac{i\pi}{3}\right)^3$ is located. Each plot, for a given n, starts on the real axis at $(1, 0)$ because $\lim_{n \to 0} \left(1 + \frac{i\pi}{n}\right)^n = 1$. I have coded this procedure in MATLAB, and Fig. 1 shows the results for nine values of n. My point here, in including the code **eulerid.m** in the box following the figure, is not to turn you into a MATLAB coder, but rather to show you how *easy* it is to write such a code. The figure generated by the code allows you to literally *see* the convergence of $e^{i\pi}$ to -1.

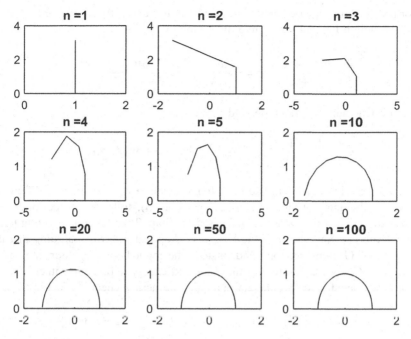

Fig. 1 Euler's wonderful formula $e^{i\pi} = -1$ as a limiting process

```
%eulerid.m
clear
n=[1 2 3 4 5 10 20 50 100];x(1)=1;y(1)=0;
for k=1:9
  N=n(k);
  c=1+(i*pi)/N;
  w(1)=c;
  for loop=2:N
  w(loop)=w(loop-1)*c;
  end
  for loop=1:N
  x(loop+1)=real(w(loop));
  y(loop+1)=imag(w(loop));
  end
  subplot(3,3,k)
  plot(x,y,'k')
  title_str=strcat('n = ',int2str(N));
  title(title_str)
end
```

For my third and final example (similar in nature to my first example), consider this claim, which I recently read in a math blog on the Web:

$$\int_0^\infty \frac{\ln\left(1+x^3\right)}{1+x^2}\,dx = \pi\frac{\ln\left(2\right)}{4} - \frac{G}{3} + \frac{2\pi\ln\left(2+\sqrt{3}\right)}{3},$$

where G is Catalan's constant, defined as

$$G = \frac{1}{1^2} - \frac{1}{3^2} + \frac{1}{5^2} - \frac{1}{7^2} + \frac{1}{9^2} - \ldots = 0.9159655\ldots.$$

It was stated that this result had been found using Feynman's trick of differentiating under the integral sign (discussed in Chap. 3), followed by performing a tedious partial fraction expansion (discussed in Chap. 2) and then integrating term-by-term. Another contributor to the blog was skeptical, however, arguing that the appearance of G seemed wrong and, instead, he argued for a contour integration (discussed in Chap. 8). Before devoting time and energy to pursuing either of those approaches, I decided to give the claim a quick numerical check. In fact,

$$integral\left(@(x)\,log\left(x.^3+1\right)./\left(x.^2+1\right),0,\,inf\right) = 2.99730487\ldots,$$

while

$$pi*log\left(2\right)/4 - \left(0.9159655/3\right) + 2*pi*log\left(2+sqrt(3)\right)/3 = 2.99730485\ldots.$$

This excellent agreement, of course, does *not* necessarily mean the original author did the integral "correctly" but, if someone bet you $100 on that point, how would you vote? I know what *I would* do!

Despite my extensive use in this book of the packaged integration software of MATLAB, there is a strong argument that can be made for "doing it yourself." That is, for writing your own numerical integration code, and this argument was ably made by the late Forman S. Acton in his book *Real Computing Made Real: Preventing Errors in Scientific and Engineering Calculations*, Princeton, 1996. All those interested in "doing integrals" should, in particular, read that book's Chap. 3 ("Quadratures"), pp. 141–171.

Acton (1920–2014) was originally trained as a chemical engineer, but his early work during the Second World War (at Oak Ridge, Tennessee, on the atomic-bomb Manhattan Project) introduced him to the then primitive computer tools of the day, and he became hooked on what we today call computer science. After the war, his computer expertise brought him onto the electrical engineering faculty at Princeton, and then later, when the computer people on that faculty split off to form their own department, he became a professor of computer science. Over his long career, Acton seemingly fell into, and then discovered a way out of, every disastrous numerical analysis horror trap one can image, and his book is a treasure trove of wisdom and escape tricks. Even purists who think the only good definite integral is one with a closed-form evaluation will find much of interest in Acton's book.

I opened this new Preface with a quotation from Feynman and so, in the interest of symmetry, I will close with a few more words about him. My occasional focus on Feynman in this book is not because his contributions to pure mathematics are noteworthy—they are not—but rather because of his tremendous enthusiasm, his sheer *joy*, in solving real problems.[9] That is the spirit in which I wrote this book, and I hope that is yours, too, in reading it. Feynman was a theoretical physicist, yes, but he was a *practical* theoretical physicist; that is, he was a very good *applied* mathematician. His writings and lectures on mathematics read and sound very much like what an engineer would write and say. What you will read in Chap. 7, on Dirac's impulse function is, for example, precisely how Feynman discussed the topic in one of his Hughes/Malibu lectures. (If Feynman had *anybody* he thought of as a hero, it was the English mathematical physicist Paul Dirac, who had an undergraduate degree in electrical engineering.)

In addition to correcting slip-ups[10] in the original discussions in this book, I have added a lot of new material, too, in the form of even more "interesting integrals" from both physics (for example, Trier's double integral that occurs in an electrical circuit containing an *infinity* of components) and mathematics (for example, the use of Ramanujan's Master Theorem, and of the Laplace transform, in doing definite integrals). One particular addition of historical interest shows how G. H. Hardy evaluated

$$\int_0^\infty \frac{\ln(x)}{1 + x^3} dx,$$

using an approach that I believe will absolutely dazzle you. *Why* did he do this integral? As you will see when you read the discussion, it was clearly simply for fun.

---

[9]I never personally talked with Feynman, but when I was in high school, my father took me to a public lecture Feynman gave one weekend at Caltech. It must have been sometime in 1957 or so. Brea, my hometown in 1950s Southern California, is only an hour's drive from Pasadena, and my father (holder of a PhD in chemistry, not physics, but with scientific interests that ranged far beyond chemistry) had heard that Feynman was "the next Einstein." So, off we went. That experience gave me a firsthand encounter with Feynman's personality. I had never before heard anyone like Feynman. He certainly was not boring, dull, or pompous! He talked like how I imagined a New York City wise guy would talk, cracked jokes, and clearly had a good time. I recall being initially shocked (and then enormously entertained) by how irreverent he was during the talk, which long after I recognized as presenting material that later appeared in the three volumes of his famous, equally irreverent *Lectures on Physics*. After the lecture, my father wanted to go up and introduce me to Feynman, but I was too shy. What a bad decision! This was, of course, my first "near miss."

[10]One special change I have made is to "correct" what was not a slip-up at the time I originally wrote: the use of *quad* (for *quadrature*) for numerical integration in MATLAB (and its multidimensional partners *dblquad* and *triplequad*). Those commands are now unsupported, and the proper commands in the newer versions of MATLAB are *integral* and *integral2* (and *integral3* for triple integrals). I have made the appropriate changes in the text, but of course the *numerical* results are unaffected.

Another of the new historical additions explains how Feynman became interested in the double integral

$$I(n) = \int_1^\infty \int_1^\infty (x^n + y^n)^{-1+\frac{1}{n}} dx\, dy,$$

an integral that for a while he thought had a connection to one of the most famous problems in number theory (which is about as pure as mathematics gets). *Why* did Feynman think this? I have no proof but, again, I think it was all just for fun for him.

Finally, I have added a large number of new challenge problems and their solutions, more or less uniformly distributed throughout the book.

Fig. 2 An elegantly attired Feynman (he would become "more casual" in later years), circa 1950, about the time he moved from snowy Cornell University physics to sunny (if occasionally smoggy) Caltech physics. Courtesy of the Archives, California Institute of Technology

As I finished writing those last words, I was reminded of a remark made by the late University of Virginia mathematician E. J. McShane (1904–1989), during his Presidential Address at the 1963 annual meeting of the American Mathematical Society: "There are in this world optimists who feel that any symbol that starts off with an integral sign must necessarily denote something that will have every property that they would like an integral to possess. This is of course quite annoying to us rigorous mathematicians: what is even more annoying is that by doing so, they often come up with the right answer."[11]

I admit that the carefree attitude displayed throughout this book would surely have had McShane place me in the box labeled *optimist*. I do think, however, that McShane would have, in turn, admitted that the way of the optimist (especially when backed-up with MATLAB or its equivalent) is a *lot* more fun than is the one demanding the citing of theorems/lemmas and proofs of uniform convergence at every step of a derivation. *That* way is a journey of hard work. Some readers may wonder at my use of the word *fun*, but what after all could *possibly* be more enjoyable (for a mathematically inclined person) than watching a seemingly impenetrable integral suddenly yield when faced with a tsunami of unimaginably powerful brilliance? (Well, perhaps that *is* just a bit over the top, but one can always hope, right?)

And so, perhaps, you can now see why I wrote this book in the way I did. As with Hardy and Feynman, it was mostly just to have fun. I do hope *you* have fun, too, while reading it.

Exeter, NH, USA   Paul J. Nahin
February 2020

---

[11] You can find McShane's complete address, "Integrals Devised for Special Purposes," in the *Bulletin of the American Mathematical Society* 69, 1963, pp.597–627.

Preface to the Original Edition

Engineering is like dancing; you don't learn it in a darkened lecture hall watching slides: you learn it by getting out on the dance floor and having your toes stepped on.

—Professor Jack Alford (1920–2006), cofounder of the Engineering Clinic at Harvey Mudd College, who hired the author in 1971 as an assistant professor. The same can be said for doing definite integrals.

To really appreciate this book, one dedicated to the arcane art of calculating definite integrals, it is necessary (although perhaps it is *not* sufficient) that you be the sort of person who finds the following question fascinating, one right up there in a fierce battle with a hot cup of coffee and a sugar donut for first place on the list of sinful pleasures:

without actually calculating x, show that

if $x + \frac{1}{x} = 1$ it then follows that $x^7 + \frac{1}{x^7} = 1$.

Okay, I know what many (but, I hope, not *you*) are thinking at being confronted with a question like this: of what Earthly significance could such a problem possibly have? Well, *none* as far as I know, but its fascination (or not) for *you* provides (I think) excellent psychological insight into whether or not *you* should spend time and/or good money on this book. If the problem leaves someone confused, puzzled, or indifferent (maybe all three) then my advice to them would be to put this book down and to look instead for a good mystery novel, the latest Lincoln biography (there seems to be a new one every year—what could *possibly* be left unsaid?), or perhaps a vegetarian cookbook.

But, if your pen is already out and scrawled pages of calculations are beginning to pile-up on your desk, then by gosh you *are* just the sort of person for whom I wrote this book. (If, after valiant effort you are still stumped but nonetheless just *have* to see how to do it—or if your pen simply ran dry—an analysis is provided at the end of the book.) More specifically, I've written with three distinct types of readers in mind: (1) physics/engineering/math students in their undergraduate years; (2) professors looking for interesting lecture material; and (3) non-academic professionals looking for a 'good technical read.'

There are two possible concerns associated with calculating definite integrals that we should address with no delay. First, do *real* mathematicians actually do that sort of thing? Isn't mere *computation* the dirty business (best done out of sight, in the shadows of back-alleys so as not to irreparably damage the young minds of impressionable youths) of grease-covered engineers with leaky pens in their shirts, or of geeky physicists in rumbled pants and chalk dust on their noses? Isn't it in the deep, clear ocean of analytical proofs and theorems where we find *real* mathematicians, swimming like powerful, sleek seals? As an engineer, myself, I find that attitude just a bit elitist, and so I am pleased to point to the pleasure in computation that many of the great mathematicians enjoyed, from Newton to the present day.

Let me give you two examples of that. First, the reputation of the greatest English mathematician of the first half of the twentieth century, G. H. Hardy (1877–1947), partially rests on his phenomenal skill at doing definite integrals. (Hardy appears in numerous places in this book.) And second, the hero of this book (Riemann) is best known today for (besides his integral) his formulation of the greatest unsolved problem in mathematics, about which I'll tell you lots more at the end of the book. But after his death, when his private notes on that very problem were studied, it was found that imbedded in all the deep theoretical stuff was a calculation of $\sqrt{2}$. To 38 decimal places!

The other concern I occasionally hear strikes me as just plain crazy; the complaint that there is no end to definite integrals. (This should, instead, be a cause for *joy*.) You can fiddle with integrands, and with upper and lower limits, in an uncountable infinity of ways,[12] goes the grumbling, so what's the point of calculating definite integrals since you can't possibly do them all? I hope writing this concern out in words is sufficient to make clear its ludicrous nature. We can never do all possible definite integrals, so why bother doing any? Well, what's next—you can't possibly add together all possible pairs of the real numbers, so why bother learning to add? Like I said—that's nuts!

What makes doing the specific integrals in this book of value aren't the specific answers we'll obtain, but rather the tricks (excuse me, the *methods*) we'll use in obtaining those answers; methods you may be able to use in evaluating the integrals you will encounter in the future in your own work. Many of the integrals I'll show you do have important uses in mathematical physics and engineering, but others are included just because they look, at first sight, to be so damn tough that it's a real kick to see how they simply crumble away when attacked with *the right trick*.

From the above you've probably gathered that I've written this book in a light-hearted manner (that's code for 'this is *not* a rigorous math textbook'). I am not going to be terribly concerned, for example, with proving the uniform convergence

---

[12]You'll see, in the next chapter, that with a suitable change of variable we can transform any integral into an integral from 0 to ∞, or from 1 to ∞, or from 0 to 1. So things aren't *quite* so bad as I've made them out.

of *anything*, and if you don't know what that means don't worry about it because I'm not going to worry about it, either. It's not that issues of rigor aren't important—*they are*—but not for us, here. When, after grinding through a long, convoluted sequence of manipulations to arrive at what we *think* is the value for some definite integral, I'll then simply unleash a wonderful MATLAB numerical integration command (*integral*) and we'll *calculate* the value. If our theoretical answer says it's $\sqrt{\pi} = 1.772453\ldots$ and *integral* says it's -9.3, we'll of course suspect that *somewhere* in all our calculations we just maybe fell off a cliff. If, however, *integral* says it is 1.77246, well then, that's good enough for me and on we'll go, happy with success and flushed with pleasure, to the next problem.

Having said that, I would be less than honest if I don't admit, right now, that such happiness *could* be delusional. Consider, for example, the following counter-example to this book's operational philosophy. Suppose you have used a computer software package to show the following:

$$\int_0^\infty \cos(x) \frac{\sin(4x)}{x} dx = 1.57079632679\ldots,$$

$$\int_0^\infty \cos(x) \cos\left(\frac{x}{2}\right) \frac{\sin(4x)}{x} dx = 1.57079632679\ldots,$$

$$\int_0^\infty \cos(x) \cos\left(\frac{x}{2}\right) \cos\left(\frac{x}{3}\right) \frac{\sin(4x)}{x} dx = 1.57079632679\ldots,$$

and so on, all the way out to

$$\int_0^\infty \cos(x) \cos\left(\frac{x}{2}\right) \cos\left(\frac{x}{3}\right) \ldots \cos\left(\frac{x}{30}\right) \frac{\sin(4x)}{x} dx = 1.57079632679\ldots.$$

One would have to be blind (as well as totally lacking in imagination) not to immediately suspect two things:

1. the consistent value of $1.57079\ldots$ is actually $\frac{\pi}{2}$, and
2. $\int_0^\infty \left\{ \prod_{k=1}^n \cos\left(\frac{x}{k}\right) \right\} \frac{\sin(4x)}{x} dx = \frac{\pi}{2}$ for *all* n.

This is exciting! But then you run the very next case, with $n = 31$, and the computer returns an answer of

$$\int_0^\infty \cos(x) \cos\left(\frac{x}{2}\right) \cos\left(\frac{x}{3}\right) \ldots \cos\left(\frac{x}{30}\right) \cos\left(\frac{x}{31}\right) \frac{\sin(4x)}{x} dx$$
$$= 1.57079632533\ldots$$

In this book I would dismiss the deviation (notice those last three digits!) as round-off error—*and I would be wrong!* It's *not* round-off error and, despite the highly suggestive numerical computations, the supposed identity "for all n" is simply *not true*. It's 'almost' true, but in math 'almost' doesn't cut-it.[13]

That's the sort of nightmarish thing that makes mathematicians feel obligated to clearly state any assumptions they make and, if they be really pure, to show that these assumptions are valid before going forward with an analysis. I will not be so constrained here and, despite the previous example of how badly things can go wrong, I'll assume just about anything that's convenient at the moment (short of something *really* absurd, like $1 + 1 = 3$), deferring the moment of truth to when we 'check' a theoretical result with MATLAB. A true mathematician would feel shame (perhaps even thinking that a state of moral degeneracy had been entered) if they should adopt such a cavalier attitude. I, on the other hand, will be immune to such soul-crushing doubts. Still, remain aware that we *will* be taking some risks.

So I *will* admit, again, that violation of one or more of the conditions that rigorous analyses have established *can* lead to disaster. Additional humorous examples of this disconcerting event can be found in a paper[14] by a mathematician with a sense of humor. The paper opens with this provocative line: "Browsing through an integral table on a dull Sunday afternoon [don't you often do the very same thing?] some time ago, I came across four divergent trigonometric integrals. I wondered how those divergent integrals [with *incorrect* finite values] ended up in a respectable table." A couple of sentences later the author writes "We have no intent to defame either the well-known mathematician who made the original error [the rightfully famous French genius Augustin-Louis Cauchy (1789–1857) that you'll get to know when we get to contour integration], or the editors of the otherwise fine tables in which the integrals appear. We all make mistakes and we're not out to point the finger at anyone. . ."

And if we *do* fall off a cliff, well, so what? Nobody need know. We'll just quietly gather-up our pages of faulty analysis, rip them into pieces, and toss the whole rotten mess into the fireplace. Our mathematical sins will be just between us and God (who is well-known for being forgiving).

Avoiding a computer is not necessarily a help, however. Here's a specific example of what I mean by that. In a classic of its genre,[15] Murray Spiegel (late professor of mathematics at Rensselaer Polytechnic Institute) asks readers to show that

$$\int_0^\infty \frac{\ln(1+x)}{1+x^2}\, dx = \frac{\pi \ln(2)}{2}$$

---

[13]For an informative discussion of the fascinating mathematics behind these calculations, see Nick Lord, "An Amusing Sequence of Trigonometrical Integrals," *The Mathematical Gazette*, July 2007, pp. 281–285.

[14]Erik Talvila, "Some Divergent Trigonometric Integrals," *The American Mathematical Monthly*, May 2001, pp. 432–436.

[15]M. R. Spiegel, *Outline of Theory and Problems of Complex Variables with an Introduction to Conformal Mapping and Its Applications*, Schaum 1964, p. 198 (problem 91).

which equals 1.088793.... One can only wonder at how many students struggled (and for how long) to do this, as the given answer is incorrect. Later in this book, in (5.1.3), we'll do this integral correctly, but a use of *integral* (*not* available to Spiegel in 1964) quickly shows the numerical value is actually the *significantly* greater 1.4603.... At the end of this Preface I'll show you two examples (including Spiegel's integral) of this helpful use of *integral*.

Our use of *integral* does prompt the question of *why*, if we can always calculate the value of any definite integral to as many decimal digits we wish, do we even care about finding exact expressions for these integrals? This is really a philosophical issue, and I think it gets to the mysterious interconnectedness of mathematics—how seemingly unrelated concepts can turn-out to actually be intimately related. The expressions we'll find for many of the definite integrals evaluated in this book will involve such familiar numbers as ln(2) and π, and other numbers that are not so well-known, like *Catalan's constant* (usually written as G) after the French mathematician Eugène Catalan (1814–1894). The common thread that stitches these and other numbers together is that all can be written as infinite series that can, in turn, be written as definite integrals[16]:

$$\ln(2) = 1 - \frac{1}{2} + \frac{1}{3} - \frac{1}{4} + \frac{1}{5} - \ldots = \int_0^1 \frac{2x}{1+x^2}\ dx = 0.693147\ldots$$

$$\frac{\pi}{4} = 1 - \frac{1}{3} + \frac{1}{5} - \frac{1}{7} + \frac{1}{9} - \ldots = \int_0^1 \frac{1}{1+x^2}\ dx = 0.785398\ldots$$

$$G = \frac{1}{1^2} - \frac{1}{3^2} + \frac{1}{5^2} - \frac{1}{7^2} + \frac{1}{9^2} - \ldots = \int_1^\infty \frac{\ln(x)}{1+x^2}\ dx = 0.9159655\ldots.$$

And surely it is understanding at a far deeper level to know that the famous Fresnel integrals $\int_0^\infty \cos(x^2)dx$ and $\int_0^\infty \sin(x^2)dx$ are *exactly* equal to $\frac{1}{2}\sqrt{\frac{\pi}{2}}$, compared to knowing only that they are 'pretty near' 0.6267.

In 2004 a wonderful book, very much like this one in spirit, was published by two mathematicians, and so I hope my cavalier words will appear to be appalling ones only to the most rigid of hard-core purists. That book, *Irresistible Integrals* (Cambridge University Press) by the late George Boros, and Victor Moll at Tulane University, is not quite as willing as this one is to return to the devil-may-care, eighteenth century mathematics of Euler's day, but I strongly suspect the authors were often tempted. Their sub-title gave them away: *Symbolics, Analysis and Experiments* [particularly notice this word!] *in the Evaluation of Integrals*. Being mathematicians, their technical will-power was stronger than is my puny electrical

---

[16]This integral is $\tan^{-1}(x)\big|_0^1 = \tan^{-1}(1) - \tan^{-1}(0) = \frac{\pi}{4}$. If we expand the integrand $\frac{1}{1+x^2}$ as a power series we get $1 - x^2 + x^4 - x^6 + x^8 - \ldots$ (perform the long division, or just verify the series by cross multiplying) and then integrate term-by-term between 0 and 1, the series for $\frac{\pi}{4}$ immediately follows. It was discovered by the French mathematician Gottfried Leibniz (1646–1716) in 1674.

engineer's dedication to rigor, but every now and then even they could not totally suppress their sheer pleasure at doing definite integrals.

And then three years later, in another book co-authored by Moll, we find a statement of philosophy that exactly mirrors my own (and that of this book): "Given an interesting identity buried in a long and complicated paper on an unfamiliar subject, which would give you more confidence in its correctness: staring at the proof, or confirming computationally that it is correct to 10,000 decimal places?"[17] That book, and *Irresistible Integrals*, are really fun math books to read.

Irresistible Integrals is different from this one, though, in that Boros and Moll wrote for a more mathematically sophisticated audience than I have, assuming a level of knowledge equivalent to that of a junior/senior college math major. They also use *Mathematica* much more than I use MATLAB. I, on the other hand, have assumed far less, just what a good student would know—with one BIG exception— after the first year of high school AP calculus, plus just a bit of exposure to the concept of a differential equation. That big exception is contour integration, which Boros and Moll avoided in their book because "not all [math majors] (we fear, few) study complex analysis."

Now *that*, I have to say, caught me by surprise. For a modern undergraduate math major not to have ever had a course in complex analysis seems to me to be shocking. As an electrical engineering major, *sixty years ago*, I took complex analysis up through contour integration (from Stanford's math department) at the start of my junior year using R. V. Churchill's famous book *Complex Variables and Applications*. (I still have my beat-up, coffee-stained copy.) I think contour integration is just too beautiful and powerful to be left out of this book but, recognizing that my assumed reader may not have prior knowledge of complex *analysis*, all the integrals done in this book by contour integration are gathered together in their own chapter at the end of the book. Further, in that chapter I've included a 'crash mini-course' in the theoretical complex analysis required to understand the technique (assuming only that the reader has already encountered complex *numbers* and their manipulation).

Irresistible Integrals contains many beautiful results, but a significant fraction of them are presented mostly as 'sketches,' with the derivation details (often presenting substantial challenges) left up to the reader. In this book *every* result is *fully* derived. Indeed, there are results here that are not in the Boros and Moll book, such as the famous integral first worked out in 1697 by the Swiss mathematician John Bernoulli (1667–1748), a result that so fascinated him he called it his "series mirabili" ("marvelous series"):

$$\int_0^1 x^x \, dx = 1 - \frac{1}{2^2} + \frac{1}{3^3} - \frac{1}{4^4} + \frac{1}{5^5} - \ldots = 0.78343\ldots$$

or its variant

---

[17] *Experimental Mathematics in Action*, A. K. Peters 2007, pp. 4–5. Our calculations here with *integral* won't be to 10,000 decimal places, but the idea is the same.

$$\int_0^1 x^{-x}\, dx = 1 + \frac{1}{2^2} + \frac{1}{3^3} + \frac{1}{4^4} + \frac{1}{5^5} + \ldots = 1.29128\ldots.$$

Also derived here are the equally exotic integrals

$$\int_0^1 x^{x^2}\, dx = 1 - \frac{1}{3^2} + \frac{1}{5^3} - \frac{1}{7^4} + \frac{1}{9^5} - \ldots = 0.89648\ldots$$

and

$$\int_0^1 x^{\sqrt{x}}\, dx = 1 - \left(\frac{2}{3}\right)^2 + \left(\frac{2}{4}\right)^3 - \left(\frac{2}{5}\right)^4 + \left(\frac{2}{6}\right)^5 - \ldots = 0.65858\ldots.$$

I don't believe either of these last two integrals has appeared in *any* book before now.

One famous integral that is also not in *Irresistible Integrals* is particularly interesting, in that it *seemingly* (I'll explain this in just a bit) owed its evaluation to a mathematician at Tulane University, Professor Moll's home institution. The then head of Tulane's math department, Professor Herbert Buchanan (1881–1974), opened a 1936 paper[18] with the following words: "In the consideration of a research problem in quantum mechanics, Professor J. C. Morris of Princeton University recently encountered the integral

$$I = \int_0^\infty \frac{x^3}{e^x - 1}\, dx.$$

Since the integral does not yield to any ordinary methods of attack, Professor Morris asked the author to evaluate it [Joseph Chandler Morris (1902–1970) was a graduate of Tulane who did his PhD in physics at Princeton; later he was head of the physics department, and then a Vice-President, at Tulane]." Professor Buchanan then showed that the integral is equal to an infinite series that sums to 6.49. . ., and just after arriving at that value he wrote "It had been found from other considerations [the details of which are not mentioned, but which I'm guessing were the results of either numerical calculations or even, perhaps, of *physics* experiments done at Princeton by Morris] that the integral should have a value between 6.3 and 6.9. Thus the value above [6.4939. . . = $\frac{\pi^4}{15}$] furnishes a theoretical verification of experimental results."

So here we have an important definite integral apparently 'discovered' by a physicist and solved by a mathematician. In fact, as you'll learn in Chap. 5, Buchanan was *not* the first to do this integral; it had been evaluated by Riemann in

---

[18]H. E. Buchanan, "On a Certain Integral Arising in Quantum Mechanics," *National Mathematics Magazine*, April 1936, pp. 247–248. I treated this integral, in a way different from Buchanan, in my book *Mrs. Perkins's Electric Quilt*, Princeton 2009, pp. 100–102, and in Chap. 5 we'll derive it in yet a different way, as a special case of (5.3.4).

1859, long before 1936. Nonetheless, this is a nice illustration of the fruitful co-existence and positive interaction of experiment and theory, and it is perfectly aligned with the approach I took while writing this book.

There is one more way this book differs from *Irresistible Integrals*, that reflects my background as an engineer rather than as a professional mathematician. I have, all through the book, made an effort to bring into many of the discussions a variety of physical applications, from such diverse fields as radio theory and theoretical mechanics. In all such cases, however, math plays a central role. So, for example, when the topic of elliptic integrals comes up (at the end of Chap. 6), I do so in the context of a famous *physics* problem. The origin of that problem is due, however, not to a physicist but to a nineteenth century *mathematician*.

Let me close this Preface on the same note that opened it. Despite all the math in it, this book has been written in the spirit of 'let's have fun.' That's the same attitude Hardy had when, in 1926, he replied to a plea for help from a young undergraduate at Trinity College, Cambridge. That year, while he was still a teenager, H. S. M. Coxeter (1907–2003) had undertaken a study of various four-dimensional shapes. His investigations had *suggested* to him ("by a geometrical consideration and verified graphically") several quite spectacular definite integrals, like[19]

$$\int_0^{\pi/2} \cos^{-1}\left\{ \frac{\cos(x)}{1 + 2\cos(x)} \right\} dx = \frac{5\pi^2}{24}.$$

In a letter to the *Mathematical Gazette* he asked if any reader of the journal could show him how to *derive* such an integral (we'll calculate the above so-called *Coxeter's integral* later, in the longest derivation in this book). Coxeter went on to become one of the world's great geometers and, as he wrote decades later in the Preface to his 1968 book *Twelve Geometric Essays*, "I can still recall the thrill of receiving [solutions from Hardy] during my second month as a freshman at Cambridge." Accompanying Hardy's solutions was a note scribbled in a margin declaring that "I tried very hard not to spend time on your integrals, but to me the challenge of a definite integral is irresistible."[20]

---

[19]MATLAB's *integral* says this integral is 2.0561677..., which agrees quite nicely with $\frac{5\pi^2}{24} =$ 2.0561675 . . .. The code is: *integral(@(x)acos(cos(x)./(1+2*cos(x))),0,pi/2)*.

For the integral I showed you earlier, from Spiegel's book, the *integral* code is (I've used *inf* for the upper limit of infinity): *integral(@(x)log(1+x)./(1+x.^2),0,inf)*. Most of the integrals in this book are one-dimensional but, for those times that we will encounter higher dimensional integrals there is *integral2* and *integral3*. The syntax for those cases, particularly for non-rectangular regions of integration, will be explained when we first encounter multi-dimensional integrals.

[20]And so we see where Boros and Moll got the title of their book. Several years ago, in my book *Dr. Euler's Fabulous Formula* (Princeton 2006, 2011), I gave another example of Hardy's fascination with definite integrals: see that book's Sect. 5.7, "Hardy and Schuster, and their optical integral," pp. 263–274. There I wrote "displaying an unevaluated definite integral to Hardy was very much like waving a red flag in front of a bull." Later in this book I'll show you a 'first principles' derivation of the optical integral (Hardy's far more sophisticated derivation uses Fourier transforms).

If you share Hardy's (and my) fascination for definite integrals, then this is a book for you. Still, despite my admiration for Hardy's near magical talent for integrals, I don't think he was *always* correct. I write that nearly blasphemous statement because, in addition to Boros and Moll, another bountiful source of integrals is *A Treatise on the Integral Calculus*, a massive two-volume work of nearly 1,900 pages by the English educator Joseph Edwards (1854–1931). Although now long out-of-print, both volumes are on the Web as Google scans and available for free download. In an April 1922 review in *Nature* that stops just short of being a sneer, Hardy made it quite clear that he did not like Edwards' work ("Mr. Edwards's book may serve to remind us that the early nineteenth century is not yet dead," and "it cannot be treated as a serious contribution to analysis"). Finally admitting that there is *some* good in the book, even then Hardy couldn't resist tossing a cream pie in Edwards' face with his last sentence: "The book, in short, may be useful to a sufficiently sophisticated teacher, provided he is careful not to allow it to pass into his pupil's hands." Well, I disagree. I found Edwards' *Treatise* to be a terrific read, a treasure chest absolutely *stuffed* with mathematical gems.

You'll find some of them in this book. Also included are dozens of challenge problems, with complete, detailed solutions at the back of the book if you get stuck. Enjoy!

Lee, NH, USA   Paul J. Nahin
August 2014

In support of the theoretical calculations performed in this book, numerical "confirmations" are provided by using several of the integration commands available in software packages developed by The MathWorks, Inc. of Natick, MA. Specifically, MATLAB® 8.4 (Release 2014b) and Symbolic Math Toolbox 6.10, with both packages running on a Windows 7 PC. These versions are now several releases old, but all the commands used in this book work with newer versions (as I write, Release 2019b and Symbolic Math Toolbox 8.3, both running on a Windows 10 PC), and are likely to continue to work for subsequent versions for several years more. (The *numerical* results should, of course, be essentially independent of the details of both the software used and of the machine hardware running that software.) MATLAB® is the registered trademark of The MathWorks, Inc. The MathWorks, Inc. does not warrant the accuracy of the text in this book. This book's use or discussion of MATLAB® and of the Symbolic Math Toolbox does not constitute an endorsement or sponsorship by The MathWorks, Inc. of a particular pedagogical approach or particular use of the MATLAB® and the Symbolic Math Toolbox software.

List of Author's Previous Books

- *Oliver Heaviside* (1988, 2002)
- *Time Machines* (1993, 1999)
- *The Science of Radio* (1996, 2001)
- *An Imaginary Tale* (1998, 2007, 2010)
- *Duelling Idiots* (2000, 2002)
- *When Least Is Best* (2004, 2007)
- *Dr. Euler's Fabulous Formula* (2006, 2011)
- *Chases and Escapes* (2007, 2012)
- *Digital Dice* (2008, 2013)
- *Mrs. Perkins's Electric Quilt* (2009)
- *Time Travel* (1997, 2011)
- *Number-Crunching* (2011)
- *The Logician and the Engineer* (2012, 2017)
- *Will You Be Alive Ten Years From Now?* (2013)
- *Holy Sci-Fi!* (2014)
- *In Praise of Simple Physics* (2016, 2017)
- *Time Machine Tales* (2017)
- *How to Fall Slower Than Gravity* (2018)
- *Transients for Electrical Engineers* (2019)
- *Hot Molecules, Cold Electrons* (2020)

Contents

1 Introduction 1
 1.1 The Riemann Integral 1
 1.2 An Example of Riemann Integration 5
 1.3 The Lebesgue Integral 7
 1.4 'Interesting' and 'Inside' 11
 1.5 Some Examples of Tricks 12
 1.6 Singularities 18
 1.7 Dalzell's Integral 23
 1.8 Where Integrals Come From 25
 1.9 Last Words 52
 1.10 Challenge Problems 55

2 'Easy' Integrals 59
 2.1 Six 'Easy' Warm-Ups 59
 2.2 A New Trick 64
 2.3 Two Old Tricks, Plus a New One 70
 2.4 Another Old Trick: Euler's Log-Sine Integral 79
 2.5 Trier's Double Integral 85
 2.6 A Final Example of a 'Hard' Integral Made 'Easy' 94
 2.7 Challenge Problems 97

3 Feynman's Favorite Trick 101
 3.1 Leibniz's Formula 101
 3.2 An Amazing Integral 110
 3.3 Frullani's Integral 112
 3.4 The Flip-Side of Feynman's Trick 115
 3.5 Combining Two Tricks and Hardy's Integral 124
 3.6 Uhler's Integral and Symbolic Integration 131
 3.7 The Probability Integral Revisited 134
 3.8 Dini's Integral 140

3.9   Feynman's Favorite Trick Solves a Physics Equation   142
3.10   Challenge Problems   144

4   Gamma and Beta Function Integrals   149
4.1   Euler's Gamma Function   149
4.2   Wallis' Integral and the Beta Function   151
4.3   Double Integration Reversal   162
4.4   The Gamma Function Meets Physics   172
4.5   Gauss 'Meets' Euler   176
4.6   Challenge Problems   179

5   Using Power Series to Evaluate Integrals   181
5.1   Catalan's Constant   181
5.2   Power Series for the Log Function   185
5.3   Zeta Function Integrals   193
5.4   Euler's Constant and Related Integrals   204
5.5   Generalizing Catalan's Constant   218
5.6   Challenge Problems   222

6   Some Not-So-Easy Integrals   227
6.1   Bernoulli's Integral   227
6.2   Ahmed's Integral   230
6.3   Coxeter's Integral   234
6.4   The Hardy-Schuster Optical Integral   241
6.5   The Watson/van Peype Triple Integrals   245
6.6   Elliptic Integrals in a Physical Problem   252
6.7   Ramanujan's Master Theorem   258
6.8   Challenge Problems   269

7   Using $\sqrt{-1}$ to Evaluate Integrals   275
7.1   Euler's Formula   275
7.2   The Fresnel Integrals   277
7.3 $\zeta(3)$ and More Log-Sine Integrals   280
7.4 $\zeta(2)$, at Last (Three Times!)   288
7.5   The Probability Integral *Again*   298
7.6   Beyond Dirichlet's Integral   302
7.7   Dirichlet Meets the Gamma Function   309
7.8   Fourier Transforms and Energy Integrals   312
7.9   'Weird' Integrals from Radio Engineering   317
7.10   Causality and Hilbert Transform Integrals   326
7.11   Laplace Transform Integrals   334
7.12   Challenge Problems   345

8   Contour Integration   351
8.1   Prelude   351
8.2   Line Integrals   352
8.3   Functions of a Complex Variable   355

8.4   The Cauchy-Riemann Equations and Analytic Functions   361
8.5   Green's Integral Theorem   364
8.6   Cauchy's First Integral Theorem   368
8.7   Cauchy's Second Integral Theorem   381
8.8   Singularities and the Residue Theorem   399
8.9   Integrals with Multi-Valued Integrands   407
8.10   A Final Calculation   415
8.11   Challenge Problems   419

9   Epilogue   423
9.1   Riemann, Prime Numbers, and the Zeta Function   423
9.2   Deriving the Functional Equation for $\zeta(s)$ 432
9.3   Our Final Calculation   442
9.4   Adieu   445
9.5   Challenge Questions   447

Solutions to the Challenge Problems   449

Index   499

About the Author

Paul J. Nahin is professor emeritus of electrical engineering at the University of New Hampshire. He is the author of 21 books on mathematics, physics, and the history of science, published by Springer, and the university presses of Princeton and Johns Hopkins. He received the 2017 Chandler Davis Prize for Excellence in Expository Writing in Mathematics (for his paper "The Mysterious Mr. Graham," The Mathematical Intelligencer, Spring 2016). He gave the invited 2011 Sampson Lectures in Mathematics at Bates College, Lewiston, Maine.

Chapter 1
Introduction

1.1 The Riemann Integral

The immediate point of this opening section is to address the question of whether
you will be able to understand the *technical commentary* in the book. To be blunt, do
you know what an integral *is*? You can safely skip the next few paragraphs if this
proves to be old hat, but perhaps it will be a useful over-view for some. It's far less
rigorous than a pure mathematician would like, and my intent is simply to define
terminology.

If $y = f(x)$ is some (any) 'sufficiently well-behaved' function (if you can draw it,
then for us it is 'sufficiently well-behaved') then the *definite integral of f(x)*, as x
varies from $x = a$ to $x = b \geq a$, is the area (a *definite* number, completely determined
by a, b, and f(x)) bounded by the function f(x) and the x-axis. It is, in fact, the shaded
area shown in Fig. 1.1.1. That's why you'll often see the phrase 'area under the
curve' in this and in other books on integrals. (We'll deal mostly with real-valued
functions in this book, but there will be, towards the end, a fair amount of discussion
dealing with complex-valued functions as well). In the figure I've shown f(x) as
crossing the x-axis at $x = c$; area *above* the x-axis (from $x = a$ to $x = c$) is *positive*
area, while area *below* the x-axis (from $x = c$ to $x = b$) is *negative* area.

We write this so-called *Riemann integral*—after the genius German mathematician Bernhard Riemann (1826–1866)—in mathematical notation as $\int_a^b f(x)dx$, where
the elongated s (that's what the integral sign *is*) stands for *summation*. It is worth
taking a look at how the Riemann integral is constructed, because *not all functions
have a Riemann integral* (these are functions that are *not* 'well-behaved'). I'll show
you such a function in Sect. 1.3. Riemann's ideas date from an 1854 paper.

Summation comes into play because the integral is actually the limiting value of
an *infinite* number of terms in a sum. Here's how that happens. To calculate the area
under a curve we imagine the integration interval on the x-axis, from a to b, is
divided into n sub-intervals, with the i-th sub-interval having length Δx_i. That is, if
the end-points of the sub-intervals are

© Springer Nature Switzerland AG 2020

P. J. Nahin, *Inside Interesting Integrals*, Undergraduate Lecture Notes in Physics,
https://doi.org/10.1007/978-3-030-43788-6_1

Fig. 1.1.1 The Riemann
definite integral $\int_a^b f(x)dx$

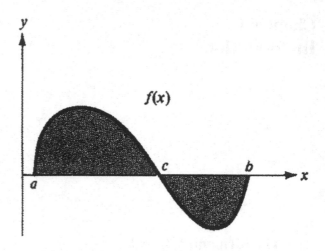

$$a = x_0 < x_1 < x_2 < \ldots < x_{i-1} < x_i < \ldots < x_{n-1} < x_n = b$$

then

$$\Delta x_i = x_i - x_{i-1}, 1 \leq i \leq n.$$

If we make the sub-intervals of equal length then

$$x_i = a + \frac{i}{n}(b - a)$$

which says

$$\Delta x_i = \Delta x = \frac{b - a}{n}.$$

We imagine that, for now, n is some finite (but 'large') integer which means that Δx is 'small.'

Indeed, we imagine that Δx is sufficiently small that f(x), from the beginning to the end of any given sub-interval, changes only slightly over the entire sub-interval. Let ζ_i be any value of x in the interval $x_{i-1} \leq \zeta_i \leq x_i$. Then, the area bounded by f(x) over that sub-interval is just the area of a very thin, vertical rectangle of height $f(\zeta_i)$ and horizontal width Δx, that is, an area of $f(\zeta_i)\Delta x$. (The shaded vertical strip in Fig. 1.1.2.) If we add all these rectangular areas, from x = a to x = b, we will get a very good approximation to the total area under the curve from x = a to x = b, an approximation that gets better and better as we let $n \to \infty$ and so $\Delta x \to 0$ (the individual rectangular areas become thinner and thinner). That is, the value of the integral, I, is given by

Fig. 1.1.2 Approximating
the 'area under the curve'

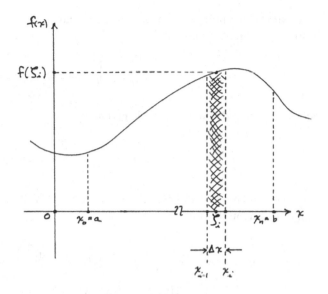

$$I = \lim_{n \to \infty} \sum_{i=1}^{n} f(\zeta_i) \, \Delta x = \int_a^b f(x) \, dx$$

where the summation symbol has become the integral symbol and Δx has become
the *differential* dx. We call $f(x)$ the *integrand* of the integral, with a and b the *lower*
and *upper* integration limits, respectively.

If a and b are both finite ($-\infty < a < b < \infty$), it is worth noting that we can always
write the definite integral $\int_a^b f(x)$ dx as a definite integral $\int_0^\infty g(t)$ dt. That is, we can
normalize the integration interval. Simply make the change of variable $t = \frac{x-a}{b-x}$. If a
and b are not both finite, there is always still some change of variable that normalizes
the integration interval. For example, suppose we have $\int_{-\infty}^\infty f(x)$ dx. Write this as
$\int_{-\infty}^0 f(x)dx + \int_0^\infty f(x)$ dx and make the change of variable $t = -x$ in the first integral.
Notice, too, that the integral $\int_0^\infty$ can always be rewritten in the form $\int_0^1$ by writing
$\int_0^\infty = \int_0^1 + \int_1^\infty$, and then changing variable in the last integral to $y = \frac{1}{x}$ which
changes the limits to 0 and 1 on that integral, too. Now, having noted this, for the
rest of this book I'll *not* bother with normalization.

We can write

$$I = \int_a^b f(x) \, dx = \int_a^b f(u) \, du$$

since the symbol we use to label the horizontal axis is x *only* because of tradition.
Labeling it u makes no difference in the numerical value of I. We call the integration
variable the *dummy* variable of integration. Suppose that we use u as the dummy
variable, and replace the upper limit on the integral with the variable x and the lower

limit with minus infinity. Now we no longer have a *definite* integral with a specific numerical value, but rather a *function* of x. That is,

$$F(x) = \int_{-\infty}^{x} f(u)\,du$$

and our original definite integral is given by

$$I = F(b) - F(a) = \int_{a}^{b} f(u)\,du.$$

What's the relationship between F(x) and f(x)? Well, recall that from the definition of a derivative we have

$$\frac{dF}{dx} = \lim_{\Delta x \to 0} \frac{F(x + \Delta x) - F(x)}{\Delta x} = \lim_{\Delta x \to 0} \frac{\int_{-\infty}^{x+\Delta x} f(u)\,du - \int_{-\infty}^{x} f(u)\,du}{\Delta x}$$

$$= \lim_{\Delta x \to 0} \frac{\int_{x}^{x+\Delta x} f(u)\,du}{\Delta x} = \frac{f(x)\Delta x}{\Delta x} = f(x),$$

where the last line follows because f(x) is essentially a constant over an integration interval of length Δx. Integration and differentiation are each the inverse operation of the other. Since the derivative of any constant C is zero, we write the so-called *indefinite* integral (an integral with no upper and lower limits) as

$$F(x) + C = \int f(x)dx.$$

So, one way to do integrals is to simply find an F(x) that, when differentiated, gives the integrand f(x). This is called 'look it up in a table' and it is a *great* way to do integrals *when you have a table that has the entry you need*.

What do you do when there is no such entry available? Well, that is when matters become 'interesting'! You'll have to either get a more extensive table of F(x) ↔ f(x) pairs, or work out F(x) for yourself. Or, perhaps, you'll grudgingly have to accept the fact that maybe, for the particular f(x) you have, there simply *is* no F(x). Amazingly, though, it can happen in that last case that while there is no F(x) there may still be a computable expression to the *definite* integral for specific values of the integration limits. For example, there is no F(x) for $f(x) = e^{-x^2}$, and all we can write for the *indefinite* integral is the admission

$$\int e^{-x^2}\,dx = ?$$

and yet (as will be shown later in this book), we can still write the *definite* integral

Fig. 1.1.3 Drawing by arrangement with Sidney Harris, ScienceCartoonsPlus.com

"I THINK YOU SHOULD BE MORE EXPLICIT HERE IN STEP TWO."

$$\int_0^\infty e^{-x^2}\, dx = \frac{1}{2}\sqrt{\pi}.$$

This is about as close to a miracle as you'll get in mathematics (Fig. 1.1.3).

1.2   An Example of Riemann Integration

Here's a pretty little calculation that nicely demonstrates the area interpretation of the Riemann integral. Consider the collection of all points (x, y) that, together, are the points in that region **R** of the x,y-plane where $|x| + |y| < 1$, where $-1 < x < 1$. What's the area of **R**? Solving for y gives

$$|y| < 1 - |x|$$

which is a condensed version of the double inequality

$$-[1 - |x|] = y_2 < y < 1 - |x| = y_1.$$

Let's consider the cases of $x > 0$ and $x < 0$ separately.

Case 1: if $x > 0$ then $|x| = x$ and the double inequality becomes

$$-(1 - x) = y_2 < y < 1 - x = y_1$$

and the area for the $x > 0$ portion of $\mathbf{R}$ is

$$\int_0^1 (y_1 - y_2)\, dx = \int_0^1 \{(1 - x) + (1 - x)\}\, dx = \int_0^1 (2 - 2x)\, dx = (2x - x^2)\big|_0^1$$
$$= 2 - 1 = 1$$

Case 2: if $x < 0$ then $|x| = -x$ and the double inequality becomes

$$-(1 + x) = y_2 < y < 1 + x = y_1$$

and the area for the $x < 0$ portion of $\mathbf{R}$ is

$$\int_{-1}^0 (y_1 - y_2)\, dx = \int_{-1}^0 \{(1 + x) + (1 + x)\}\, dx = \int_{-1}^0 (2 + 2x)\, dx$$
$$= (2x + x^2)\big|_{-1}^0 = 2 - 1 = 1.$$

So, the total area of $\mathbf{R}$ is 2.

Notice that we did the entire calculation without any concern about the shape of $\mathbf{R}$. So, what *does* $\mathbf{R}$ look like? If we knew that, then maybe the area would be obvious (it will be!). For $x > 0$ we have

$$|y| < 1 - |x|$$

which says that one edge of $\mathbf{R}$ is

$$y_a(x) = 1 - x, \quad x > 0$$

and another edge of $\mathbf{R}$ is

$$y_b(x) = -(1 - x) = -1 + x, \quad x > 0.$$

For $x < 0$ we have

$$|x| + |y| < 1$$

or, as $|x| = -x$ for $x < 0$,

Fig. 1.2.1 The shaded,
rotated square is **R**

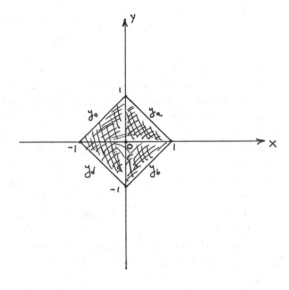

$$|y| < 1 + x.$$

Thus, a third edge of **R** is

$$y_c(x) = 1 + x, \quad x < 0$$

and a fourth edge of **R** is

$$y_d(x) = -(1 + x) = -1 - x, \quad x < 0.$$

Figure 1.2.1 shows these four edges plotted, and we see that **R** is a rotated (through 45°) square centered on the origin, with a side length of $\sqrt{2}$. That is, it has an area of 2, just as we calculated with the Riemann integral.

1.3   The Lebesgue Integral

Now, in the interest of honesty for the engineers reading this (and to avoid scornful denunciation by the mathematicians reading this!), I must take a time-out here and tell you that the Riemann integral (and its area interpretation) is not the end of the line when it comes to integration. In 1902 the French mathematician Henri Lebesgue (1875–1941) extended the Riemann integral to be able to handle integrand functions that, in no obvious way, bound an area. There *are* such functions; probably the most famous is the one cooked-up in 1829 by the German mathematician Lejeune Dirichlet (1805–1859):

$$\phi(x) = \left\{ \begin{array}{l} 1 \text{ when x is rational} \\ 0 \text{ when x is irrational} \end{array} \right.$$

Try drawing a sketch of $\phi(x)$—and I'll bet you can't. In any interval of finite length there are an infinity of rationals and irrationals, *each*, and Dirichlet's function is a very busy one, jumping wildly back-and-forth an infinity of times between 0 and 1 like an over-caffeinated frog on a sugar-high. (That the rationals are a *countable* infinity and the irrationals are an *uncountable* infinity are two famous results in mathematics, both due to the Russian-born German mathematician Georg Cantor (1845–1918).)[1] And if you can't even draw it, how can you talk of Dirichlet's function as 'bounding an area'? You can't, and Dirichlet's function is an example of a function that isn't integrable in the Riemann sense.

You might suspect that it is the infinite number of discontinuous jumps within a finite interval that makes $\phi(x)$ non-Riemann integrable, but in fact it is possible to have a similarly wildly discontinuous function that remains Riemann integrable. Indeed, in 1854 Riemann himself created such a function. Define [x] as the integer nearest to x. For example, [9.8] = 10 and [−10.9] = −11. If x is exactly between two integers, then [x] is defined to be zero; [3.5] = 0. Riemann's infinitely discontinuous function is then

$$r(x) = \sum_{k=1}^{\infty} \frac{kx - [kx]}{k^2}.$$

In Fig. 1.3.1 I've plotted an *approximation* (using the first 20 terms of the sum) to $r(x)$ at 100,000 points uniformly spaced over the interval 0–1. That figure will, I think, just *start* to give you a sense of how crazy-wild $r(x)$ is; it has an infinite number of discontinuities in the interval 0–1, and yet Riemann showed that $r(x)$ is still integrable in the Riemann sense even while $\phi(x)$ is not.[2]

But Dirichlet's function *is* integrable in the Lebesgue sense. Rather than sub-dividing the integration interval into sub-*intervals* as in the Riemann integral, the Lebesgue integral divides the integration interval into *sets of points*. Lebesgue's central intellectual contribution was the introduction of the concept of the *measure* of a set. For the Riemann sub-interval, its measure was simply its length. The Riemann integral is thus just a special case of the Lebesgue integral, as a sub-interval is just one *particular* way to define a set of points; but there are other ways, too. When the Riemann integral exists, so does the Lebesgue integral, but the converse is not true. When both integrals exist they are equal.

---

[1] You can find high school level proofs of Cantor's results in my book *The Logician and the Engineer*, Princeton 2012, pp. 168–172.

[2] For more on r(x), see E. Hairer and G. Wanner, *Analysis by Its History*, Springer 1996, p. 232, and William Dunham, *The Calculus Gallery*, Princeton 2005, pp. 108–112.

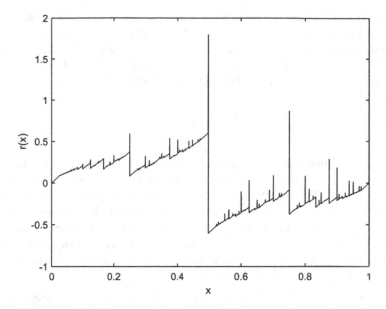

Fig. 1.3.1 Ricmann's weird function

To see how this works, let's calculate the Lebesgue integral of $\phi(x)$ over the interval 0–1. In this interval, focus first on all the rational values of x that have the particular integer n in the denominator of the fraction $\frac{m}{n}$ (by definition, this is the form of a rational number). Given the n we've chosen, we see that m can vary from 0 to n, that is, there are $n + 1$ such rational values (points) along the x-axis from 0 to 1. Now, imagine each of these points embedded in an interval of length $\frac{\varepsilon}{n^3}$, where ε is an arbitrarily small (but non-zero) positive number. That means we can imagine as tiny an interval as we wish, as long as it has a non-zero length. The total length of all the $n + 1$ intervals is then

$$(n + 1)\frac{\varepsilon}{n^3} = \frac{\varepsilon}{n^2} + \frac{\varepsilon}{n^3}.$$

Remember, this is for a *particular* n.

Now, sum over all possible n, that is, let n run from 1 to infinity. There will, of course, be a lot of repetition: for example, $n = 2$ and $m = 1$, and $n = 26$ and $m = 13$, define the same point. So, the *total* length of all the intervals that cover *all* the rational numbers from 0 to 1 is *at most*

$$\varepsilon\left\{\sum_{n=1}^{\infty}\frac{1}{n^2} + \sum_{n=1}^{\infty}\frac{1}{n^3}\right\}.$$

As is well-known, both sums have a finite value (the first is, of course, Euler's famous result of $\frac{\pi^2}{6}$ which will be derived in Chap. 7, and the second sum is obviously even smaller). The point is that the total sum in the braces has some finite value S, and so the *total* length of *all* the intervals that cover *all* the rational numbers from 0 to 1 is at most εS, and we can make this as small as we wish by picking ever smaller values for ε. Lebesgue therefore says that the *measure* of the set of all the rationals from 0 to 1 is zero and so, in the Lebesgue sense, we have

$$\int_0^1 \phi(x) \, dx = 0.$$

Now, with all that said, I'll simultaneously admit to the beauty of the Lebesgue integral as well as admit to the 'scandalous' fact that in this book I'm not going to worry about it. In 1926 the President of the Mathematical Association (England) sternly stated "To be a serious mathematician and not to use the Lebesgue integral is to adopt the attitude of the old man in a country village who refuses to travel in a train."[3] On the other hand, the American electrical engineer and computer scientist Richard Hamming (1915–1998) somewhat cavalierly rebutted that when he declared (in a 1997 address[4] to *mathematicians!*)

> ... for more than 40 years I have claimed that if whether an airplane would fly or not depended on whether some function that arose in its design was Lebesgue but *not Riemann integrable*, then I would not fly in it. Would you? Does Nature recognize the difference? I doubt it! You may, of course, choose as you please in this matter, but I have noticed that year by year the Lebesgue integration, and indeed all of measure theory, seems to be playing a smaller and smaller role in other fields of mathematics, and *none at all in fields that merely use mathematics* [my emphasis].

I think Hamming has the stronger position in this debate, and all the integrals you'll see from now on in this book are to be understood as Riemann integrals.

For mathematicians who might be tempted to dismiss Hamming's words with 'Well, what else would you expect from an *engineer!*,' let me point out that the year before Hamming's talk a mathematician had said essentially the same thing in a paper that Hamming had surely read. After admitting that the Lebesgue integral "has become the 'official' integral in mathematical research," Robert Bartle (1927–2003) then stated *"the time has come to discard the Lebesgue integral as the primary integral* [Bartle's emphasis]."[5]

---

[3]Comment made after the presentation by E. C. Francis of his paper "Modern Theories of Integration," *The Mathematical Gazette*, March 1926, pp. 72–77.

[4]In Hamming's paper "Mathematics On a Distant Planet," *The American Mathematical Monthly*, August–September 1998, pp. 640–650.

[5]See Bartle's award-winning paper "Return to the Riemann Integral," *The American Mathematical Monthly*, October 1996, pp. 625–632. He was professor of mathematics at the University of Illinois for many years, and then at Eastern Michigan University.

1.4 'Interesting' and 'Inside'

So, what's an interesting integral, and what does it mean to talk of being 'inside' it? I suppose the honest answer about the 'interesting' part is sort of along the lines of Supreme Court Associate Justice Potter Stewart's famous 1964 comment on the question of "what is pornography?" He admitted it was hard to define "but I know it when I see it." It's exactly the same with an interesting integral. Here's what I mean by that.

In 1957, in the summer between my junior and senior years in high school, I bought a copy of the second edition of MIT professor George B. Thomas' famous textbook *Calculus and Analytic Geometry* for a summer school class at the local junior college. I still remember the thrill I felt when, flipping through the pages for the first time, I happened across (on p. 369) the following (Thomas says it's from the "lifting theory" of aerodynamics):

$$\int_{-1}^{1} \sqrt{\frac{1+x}{1-x}}\, dx = \pi.$$

Why did I, a kid not yet 17 who had only recently still been building model airplanes from balsa wood kits and leaky tubes of glue, find that 'interesting'? I didn't know then—and I'm not really sure I do even now—except that it just looks 'mysterious and exotic.'

Now some may snicker at reading that, but I don't care—by god, to me that aerodynamic integral *still* looks mysterious and exotic! It's a line of writing as wonderful as anything you'll find in Hemingway or Dostoevsky and, unlike with even the greatest fiction, it's a line that nobody could just make-up. What, after all, does all that strange (to me, then) stuff on the left-hand-side of the equal sign have to do with π? I of course knew even in those long-ago days that π was intimately connected to circles, but I didn't see any circles on the left. Interesting! (Later, in Challenge Problem (C3.6), I'll ask you to evaluate the "lifting theory" integral.)

This sort of emotional reaction isn't limited to just amateur mathematicians; professionals can be swept-up by the same euphoria. In 1935, for example, at the end of his presidential address to the London Mathematical Society, the English mathematician G. N. Watson (1886–1965) mentioned this astonishing definite integral:

$$\int_{0}^{\infty} e^{-3\pi x^2} \frac{\sinh{(\pi x)}}{\sinh{(3\pi x)}}\, dx = \frac{1}{e^{2\pi/3}\sqrt{3}} \sum_{n=0}^{\infty} \frac{e^{-2n(n+1)\pi}}{(1+e^{-\pi})^2(1+e^{-3\pi})^2 \ldots (1+e^{-(2n+1)\pi})^2}.$$

Of it he declared that it gave him "a thrill which is indistinguishable from the thrill which I feel when I enter the Sagrestia Nuova of the Capelle Medicee and see before me the austere beauty of the four statues representing Day, Night, Evening, and

Dawn which Michelangelo has set over the tombs of Guiliano de' Medici and Lorenzo de' Medici."

Wow.

Alright, now what does being 'inside' an integral mean? I'll try to answer that by example. Suppose I tell you that

$$\int_0^1 \frac{\ln(1+x)}{1+x^2}\,dx = \frac{\pi}{8}\ln(2).$$

You'd probably just shrug your shoulders and say (or at least think it if too polite to actually say it) "Okay. Where did you get that from, a good table of integrals? Which one did you use? I'll look it up, too." And so you could, but could you *derive* the result? That's what I mean by getting 'inside' an integral. It's the art of starting with the integral on the left side of the equal sign and somehow getting to a computable expression on the right side. (We'll do this integral later in the book— see (2.2.4).

I use the word *art* with intent. There is no theory to doing definite integrals. Each new integral is a brand new challenge and each new challenge almost always demands a new trick or, at least, a new twist or two to a previous trick. That's right, a *trick*. Some people might recoil in horror at that, but a real analyst simply smiles with anticipation at the promise of a righteous struggle. And even if the integral wins for the moment, leaving the analyst exhausted and with a pounding headache, the analyst knows that there is always tomorrow in which to try again.

1.5   Some Examples of Tricks

Now, just so you know what I'm talking about when I use the word *trick*, here's the first example of a trick in this book (the rest of this book is, essentially, just one new trick after the next). What goes on the right-hand side of

$$\int_{-1}^1 \frac{\cos(x)}{e^{(1/x)}+1}\,dx = ?$$

This integral almost surely looks pretty scary to nearly everybody at first sight; trying to find an F(x) that, when differentiated gives the integrand, certainly doesn't seem very promising. But in fact the *definite* integral *can* be done by a first-year calculus student *if that student sees the trick*! Play around with this for a while as I work my way towards the stunning revelation of the answer, and see if you can beat me to it.

Consider the more general integral

$$\int_{-1}^{1} \frac{\cos(x)}{d(x) + 1}\, dx$$

where $d(x)$ is just about any function of x that you wish.
(For the original integral, $d(x) = e^{(1/x)}$.) I'll impose a restriction on $d(x)$ in just a bit, but not right now. If we then write

$$g(x) = \frac{\cos(x)}{d(x) + 1}$$

then we arrive (of course!) at the integral

$$\int_{-1}^{1} g(x)\, dx.$$

No matter what $g(x)$ is, we can always write it as the sum of an even function $g_e(x)$ and an odd function $g_o(x)$. That is,

$$g(x) = g_e(x) + g_o(x).$$

How do we know we can do this? Because we can actually write down what $g_e(x)$ and $g_o(x)$ are. By the definitions of even and odd we have

$$g(-x) = g_e(-x) + g_o(-x) = g_e(x) - g_o(x)$$

and so it is simple algebra to arrive at

$$g_e(x) = \frac{g(x) + g(-x)}{2}$$

and

$$g_o(x) = \frac{g(x) - g(-x)}{2}.$$

Now,

$$\int_{-1}^{1} g(x)\, dx = \int_{-1}^{1} g_e(x)\, dx + \int_{-1}^{1} g_o(x)\, dx - \int_{-1}^{1} g_e(x)\, dx$$

because by the 'symmetry' of an odd function around $x = 0$ we have (think 'area under the curve')

$$\int_{-1}^{1} g_o(x)\, dx = 0.$$

From our original integral, we have

$$g(x) = \frac{\cos{(x)}}{d(x) + 1}$$

and so

$$g_e(x) = \frac{1}{2}\left[\frac{\cos{(x)}}{d(x)+1} + \frac{\cos{(-x)}}{d(-x)+1}\right]$$

or, because $\cos(-x) = \cos(x)$, that is, because the cosine is an even function,

$$g_e(x) = \frac{\cos{(x)}}{2}\left[\frac{1}{d(x)+1} + \frac{1}{d(-x)+1}\right]$$

$$= \frac{\cos{(x)}}{2}\left[\frac{d(-x)+1+d(x)+1}{d(x)d(-x)+d(x)+d(-x)+1}\right]$$

or,

$$g_e(x) = \frac{\cos{(x)}}{2}\left[\frac{2+d(-x)+d(x)}{d(x)d(-x)+d(x)+d(-x)+1}\right].$$

Okay, suppose we now put a restriction on $d(x)$. Suppose $d(x)d(-x) = 1$. This is the case, you'll notice, for the $d(x)$ in our original integral ($= e^{(1/x)}$) because

$$d(x)d(-x) = e^{\left(\frac{1}{x}\right)}e^{\left(-\frac{1}{x}\right)} = e^0 = 1.$$

Well, then, you should now see that the numerator and the denominator of all that stuff in the brackets on the right-hand-side of the last expression for $g_e(x)$ are equal! That is, everything in the brackets reduces to 1 and so

$$g_e(x) = \frac{\cos{(x)}}{2}$$

and our scary integral has vanished like a balloon pricked with a pin:

$$\int_{-1}^{1}\frac{\cos{(x)}}{e^{(1/x)}+1}\,dx = \int_{-1}^{1}\frac{\cos{(x)}}{2}\,dx = \frac{1}{2}\int_{-1}^{1}\cos{(x)}\,dx = \frac{1}{2}\{\sin{(x)}\}\Big|_{-1}^{1}$$

$$= \frac{\sin(1) - \sin(-1)}{2} = \sin(1) = 0.8414709\ldots.[6]$$

Now *that's* a trick! MATLAB's *integral* agrees, computing a value of 0.8414623 .... The code syntax is: *integral(@(x)cos(x)./(exp(1./x)+1),-1,1)*.

This was a fairly sophisticated trick, but sometimes even pretty low-level approaches can achieve 'tricky' successes. For example, what is the value of

$$\int_0^\infty \frac{\ln(x)}{1 + x^2}\, dx = ?$$

Let's start with the obvious $\int_0^\infty = \int_0^1 + \int_1^\infty$. In the first integral on the right, make the change of variable $t = 1/x$ (and so $dx = -dt/t^2$). Then,

$$\int_0^1 \frac{\ln(x)}{1 + x^2}\, dx = \int_\infty^1 \frac{\ln\left(\frac{1}{t}\right)}{1 + \frac{1}{t^2}}\left(-\frac{1}{t^2}dt\right) = -\int_\infty^1 \frac{\ln\left(\frac{1}{t}\right)}{t^2 + 1}\, dt = \int_1^\infty \frac{\ln\left(\frac{1}{t}\right)}{t^2 + 1}\, dt$$

$$= -\int_1^\infty \frac{\ln(t)}{t^2 + 1}\, dt.$$

That is, recognizing that t and x are just dummy variables of integration, we have $\int_0^1 = -\int_1^\infty$ and so we immediately have our result (first derived by Euler):

$$\int_0^\infty \frac{\ln(x)}{1 + x^2}\, dx = 0 \tag{1.5.1}$$

This is, in fact, a special case of the more general integral

$$\int_0^\infty \frac{\ln(x)}{b^2 + x^2}\, dx.$$

See if you can calculate this integral (your answer should, of course, reduce to zero when $b = 1$); if you have trouble with it we'll evaluate this integral in the next chapter—in (2.1.3)—where you'll see that knowing the special case helps (a *lot*!) in doing the general case.

For a third example of a trick, let me show you one that illustrates just how clever the early students of the calculus could be in teasing-out the value of a definite integral. This trick dates back to 1719, when the Italian mathematician Giulio Fagnano (1682–1766) calculated $\int_0^\infty \frac{dx}{1+x^2}$. Today, of course, a first-year calculus freshman would recognize the *indefinite* integral to be $\tan^{-1}(x)$, and so the answer is

---

[6]Writing $\sin(1)$ means, of course, the sine of 1 *radian* $= \frac{180°}{\pi} = 57.3°$ (*not* of 1°).

$\tan^{-1}(\infty) - \tan^{-1}(0) = \frac{\pi}{2}$. But Fagnano's clever trick does not require knowledge of the indefinite integral,[7] but only how to differentiate. Here's how Fagnano did it.

Imagine a circle with radius r, centered on the origin of the xy-plane. The arc-length L along the circumference of that circle that is subtended by a central angle of θ is $L = r\theta$. Now, suppose that $r = 1$. Then $L = \theta$ and so, with t as a dummy variable of integration,

$$L = \theta = \int_0^\theta dt.$$

Fagnano then played around with this very simple integrand (indeed, how could it possibly be any simpler!?) to make it *more* complicated! Specifically,

$$L = \int_0^\theta \frac{\frac{1}{\cos^2(t)}}{\frac{1}{\cos^2(t)}} \, dt = \int_0^\theta \frac{\frac{1}{\cos^2(t)}}{\frac{\cos^2(t) + \sin^2(t)}{\cos^2(t)}} \, dt = \int_0^\theta \frac{\frac{1}{\cos^2(t)}}{1 + \tan^2(t)} \, dt.$$

Next, change variable to $x = \tan(t)$, which says

$$dx = \frac{dt}{\cos^2(t)}.$$

Thus,

$$L = \int_0^{\tan(\theta)} \frac{dx}{1 + x^2}.$$

Suppose $\theta = \frac{\pi}{2}$. Then $\tan(\theta) = \tan(\frac{\pi}{2}) = \infty$, and of course L is one-fourth the circle's circumference and so equals $\frac{\pi}{2}$ because $\frac{\pi}{2}$ is one-fourth of 2π, the circumference of a unit radius circle. Thus, *instantly*,

$$\frac{\pi}{2} = \int_0^\infty \frac{dx}{1 + x^2}.$$

By the same reasoning, we *instantly* have[8]

---

[7] I am assuming that when you see $\int \frac{1}{a^2 + x^2} \, dx$ you immediately recognize it as $\frac{1}{a} \tan^{-1}\left(\frac{x}{a}\right)$. This is one of the few 'fundamental' indefinite integrals I'm going to assume you've seen previously from a first course of calculus. Others are: $\int \frac{1}{x} dx = \ln(x)$, $\int e^x \, dx = e^x$, $\int x^n \, dx = \frac{x^{n+1}}{n+1}$ $(n \neq -1)$, $\int \frac{dx}{\sqrt{a^2 - x^2}} = \sin^{-1}\left(\frac{x}{a}\right)$, and $\int \ln(x) dx = x\ln(x) - x$.

[8] This result says, if we expand the integrand as a power series,
$\frac{\pi}{4} = \int_0^1 \left(1 - x^2 + x^4 - x^6 + \ldots\right) dx = \left(x - \frac{1}{3}x^3 + \frac{1}{5}x^5 - \frac{1}{7}x^7 + \ldots\right)\big|_0^1$, or
$\frac{\pi}{4} = 1 - \frac{1}{3} + \frac{1}{5} - \frac{1}{7} + \ldots$, a famous formula discovered (by other means) decades earlier by Leibniz (see note 5 in the original Preface).

$$\frac{\pi}{4} = \int_0^1 \frac{dx}{1+x^2},$$

and

$$\frac{\pi}{3} = \int_0^{\sqrt{3}} \frac{dx}{1+x^2},$$

and

$$\frac{\pi}{6} = \int_0^{\sqrt{3}/3} \frac{dx}{1+x^2}.$$

Pretty clever.

Finally, to end this 'sampler of tricks,' consider

$$I = \int_0^{\pi/2} \frac{dx}{1+\tan^m(x)},$$

where m is an arbitrary constant. The trick here is to think of x as one of the acute angles of a right triangle, and so the other acute angle is $y = \frac{\pi}{2} - x$. With that change of variable, $dx = -dy$. Now, with the triangle imagery still in mind, we can immediately write

$$\tan(x) = \cot(y) = \frac{1}{\tan(y)}.$$

Thus,

$$I = \int_{\pi/2}^0 \frac{-dy}{1+\left\{\frac{1}{\tan(y)}\right\}^m} = \int_0^{\pi/2} \frac{\tan^m(y)}{1+\tan^m(y)} dy.$$

Adding this expression to our original expression for I, we have

$$I + I = 2I = \int_0^{\pi/2} \frac{dx}{1+\tan^m(x)} + \int_0^{\pi/2} \frac{\tan^m(y)}{1+\tan^m(y)} dy$$

and realizing that x and y are simply dummy variables of integration, we have

$$2I = \int_0^{\pi/2} \frac{1 + \tan^m(y)}{1 + \tan^m(y)} dy = \int_0^{\pi/2} dy = \frac{\pi}{2}$$

and so, independent of m (which may be a bit of a surprise),

$$\int_0^{\pi/2} \frac{dx}{1 + \tan^m(x)} = \frac{\pi}{4}. \tag{1.5.2}$$

The claim of independence might seem to be a pretty strong assertion, so let's use MATLAB to do a quick check, with m = 11 and m = − 7 as randomly selected test values:

integral(@(x)1./(1+tan(x).^11),0,pi/2) = 0.7854
integral(@(x)1./(1+tan(x).^(−7)),0,pi/2) = 0.7854.

So, the claim looks like it's valid.

1.6   Singularities

Tricks are neat, and the discovery of each new one is like the thrill you get when digging through a box of raisins (think 'routine' integration methods, good for you, yeah, but still sorta boring) and every now and then coming across a chocolate-covered peanut (think 'fantastic new trick'). But—one does have to be ever alert for pitfalls that, we hope, *integral* will be our last-ditch defense against. Here's an example of what I'm getting at, in which that most common of operations—blindly plugging into a standard integration formula—can lead to disaster.

Suppose we have the integral $I = \int_{-1}^1 \frac{dx}{x^2}$. The integrand $f(x) = \left(\frac{1}{x}\right)^2$ is *never* negative because it's the square of a real value (from −1 to 1). Thus, we know immediately, from the area interpretation of the Riemann integral, that I > 0. However, from differential calculus we also know that

$$\frac{d}{dx}\left(\frac{1}{x}\right) = -\frac{1}{x^2}$$

and so, upon integrating, we have

$$I = \left(-\frac{1}{x}\right)\Big|_{-1}^1 = \left(-\frac{1}{1}\right) - \left(-\frac{1}{-1}\right) = -1 - 1 = -2.$$

That's right, *minus* 2, which is certainly <0. What's going on with this?

The problem is that f(x) blows-up at x = 0, right in the middle of the integration interval. The integral is called *improper*, and x = 0 is called a *singularity*. You have to be ever alert for singularities when you are doing integrals; always, *stay away* from singularities. Singularities are the black holes of integrals; don't 'fall into' one (don't integrate across a singularity). You'll find, when we get to contour integration, this will be *very* important to keep in mind. Here's how to do that for our integral. We'll write I as follows, where ε is an arbitrarily small, *positive* quantity:

$$I = \int_{-1}^{1} \frac{dx}{x^2} = \lim_{\varepsilon \to 0} \int_{-1}^{-\varepsilon} \frac{dx}{x^2} + \lim_{\varepsilon \to 0} \int_{\varepsilon}^{1} \frac{dx}{x^2}.$$

Then,

$$I = \lim_{\varepsilon \to 0} \left(-\frac{1}{x}\right)\Big|_{-1}^{-\varepsilon} + \lim_{\varepsilon \to 0} \left(-\frac{1}{x}\right)\Big|_{\varepsilon}^{1}$$

$$= \lim_{\varepsilon \to 0} \left[\left(-\frac{1}{-\varepsilon}\right) - \left(-\frac{1}{-1}\right)\right] + \lim_{\varepsilon \to 0} \left[\left(-\frac{1}{1}\right) - \left(-\frac{1}{\varepsilon}\right)\right]$$

$$= \lim_{\varepsilon \to 0} \left(\frac{1}{\varepsilon} - 1 - 1 + \frac{1}{\varepsilon}\right) = \lim_{\varepsilon \to 0} \left(\frac{2}{\varepsilon}\right) - 2 = +\infty.$$

The integral is (as we expected) positive—in fact, it's *infinitely* positive! It certainly is *not* negative. Notice, too, that our first (incorrect) result of -2 is now understandable—it's right there in the correct answer, *along with the infinite contribution from the singularity* that we originally missed.

Now, to see if you've *really* grasped the problem with the above integral, consider *this* integral, one made famous in mathematical physics by the Nobel prize-winning physicist Richard Feynman (1918–1988): if a and b are real-valued, arbitrary constants, then

$$\int_{0}^{1} \frac{1}{[ax + b(1-x)]^2} \, dx = \frac{1}{ab}. \tag{1.6.1}$$

Do you see the issue here? If a and b have opposite algebraic signs then the right-hand side of the above formula is negative. But on the left-hand-side, the integrand is always something *squared*, no matter what a and b may be, and so the integrand is *never* negative. We appear to have a conflict. What's going on with this? Hint: ask yourself if the integrand has a singularity and, if so, where is it located?[9] (See Challenge Problem (C1.3) at the end of this chapter.)

Before leaving the subject of singularities, I should tell you that there are other infinity concerns besides the blowing-up of the integrand that can occur when doing integrals. What do we mean, for example, when we write $\int_{-\infty}^{\infty} f(x)\,dx$? The area interpretation of the Riemann integral can fail us in this case, even when f(x) is a 'nice' function. For example, how much area is under the curve f(x) = sin(x),

---

[9]This integral appeared in Feynman's famous paper "Space-Time Approach to Quantum Electrodynamics," *Physical Review*, September 15, 1949, pp. 769–789. Some historical discussion of the integral is in my book *Number–Crunching*, Princeton 2011, pp. xx–xxi.

$-\infty < x < \infty$? Since sin(x) is an odd function it seems we should be able to argue that there is always a piece of negative area that cancels every piece of positive area, and so we'd like to write

$$\int_{-\infty}^{\infty} \sin(x)dx = 0.$$

But what then about the area under the curve f(x) = cos(x), $-\infty < x < \infty$? Now we have an even function, but it's natural to wonder if we can continue to make the negative/positive area cancellation argument since cos(x) is just a shifted sine function; that is, is it true that

$$\int_{-\infty}^{\infty} \cos(x)dx = 2\int_{0}^{\infty} \cos(x)dx = 0?$$

That is, is it true that

$$\int_{0}^{\infty} \cos(x)dx = 0?$$

The answer is *no*, neither $\int_{-\infty}^{\infty} \sin(x)dx$ or $\int_{-\infty}^{\infty} \cos(x)dx$ exist. We can, however, write what mathematicians call the *Cauchy Principal Value* of the integral:

$$\int_{-\infty}^{\infty} f(x)\, dx = \lim_{R\to\infty} \int_{-R}^{R} f(x)\, dx,$$

an integral that is zero if f(x) is odd. This approach, using a *symmetrical* limiting operation, means that the Cauchy Principal Value for $\int_{-\infty}^{\infty} \sin(x)dx$ *does* exist (it is zero), even though the integral does not.

If, however, we have an f(x) integrand such that $\lim_{x\to\pm\infty} f(x) = 0$ 'fast enough'—that means faster than $\frac{1}{x}$ (for example, $\frac{\sin(x)}{x}$)—then we don't have these conceptual difficulties. But even this isn't enough to fully capture all the subtle problems that can come with infinity. That's because an integrand like f(x) = cos (x^2), which *doesn't* go to zero as x → ±∞, *does* have a definite integral over $-\infty < x < \infty$ (you'll recall this integral—called a *Fresnel integral*—from the original Preface). That's because cos(x^2) oscillates faster and faster between ±1 as x increases in both directions, and so the positive and negative areas above and below the x-axis *individually* go to zero faster and faster and so contribute less and less to the *total* area under the curve (an area which is finite).

Our little trick of 'sneaking-up' on a singularity can be quite powerful. Consider, for example, the interesting integral

$$\int_0^\infty \frac{dx}{x^3 - 1}.$$

When x is between 0 and 1 the integrand is negative, while when x is >1 the integrand is positive. There is obviously a singularity *at* $x = 1$, with the integrand blowing-up to minus-infinity as x approaches 1 from values <1, and blowing-up to plus-infinity as x approaches 1 from values >1. Is it possible, we might wonder, for these two infinite explosions (with opposite signs) to perhaps *cancel* each other? In fact they do, and to convince you of that I'll use the 'sneak' trick to write our integral as

$$\int_0^{1-\varepsilon} \frac{dx}{x^3 - 1} + \int_{1+\varepsilon}^\infty \frac{dx}{x^3 - 1}$$

and then explore what happens as $\varepsilon \to 0$.

In the spirit of this book I'll first use *integral* to *experimentally* study what happens as $\varepsilon \to 0$. In the following table I've listed the result of running the following MATLAB command for various, ever smaller, values of ε (stored in the vector e(j)):

$integral(@(x)1./(x.\wedge 3-1),0,1-e(j))+integral(@(x)1./(x.\wedge 3-1),1+e(j),inf)$

ε	$\int_0^{1-\varepsilon} \frac{dx}{x^3-1} + \int_{1+\varepsilon}^\infty \frac{dx}{x^3-1}$
0.1	−0.53785
0.01	−0.59793
0.001	−0.60393
0.0001	−0.60453
0.00001	−0.60459
0.000001	−0.60459
0.0000001	−0.60459

So, from this numerical work it would appear that

$$\int_0^\infty \frac{dx}{x^3 - 1} = -0.60459.$$

Well, what could this curious number *be*? As it turns out we can answer this question, *exactly*, because it proves possible to actually find the *indefinite* integral! That's because we can write the integrand as the partial fraction expansion

$$\frac{1}{x^3 - 1} = \frac{1}{3(x - 1)} - \frac{2x + 1}{6(x^2 + x + 1)} - \frac{1}{2(x^2 + x + 1)}$$

or, completing the square in the denominator of the last term,

$$\frac{1}{x^3 - 1} = \frac{1}{3(x - 1)} - \frac{2x + 1}{6(x^2 + x + 1)} - \frac{1}{2\left[(x + \frac{1}{2})^2 + \frac{3}{4}\right]}.$$

Each of the individual terms on the right is easily integrated, giving

$$\int \frac{dx}{x^3 - 1} = \frac{1}{3} \ln (x - 1) - \frac{1}{6} \ln (x^2 + x + 1) - \frac{1}{\sqrt{3}} \tan^{-1}\left(\frac{2x + 1}{\sqrt{3}}\right)$$

$$= \frac{1}{6} \{2 \ln (x - 1) - \ln (x^2 + x + 1)\} - \frac{1}{\sqrt{3}} \tan^{-1}\left(\frac{2x + 1}{\sqrt{3}}\right)$$

$$= \frac{1}{6} \ln \left\{\frac{(x - 1)^2}{x^2 + x + 1}\right\} - \frac{1}{\sqrt{3}} \tan^{-1}\left(\frac{2x + 1}{\sqrt{3}}\right)$$

$$= \frac{1}{6} \ln \left\{\frac{x^2 - 2x + 1}{x^2 + x + 1}\right\} - \frac{1}{\sqrt{3}} \tan^{-1}\left(\frac{2x + 1}{\sqrt{3}}\right).$$

The argument of the log function is well-behaved for all x in the integration interval *except* at x = 1 where we get log(0), and so let's again use the sneak trick, but this time *analytically*. That is, we'll integrate from 0 to 1 − ε and add it to the integral from 1 + ε to ∞. Then we'll let ε → 0. So, noticing that the log function vanishes at both x = 0 and at x = ∞ (each give log(1)), we have

$$\int_0^\infty \frac{dx}{x^3 - 1} = \lim_{\varepsilon \to 0} \left[\frac{1}{6} \ln \left\{\frac{(1-\varepsilon)^2 - 2(1-\varepsilon) + 1}{(1-\varepsilon)^2 + (1-\varepsilon) + 1}\right\} - \frac{1}{\sqrt{3}} \tan^{-1}\left\{\frac{2(1-\varepsilon) + 1}{\sqrt{3}}\right\} + \frac{1}{\sqrt{3}} \tan^{-1}\left(\frac{1}{\sqrt{3}}\right)\right]$$

$$+ \lim_{\varepsilon \to 0} \left[-\frac{1}{\sqrt{3}} \tan^{-1}(\infty) - \frac{1}{6} \ln \left\{\frac{(1+\varepsilon)^2 - 2(1+\varepsilon) + 1}{(1+\varepsilon)^2 + (1+\varepsilon) + 1}\right\} + \frac{1}{\sqrt{3}} \tan^{-1}\left\{\frac{2(1+\varepsilon) + 1}{\sqrt{3}}\right\}\right].$$

The log terms expand as

$$\frac{1}{6} \ln \left\{\frac{1 - 2\varepsilon + \varepsilon^2 - 2 + 2\varepsilon + 1}{1 - 2\varepsilon + \varepsilon^2 + 1 - \varepsilon + 1}\right\} - \frac{1}{6} \ln \left\{\frac{1 + 2\varepsilon + \varepsilon^2 - 2 - 2\varepsilon + 1}{1 + 2\varepsilon + \varepsilon^2 + 1 + \varepsilon + 1}\right\}$$

$$= \frac{1}{6} \ln \left\{\frac{\varepsilon^2}{\varepsilon^2 - 3\varepsilon + 3}\right\} - \frac{1}{6} \ln \left\{\frac{\varepsilon^2}{\varepsilon^2 + 3\varepsilon + 3}\right\} = \frac{1}{6} \ln \left\{\frac{\varepsilon^2 + 3\varepsilon + 3}{\varepsilon^2 - 3\varepsilon + 3}\right\}.$$

As ε → 0 this last log term obviously vanishes. The tan^{-1} terms expand as

$$-\frac{1}{\sqrt{3}} \tan^{-1}\left\{\frac{2 - 2\varepsilon + 1}{\sqrt{3}}\right\} + \frac{1}{\sqrt{3}} \tan^{-1}\left(\frac{1}{\sqrt{3}}\right) - \frac{1}{\sqrt{3}} \tan^{-1}(\infty) + \frac{1}{\sqrt{3}} \tan^{-1}\left\{\frac{2 + 2\varepsilon + 1}{\sqrt{3}}\right\}$$

$$= -\frac{1}{\sqrt{3}} \tan^{-1}\left\{\frac{3 - 2\varepsilon}{\sqrt{3}}\right\} + \frac{1}{\sqrt{3}} \tan^{-1}\left(\frac{1}{\sqrt{3}}\right) - \frac{1}{\sqrt{3}} \left(\frac{\pi}{2}\right) + \frac{1}{\sqrt{3}} \tan^{-1}\left\{\frac{3 + 2\varepsilon}{\sqrt{3}}\right\}$$

or, as ε → 0, this reduces to

$$-\frac{1}{\sqrt{3}}\tan^{-1}\left\{\frac{3}{\sqrt{3}}\right\} + \frac{1}{\sqrt{3}}\tan^{-1}\left(\frac{1}{\sqrt{3}}\right) - \frac{1}{\sqrt{3}}\left(\frac{\pi}{2}\right) + \frac{1}{\sqrt{3}}\tan^{-1}\left\{\frac{3}{\sqrt{3}}\right\}$$

$$= \frac{1}{\sqrt{3}}\tan^{-1}\left(\frac{1}{\sqrt{3}}\right) - \frac{1}{\sqrt{3}}\left(\frac{\pi}{2}\right) = \frac{1}{\sqrt{3}}\left[\tan^{-1}\left(\frac{1}{\sqrt{3}}\right) - \frac{\pi}{2}\right] = \frac{1}{\sqrt{3}}\left[\frac{\pi}{6} - \frac{\pi}{2}\right] = -\frac{\pi}{3\sqrt{3}} = -\frac{\pi\sqrt{3}}{9}.$$

So, at last,

$$\int_0^\infty \frac{dx}{x^3 - 1} = -\frac{\pi\sqrt{3}}{9} \tag{1.6.2}$$

which is, indeed, the curious -0.60459.

1.7   Dalzell's Integral

The rest of this book is simply a lot more tricks, some even more spectacular than the one I showed you in Sect. 1.5. But why, some may ask, should we study *tricks*? After all, with modern computer software even seemingly impossible integrals can be done far faster than a human can work. Please understand that I'm *not* talking about *numerical* integrators, like MATLAB's *integral* (although far less rapidly, mathematicians could do *that* sort of thing *centuries* ago!). I'm talking about *symbolic* integrators like, for example, the on-line (available for free) *Mathematica* symbolic integrator software that needs only a fraction of a second to evaluate the *indefinite* aerodynamic "lifting theory" integral from earlier in this Introduction (Sect. 1.4):

$$\int \sqrt{\frac{1+x}{1-x}}\,dx = -\sqrt{1-x^2} + 2\sin^{-1}\left(\sqrt{\frac{x+1}{2}}\right).$$

You can verify that this is indeed correct by simply differentiating the right-hand-side and observing the integrand on the left-hand-side appear. And when the lower and upper limits of -1 and $+1$, respectively, are plugged-in, we get π, just as Professor Thomas wrote in his book.

Now, I'd be the first to admit that there is a *lot* of merit to using automatic, computer integrators. If *I* had to do a tough integral as part of a job for which I was getting paid, the *first* thing *I* would do is go to Wolfram (although, I should tell you, Wolfram *fails* on the indefinite $\int \frac{\cos(x)}{d(x)+1}\,dx$ that we solved for a *definite* case with our little even/odd trick—fails, almost surely, since there is no *indefinite* integral). That admission ignores the *fun* of doing definite integrals, however, the same sort of fun that so many people enjoy when they do Sudoku puzzles. Battling Sudoku puzzles *or* integrals is a combat of wits with an 'adversary' (the rules of math) that tolerates *zero* cheating. If you succeed at either, it ain't luck—it's skill.

Now, just to show you I'm serious when I say doing definite integrals can be fun, consider

$$I = \int_0^1 \frac{x^4(1-x)^4}{1+x^2} \, dx,$$

which first appeared (in 1944) on the pages of the *Journal of the London Mathematical Society*. What's so 'fun' about this, you ask? Well, look at what we get when I is evaluated. Multiplying out the numerator, and then doing the long division of the result by the denominator, we get

$$I = \int_0^1 \left(x^6 - 4x^5 + 5x^4 - 4x^2 + 4 - \frac{4}{1+x^2} \right) dx,$$

integrations that are all easily done to give

$$I = \int_0^1 \frac{x^4(1-x)^4}{1+x^2} \, dx = \left(\frac{x^7}{7} - \frac{2x^6}{3} + x^5 - \frac{4x^3}{3} + 4x - 4\tan^{-1}(x) \right) \Bigg|_0^1$$

$$= \frac{1}{7} - \frac{2}{3} + 1 - \frac{4}{3} + 4 - \pi.$$

That is,

$$\int_0^1 \frac{x^4(1-x)^4}{1+x^2} \, dx = \frac{22}{7} - \pi. \tag{1.7.1}$$

Since the integrand is never negative, we know that $I > 0$ and so we have the sudden (and, I think, totally unexpected) result that $\frac{22}{7} > \pi$. That is, the classic schoolboy approximation to π is an *over*estimate, a fact that is not so easy to otherwise establish.

Our calculations actually give us more information than just $\frac{22}{7} > \pi$; we can also get an idea of just how good an approximation $\frac{22}{7}$ is to π. That's because if we replace the denominator of the integrand with 1 we'll clearly get a bigger value for the integral, while if we replace the denominator with 2 we'll get a smaller value for the integral. That is,

$$\int_0^1 \frac{x^4(1-x)^4}{2} \, dx < \int_0^1 \frac{x^4(1-x)^4}{1+x^2} \, dx = \frac{22}{7} - \pi < \int_0^1 x^4(1-x)^4 \, dx$$

or, multiplying out, integrating, and plugging-in the limits,

$$\frac{1}{2}\left(\frac{1}{5}-\frac{2}{3}+\frac{6}{7}-\frac{1}{2}+\frac{1}{9}\right)=\frac{1}{1,260}<\frac{22}{7}-\pi<\left(\frac{1}{5}-\frac{2}{3}+\frac{6}{7}-\frac{1}{2}+\frac{1}{9}\right)=\frac{1}{630}$$

or, multiplying through by -1 (which reverses the sense of the inequalities),

$$-\frac{1}{630}<\pi-\frac{22}{7}<-\frac{1}{1,260}$$

or, at last,

$$\frac{22}{7}-\frac{1}{630}<\pi<\frac{22}{7}-\frac{1}{1,260}.$$

To five decimal places, this says $3.14127 < \pi < 3.14206$, which nicely and fairly tightly bounds π ($= 3.14159 \ldots$). Now who would deny that this sort of thing is *fun*?!

The author of the 1944 paper that first published this gem was D. P. Dalzell, a curious fellow who is mostly a ghost in the history of mathematics. All of the modern references to Dalzell's integral make no mention of the man, himself, even though he wrote a number of high quality mathematical papers and had an excellent reputation among mathematicians. Dalzell didn't help his cause by his habit of always using his initials. In fact, he was Donald Percy Dalzell (1898–1988), who graduated in 1921 from St. John's College, Cambridge, in mathematics and mechanical sciences. He received an MA degree in 1926, and his career was *not* as a mathematician but rather as a chartered engineer (a term used in England for a masters level professional engineer). He worked for a number of years for the Standard Telephones and Cables Company in London, and had two patents on electrical communication cables. The only known photograph of him is the one on the MacTutor math web-site (taken at the 1930 Edinburgh Mathematical Society Colloquium at St. Andrews).

1.8   Where Integrals Come From

Just about all of the discussions in this book are in the form of 'here's an integral, how can we evaluate it?' Mathematicians, of course, can simply 'make-up' integrals off the tops of their heads, but engineers and physicists generally encounter integrals resulting from the analysis of a physical problem. I thought, therefore, that I should include examples of that sort of origin of integrals, too, in addition to the pure imagination of mathematicians. Still, with each of the seven examples in this section I have made an effort to keep the interest of mathematicians, too, by selecting *physical* problems (each giving birth to integrals) that are *mathematical* problems at heart; the first two of these examples come from probability. For a third example of where interesting integrals 'come from,' I've selected a problem that I first came across while reading *Irresistible Integrals*, a book I mentioned in both the original

Preface and the new Preface. I think that problem nicely illustrates how mathematicians can need motivation, too (just like physicists), for some of the 'weird' integrals *they* conjure-up! The next three examples are from engineering physics, pure math *by a physicist*, and electronics. The final example returns to probability. Later in the book I'll show you some additional examples of interesting integrals occurring in the analysis of physical problems (see, for example, the discussion of Trier's double integral in Chap. 2).

So, to start, here is what I call 'The Circle in a Circle' problem. Imagine a circle (let's call it C_1) that has radius a. We then chose *at random*,[10] *and independently*, three points from the interior of that circle. These three points, *if non-collinear*, uniquely determine another circle, C_2. C_2 may or may not be totally contained within C_1. What is the probability that C_2 lies totally inside C_1?[11]

To answer this, imagine that after picking the three points we've drawn C_2 as shown in Fig. 1.8.1. There I've shown C_2 as totally inside C_1, but that's just because

Fig. 1.8.1 A 'circle inside a circle'

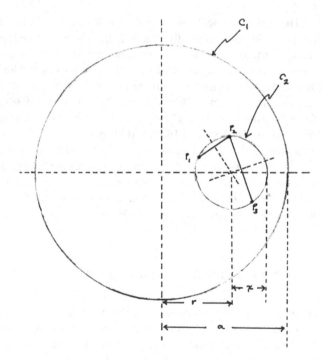

[10] 'At random' has the following meaning. If we look at any tiny patch of area dA in the interior of C_1, a patch of any shape, then the probability a point is selected from that area patch is dA divided by the area of C_1. We say that each of the three points is selected *uniformly* from the interior of C_1.

[11] I think it almost intuitively obvious that the probability is *scale-invariant* (the same for any value of a), but just in case it isn't obvious for you I'll carry the radius of C_1 along explicitly. At the end of our analysis you'll see that the scale-setting parameter a has disappeared, proving my claim.

I arbitrarily decided to do it that way instead of showing the alternative. Without any loss of generality, we can further imagine that the horizontal axis passes through the center of C_2 (as shown) because we can always rotate the figure to make that so. The center of C_2 is taken to be distance r from the center of C_1, and the radius of C_2 is taken to be x.

Next, imagine a thin, circular band of width Δx is drawn that encloses the circumference of C_2. The area of this band is, to a first approximation that gets better and better as $\Delta x \to 0$, given by $2\pi x \, \Delta x$. The probability a point selected at random from the interior of C_1 is from this band is, therefore, the ratio of the area of the band to the area of C_2 (see note 10 again):

$$\frac{2\pi x \, \Delta x}{\pi a^2} = \frac{2}{a^2} x \, \Delta x.$$

(I'll comment further on this claim, in just a bit.) Since the *three* points that determine C_2 all fall inside the band (*by definition!*), and they are *independently* selected, then the probability of that event is

$$\left\{ \frac{2}{a^2} x \, \Delta x \right\}^3 = \frac{8}{a^6} x^3 \{\Delta x\}^2 \Delta x = \frac{8}{a^6} x^3 \Delta A \, \Delta x$$

where $\Delta A = \{\Delta x\}^2$ is the differential area in rectangular coordinates. (I'll elaborate on this claim, in just a bit.) I'll write ΔA and Δx as dA and dx, respectively, from now on.

Obviously if $x = 0$ (that is, if C_2 is a degenerate circle that is actually a point) we see that C_2 is necessarily inside C_1. In fact, as x increases from 0, it can get as large as $a - r$ before C_2 penetrates C_1. So, the probability the C_2 circle, centered on a particular line (the horizontal axis), is totally within C_1 is

$$\frac{8 \, dA}{a^6} \int_0^{a-r} x^3 dx = \frac{8 \, dA}{a^6} \left\{ \frac{x^4}{4} \right\} \Big|_0^{a-r} = \frac{2 \, dA}{a^6} (a - r)^4.$$

In general, of course, the center of C_2 could fall anywhere inside C_1, with any value of r from 0 to a occurring, and so we need to integrate the above differential probability (*differential* because of the dA) over the entire interior of C_1. This is most easily done by writing dA in polar coordinates, as r dr dθ, and so the probability we are after is

$$\int_0^{2\pi} \int_0^a \frac{2}{a^6} (a - r)^4 r \, dr \, d\theta = \frac{2}{a^6} \int_0^{2\pi} \left\{ \int_0^a (a - r)^4 r \, dr \right\} d\theta = \frac{4\pi}{a^6} \int_0^a (a - r)^4 r \, dr,$$

since the θ-integral is obviously just 2π (there is no θ-dependency in the integrand).

The remaining r-integral is easily done by making the change of variable $u = a - r$ (and so $du = -dr$). So,

$$\int_0^a (a-r)^4 r\, dr = \int_a^0 u^4(a-u)(-du) = \int_0^a u^4(a-u)du = a\int_0^a u^4 du - \int_0^a u^5 du$$

$$= a\left\{\frac{u^5}{5}\right\}\Big|_0^a - \left\{\frac{u^6}{6}\right\}\Big|_0^a = \frac{a^6}{5} - \frac{a^6}{6} = \frac{a^6}{30}.$$

Thus, the probability that C_2 lies totally inside C_1 is

$$\frac{4\pi}{a^6}\left(\frac{a^6}{30}\right) = \frac{4\pi}{30} = \frac{2\pi}{15} = 0.418879\ldots.$$

Notice that the radius of C_1, a, has cancelled away, which supports my earlier claim that the probability is scale-invariant.

Well, all this is fine as it stands, BUT—how do we *really know* that one of our admittedly casual manipulations along the way didn't have a hidden flaw in it?[12] (Like the claim, for example, that $\{\Delta x\}^2$ is the differential area in rectangular coordinates, in which I've replaced the Δy in the usual $dA = \Delta x \Delta y$ with another Δx. Or, what about the claim a randomly selected point from the interior of C_1 is in the Δx band around C_2, as that *assumes* C_2 is totally inside C_1?) The integral we evaluated was not a technically difficult one to do, but how do we know we arrived at the *correct* integral? This is a question, often confronting the engineering analyst, which might not be of great concern to a pure mathematician who is simply looking for an 'interesting integral.'

When we 'check' integrals that are just given to us, we'll use *integral*, but when we are faced with this new type of question we need to do something different. What we'll do still uses a computer, but now we'll *simulate* the physical process of drawing a circle C_2 using points chosen randomly from the interior of a given circle, C_1. That is, our computer will 'draw' many such random C_2 circles and literally count the fraction of them that are totally inside C_1. The computer code that accomplishes this will be developed by an analysis that is distinct, separate, and independent of the mathematical arguments used to arrive at the integral in our theoretical result. So, here's how to create what physicists call a *Monte Carlo simulation* of the problem, a technique that in the pre-computer days of the 1920s Edwards could only have imagined in a science fiction fantasy.

Given three points that are not collinear, p_1, p_2, and p_3, where $p_1 = (x_1, y_1)$, $p_2 = (x_2, y_2)$, and $p_3 = (x_3, y_3)$, we form two chords: chord a as $p_1 p_2$ and chord b as $p_2 p_3$, with centers as $\left(\frac{x_1+x_2}{2}, \frac{y_1+y_2}{2}\right)$ and $\left(\frac{x_2+x_3}{2}, \frac{y_2+y_3}{2}\right)$, respectively. The equations of these two chords are

---

[12]The analysis I've just taken you through is the one given on pp. 817–818 of Edwards' book that I mentioned in the original Preface.

$$y_a = m_a(x - x_1) + y_1$$

and

$$y_b = m_b(x - x_2) + y_2$$

where m_a and m_b are the slopes of chord a and chord b, respectively. In fact,

$$m_a = \frac{y_2 - y_1}{x_2 - x_1}$$

and

$$m_b = \frac{y_3 - y_2}{x_3 - x_2}.$$

The center of C_2 is the intersection point of the perpendicular bisectors of the two chords (dashed lines in Fig. 1.8.1). The slope of a line perpendicular to a line with slope m is the negative reciprocal of m, and so the equations of the perpendicular bisectors are

$$y_A = -\frac{1}{m_a}\left(x - \frac{x_1 + x_2}{2}\right) + \frac{y_1 + y_2}{2}$$

and

$$y_B = -\frac{1}{m_b}\left(x - \frac{x_2 + x_3}{2}\right) + \frac{y_2 + y_3}{2}.$$

The intersection point of the bisectors (the center of C_2) is such that $y_A = y_B$ and so, solving for the x-ordinate of the center, we have

$$x_c = \frac{m_a m_b(y_1 - y_3) + m_b(x_1 + x_2) - m_a(x_2 + x_3)}{2(m_b - m_a)}.$$

The value of y_c, the y-ordinate of the center, is found by substituting x_c into either of the y_A, y_B equations. Finally, the distance of the center of C_2 from the center of C_1 is $\sqrt{x_c^2 + y_c^2}$. The radius of C_2 is the distance between the center of C_2 and any one of the three points p_1, p_2, p_3. As long as the sum of these two distances is no more than 1, C_2 is inside C_1. Otherwise, C_2 penetrates C_1.

The MATLAB code **circles.m** performs all of this grubby number-crunching, over and over, a total of one million times, keeping track of how many of those times

C_2 is inside C_1. Running **circles.m** numerous times produced estimates[13] for the probability C_2 is inside C_1 ranging over the interval 0.39992–0.400972. Comparing this interval with the theoretical result we computed earlier, 0.418879, leaves one (me, anyway) with a feeling of concern. A million samples is a lot of samples, and yet we have a disagreement between theory and 'experiment' of about 5%. That's not an insignificant difference. In addition, you'll notice that the code's interval of estimates does not include the theoretical result while, in general, multiple Monte Carlo computer simulations will nearly always *bound* a theoretical result, sometimes overestimating and other times underestimating the theoretical value.

It is difficult to look at the simulation results and not to come away with the feeling that the actual probability C_2 is inside C_1 is *exactly* 0.4. That is, $\frac{2}{5} = \frac{2}{15/3}$ rather than $\frac{2\pi}{15} = \frac{2}{15/\pi}$. But that's simply speculation on my part. *Why* the lack of better agreement between theory and experiment? I don't know. *Perhaps* there is a subtle error in the theoretical analysis. Or *perhaps* I made an error in the code. But it doesn't matter, for this book. My only point with this example is to show you how a theoretical analysis (involving an integral) and a computer can work together.

circles.m

```
% Created by P.J.Nahin (9/12/2012) for Inside Interesting Integrals.
% This code performs simulations of the 'circle in a circle' problem
% one million times.
inside=0;
for loop1=1:1000000
   for loop2=1:3
     done=0;
     while done==0
       x=-1+2*rand;y=-1+2*rand;
       if x^2+y^2<=1;
         px(loop2)=x;py(loop2)=y;done=1;
       end
     end
   end
   ma=(py(2)-py(1))/(px(2)-px(1));
   mb=(py(3)-py(2))/(px(3)-px(2));
   xc=ma*mb*(py(1)-py(3))+mb*(px(1)+px(2))-ma*(px(2)+px(3));
   xc=xc/(2*(mb-ma));
   yc=(-1/ma)*(xc-((px(1)+px(2))/2));
   yc=yc+(py(1)+py(2))/2;
   center=sqrt(xc^2+yc^2);
   radius=sqrt((xc-px(1))^2+(yc-py(1))^2);
   if center+radius<=1
     inside=inside+1;
   end
end
inside/loop1
```

---

[13]This *interval* of estimates is a result of the code's use of a random number generator (with the *rand* command)—every time we run the code we get a new estimate that is (slightly) different from the estimates produced by previous runs.

This is where the discussion of the problem ended in the first edition of the book, but I have since learned that my suspicion (from the simulation result) that the correct answer is 2/5 = 0.4 was, in fact, well-founded. The history of this problem goes back to mid-Victorian times, to its original proposal by the editor of *The Educational Times* in England.[14] The solution in Edwards book that I've reproduced here (see note 12) was the work of the British actuary Wesley Woolhouse (1809–1893), dating from 1867, who acknowledged that his result of $2\pi/15$ was only approximate. Another correspondent to *The Educational Times* (writing under the pen-name of *N'Importe*, French for 'it matters not') first published the correct 0.4 result in 1870. In the January 1879 issue of *The Mathematical Visitor* the American amateur mathematician Enoch Seitz (1846–1883) solved a more general problem, of which the circle-in-a-circle problem is a special case, thus firmly establishing the 0.4 result.[15] I particularly like this problem because it illustrates just how powerful the *combined* use of analysis and electronic computation can be. (The final discussion of this section will give you yet another example of the Monte Carlo method providing support to a theoretical analysis.)

For a second example of where integrals come from, let's go back to June 1827, when the Scottish botanist Robert Brown (1773–1858) observed (through a one-lens microscope, that is, a magnifying glass, with a magnification in excess of 300) the chaotic motion of tiny grains of plant pollen suspended in water drops. This motion had earlier been noticed in passing by others, but Brown took the time to publish what he saw in a September 1828 paper in the *Philosophical Magazine*, and thereby initiated a search for what was going on. The German-born physicist Albert Einstein (1879–1955), in a series of papers published between 1905 and 1908, applied statistical mechanics to show that what is now called *Brownian motion* is the result of the random bombardment of the particles by the molecules of the suspension medium. Indeed, Brownian motion is viewed as strong macroscopic experimental evidence for the reality of molecules (and so of atoms, too).

Figure 1.8.2 shows four typical paths of particles executing Brownian motion in two dimensions,[16] and they *are* pretty erratic. In fact, in the early 1920s the American mathematician Norbert Wiener (1894–1964) made a deep analysis of the *mathematics* of Brownian motion (Einstein explained the *physics*), studying in

---

[14]*The Educational Times* was a self-described "monthly journal of education, science and literature," published by the College of Preceptors in London, to deal with general educational questions at the school and university level. In the years it appeared, 1847–1918, over 18,000 questions on mathematics were presented and solved.

[15]You can find Seitz's solution reprinted in *Problems and Solutions from The Mathematical Visitor 1877–1896* (Stanley Rabinowitz, editor), MathPro Press 1996, pp. 125–126.

[16]Imagine that we have defined the maximum absolute value of a *step*, which will be our unit distance. Then, in one dimension (call it x) the particle moves, after each molecular hit, a distance randomly selected from the interval −1 to +1. In a second, perpendicular direction (call it y) the particle moves, after each molecular hit, a distance randomly selected from the interval −1 to +1. Figure 1.8.2 shows the combined result of these two *independent* motions for four particles, each for 500 hits (the four curves are MATLAB simulations).

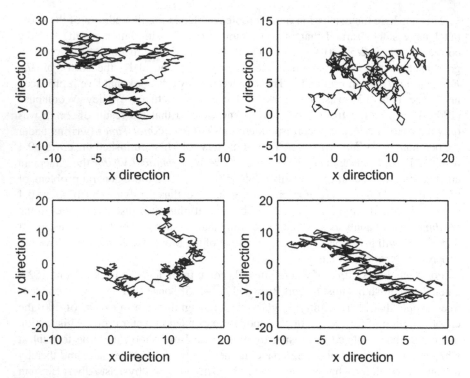

Fig. 1.8.2 Typical paths of two-dimensional Brownian motion (each 500 steps)

particular what happens in the limit of a particle being *continuously* hit by molecules (that is, successive hits are separated by vanishingly small time intervals). In that limit, Brownian motion becomes what mathematicians somewhat forbiddingly call a *Wiener stochastic random process*. Each realization of such a process is called a 'Wiener random walk,' and they are continuous curves that are so kinky they *fail*, at almost every point, to have a derivative (that is, to have a direction)!

Now, suppose at time $t = 0$ we put a particle at the origin of a plane and then watch it execute a one-dimensional Wiener walk on the x-axis. Where will it be at some later time $t > 0$? Certainly it will always be on the x-axis, but *where*? Since the walk is random, the best we can do is give the *probability* the particle will be somewhere, and this was one of the first things Einstein calculated in 1905. To be specific, let's write $W(x, t)$ as the probability the particle will be somewhere in the interval $(x, x + \Delta x)$ at time t. (The W is, of course, in honor of Wiener.) Let's assume this probability is, for very short intervals, *linear* in Δx (i.e., double the length the of the interval and so double the probability the particle is there), and so

$$W(x, t) = f(x, t)\Delta x$$

where $f(x, t)$ is called the *probability density function* of the Wiener walk (a *density* because we multiply it by the interval length Δx to get the probability).

Einstein showed, through ingenious physical arguments (which we can ignore here), that f(x, t) is the solution to the second-order partial differential equation

$$\frac{\partial f}{\partial t} = D \frac{\partial^2 f}{\partial x^2}$$

where D is a constant. (D is determined by a number of physical constants, like the mass of the particle and the temperature, and Einstein derived a formula for D, but for us it is sufficient to just write D.) This differential equation is, in fact, a famous one in mathematical physics, called the *heat* or *diffusion* equation, and so D is called the *diffusion constant*. In the usual appearance of this equation, the function being solved for is the temperature in a solid as a function of location and time, but Einstein showed that the probability density function for the Wiener walk satisfies the very same equation. Amazing!

The solution for f(x, t) is

$$f(x, t) = \frac{e^{-x^2/4Dt}}{2\sqrt{\pi Dt}},$$

which is easily verified by direct substitution, although it can be formally derived using nothing but simple combinatorial (that is, purely *mathematical*) arguments.[17] Since the particle has to be *somewhere* (I do hope that is obvious!) it must be true that if we integrate the probability density over the *entire* x-axis we have to get a probability of 1. That is,

$$\int_{-\infty}^{\infty} f(x, t)dx = 1$$

for all $t \geq 0$.[18]

From elementary probability theory we know the average value of x, at any time t, is

$$< x > = \int_{-\infty}^{\infty} x\, f(x, t)dx = 0.$$

This is *mathematically* so because f(x, t) is even and so the integrand is an odd function of x, and it is *physically* so because a one-dimensional Wiener walk is equally likely to move in either direction at every instant in time. And yet, as Fig. 1.8.2 clearly illustrates, a *two*-dimensional Wiener walk does tend to slowly migrate away in absolute distance from the origin at t increases. A measure of this

---

[17]If you are curious about the details of such a derivation, you can find them in my book *Mrs. Perkins's Electric Quilt*, Princeton University Press 2009, pp. 263–267.

[18]For a proof of this, see *Mrs. Perkins's*, pp. 282–283.

drift is the average *squared* value of x (because then the individual horizontal (and vertical) movements of the particle that have opposite signs don't tend to cancel each other). So, like Einstein, let's calculate

$$< x^2 >= \int_{-\infty}^{\infty} x^2 \, f(x,t)dx = \frac{1}{2\sqrt{\pi Dt}} \int_{-\infty}^{\infty} x^2 \, e^{-x^2/4Dt} dx$$

or, since the integrand is an even function of x,

$$< x^2 >= \frac{1}{\sqrt{\pi Dt}} \int_{0}^{\infty} x^2 \, e^{-x^2/4Dt} dx.$$

Making the obvious change of variable $u = \frac{x^2}{4Dt}$, we have

$$\frac{du}{dx} = \frac{2x}{4Dt} = \frac{x}{2Dt}$$

and so

$$dx = \frac{2Dt}{x} \, du.$$

Or, since $x = 2\sqrt{Dt}\sqrt{u}$, we have

$$dx = \frac{2Dt}{2\sqrt{Dt}} \left(\frac{du}{\sqrt{u}} \right) = \sqrt{Dt} \, \frac{du}{\sqrt{u}}.$$

Thus,

$$< x^2 >= \frac{1}{\sqrt{\pi Dt}} \int_{0}^{\infty} 4Dt \, u \, e^{-u} \sqrt{Dt} \, \frac{du}{\sqrt{u}} = Dt \frac{4}{\sqrt{\pi}} \int_{0}^{\infty} \sqrt{u} \, e^{-u} \, du.$$

At this point we actually have Einstein's basic result that $<x^2>$ varies *linearly* with t because, whatever the value of the integral, we know it is simply a *number*. For us *in this book*, however, the calculation of the integral (which has appeared in a natural way in a physical problem) is the challenge. In Chap. 4 we'll do this calculation—see (4.2.8)—and find that the integral's value is $\frac{1}{2}\sqrt{\pi}$. So (and just as Einstein wrote), $<x^2> = 2Dt$.

For my third example of where integrals come from, this time from pure mathematics, just imagine someone has just dropped *this* on your desk: show that

$$\int_{1}^{\infty} \frac{\{x\} - \frac{1}{2}}{x} \, dx = -1 + \ln\left(\sqrt{2\pi}\right) = -0.08106\ldots,$$

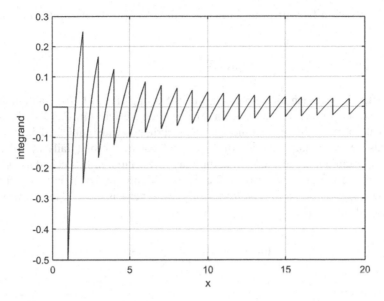

Fig. 1.8.3 The *Irresistible Integrals* integrand

where {x} denotes the fractional part of x (for example, {5.619} = 0.619 and
{7} = 0). Holy cow!, I can just hear most readers exclaim, how do you prove
something like *that*? The authors of *Irresistible Integrals* outline a derivation (see
their pp. 92–93) but skip over a *very* big step, with just a tiny hint at how to proceed.
At least as interesting as the derivation, itself, is that the integral was not simply
'made-up out of the blue,' but is actually the result of a preliminary analysis.

Before getting into the derivation details, it is useful to get a 'physical feel' for the
integral. In Fig. 1.8.3 the integrand is plotted over the interval $1 \leq x \leq 20$. You can
see from that plot that the integrand starts off with a negative area over the interval
$1 \leq x \leq 1.5$, which is then partially cancelled with positive area from $1.5 \leq x \leq 2$,
followed by a negative area from $2 \leq x \leq 2.5$, which is then *almost* cancelled by
positive area from $2.5 \leq x \leq 3$, and so on. These alternating area cancellations are
never quite total, but as $x \to \infty$ the area cancellations do become ever closer to total,
which physically explains why the integral exists (doesn't blow-up). Since the
negative areas are always slightly bigger in magnitude than the positive ones, the
final slightly negative value for the integral does make sense.

We can numerically check these observations with *integral*, over the integration
interval in Fig. 1.8.3. Thus, *integral(@(x)(x-floor(x)-0.5)./x,1,20) = −0.0769...*,
where *floor(x)* computes the largest integer <x. That is, *floor(x)* rounds x down
and so $x - floor(x) = \{x\}$. This is in pretty good agreement with the *Irresistible
Integrals* theoretical result and, if we increase the integration interval, MATLAB's
value does seem to approach the theoretical value (for example, using an upper limit
of 100, MATLAB computes −0.08017). In any case, how do we *derive* the value of
the integral?

We start with a famous result in mathematics, *Stirling's asymptotic formula* for n!:

$$n! \tilde{\sqrt{2\pi}}\, n^{n+\frac{1}{2}}\, e^{-n},$$

named after the Scottish mathematician James Stirling (1692–1770)—although it is known that the French mathematician Abraham de Moivre (1667–1754) knew an equivalent form at the same time (or even earlier)—who published it in 1730. Factorials get very large, very fast (my hand calculator first fails at 70!), and Stirling's formula is quite useful in computing n! for large n. It is called *asymptotic* because, while the *absolute* error in the right-hand side in evaluating the left-hand side blows-up as n $\rightarrow \infty$ the *relative* error goes to zero as n $\rightarrow \infty$ (that's why $\sim$ is used instead of $=$). That is,

$$\lim_{n \to \infty} \frac{n!}{\sqrt{2\pi}\, n^{n+\frac{1}{2}}\, e^{-n}} = 1$$

or, alternatively,

$$\lim_{n \to \infty} \frac{n!}{n^{n+\frac{1}{2}}\, e^{-n}} = \sqrt{2\pi}.$$

Tuck this away in your memory because we'll use it at the end of our derivation of the above integral. So, here we go.

We start by writing

$$\ln\{n!\} = \ln\{n(n-1)(n-2)\ldots(3)(2)(1)\}$$
$$= \ln(n) + \ln(n-1) + \ldots + \ln(2) + \ln(1)$$

or, since $\ln(1) = 0$,

$$\ln\{n!\} = \sum_{k=2}^{n} \ln(k).$$

Next, first notice that since

$$\int_{1}^{k} \frac{dx}{x} = \{\ln(x)\}\Big|_{1}^{k} = \ln(k)$$

then

$$\ln\{n!\} = \sum_{k=2}^{n} \int_{1}^{k} \frac{dx}{x},$$

and then further notice that

$$\int_1^k \frac{dx}{x} = \sum_{j=1}^{k-1} \int_j^{j+1} \frac{dx}{x}.$$

Thus,

$$\ln\{n!\} = \sum_{k=2}^n \left\{ \sum_{j=1}^{k-1} \int_j^{j+1} \frac{dx}{x} \right\}.$$

So far, so good. It is at this point, however, that the authors of *Irresistible Integrals* write that the next thing to do is "exchange the order of the two sums," and then they *immediately* write

$$\ln\{n!\} = \int_1^n \frac{n - \lfloor x \rfloor}{x} dx$$

where the notation $\lfloor x \rfloor$ means the integer part of x (for example, $\lfloor 5.619 \rfloor = 5$). Clearly,

$$x = \lfloor x \rfloor + \{x\}.$$

This 'exchange' step may look mysterious to you (it did to me, at first!), but here's how to see it. Write down the terms of the inner sum for each value of k as that index goes from 2 to n (the outer sum):

$k = 2$: $\int_1^2$

$k = 3$: $\int_1^2 + \int_2^3$

$k = 4$: $\int_1^2 + \int_2^3 + \int_3^4$

. . . .

$k = n$: $\int_1^2 + \int_2^3 + \int_3^4 + \ldots + \int_{n-1}^n$.

As written, the double summation adds all these integrals *horizontally*, across row by row. Exchanging the order of the sums simply means to add all the integrals *vertically*, down column by column. You get the same answer, either way! Thus,

$$\ln\{n!\} = (n-1)\int_1^2 \frac{dx}{x} + (n-2)\int_2^3 \frac{dx}{x} + (n-3)\int_3^4 \frac{dx}{x} + \ldots + \int_{n-1}^n \frac{dx}{x}.$$

$$= \int_1^2 \frac{(n-1)}{x} dx + \int_2^3 \frac{(n-2)}{x} dx + \int_3^4 \frac{(n-3)}{x} dx + \ldots + \int_{n-1}^n \frac{1}{x} dx.$$

The general form of the terms in the last summation is, with $1 \le j \le n - 1$,

$$\int_j^{j+1} \frac{(n-j)}{x} dx = \int_j^{j+1} \frac{n - \lfloor x \rfloor}{x} dx$$

because, for the integration interval $j \leq x < j+1$, we have *by definition* that $\lfloor x \rfloor = j$. Thus,

$$\ln\{n!\} = \int_1^n \frac{n - \lfloor x \rfloor}{x} dx$$

as claimed in *Irresistible Integrals*.

Now, since

$$\lfloor x \rfloor = x - \{x\}$$

then

$$n - \lfloor x \rfloor = n - [x - \{x\}] = n - x + \{x\}.$$

So,

$$\ln\{n!\} = \int_1^n \frac{n - x + \{x\}}{x} dx = \int_1^n \frac{n}{x} dx - \int_1^n dx + \int_1^n \frac{\{x\}}{x} dx$$

$$= n \ln(n) - (n-1) + \int_1^n \frac{\{x\}}{x} dx = n \ln(n) - n + 1 + \frac{1}{2}\ln(n) + \int_1^n \frac{\{x\} - \frac{1}{2}}{x} dx$$

$$= \left(n + \frac{1}{2}\right) \ln(n) - n + 1 + \int_1^n \frac{\{x\} - \frac{1}{2}}{x} dx$$

$$= \ln\left(n^{n+\frac{1}{2}}\right) + \ln(e^{-n}) + 1 + \int_1^n \frac{\{x\} - \frac{1}{2}}{x} dx$$

$$= \ln\left(n^{n+\frac{1}{2}} e^{-n}\right) + \ln\left(e^{1+\int_1^n \frac{\{x\}-\frac{1}{2}}{x} dx}\right) = \ln\left(n^{n+\frac{1}{2}} e^{-n} e^{1+\int_1^n \frac{\{x\}-\frac{1}{2}}{x} dx}\right).$$

So,

$$n! = n^{n+\frac{1}{2}} e^{-n} e^{1 + \int_1^n \frac{\{x\} - \frac{1}{2}}{x} dx}$$

or,

$$e^{1 + \int_1^n \frac{\{x\} - \frac{1}{2}}{x} dx} = \frac{n!}{n^{n+\frac{1}{2}} e^{-n}}$$

or, if we now let $n \to \infty$ and recall Stirling's formula, we have

$$e^{1 + \int_1^\infty \frac{\{x\} - \frac{1}{2}}{x} dx} = \sqrt{2\pi}.$$

Thus,

$$1 + \int_1^\infty \frac{\{x\} - \frac{1}{2}}{x} dx = \ln\left(\sqrt{2\pi}\right)$$

and so, just as *Irresistible Integrals* claimed,

$$\int_1^\infty \frac{\{x\} - \frac{1}{2}}{x} dx = -1 + \ln\left(\sqrt{2\pi}\right). \tag{1.8.1}$$

To really be sure you've understood the derivation of (1.8.1), try your hand at the second challenge problem of Chap. 5 (when you get there).

For my next example of this section, let me take you back in time to the morning of January 6, 1876, to the ancient Victorian campus of Cambridge University. In an examination room there, approximately 100 extremely nervous math students are in the fourth day of a *nine* (!) day test. Their state of anxiety has not been helped as they stare at the following question, one which (as they listen to a ticking clock on the wall) they have just 18 min to answer before having to move on to the next problem:

"A uniform steel wire in the form of a circular ring is made to revolve in its own plane about its centre of figure. Show that the greatest possible linear velocity is independent both of the [cross]section of the wire and the radius of the ring, and find roughly this velocity, the breaking strength of the wire being given as 90,000 lbs per square inch, and the weight of a cubic foot [of steel] as 490 lbs." This problem was the creation of Professor John William Strutt (1842–1919), better known in the world of mathematical physics as Lord Rayleigh (three decades later he would win the 1904 Nobel Prize in physics). The problem, part of the annual Cambridge University Mathematical Tripos exam, stunned all of the students. As the eventually top-scoring man (the so-called *Senior Wrangler*) remembered it, Rayleigh's ring problem was "uncommonly high" (Victorian code for 'damn hard'). Here's how to do it.

A modern engineering student quickly "sees" the ultimate fate of the ring as its rate of rotation increases: each portion of the ring experiences an outward force (the centrifugal force, called a fictitious force by physicists, but a seemingly quite real force to all who have felt it on a merry-go-round) that increases with the spin rate until the ring eventually explosively disintegrates. In particular, an arbitrary half of the ring—see Fig. 1.8.4—experiences a force that is attempting to separate it from the other half (which is 'enjoying' the same experience).

We imagine the ring, with radius r, rotating in a vertical plane, where dm denotes a differential mass of the ring. Fig. 1.8.4 shows two such masses, one at angle θ and a matching mass at angle $180^\circ - \theta$. The radially-outward centrifugal force dF on each of the dm's can be resolved into horizontal and vertical components, where it is clear

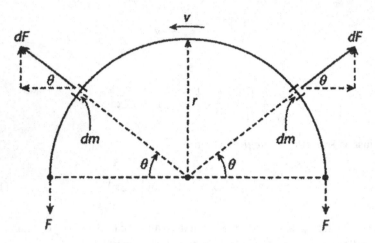

Fig. 1.8.4 Arbitrary half of Rayleigh's rotating ring

by symmetry that the two horizontal components cancel. The two equal vertical components, however, add, and so if we sum (integrate) the vertical components over $0 \le \theta \le \pi/2$ radians, all we need do is double the result to get the total vertical centrifugal force acting on the half-ring.

The differential length of each differential mass is $dl = rd\theta$ and so, with a uniform cross-sectional area of a, the volume of a differential mass is adl. If the ring has density ρ, then the differential mass is $dm = adl\rho = a\rho rd\theta$. If the tangential speed of the ring is v, then the radially outward differential centrifugal force on a differential mass is, from the well-known formula for centrifugal force,

$$dF = \frac{v^2 dm}{r} = a\rho v^2 d\theta,$$

and so the *vertical, upward* component of the differential force on dm is given by

$$dF\sin(\theta) = a\rho v^2 \sin(\theta)d\theta.$$

Thus, the total integrated vertical force on the half-ring is given by the freshman calculus integral (which makes it no less important)

$$2\int_0^{\pi/2} a\rho v^2 \sin(\theta)d\theta = 2a\rho v^2 [-\cos(\theta)]\big|_0^{\pi/2} = 2a\rho v^2.$$

This is the force that is balanced—until the ring comes apart—by the tensions in the two ends of the half-ring. So,

$$2F = 2a\rho v^2$$

or,

$$F = a\rho v^2$$

Assuming the wire's cross-sectional dimensions are 'small' compared to r, the stress in the ring—at *every* point in the ring, not just at the ends of the imagined half-ring—is given by

$$\sigma = \frac{F}{a} = \rho v^2.$$

Thus,

$$v = \sqrt{\frac{\sigma}{\rho}}.$$

This last expression is fine and good as it stands, but what's the *numerical* value of v, which is what Rayleigh specifically asked for? Surviving exam sheets show that some of the students correctly derived the symbolic expression, but also that *nobody* was able to use it to compute the correct numerical value for the maximum v before explosive disintegration destroyed the ring. How, indeed, do you get something with the units of feet/second (v) from lbs per square inch (σ) and lbs per cubic foot (ρ)? To answer that would take us too far away from the mathematical theme of this book and into engineering physics, but if you're interested take a look at ...[19]

As the next example of how integrals arise in scientific work, I'll use a Feynman (who, you'll recall, opened the new Preface) curiosity. While remembered as a physicist, Feynman was also a very good mathematician, and a two-page, unpublished manuscript dealing with Fermat's Last Theorem[20] is among his papers.

---

[19]Paul J. Nahin, "Rayleigh's Rotating Ring," *The Mathematical Intelligencer*, Summer 2017, pp. 72–75. The answer to Rayleigh's question is that the ring reaches its structural limit at just a bit <2700 rpm. This problem isn't as theoretical as it might appear. A specific concern over high-speed rotational disintegration occurred when Rayleigh's friend, the English physicist Oliver Lodge (1851–1940), conducted experiments in a study of the ether (the mythical substance Victorian physicists at one time believed filled all of what appeared to be empty space, through which electromagnetic waves, light, could travel). Lodge's experimental setup involved the high-speed rotation of massive steel plates and, by Christmas of 1891, he was operating them at 2800 rpm. At that speed, one of Lodge's friends worried that if there were any flaws in the plates they could disintegrate, and as Lodge himself wrote in a lab notebook, "we should have our heads cut off."

[20]Named after the French lawyer Pierre de Fermat (1601–1665), who sometime around 1637 wrote in the margin of one of his private books that he had discovered a truly wonderful proof that there are no integer solutions to $x^n + y^n = z^n$ for any $n \geq 3$. As is well-known, Fermat didn't provide that proof, claiming the margin too small to hold it. It wasn't until more than three centuries later that a proof was finally found (mathematicians believe, today, that whatever Fermat *thought* he had in 1637, he later realized was flawed).

This paper is undated, but it must have been written no later than 1963 because it is known that he discussed his results with a Caltech mathematician who died that year. At that time, the famous Fermat theorem was still unproven, and so Feynman came-up with the following quite curious *probabilistic* analysis.

He started by asking himself 'What is the prob[ability] N is a perfect n-*th* power?' He answered[21] this by mapping the unit interval defined by two large consecutive integers (N and N + 1) into the shorter interval of the n-*th* roots of the unit interval (you'll see why he did that, in just a moment). Such a mapping is monotonic, with the end-points mapping into the end-points, and all the interior points in the unit interval mapping into all the interior points of the smaller interval, with the ordering preserved. To determine the length l of the shorter interval, Feynman wrote

$$l = (N+1)^{1/n} - N^{\frac{1}{n}} = \left[N\left(1+\frac{1}{N}\right)\right]^{\frac{1}{n}} - N^{\frac{1}{n}} = N^{\frac{1}{n}}\left(1+\frac{1}{N}\right)^{\frac{1}{n}} - N^{\frac{1}{n}}$$

or

$$l = N^{\frac{1}{n}}\left[\left(1+\frac{1}{N}\right)^{\frac{1}{n}} - 1\right].$$

He then recalled the power-series expansion (the binomial theorem)

$$(1+x)^a = 1 + ax + \frac{a(a-1)}{2}x^2 + \cdots$$

and set $x = 1/N$ and $a = 1/n$. Thus,

$$l = N^{\frac{1}{n}}\left[\left(1+\frac{1}{n}\cdot\frac{1}{N}+\frac{\frac{1}{n}\left(\frac{1}{n}-1\right)}{2}\cdot\frac{1}{N^2}+\cdots\right)-1\right]$$

which, for large N, becomes (if we keep just the first surviving term),

$$l \approx \frac{N^{\frac{1}{n}}}{nN}.$$

For example, if $N = 1000$ and $n = 4$ then this *approximate* expression says $l = 0.0014059$ while the exact value is $(1,001)^{1/4} - (1,000)^{\frac{1}{4}} = 0.0014053$. Pretty close.

---

[21]I've taken the Feynman quotes from Silvan S. Schweber, *QED and the Men Who Made It: Dyson, Feynman, Schwinger, and Tomonaga*, Princeton 1994, p. 464. 'QED' is quantum electrodynamics, the theory that won Feynman his share of the 1965 Nobel Prize in physics.

Feynman then wrote "Probability that N is a perfect n-*th* power is $\frac{N^{\frac{1}{n}}}{nN}$ [and]
therefore probability $x^n + y^n$ is a perfect n-*th* power is $\frac{(x^n+y^n)^{1/n}}{n(x^n+y^n)}$." The second half
of this claim certainly follows from the first half (just set $N = x^n + y^n$), but where
does the first half come from? Feynman himself left no record, but what I think he
had in mind is simply this: the shorter, mapped interval of length *l* includes an integer
with probability $l/1$ because the integers are spaced—dare I actually write this?—
unit distance apart! Feynman then continues with "Therefore the total probability
any $x^n + y^n$ is a perfect n-*th* power [that is, is z^n] for $x > x_0$ and $y > y_0$ is equal to

$$\int_{x_0}^{\infty} \int_{x_0}^{\infty} \frac{1}{n} (x^n + y^n)^{-1+1/n} dxdy = \frac{2}{nx_0^{n-3}} c_n$$

where
$c_n = \frac{1}{2} \int_0^{\infty} \int_0^{\infty} \frac{1}{n} (u^n + v^n)^{-1+1/n} dudv$."
    Feynman *integrated* over all x and y rather than *summing* over all integer x and y
simply (I believe) because integrals are generally easier to do than are sums. Most
analysts would have no problem with doing that. But what happened to y_0? Why
isn't the double integral actually $\int_{x_0}^{\infty} \int_{y_0}^{\infty}$ instead of $\int_{x_0}^{\infty} \int_{x_0}^{\infty}$? This is explained
(I think) by imagining that Feynman thought there to be no reason why y couldn't
range over the same interval as does x, and so he took $y_0 = x_0$. So far, so good. Now,
where do Feynman's double integral and c_n expressions come from?
    If we make the obvious change of variables

$$s = \frac{x - x_0}{x_0}, t = \frac{y - x_0}{x_0},$$

which maps x_0 into 0 and ∞ into ∞, then (with $dx = x_0 ds$, $dy = x_0 dt$, $x = (s+1) x_0$,
and $y = (t + 1) x_0$) we have

$$\int_{x_0}^{\infty} \int_{x_0}^{\infty} \frac{1}{n} (x^n + y^n)^{-1+1/n} dxdy = \int_0^{\infty} \int_0^{\infty} \frac{1}{n} \{(s+1)^n x_0^n + (t+1)^n x_0^n\}^{\frac{1-n}{n}} x_0 dsx_0 dt$$

$$= \frac{x_0^{1-n} x_0^2}{n} \int_0^{\infty} \int_0^{\infty} \{(s+1)^n + (t+1)^n\}^{\frac{1-n}{n}} dsdt$$

$$= \frac{1}{nx_0^{n-3}} \int_0^{\infty} \int_0^{\infty} \{(s+1)^n + (t+1)^n\}^{\frac{1-n}{n}} dsdt.$$

    Next, let $u = s + 1$ and $v = t + 1$. Then, finally, we have for the probability of a
solution to $x^n + y^n = z^n$ the value of the integral

$$\frac{1}{nx_0^{n-3}} \int_1^{\infty} \int_1^{\infty} \{u^n + v^n\}^{-1+\frac{1}{n}} dudv.$$

This is almost Feynman's result (except for the lower limits on the integrals), with one final touch. Feynman included the factor of $\frac{1}{2}$ in front of the c_n integral because integration over all u and v counts any possible solution twice (swapping the values of u and v does not give distinct solutions), which explains (I think) why he included the 2 in his first expression. From all of this Feynman somehow concluded that 'the probability of Fermat being wrong was $<10^{-200}$.' Well, whatever you might think of Feynman's double integral, and of how he arrived at it, you have to admit it's 'interesting'!

For my next example of the appearance of integrals in the study of physical problems, consider the vast topic of inequalities. Inequalities are one of the most powerful tools in mathematics, in that they make general statements that cover classes of functions, and are not limited to specific cases. There are an enormous number of mathematical inequalities, but I'd bet if you asked a 1000 professional mathematicians to each list their choice for the top ten, *every last one of them* would put the Cauchy-Schwarz inequality as one of their top two.[22] That inequality gets its name from the French mathematician Augustin-Louis Cauchy (who gets a whole chapter at the end of this book), and the German mathematician Hermann Schwarz (1843–1921). Each discovered different forms of the inequality. The Russian mathematician V. Y. Bunyakovsky (1804–1889), however, was the first to publish it (in 1859), in the form we'll develop here (in terms of integrals).

The derivation of the Cauchy-Schwarz inequality uses nothing beyond high school algebra and a touch of AP-calculus. Suppose f(t) and g(t) are any two real-valued functions of the real variable t. Then, for λ any real-valued parameter, and a and b as two constants (each of which may take their values from minus infinity to plus infinity), it is certainly true that

$$\int_a^b \{f(t) + \lambda g(t)\}^2 dt \geq 0 \qquad (1.8.2)$$

because, as something real and *squared*, the integrand in (1.8.2) is nowhere negative. Expanding (1.8.2), we have

$$\lambda^2 \int_a^b g^2(t)dt + 2\lambda \int_a^b f(t)g(t)dt + \int_a^b f^2(t)dt \geq 0. \qquad (1.8.3)$$

Since the definite integrals in (1.8.3) are constants (call their values A, B, and C, respectively), (1.8.3) is simply a quadratic in λ. Writing $h(\lambda)$ for the left-hand-side of (1.8.3), we have

---

[22]In Ralph Palmer Agnew's 1960 *Differential Equations* we read the following words (page 370): "This [the Cauchy-Schwarz inequality] is an exceptionally potent weapon. There are many occasions on which persons who know and think of using this formula can shine while their less fortunate brethren flounder."

$$h(\lambda) = A\lambda^2 + 2B\lambda + C \geq 0 \qquad (1.8.4)$$

where

$$A = \int_a^b g^2(t)dt, B = \int_a^b f(t)g(t)dt, C = \int_a^b f^2(t)dt.$$

The inequality in (1.8.4) has a simple geometric interpretation: a plot of $h(\lambda)$ versus λ does *not* cross the horizontal (λ) axis (that, after all, is what the 'greater than or equal to' sign *means*). At most, such a plot (a parabola) may just *touch* the λ axis, if the 'greater than or equal to' condition is actually one of equality, that is if $h(\lambda) = 0$. All of this means that there can be no *real* solutions (other than the possibility there is a double root to $A\lambda^2 + 2B\lambda + C = 0$) because a *real* solution would be the location of a λ-axis crossing. In other words, the two solutions to the quadratic $A\lambda^2 + 2B\lambda + C = 0$ must be the *complex* conjugate pair

$$\lambda = \frac{-2B \pm \sqrt{4B^2 - 4AC}}{2A} = \frac{-B \pm \sqrt{B^2 - AC}}{A}$$

where $B^2 < AC$ is the condition that gives complex values for λ (or a real double root if $B^2 - AC = 0$). So, the inequality $h(\lambda) \geq 0$ requires that

$$\left\{ \int_a^b f(t)g(t)dt \right\}^2 \leq \left\{ \int_a^b g^2(t)dt \right\} \left\{ \int_a^b f^2(t)dt \right\} \qquad (1.8.5)$$

and (1.8.5) is the Cauchy-Schwarz inequality, which holds for any real $f(t)$ and $g(t)$.

An elegant illustration of an illustration of the Cauchy-Schwarz inequality is its use in studying the autocorrelation function $R_f(\tau)$ of the real-valued function $f(t)$:

$$R_f(\tau) = \int_{-\infty}^{\infty} f(t)f(t - \tau)dt. \qquad (1.8.6)$$

Notice, carefully, that $R_f(\tau)$ is a function of τ, *not* t. The usual interpretation of t is as *time*, so what is τ? Dimensionally, of course, τ does have the units of time because you can't subtract apples from oranges and so $t - \tau$ means t and τ have the same units. But if τ isn't time, then what *is* it? We can answer that question by noticing that $R_f(\tau)$ is a measure of the similarity of $f(t)$ with a *shifted* (so τ is a time *shift*) version of itself: hence the name *correlation*, with *auto* coming from $f(t)$ being compared against itself.

The autocorrelation function has enormous application in the construction of electronic signal processing circuitry that can 'extract' an information-bearing signal that is literally *buried* in random noise. (I tell you all this just so you'll appreciate that the mathematical study of $R_f(\tau)$ is not some idle exercise in symbol-pushing. At the end of this discussion, I'll give you an elementary illustration of such a signal processing operation.)

The autocorrelation function has a number of fundamental, general properties, and here are three of them, each established by considering the nature of the *integral* definition of $R_f(\tau)$.

(a) $R_f(0) \geq 0$. This follows by simply inserting $\tau = 0$ into (1.8.6) and observing that $R_f(0) = \int_{-\infty}^{\infty} f^2(t)dt$, which is certainly never negative.

(b) $R_f(-\tau) = R_f(\tau)$. That is, $R_f(\tau)$ is an even function. This follows by first writing $R_f(-\tau) = \int_{-\infty}^{\infty} f(t)f(t+\tau)dt$ and then changing variable to $s = t + \tau$ ($ds = dt$). Then, $R_f(-\tau) = \int_{-\infty}^{\infty} f(s-\tau)f(s)ds$ or, making the trivial change in the dummy variable of integration from s to t, $R_f(-\tau) = \int_{-\infty}^{\infty} f(t-\tau)f(t)dt = R_f(\tau)$.

(c) $R_f(0) \geq |R_f(\tau)|$. That is, $R_f(\tau)$ attains its *maximum* value with a zero time shift. To establish this, first write $\int_{-\infty}^{\infty} \{f(t) \pm f(t-\tau)\}^2 dt \geq 0$ which (by an argument we've used now already twice before) is obviously true. Then, expanding,

$$\int_{-\infty}^{\infty} f^2(t)dt \pm 2\int_{-\infty}^{\infty} f(t)f(t-\tau)dt + \int_{-\infty}^{\infty} f^2(t-\tau)dt \geq 0$$

or,

$$R_f(0) \pm 2R_f(\tau) + R_f(0) \geq 0$$

or,

$$R_f(0) \geq \pm R_f(\tau)$$

from which $R_f(0) \geq |R_f(\tau)|$ immediately follows.

Now, how does the Cauchy-Schwarz inequality come into all of this? The last property, (c), has an important qualification. If f(t) is *periodic* (as is the sine-wave of a radar signal that has bounced off of a distant target and which has been received back at the transmitter site) with period T, then it seems clear that $R_f(\tau)$ will also be periodic with period T. That is,

$R_f(\tau)$ will achieve its maximum value not only at $\tau = 0$, but at $\tau = kT$ where k is *any* integer (not just k = 0). This is generally stated as follows, in the reverse: If there is a constant T > 0 such that $R_f(0) = R_f(T)$ then $R_f(\tau)$ is periodic with period T. Here's how to prove that with the Cauchy-Schwarz inequality.

Define $g(t) = f(t - \tau + T) - f(t - \tau)$. Then, the Cauchy-Schwarz inequality of (1.8.5) says

$$\left\{ \int_{-\infty}^{\infty} f(t)[f(t-\tau+T) - f(t-\tau)]dt \right\}^2 \tag{1.8.7}$$

$$\le R_f(0)\int_{-\infty}^{\infty}[f(t-\tau+T)-f(t-\tau)][f(t-\tau+T)-f(t-\tau)]dt.$$

The left-hand-side of (1.8.7) is

$$\left\{\int_{-\infty}^{\infty}f(t)f(t-\tau+T)dt - \int_{-\infty}^{\infty}f(t)f(t-\tau)dt\right\}^2 = \{R_f(\tau-T)-R_f(\tau)\}^2.$$

The integral on the right-hand-side of (1.8.7) is

$$\int_{-\infty}^{\infty}f(t-\tau+T)f(t-\tau+T)dt - \int_{-\infty}^{\infty}f(t-\tau)f(t-\tau+T)dt$$

$$-\int_{-\infty}^{\infty}f(t-\tau+T)f(t-\tau)dt + \int_{-\infty}^{\infty}f(t-\tau)f(t-\tau)dt$$

$$= R_f(0) - R_f(T) - R_f(T) + R_f(0) = 2R_f(0) - 2R_f(T).$$

Thus,

$$\{R_f(\tau-T)-R_f(\tau)\}^2 \le R_f(0)2[R_f(0)-R_f(T)]. \tag{1.8.8}$$

Now, remember we are *given* that $R_f(0) = R_f(T)$, and so (1.8.8) reduces to

$$\{R_f(\tau-T)-R_f(\tau)\}^2 \le 0$$

and we notice that the inequality must actually be equality. That is,

$$\{R_f(\tau-T)-R_f(\tau)\}^2 = 0 \tag{1.8.9}$$

because as something real and squared (you should know this argument by heart, by now) the left-hand-side of (1.8.9) can't possibly be negative. So, finally,

$$R_f(\tau-T) = R_f(\tau)$$

for *any* τ, which is the very *definition* of a periodic function. If $R_f(\tau)$ has a *double* maximum then it has an *infinite* number of maximums and f(t) is periodic.

This property of the autocorrelation function can be used to detect the presence of a sine-wave of known frequency (period equal to T seconds) in a received radar echo that is dominated by large-amplitude random electronic noise. Such noise has zero correlation with either itself (for $\tau \ne 0$) or with the sine-wave (for any τ) and so running the received echo (sine-wave plus noise) into a correlation circuit that 'compares' (using (1.8.6)) that received echo with a replica of the sine-wave should generate a periodic sequence of pulses spaced T seconds apart. Such a so-called 'auto-correlation detector' can work even when a direct display of the received echo (sine-wave plus noise) on an oscilloscope screen looks, to the eye, like pure noise and certainly *nothing* like a sine-wave. We'll return to the autocorrelation function in Challenge Problem (C7.16).

For a final example of how integrals occur in physical problems, let's return to probability. Suppose we have a circular dartboard with unit radius, and we toss two darts at it. The darts each land on the board at independent, random locations. As in the circle-in-a-circle problem, let's assume the randomness is *uniform* (see note 10 again). Our problem is to determine the probability that the two darts are at least unit distance apart.

This question can, via some interesting integrations, be answered exactly (I'll show you how, in just a moment): the probability is $\frac{3\sqrt{3}}{4\pi} = 0.41349\ldots$. The analytical solution is definitely not a trivial exercise, however, and so I particularly like this problem because it again dramatically demonstrates the power of a Monte Carlo computer simulation. With such a simulation, which in this case is actually quite easy to create, we can quickly get our hands on a pretty good estimate for the probability we are after, and that will give us a check on whatever we come up with theoretically.

The MATLAB code **twodart.m** simulates our problem by generating ten million double tosses on the dartboard and calculating the separation distance for each pair (the variable *total* is the number of those double tosses that are at least unit distance apart). The code achieves a uniform distribution of dart landing points in the same way **circles.m** generated its uniformly random points over a circle: the dartboard is imagined to be enclosed by a square with edge length 2, and pairs of darts are 'tossed' at the square but retained for use in the simulation only if *both* darts of a pair land within a circle of radius 1.[23]

twodart.m

```
% Created by P. J. Nahin (8/2/2019) for Inside Interesting Integrals.
% This code performs simulations of the 'two-dart' problem ten
% million times.
hits=0;total=0;
while hits<10000000
  x1=-1+2*rand;y1=-1+2*rand;
  x2=-1+2*rand;y2=-1+2*rand;
  d1=x1^2+y1^2;d2=x2^2+y2^2;
  if d1<1&&d2<1
    hits=hits+1;
    s=(x1-x2)^2+(y1-y2)^2;
    if s>1
       total=total+1;
    end
  end
end
total/hits
```

---

[23]This way of generating random locations over a circle is called the *rejection method*. It's not computationally efficient, but does have the virtue of being intuitive. A more sophisticated (but less intuitive) way to directly generate uniformly distributed points over a circular region is discussed in my book *Digital Dice*, Princeton 2008, pp. 16–18.

Fig. 1.8.5 The geometry of
the two-dart problem

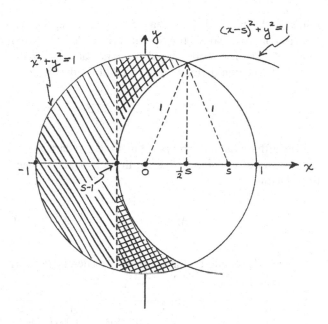

When run numerous times, **twodart.m** produced estimates for the probability of a pair of darts being at least unit distance apart that ranged from *0.41346* to *0.41369*. (You'll notice that this interval *includes* the theoretical value I gave you earlier, unlike the situation in the circle-in-a-circle simulation that prompted our suspicion of the theoretical analysis we did.) So, where does the two-dart theoretical value come from? In Fig. 1.8.5 I've drawn the dartboard as a circle centered on the origin (equation $x^2 + y^2 = 1$). You are to imagine that the board has then been rotated after the first dart has landed so that the dart is on the positive x-axis, distance s from the origin, where s is uniformly distributed from 0 to 1. Because of the symmetry of the circle, we lose no generality in doing that. Next, imagine a thin circular band of radius s and with *very small* width Δs centered on the origin. The probability that the first dart actually lands at distance s from the origin is the area of the band divided by the area of the dartboard (because of the assumption that the dart landing points are *uniformly* random over the board). That is, the probability the first dart is at distance s from the origin is

$$\frac{\pi(s + \Delta s)^2 - \pi s^2}{\pi(1)^2} = \frac{2\pi s \Delta s + \pi(\Delta s)^2}{\pi}$$

which, as $\Delta s \to 0$ becomes 2sΔs (and, of course, this eventually becomes 2sds):

Now, superimposed on the dartboard circle in Fig. 1.8.5 is another circle, also with unit radius, that is centered on the first dart. The equation of this circle is, therefore, $(x - s)^2 + y^2 = 1$. If the second dart lands outside this second circle (but inside the first circle, of course, as *both* darts are given as landing on the dartboard), then the second dart is at least unit distance from the first dart. That is, the second dart lands in the hatched, lune-shaped region (a region with an area that obviously depends on s, and so we'll write that area as A(s)), where it is geometrically obvious that $A(0) = 0$. The probability the second dart lands in the lune-shaped region is the area of that region divided by the area of the dartboard, and so the probability is

$$\frac{A(s)}{\pi}.$$

The differential probability dP that the two darts are at least unit distance apart is the *product* of the probabilities of these two *independent* (this is a crucial assumption) landing events:

$$dP = (2sds)\left(\frac{A(s)}{\pi}\right) = \frac{2sA(s)}{\pi}ds.$$

The *total* probability, the value we are after, is the integral of dP over all possible s:

$$\int_{all\ s} dP = P = \frac{2}{\pi}\int_0^1 sA(s)ds.$$

To do this integral we obviously now need to find A(s), which itself is given by an integral.

Writing the A(s) integral as twice the area of that portion of the hatched, lune region that is above the x-axis, we have (the integrals are the sums of the areas of vertical strips):

$$A(s) = \left\{2\int_{-1}^{s-1}\sqrt{1 - x^2}\,dx\right\} + \left\{2\int_{s-1}^{s/2}\left[\sqrt{1 - x^2} - \sqrt{1 - (x - s)^2}\right]dx\right\}$$

where the term in the first pair of curly brackets is the area of the single-hatched part of the lune, and the term in the second pair of curly brackets is the area of the double-hatched part of the lune.[24] A simple change of variable in the second integral in the second pair of curly brackets, to $u = x - s$, gives us

---

[24]Don't continue reading until you fully understand the limits on the two integrals. Since the second circle in Fig. 1.8.5 has radius 1, and is centered on $x = s$, then that circle crosses the x-axis at $s - 1$ (which is, of course, non-positive as s is, at most, 1). Also, since both circles have radius 1, the isosocles triangle in Fig. 1.8.5 shows that the two circles intersect at $x = \frac{1}{2}s$.

$$A(s) = 2\left[\int_{-1}^{s-1}\sqrt{1-x^2}\;dx + \int_{s-1}^{s/2}\sqrt{1-x^2}\;dx - \int_{-1}^{-s/2}\sqrt{1-u^2}\;du\right]$$

or,

$$A(s) = 2\int_{-s/2}^{s/2}\sqrt{1-x^2}\;dx.$$

From any good set of integral tables we find the indefinite integration formula

$$\int\sqrt{1-x^2}dx = \frac{x\sqrt{1-x^2}}{2} + \frac{1}{2}\sin^{-1}(x)$$

and so

$$A(s) = 2\left\{\frac{x\sqrt{1-x^2}}{2} + \frac{1}{2}\sin^{-1}(x)\right\}\Big|_{-s/2}^{s/2} = \left\{\frac{s}{2}\sqrt{1-\frac{s^2}{4}} + \sin^{-1}\left(\frac{s}{2}\right)\right\}$$

$$+\left\{\frac{s}{2}\sqrt{1-\frac{s^2}{4}} + \sin^{-1}\left(\frac{s}{2}\right)\right\}$$

$$= s\sqrt{1-\frac{s^2}{4}} + 2\sin^{-1}\left(\frac{s}{2}\right).$$

Thus, finally,

$$A(s) = \frac{1}{2}s\sqrt{4-s^2} + 2\sin^{-1}\left(\frac{s}{2}\right).$$

Notice that $A(0) = 0$, just as we argued it must be, earlier, from the geometry of Fig. 1.8.5. So, putting this $A(s)$ into our integral for P, we have

$$P = \frac{2}{\pi}\int_0^1\left\{\frac{1}{2}s^2\sqrt{4-s^2} + 2s\;\sin^{-1}\left(\frac{s}{2}\right)\right\}ds$$

Turning once again to our good set of integral tables, we find these two formulas:

$$\int x^2\sqrt{4-x^2}dx = -\frac{x(4-x^2)^{\frac{3}{2}}}{4} + \frac{4x\sqrt{4-x^2}}{8} + 2\sin^{-1}\left(\frac{x}{2}\right)$$

and

$$\int x \sin^{-1}\left(\frac{x}{2}\right) dx = \left(\frac{x^2}{2} - 1\right) \sin^{-1}\left(\frac{x}{2}\right) + \frac{x\sqrt{4 - x^2}}{4}.$$

So,

$$P = \frac{1}{\pi}\left\{ -\frac{x(4 - x^2)^{\frac{3}{2}}}{4} + \frac{4x\sqrt{4 - x^2}}{8} + 2\sin^{-1}\left(\frac{x}{2}\right) \right\}\Big|_0^1 + \frac{4}{\pi}$$

$$\times \left\{ \left(\frac{x^2}{2} - 1\right) \sin^{-1}\left(\frac{x}{2}\right) + \frac{x\sqrt{4 - x^2}}{4} \right\}\Big|_0^1$$

which I'll let you confirm (with just a few lines of easy arithmetic) reduces to $\frac{3\sqrt{3}}{4\pi}$.

Well, perhaps that is enough of a pep-talk from me on how integrals occur in real problems (or, at least 'more real' than just making integrals up off the tops of our heads). In the following, final section of this chapter, I'll let a famous physicist take over.

1.9 Last Words

In a rightfully famous (and very funny) collection of autobiographical essays, Richard Feynman writes[25]

> "I had learned to do integrals by various methods shown in a book that my high school teacher Mr. Bader had given me [Abram Bader (1908-1989) was Feynman's physics teacher during his senior year in high school, and the book was the classic *Advanced Calculus* by MIT math professor Frederick Woods]. [I]t was for a junior or senior course in college. It had Fourier series, Bessel functions, determinants, elliptic functions—all kinds of wonderful stuff that I didn't know anything about. That book also showed how to differentiate parameters under the integral sign—it's a certain operation. It turns out that's not taught very much in the universities; they don't emphasize it. But I caught on how to use that method, and I used that one damn tool again and again. So, because I was self-taught using that book, I had peculiar methods of doing integrals."

Upon graduation from Far Rockaway High School in New York, in 1935, Feynman spent the next 4 years at MIT, first as a math major, then as an electrical engineering major (having found math too esoteric), and then finally as a physics major (having found electrical engineering not esoteric enough).

Although no longer a math major, the MIT math people hadn't forgotten him, and so in his senior year he was asked to become a member of the MIT math team for the 1939 Putnam Math Competition. The Putnam is an extremely challenging test given nation-wide, yearly, with the median score typically being *one*. That's right, half the

---

[25]In the essay titled "A Different Box of Tools," in *Surely You're Joking, Mr. Feynman!*, W. W. Norton 1985, pp. 84–87.

people taking the exam don't get even a single point.[26] Feynman did a bit better than that, and he was ranked in the top five of all the participants that year (see Challenge Problem (C1.7)). The Putnam was mostly an amusing, temporary diversion for Feynman, however, and he turned down an offer (due to his outstanding test performance) to do graduate work in math at Harvard in favor of studying physics at Princeton.

Feynman continued his essay of note 25 with

> When guys at MIT or Princeton had trouble doing a certain integral, it was because they couldn't do it with the standard methods they had learned in school. If it was contour integration, they would have found it; if it was a simple series expansion, they would have found it. Then I come along and try differentiating under the integral sign, and often it worked. So I got a great reputation for doing integrals, only because my box of tools was different from everybody else's, and they had tried all their tools on it before giving the problem to me.

Feynman was writing of his experiences in the late 1930s, long before Wolfram's on-line symbolic integrator was even just a science fiction fantasy, much less actually available. And so while he was clearly having fun doing integrals, there was also a real pay-off to his knowing a 'trick' that others didn't know. Later in the same book he writes[27] of an encounter he had with another analyst who was stumped by an integral that had appeared in his work during the atom bomb project in the Second World War:

> When one of the guys was explaining [his] problem, I said, 'Why don't you do it by differentiating under the integral sign?' In half an hour he had it solved, and they'd been working on it for three weeks. So, I did something, using my 'different box of tools.'

We started this chapter with a discussion of the area interpretation of the integral, and so let me jump in here with an elaboration on Feynman's last comment that returns to the area idea. I'll finish this chapter with one last example of how that concept can crack open what might appear at first glance to be a pretty tough problem. For any real values of a $\leq$ b, what is

$$\int_a^b \sqrt{(x-a)(b-x)}\, dx = ?$$

This problem appears in an excellent book devoted entirely to complex function theory and contour integration, where it is solved using a fairly advanced method involving successive applications of L'Hospital's rule.[28] What I'll show you here is how to do the integral using nothing but the area interpretation of the integral and

---

[26]I sat for the Putnam in 1959, and I'm pretty sure I know on which side of the median score *I* fell!

[27]In the essay "Los Alamos from Below" in *Surely You're Joking, Mr. Feynman!*, pp. 107–136.

[28]D. S. Mitrinović and J. D. Kečkić, *The Cauchy Method of Residues*, D. Reidel 1984 (originally published in 1978, in Serbian), pp. 191–192. This outstanding book is, alas, out-of-print and available today only through used book dealers on the Web (at outrageous prices). Your best bet is to check the holdings of a local college library.

high school geometry (see Challenge Problem (C1.13) for an interesting twist on this integral). That is, I'm going to show you a trick that Feynman would have loved nearly as much as he did differentiating an integral: almost certainly it is one that he had in his 'box of tools.'

The integrand is

$$y(x) = \sqrt{(x-a)(b-x)}$$

which is clearly never negative over the entire interval of integration. That is, $y(x)$ lies entirely above or on the x-axis for $a \le x \le b$, and $y(a) = y(b) = 0$. The value of the integral is the area enclosed between $y(x)$ and the x-axis. Now,

$$y^2 = (x-a)(b-x) = xb - x^2 - ab + ax = x(a+b) - x^2 - ab.$$

Completing the square,

$$y^2 = -x^2 + x(a+b) + \frac{(a+b)^2}{4} - ab - \frac{(a+b)^2}{4},$$

and then rearranging,

$$y^2 + x^2 - x(a+b) + \frac{(a+b)^2}{4} = \frac{(a+b)^2}{4} - ab$$

and so

$$y^2 + \left(x - \frac{a+b}{2}\right)^2 = \frac{(a+b)^2 - 4ab}{4} = \frac{a^2 - 2ab + b^2}{4} = \left(\frac{b-a}{2}\right)^2,$$

the equation of a circle.

That is, the integral $\int_a^b y(x)dx$ is the area of the upper-half of the circle centered on $\left(\frac{a+b}{2}, 0\right)$ with radius $\frac{b-a}{2}$. The area of that circle is $\pi\left(\frac{b-a}{2}\right)^2$, and so the integral, just like that, is one-half of that area:

$$\int_a^b \sqrt{(x-a)(b-x)}dx = \frac{1}{2}\pi\left(\frac{b-a}{2}\right)^2 = \frac{\pi}{8}(b-a)^2. \tag{1.9.1}$$

Later (in Sect. 4.2) you'll see this same trick again, when we discuss an historically important integral. And you'll see lots more even more wonderful tricks, too, as we work our way through the book. I'll show you, for example, numerous examples of Feynman's favorite trick of 'differentiating under the integral sign' (which can be traced all the way back to Leibniz—as I mentioned at the start of this chapter, he was the man who, in 1686, introduced the elongated-S integral sign that we use today—in the late seventeenth century), although it seems the final form of it that we use today is really a post-Leibniz development. We'll also use other tricks involving the more familiar operations of change of variable, power series, and integration-by-parts. And, as I promised in the original Preface, we'll explore contour integration, too. In fact, why wait? Let's get started right now.

1.10   Challenge Problems

Before moving on to the next chapter, however, here are some challenge problems for your amusement.

(C1.1) Use the 'sneaking up on a singularity' trick to evaluate

$$\int_0^8 \frac{dx}{x - 2}.$$

This is an improper integral because of the singularity in the integration interval at $x = 2$, but it does have a value. Repeat for the integral

$$\int_0^3 \frac{dx}{(x - 1)^{2/3}}$$

and show that, despite the singularity at $x = 1$, this integral also has a value. Notice that unlike in the first integral, the integrand is always positive (there is no cancellation between negative and positive areas). Nevertheless, the integral still has a value.

(C1.2) Show that

$$\int_1^\infty \frac{dx}{\sqrt{x^3 - 1}}$$

exists because there is a finite upper-bound on its value. In particular, show that the integral[29] is < 4. (The solution I give at the back of the book is perfectly okay but, as I discussed in the new Preface, a clever reader of the original edition found an even better—that is, smaller—upper-bound.)

(C1.3) What is the answer to the Feynman integral puzzle in (1.6.1)? Start by deriving (1.6.1) and think about just how arbitrary the constants a and b really are.

---

[29]For a discussion of how this integral appears in a physics problem, see my book *Mrs. Perkins's Electric Quilt*, Princeton 2009, pp. 2–3 (and also that book's p. 4, for how to attack the challenge question—but try on your own before looking there or at the solutions).

(C1.4) For c *any* positive constant, start by confirming that the integral

$$\int_0^\infty \frac{e^{-cx}}{x}\,dx$$

is transformed by the change of variable $y = cx$ into the integral

$$\int_0^\infty \frac{e^{-y}}{y}\,dy.$$

Now if this is so then with a and b any two positive constants it would seem we could argue that

$$I = \int_0^\infty \frac{e^{-ax} - e^{-bx}}{x}\,dx = \int_0^\infty \frac{e^{-ax}}{x}\,dx - \int_0^\infty \frac{e^{-bx}}{x}\,dx = \int_0^\infty \frac{e^{-y}}{y}\,dy - \int_0^\infty \frac{e^{-y}}{y}\,dy = 0.$$

But, as we'll see in Chap. 3, in (3.3.3),

$$I = \int_0^\infty \frac{e^{-ax} - e^{-bx}}{x}\,dx = \ln\left(\frac{b}{a}\right)$$

which is zero only for the special case of $a = b$. So, there's the puzzle—what's wrong with the first argument that claims I is zero for *any* positive a and b?

(C1.5) Here's one more example of an integral that appears in a real-world situation, and of how MATLAB makes short work of it. *Mercator's integral*, named after the inventor—Gerardus Mercator (1512–1594), born in what is today Belgium—of the famous *Mercator map* that renders the spherical surface of the Earth on a planar map, is $\int_{\theta_1}^{\theta_2} \frac{d\theta}{\cos(\theta)}$. It appears as a 'distortion' or 'warping' factor in the construction of a 'flat map' (θ_1 and θ_2 are the latitude extremes of the map). Mercator (who was a cartographer and not a mathematician) encountered this distortion in 1569, long before analytical integration techniques were developed, and he was forced to deal with it by other means (wouldn't he have loved MATLAB!). Today, of course, he would just use a table of integrals. What is the value of Mercator's integral if $\theta_1 = 0$ and $\theta_2 = \frac{\pi}{3}$ (60°)? MATLAB computes *integral(@(x)1./cos(x),0,pi/ 3)* = 1.3169579.... (If you have a table of integrals handy, then you should find that the theoretical answer is $\ln\left(2 + \sqrt{3}\right) = 1.31695789....$)

(C1.6) In a book discussion of how G. H. Hardy evaluated the integral $I = \int_0^1 \frac{\ln(x)}{1+x^3}\,dx$, it is stated[30] that he made use of the auxiliary integral $\int_0^1 x^r \ln(x)\,dx = -\frac{1}{(r+1)^2}$, followed by the claim that this result holds for $r > 0$. It is further stated that

---

[30]*The G. H. Hardy Reader*, Cambridge University Press 2015, p. 228.

"this is easily confirmed using integration-by-parts." Confirm that claim and, as a result, show that it continues to hold true for the extended condition $r > -1$. (Later, in Chap. 3, we'll go through the details of Hardy's evaluation of I.)

(C1.7) The 1939 Putnam Exam that Feynman did so well on asked for the value of this definite integral: $\int_1^\infty \frac{dx}{e^{x+1}+e^{3-x}} = ?$ Can you do it? Hint: no clever tricks are required, just routine freshman calculus methods.

(C1.8) Consider the definite integral $I(t) = \int_0^{2\pi} \{\cos(\varphi) + \sin(\varphi)\}\sqrt{\frac{1-\sqrt{t}\sin(\varphi)}{1-\sqrt{t}\cos(\varphi)}}d\varphi$,

which appears in a quantum mechanical analysis of the helium atom.[31] As part of that analysis it was important to show that $I(t) \geq 0$ for $0 \leq t \leq 1$ (obviously, $I(0) = 0$). The author of that analysis had no success in formally doing that, and was reduced to evaluating I(t) numerically at four values of t from 0.09 to 0.95. He found all four values of I(t) to be >0, and so argued that $I(t) \geq 0$ for *all* t over the interval from 0 to 1. Such an argument wouldn't mean much to a mathematician, of course, and even a less rigorous physicist would feel better about it if a *lot* more values of t were used. In 1937 the tedious numerical integration work was performed by hand (and, after doing four cases, perhaps the author's endurance simply failed), but with today's electronic computers and software like MATLAB it is duck soup to generate Figure C1.8 (in mere seconds), which shows I(t) as a semilog plot for 1000 values of t from 0 to 1, in steps of 0.001.

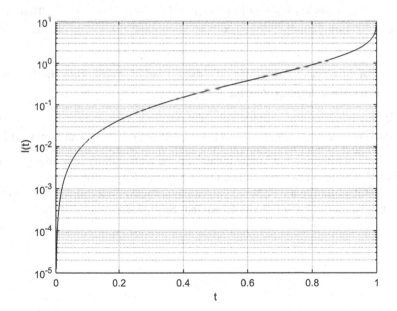

Fig. C1.8 The integral I(t)

---

[31]J. H. Bartlett, Jr., "The Helium Wave Equation," *Physical Review*, April 15, 1937, pp. 661–669.

The figure is certainly reassuring but, still, mathematicians would like to see a formal proof. And, in 1978, a proof showing that $I(t) \geq 0$ for $0 \leq t \leq 1$ finally did appear, a proof using nothing but high school algebra and trigonometry. Can you rediscover that proof? Hint: Start with a change of variable to $\theta = \varphi + \frac{\pi}{4}$.

(C1.9) The following definite integral appeared in the theoretical analysis of a famous electrical circuit involving an infinite number of resistors: $\int_0^{\pi/2} \frac{\sin^2(x)}{\sqrt{\{2-\cos(2x)\}^2-1}} dx$. The author[32] claimed the integral equals $\frac{\pi}{8} = 0.392699\ldots$, and MATLAB agrees, as *integral(@(x)(sin(x).^2)./sqrt((2-cos(2*x)).^2-1),0,pi/2) = 0.392699....* See if you can derive the exact result. Hint: Try the change of variable $2y = 1 - \cos(2x)$, and remember the double angle formulas from trigonometry.

(C1.10) Calculate the *exact* value of $\int_0^1 \frac{dx}{\sqrt{x(1-x)}}$. MATLAB's estimation is *integral (@(x)1./sqrt(x.*(1-x)),0,1) = 3.141592653589...* and this should be a *big* hint as to the *exact* value. This is a very easy problem *if* you see the 'trick,' and I include it here mostly as an example of how a preliminary computer study can potentially add guidance to a deeper theoretical analysis. Hint: The appearance of pi in the value of an integral 'suggests' that a trigonometric substitution might lead to something.

(C1.11) Calculate the autocorrelation function $R_f(\tau)$ for $f(t) = e^{-|t|}$, $-\infty < t < \infty$. Hint: Assume $\tau > 0$ and, after making a sketch of $f(t)$ and $f(t - \tau)$, notice that you can write the autocorrelation integral in (1.8.6) as $\int_{-\infty}^{\infty} = \int_{-\infty}^0 + \int_0^{\tau} + \int_{\tau}^{\infty}$. Then, re-do for $\tau < 0$. Be sure your $R_f(\tau)$ is even. As a numerical check, you should find that $R_f(0) = 1$.

(C1.12) Calculate the value of $\int_0^1 e^{\frac{1}{2}\left(x-\frac{1}{x}\right)} dx + \int_1^{\infty} e^{-\frac{1}{2}\left(x-\frac{1}{x}\right)} dx$. Hint: In the second integral, make the change of variable $y = \frac{1}{x}$.

(C1.13) Show, for *any* choice of non-negative values for $b > a$, that $\int_a^b \frac{dx}{\sqrt{(x-a)(b-x)}} = \pi$. The authors of the book cited in note 28 use an argument from complex function theory, but in fact it can be done with nothing beyond high school algebra and freshman calculus. Notice that if $a = 0$ and $b = 1$ the integral reduces to the one in (C1.10). (The $a = 1$, $b = 3$ case was yet another integral, in addition to the one in (C1.7), on the 1939 Putnam Exam that Feynman aced.)

---

[32]Leo Lavatelli, "The Resistive Net and Finite-Difference Equations," *American Journal of Physics*, September 1972, pp. 1246–1257. I'll tell you more about this famous circuit (involving a spectacular *double* integral) in the next chapter.

Chapter 2
'Easy' Integrals

2.1 Six 'Easy' Warm-Ups

You should always be alert, when confronted by a definite integral, for the happy possibility that although the integral might look 'interesting' (that is, hard!) just *maybe* it will still yield to a direct, frontal attack. The first six integrals in this chapter are in that category. If a and b are positive constants, calculate:

$$\int_1^\infty \frac{1}{(x+a)\sqrt{x-1}}\, dx \qquad (2.1.a)$$

and

$$\int_0^\infty \ln\left(1+\frac{a^2}{x^2}\right) dx \qquad (2.1.b)$$

and

$$\int_0^\infty \frac{\ln(x)}{x^2+b^2}\, dx \qquad (2.1.c)$$

and

$$\int_0^\infty \frac{1}{1+e^{ax}}\, dx. \qquad (2.1.d)$$

Finally, calculate

© Springer Nature Switzerland AG 2020
P. J. Nahin, *Inside Interesting Integrals*, Undergraduate Lecture Notes in Physics,
https://doi.org/10.1007/978-3-030-43788-6_2

$$\int_{\sqrt{2}}^{\infty} \frac{1}{\sqrt{2}\,x \ + \ x^{\sqrt{2}}}\, dx \qquad\qquad (2.1.e)$$

and

$$\int_{-\infty}^{\infty} \frac{dx}{\cosh(x)}. \qquad\qquad (2.1.f)$$

For (2.1.a) make the change of variable $x - 1 = t^2$ and so

$$\frac{dx}{dt} = 2t$$

or

$$dx = 2t\, dt = 2\sqrt{x-1}\, dt.$$

Since

$$x = 1 + t^2$$

then we have

$$\int_{1}^{\infty} \frac{1}{(x+a)\sqrt{x-1}}\, dx = \int_{0}^{\infty} \frac{2\sqrt{x-1}\, dt}{(1+t^2+a)\sqrt{x-1}} = 2\int_{0}^{\infty} \frac{dt}{(a+1)+t^2}.$$

We immediately recognize this last integral as being of the form

$$\int \frac{dt}{c^2+t^2} = \frac{1}{c} \tan^{-1}\left(\frac{t}{c}\right)$$

and so

$$\int_{1}^{\infty} \frac{1}{(x+a)\sqrt{x-1}}\, dx = 2\left\{\frac{1}{\sqrt{a+1}} \tan^{-1}\left(\frac{t}{\sqrt{a+1}}\right)\right\}\Bigg|_{0}^{\infty}$$

$$= \frac{2}{\sqrt{a+1}} \tan^{-1}(\infty) = \frac{2}{\sqrt{a+1}}\left(\frac{\pi}{2}\right)$$

which gives us

$$\int_1^\infty \frac{1}{(x+a)\sqrt{x-1}} \, dx = \frac{\pi}{\sqrt{a+1}}. \tag{2.1.1}$$

As a check, for a = 99 we have the value of the integral equal to $\frac{\pi}{10} = 0.31415\ldots$, while MATLAB says *integral(@(x)1./((x+99).*sqrt(x-1)),1,inf) = 0.31415....*

For (2.1.b) integration-by-parts will do the job. That is, we'll use

$$\int_0^\infty u \, dv = (uv)\Big|_0^\infty - \int_0^\infty v \, du,$$

where

$$u = \ln\left(1 + \frac{a^2}{x^2}\right)$$

and dv = dx. Then, v = x and

$$du = \left(-\frac{2a^2}{x}\right)\left(\frac{1}{x^2 + a^2}\right) dx.$$

So,

$$\int_0^\infty \ln\left(1 + \frac{a^2}{x^2}\right) dx = \left\{ x \ln\left(1 + \frac{a^2}{x^2}\right)\right\}\Big|_0^\infty - \int_0^\infty x\left(-\frac{2a^2}{x}\right)\left(\frac{1}{x^2 + a^2}\right) dx$$

$$= 2a^2 \int_0^\infty \frac{dx}{x^2 + a^2} = 2a^2\left\{\frac{1}{a}\tan^{-1}\left(\frac{x}{a}\right)\right\}\Big|_0^\infty = 2a \tan^{-1}(\infty)$$

and thus[1]

$$\int_0^\infty \ln\left(1 + \frac{a^2}{x^2}\right) dx = \pi a. \tag{2.1.2}$$

---

[1] In this derivation we've assumed $\lim_{x \to 0} x \ln\left(1 + \frac{a^2}{x^2}\right) = \lim_{x \to \infty} x \ln\left(1 + \frac{a^2}{x^2}\right) = 0$. To see that these two assumptions are correct, recall the power series expansion for the log function when $p \approx 0$: $\ln(1 + p) = p - \frac{1}{2}p^2 + \frac{1}{3}p^3 - \ldots$. So, with $p = \frac{a^2}{x^2}$ (which $\to 0$ as $x \to \infty$), we have $x \ln\left(1 + \frac{a^2}{x^2}\right) = x\left[\frac{a^2}{x^2} - \frac{1}{2}\left(\frac{a^2}{x^2}\right)^2 + \frac{1}{3}\left(\frac{a^2}{x^2}\right)^3 - \ldots\right] = \frac{a^2}{x} - \frac{1}{2}\frac{a^4}{x^3} + \ldots$ which $\to 0$ as $x \to \infty$. On the other hand, as $x \to 0$ we have $x \ln\left(1 + \frac{a^2}{x^2}\right) \approx x \ln\left(\frac{a^2}{x^2}\right) = x \ln(a^2) - x \ln(x^2) = x \ln(a^2) - 2x \ln(x)$ and both of these terms go to zero as x goes to zero (the first term is obvious, and in the second term x vanishes faster than ln(x) blows-up).

As a check, for a $= 10$ we have the value of the integral equal to $31.415926\ldots$, while MATLAB says $integral(@(x)log(1+(100./x.^2)),0,inf) = 31.415926\ldots$.

For (2.1.c) let $x = \frac{1}{t}$ and so $dx = -\frac{1}{t^2}\, dt$. Our integral then becomes

$$I = \int_0^\infty \frac{\ln(x)}{x^2 + b^2}\, dx = \int_\infty^0 \frac{\ln\left(\frac{1}{t}\right)}{\frac{1}{t^2} + b^2}\left(-\frac{1}{t^2}\, dt\right) = -\int_\infty^0 \frac{\ln\left(\frac{1}{t}\right)}{1 + b^2 t^2}\, dt$$

$$= -\int_0^\infty \frac{\ln(t)}{1 + b^2 t^2}\, dt.$$

Let $s = bt$ (and so $dt = \frac{1}{b}\, ds$). Then,

$$I = -\int_0^\infty \frac{\ln(t)}{b^2 t^2 + 1}\, dt = -\int_0^\infty \frac{\ln\left(\frac{s}{b}\right)}{s^2 + 1}\left(\frac{1}{b}\right) ds = \frac{1}{b}\left[-\int_0^\infty \frac{\ln(s)}{s^2 + 1}\, ds + \int_0^\infty \frac{\ln(b)}{s^2 + 1}\, ds\right]$$

or, as the first integral in the brackets is zero—we showed this in (1.5.1)—then we have

$$I = \frac{\ln(b)}{b}\int_0^\infty \frac{1}{s^2 + 1}\, ds = \frac{\ln(b)}{b}\left\{\tan^{-1}(s)\right\}\Big|_0^\infty$$

and thus

$$\int_0^\infty \frac{\ln(x)}{x^2 + b^2}\, dx = \frac{\pi}{2b}\ln(b). \tag{2.1.3}$$

Notice that this reduces to zero (as it should) when $b = 1$. For $b = 2$ our formula says the integral is equal to $\frac{\pi}{4}\ln(2) = 0.544396\ldots$ and MATLAB agrees: $integral(@(x)log(x)./(4+x.^2),0,inf) = 0.544396\ldots$.

For (2.1.d) a simple substitution is all we need. Letting $u = e^{ax}$ (and so $\frac{du}{dx} = ae^{ax}$ and thus $dx = \frac{du}{ae^{ax}} = \frac{du}{a\,u}$) we have

$$\int_0^\infty \frac{1}{1 + e^{ax}}\, dx = \int_1^\infty \frac{1}{1+u}\left(\frac{du}{a\,u}\right) = \frac{1}{a}\int_1^\infty \frac{du}{u(1+u)} = \frac{1}{a}\int_1^\infty \left\{\frac{1}{u} - \frac{1}{1+u}\right\} du$$

$$= \frac{1}{a}\left\{\ln(u) - \ln(1+u)\right\}\Big|_1^\infty = \frac{1}{a}\ln\left(\frac{u}{1+u}\right)\Big|_1^\infty = -\frac{1}{a}\ln\left(\frac{1}{2}\right)$$

or,

$$\int_0^\infty \frac{1}{1 + e^{ax}}\, dx = \frac{\ln(2)}{a}. \tag{2.1.4}$$

For $a = \pi$, for example, the integral's value is 0.220635 ..., and in agreement we have *integral($@(x)1./(1+exp(pi*x))$),0,inf) = 0.220635* ....

For (2.1.e), consider the indefinite integral

$$\int \frac{dx}{x + x^m} = \int \frac{x^{-m}}{x^{1-m} + 1}\, dx.$$

Notice that

$$\frac{d}{dx} \ln\left(x^{1-m} + 1\right) = \frac{(1-m)x^{1-m-1}}{x^{1-m} + 1} = \frac{(1-m)x^{-m}}{x^{1-m} + 1}$$

and so

$$\int \frac{dx}{x + x^m} = \frac{1}{1-m} \ln\left(x^{1-m} + 1\right) + C$$

where C is an arbitrary constant of integration. Thus, with $m = \sqrt{2}$, we have

$$\int_{\sqrt{2}}^{\infty} \frac{dx}{x + x^{\sqrt{2}}} = \frac{1}{1 - \sqrt{2}} \left\{ \ln\left(x^{1-\sqrt{2}} + 1\right) \right\} \Big|_{\sqrt{2}}^{\infty}$$

or, because $1 - \sqrt{2} < 0$ and so $\lim_{x \to \infty} x^{1-\sqrt{2}} = 0$, a bit of elementary complex number arithmetic gives us our answer:

$$\int_{\sqrt{2}}^{\infty} \frac{dx}{x + x^{\sqrt{2}}} = \left(1 + \sqrt{2}\right) \ln\left\{1 + 2^{\frac{1}{2}(1-\sqrt{2})}\right\}. \tag{2.1.5}$$

The expression on the right is 1.5063322..., while MATLAB says the value of the integral is *integral($@(x)1./(x+x.\^sqrt(2))$),sqrt(2),inf) = 1.5063319*....

For (2.1.f) let $t = e^x$ and so $\frac{dt}{dx} = e^x$ or $dx = \frac{dt}{e^x} = \frac{dt}{t}$.

Then,

$$\int_{-\infty}^{\infty} \frac{dx}{\cosh(x)} = \int_{-\infty}^{\infty} \frac{dx}{\frac{e^x + e^{-x}}{2}} = 2\int_{-\infty}^{\infty} \frac{dx}{e^x + e^{-x}} = 2\int_{0}^{\infty} \frac{1}{t + \frac{1}{t}}\frac{dt}{t} = 2\int_{0}^{\infty} \frac{1}{t^2 + 1}\, dt$$

$$= 2\tan^{-1}(t)\Big|_{0}^{\infty} = 2\left(\frac{\pi}{2}\right)$$

and so

$$\int_{-\infty}^{\infty} \frac{dx}{\cosh(x)} = \pi. \tag{2.1.6}$$

MATLAB agrees, as *integral($@(x)1./cosh(x)$),-inf,inf) = 3.14159265*....

2.2   A New Trick

The next four examples illustrate an often powerful trick for calculating definite integrals, that of 'flipping' the integration variable's 'direction.' That is, if x goes from 0 to π, try changing to $y = \pi - x$. This may seem almost trivial, but it often works! With this idea in mind, let's calculate

$$\int_0^{\pi/2} \frac{\sqrt{\sin(x)}}{\sqrt{\sin(x)} + \sqrt{\cos(x)}} \, dx \qquad (2.2.a)$$

and

$$\int_0^{\pi} \frac{x \sin(x)}{1 + \cos^2(x)} \, dx \qquad (2.2.b)$$

and

$$\int_0^{\pi/2} \frac{\sin^2(x)}{\sin(x) + \cos(x)} \, dx \qquad (2.2.c)$$

and

$$\int_0^1 \frac{\ln(x+1)}{x^2 + 1} \, dx. \qquad (2.2.d)$$

For (2.2.a), make the substitution $x = \frac{\pi}{2} - y$. Then, $dx = -dy$ and

$$I = \int_{\pi/2}^0 \frac{\sqrt{\sin\left(\frac{\pi}{2} - y\right)}}{\sqrt{\sin\left(\frac{\pi}{2} - y\right)} + \sqrt{\cos\left(\frac{\pi}{2} - y\right)}} \, (-dy) = \int_0^{\pi/2} \frac{\sqrt{\cos(y)}}{\sqrt{\cos(y)} + \sqrt{\sin(y)}} \, dy.$$

Adding this expression to the original I (and changing the dummy variable of integration variable y back to x) gives

$$2I = \int_0^{\pi/2} \frac{\sqrt{\sin(x)} + \sqrt{\cos(x)}}{\sqrt{\sin(x)} + \sqrt{\cos(x)}} \, dx = \int_0^{\pi/2} dx = \frac{\pi}{2},$$

and so

$$\int_0^{\pi/2} \frac{\sqrt{\sin(x)}}{\sqrt{\sin(x)} + \sqrt{\cos(x)}} \, dx = \frac{\pi}{4}. \qquad (2.2.1)$$

This says the integral is equal to 0.785398..., and MATLAB agrees: *integral(@ (x)sqrt(sin(x))./(sqrt(sin(x))+sqrt(cos(x))),0,pi/2) = 0.785398* ....

In (2.2.b) make the substitution $y = \pi - x$ (and so $dx = -dy$). Then,

$$I = \int_\pi^0 \frac{(\pi - y)\sin(\pi - y)}{1 + \cos^2(\pi - y)}(-dy)$$

$$= \int_0^\pi \frac{(\pi - y)\{\sin(\pi)\cos(y) - \cos(\pi)\sin(y)\}}{1 + \{\cos(\pi)\cos(y) + \sin(\pi)\sin(y)\}^2}\, dy$$

or,

$$I = \int_0^\pi \frac{(\pi - y)\sin(y)}{1 + \cos^2(y)}\, dy = \pi\int_0^\pi \frac{\sin(y)}{1 + \cos^2(y)}\, dy - \int_0^\pi \frac{y\sin(y)}{1 + \cos^2(y)}\, dy.$$

That is,

$$I = \pi\int_0^\pi \frac{\sin(x)}{1 + \cos^2(x)}\, dx - I$$

or,

$$I = \frac{\pi}{2}\int_0^\pi \frac{\sin(x)}{1 + \cos^2(x)}\, dx.$$

Now, let $u = \cos(x)$ and so $\frac{du}{dx} = \sin(x)$ (that is, $dx = -du/\sin(x)$). Thus,

$$I = \frac{\pi}{2}\int_1^{-1} \frac{\sin(x)}{1 + u^2}\left(-\frac{du}{\sin(x)}\right) = -\frac{\pi}{2}\int_1^{-1}\frac{du}{1 + u^2} = \frac{\pi}{2}\int_{-1}^1\frac{du}{1 + u^2} = \frac{\pi}{2}\{\tan^{-1}(u)\}\Big|_{-1}^1$$

$$= \frac{\pi}{2}[\tan^{-1}(1) - \tan^{-1}(-1)] = \frac{\pi}{2}\left[\frac{\pi}{4} + \frac{\pi}{4}\right] = \frac{\pi^2}{4}.$$

So,

$$\int_0^\pi \frac{x\sin(x)}{1 + \cos^2(x)}\, dx = \frac{\pi^2}{4}. \tag{2.2.2}$$

This is equal to 2.4674 ..., and MATLAB agrees as *integral(@(x)(x.*sin(x))./ (1+(cos(x).^2)),0,pi) = 2.4674* ....

For (2.2.c), make the substitution $x = \frac{\pi}{2} - y$. Then, $dx = -dy$ and

$$I = \int_{\pi/2}^0 \frac{\sin^2(\frac{\pi}{2} - y)}{\sin(\frac{\pi}{2} - y) + \cos(\frac{\pi}{2} - y)}(-dy).$$

Since

$$\sin\left(\frac{\pi}{2} - y\right) = \sin\left(\frac{\pi}{2}\right)\cos(y) - \cos\left(\frac{\pi}{2}\right)\sin(y) = \cos(y)$$

and

$$\cos\left(\frac{\pi}{2} - y\right) = \cos\left(\frac{\pi}{2}\right)\cos(y) + \sin\left(\frac{\pi}{2}\right)\sin(y) = \sin(y),$$

we have

$$I = \int_0^{\pi/2} \frac{\cos^2(y)}{\cos(y) + \sin(y)}\, dy$$

and so, changing the dummy variable of integration back to x,

$$2I = \int_0^{\pi/2} \frac{\sin^2(x) + \cos^2(x)}{\cos(x) + \sin(x)}\, dx = \int_0^{\pi/2} \frac{1}{\cos(x) + \sin(x)}\, dx.$$

Now, change variable to

$$z = \tan\left(\frac{x}{2}\right).$$

Then we have

$$\frac{dz}{dx} = \frac{\frac{1}{2}\cos^2\left(\frac{x}{2}\right) + \frac{1}{2}\sin^2\left(\frac{x}{2}\right)}{\cos^2\left(\frac{x}{2}\right)} = \frac{\frac{1}{2}}{\cos^2\left(\frac{x}{2}\right)} = \frac{\frac{1}{2}}{\frac{1}{1 + \tan^2\left(\frac{x}{2}\right)}} = \frac{1}{2}\left[1 + \tan^2\left(\frac{x}{2}\right)\right]$$

or,

$$\frac{dz}{dx} = \frac{1 + z^2}{2}$$

and so

$$dx = \frac{2}{1 + z^2}\, dz.$$

From the double-angle formulas from trigonometry we can write

$$\sin(x) = 2\sin\left(\frac{x}{2}\right)\cos\left(\frac{x}{2}\right) = 2\frac{\sin\left(\frac{x}{2}\right)}{\cos\left(\frac{x}{2}\right)}\cos^2\left(\frac{x}{2}\right)$$

and so

$$\sin(x) = 2\tan\left(\frac{x}{2}\right)\frac{1}{1+\tan^2\left(\frac{x}{2}\right)} = \frac{2z}{1+z^2},$$

as well as

$$\cos(x) = \cos^2\left(\frac{x}{2}\right) - \sin^2\left(\frac{x}{2}\right) = \cos^2\left(\frac{x}{2}\right)\left[1 - \frac{\sin^2\left(\frac{x}{2}\right)}{\cos^2\left(\frac{x}{2}\right)}\right]$$

$$= \frac{1}{1+\tan^2\left(\frac{x}{2}\right)}\left[1 - \frac{\frac{\tan^2\left(\frac{x}{2}\right)}{1+\tan^2\left(\frac{x}{2}\right)}}{\frac{1}{1+\tan^2\left(\frac{x}{2}\right)}}\right] = \frac{1}{1+\tan^2\left(\frac{x}{2}\right)}\left[1 - \tan^2\left(\frac{x}{2}\right)\right] = \frac{1-z^2}{1+z^2}.$$

Thus,

$$2I = \int_0^1 \frac{1}{\frac{2z}{1+z^2} + \frac{1-z^2}{1+z^2}}\left(\frac{2}{1+z^2}\right)dz = 2\int_0^1 \frac{dz}{1+2z-z^2} = 2\int_0^1 \frac{dz}{2-[z^2-2z+1]}$$

$$-2\int_0^1 \frac{dz}{2-(z-1)^2}.$$

Next, writing the integrand as a partial fraction expansion we have

$$2I = \frac{2}{2\sqrt{2}}\int_0^1 \left\{\frac{1}{\sqrt{2}-(z-1)} + \frac{1}{\sqrt{2}+(z-1)}\right\}dz$$

$$-\frac{2}{2\sqrt{2}}\int_0^1 \left\{\frac{1}{1+\sqrt{2}-z} + \frac{1}{-1+\sqrt{2}+z}\right\}dz$$

$$= \frac{2}{2\sqrt{2}}\left[\int_0^1 \frac{dz}{z+\sqrt{2}-1} - \int_0^1 \frac{dz}{z-1-\sqrt{2}}\right].$$

Letting $u = z + \sqrt{2} - 1$ in the first integral in the brackets gives us

$$\int_0^1 \frac{dz}{z+\sqrt{2}-1} = \int_{\sqrt{2}-1}^{\sqrt{2}} \frac{du}{u} = \ln(u)\Big|_{\sqrt{2}-1}^{\sqrt{2}} = \ln\left(\frac{\sqrt{2}}{\sqrt{2}-1}\right).$$

Letting $u = z - 1 - \sqrt{2}$ in the second integral in the brackets gives us

$$\int_0^1 \frac{dz}{z-1-\sqrt{2}} = \int_{-1-\sqrt{2}}^{-\sqrt{2}} \frac{du}{u} = \ln(u)\Big|_{-1-\sqrt{2}}^{-\sqrt{2}} = \ln\left(\frac{\sqrt{2}}{1+\sqrt{2}}\right).$$

Thus,

$$2I = \frac{2}{2\sqrt{2}} \left[\ln\left(\frac{\sqrt{2}}{\sqrt{2}-1}\right) - \ln\left(\frac{\sqrt{2}}{1+\sqrt{2}}\right) \right] = \frac{2}{2\sqrt{2}} \ln\left(\frac{\sqrt{2}}{\sqrt{2}-1} \cdot \frac{1+\sqrt{2}}{\sqrt{2}}\right)$$

$$= \frac{1}{\sqrt{2}} \ln\left(\frac{1+\sqrt{2}}{\sqrt{2}-1}\right) = \frac{1}{\sqrt{2}} \ln\left(\frac{1+\sqrt{2}}{\sqrt{2}-1} \cdot \frac{\sqrt{2}+1}{\sqrt{2}+1}\right) = \frac{1}{\sqrt{2}} \ln\left(\frac{(1+\sqrt{2})^2}{2-1}\right)$$

$$= \frac{2}{\sqrt{2}} \ln\left(1+\sqrt{2}\right)$$

or,

$$\int_0^{\pi/2} \frac{\sin^2(x)}{\sin(x) + \cos(x)} \, dx = \frac{1}{\sqrt{2}} \ln\left(1+\sqrt{2}\right). \qquad (2.2.3)$$

The value of the integral is 0.623225..., and MATLAB agrees as *integral(@(x) sin(x).^2./(sin(x)+cos(x)),0,pi/2) = 0.623225 ....*

In (2.2.d) let

$$x = \tan(\theta) = \frac{\sin(\theta)}{\cos(\theta)}$$

and so

$$\frac{dx}{d\theta} = \frac{\cos^2(\theta) + \sin^2(\theta)}{\cos^2(\theta)} = 1 + \tan^2(\theta)$$

and thus we have $dx = \{1 + \tan^2(\theta)\} \, d\theta$ and so our integral is $I = \int_0^1 \frac{\ln(x+1)}{x^2+1} \, dx = \int_0^{\pi/4} \frac{\ln\{\tan(\theta)+1\}}{1+\tan^2(\theta)} \{1 + \tan^2(\theta)\} \, d\theta = \int_0^{\pi/4} \ln\{\tan(\theta)+1\} d\theta$. Now, make the change of variable that 'flips' the direction of integration, that is, $u = \frac{\pi}{4} - \theta$ (and so $du = -d\theta$). Then,

$$I = \int_{\pi/4}^0 \ln\left\{ \tan\left(\frac{\pi}{4} - u\right) + 1 \right\}(-du) = \int_0^{\pi/4} \ln\left\{ \tan\left(\frac{\pi}{4} - u\right) + 1 \right\} du$$

or, changing back to θ as the dummy variable of integration,

$$I = \int_0^{\pi/4} \ln\left\{ \tan\left(\frac{\pi}{4} - \theta\right) + 1 \right\} d\theta.$$

Next, recall the identity

$$\tan(\alpha - \beta) = \frac{\tan(\alpha) - \tan(\beta)}{1 + \tan(\alpha)\tan(\beta)}.$$

With $\alpha = \frac{\pi}{4}$ and $\beta = \theta$ we have

$$\tan\left(\frac{\pi}{4} - \theta\right) = \frac{\tan\left(\frac{\pi}{4}\right) - \tan(\theta)}{1 + \tan\left(\frac{\pi}{4}\right)\tan(\theta)} = \frac{1 - \tan(\theta)}{1 + \tan(\theta)}$$

and so

$$I = \int_0^{\pi/4} \ln\left\{\frac{1 - \tan(\theta)}{1 + \tan(\theta)} + 1\right\} d\theta = \int_0^{\pi/4} \ln\left\{\frac{2}{1 + \tan(\theta)}\right\} d\theta$$

$$= \int_0^{\pi/4} \ln\{2\} d\theta - \int_0^{\frac{\pi}{4}} \ln\{1 + \tan(\theta)\} d\theta.$$

But the last integral is I, and so

$$I = \frac{\pi}{4} \ln(2) - I$$

or, at last (and using x as the dummy variable of integration),

$$\int_0^1 \frac{\ln(x+1)}{x^2 + 1} dx = \int_0^{\pi/4} \ln\{1 + \tan(x)\} dx = \frac{\pi}{8} \ln(2). \qquad (2.2.4)$$

This integral is often called *Serret's integral*, after the French mathematician Joseph Serret (1819–1885) who did it in 1844. Our result says the two above integrals are both equal to 0.27219826 ... and MATLAB agrees, as *integral(@(x) log(tan(x)+1),0,pi/4)* = 0.27219826 ... and *integral(@(x)log(x+1)./(x.^2+1),0,1) = 0.27219823....*

If we make the change of variable $x = \frac{t}{a}$ we can generalize this result as follows. Since $dx = \frac{1}{a} dt$, then

$$\frac{\pi}{8}\ln(2) = \int_0^1 \frac{\ln(x+1)}{x^2+1} dx = \int_0^a \frac{\ln\left(\frac{t}{a}+1\right)}{\left(\frac{t}{a}\right)^2 + 1}\left(\frac{1}{a} dt\right) = a\int_0^a \frac{\ln(t+a) - \ln(a)}{t^2 + a^2} dt$$

$$= a\left\{\int_0^a \frac{\ln(t+a)}{t^2+a^2} dt - \ln(a)\left[\frac{1}{a}\tan^{-1}\left(\frac{t}{a}\right)\right]\Big|_0^a\right\} = a\left\{\int_0^a \frac{\ln(t+a)}{t^2+a^2} dt - \frac{1}{a}\ln(a)\left[\tan^{-1}(1)\right]\right\}$$

$$= a\left\{\int_0^a \frac{\ln(t+a)}{t^2+a^2} dt - \frac{\pi}{4a}\ln(a)\right\} = a\int_0^a \frac{\ln(t+a)}{t^2+a^2} dt - \frac{\pi}{4}\ln(a).$$

Thus,

$$\int_0^a \frac{\ln(t+a)}{t^2 + a^2}\, dt = \frac{\pi}{8a} \ln(2) + \frac{\pi}{4a} \ln(a) = \frac{\pi}{8a} \ln(2) + \frac{2\pi}{8a} \ln(a)$$

or, finally, changing the dummy variable of integration back to x,

$$\int_0^a \frac{\ln(x+a)}{x^2 + a^2}\, dx = \frac{\pi}{8a} \ln\left(2a^2\right). \tag{2.2.5}$$

2.3 Two Old Tricks, Plus a New One

The following integral, with *arbitrarily many and different* quadratic factors in the denominator of the integrand, may at first glance look impossibly difficult: $\int_0^\infty \frac{dx}{\left(x^2 + a_1^2\right)\left(x^2 + a_2^2\right)\left(x^2 + a_3^2\right) \dots \left(x^2 + a_n^2\right)}$, with *all* of the $a_i \neq 0$ *and different*.

Not so, as I'll now show you. Let's start by writing the integrand in partial fraction form (look back at how we got (2.2.3)), as

$$\frac{1}{\left(x^2 + a_1^2\right)\left(x^2 + a_2^2\right)\left(x^2 + a_3^2\right) \dots \left(x^2 + a_n^2\right)} = \frac{c_1}{x^2 + a_1^2} + \frac{c_2}{x^2 + a_2^2} + \frac{c_3}{x^2 + a_3^2} + \dots + \frac{c_n}{x^2 + a_n^2}$$

where the c's are all constants.[2] Fix your attention on any one of the terms on the right, say the k-th one. Then, multiplying through by $x^2 + a_k^2$, we have

$$\frac{x^2 + a_k^2}{\left(x^2 + a_1^2\right)\left(x^2 + a_2^2\right)\left(x^2 + a_3^2\right) \dots \left(x^2 + a_k^2\right) \dots \left(x^2 + a_n^2\right)}$$

$$= \frac{c_1\left(x^2 + a_k^2\right)}{x^2 + a_1^2} + \frac{c_2\left(x^2 + a_k^2\right)}{x^2 + a_2^2} + \dots + c_k + \dots + \frac{c_n\left(x^2 + a_k^2\right)}{x^2 + a_n^2}.$$

Thus, cancelling the $x^2 + a_k^2$ factor in the numerator and denominator on the left in the above expression, and setting x to the particular value of ia_k where $i = \sqrt{-1}$, we have[3]

---

[2]Writing the partial fraction expansion this way is where the assumption that all the a_i *are different* comes into play. If any of the a_i appears multiple times, then the correct partial fraction expansion of the integrand is *not* as I've written it.

[3]There are two points to be clear on at this point. First, since we are working with an identity it must be true for all values of x, and I've just picked a particularly convenient one. Second, if the use of an imaginary x bothers you, just remember the philosophical spirit of this book—*anything* (well, *almost* anything) goes, and we'll check our result when we get to the end.

$$\frac{1}{\left(-a_k^2 + a_1^2\right)\left(-a_k^2 + a_2^2\right)\left(-a_k^2 + a_3^2\right)\ldots\left(-a_k^2 + a_n^2\right)} = \frac{1}{\prod\limits_{j=1,j\neq k}^{n}\left(a_j^2 - a_k^2\right)} = c_k$$

where only the c_k term survives on the right since all the other terms in the partial fraction expansion end-up with a zero in their numerator. So, our original partial fraction expansion is just

$$\frac{1}{\left(x^2 + a_1^2\right)\left(x^2 + a_2^2\right)\left(x^2 + a_3^2\right)\ldots\left(x^2 + a_n^2\right)} = \sum_{k=1}^{n}\frac{c_k}{x^2 + a_k^2}$$

and thus

$$\int_0^\infty \frac{dx}{\left(x^2 + a_1^2\right)\left(x^2 + a_2^2\right)\left(x^2 + a_3^2\right)\ldots\left(x^2 + a_n^2\right)} = \sum_{k-1}^{n} c_k \int_0^\infty \frac{dx}{x^2 + a_k^2}$$

$$= \sum_{k-1}^{n} c_k \frac{1}{a_k}\left[\tan^{-1}\left(\frac{x}{a_k}\right)\right]\Big|_0^\infty = \sum_{k=1}^{n}\frac{c_k}{a_k}\left(\frac{\pi}{2}\right)$$

or, at last, with all the $a_i \neq 0$,

$$\int_0^\infty \frac{dx}{\left(x^2 + a_1^2\right)\left(x^2 + a_2^2\right)\left(x^2 + a_3^2\right)\ldots\left(x^2 + a_n^2\right)} = \left(\frac{\pi}{2}\right)\sum_{k=1}^{n}\frac{c_k}{a_k}, \qquad (2.3.1)$$

where $c_k = \dfrac{1}{\prod\limits_{j=1,j\neq k}^{n}\left(a_j^2 - a_k^2\right)}$ and $a_j \neq a_i$ if $j \neq i$.

For example, suppose $a_1^2 = 1$, $a_2^2 = 4$, and $a_3^2 = 9$. Then, $a_1 = 1$, $a_2 = 2$, and $a_3 = 3$. This gives the values of the c's as

$$c_1 = \frac{1}{(4-1)(9-1)} = \frac{1}{24}, c_2 = \frac{1}{(1-4)(9-4)} = -\frac{1}{15}, c_3 = \frac{1}{(1-9)(4-9)}$$

$$-\frac{1}{40}.$$

Then the value of the integral is

$$\left(\frac{\pi}{2}\right)\left[\frac{\frac{1}{24}}{1} - \frac{\frac{1}{15}}{2} + \frac{\frac{1}{40}}{3}\right] = \left(\frac{\pi}{2}\right)\left[\frac{1}{24} - \frac{1}{30} + \frac{1}{120}\right] = \left(\frac{\pi}{2}\right)\left[\frac{2}{120}\right] = \frac{\pi}{120} = 0.0261799\ldots$$

and MATLAB agrees, as *integral(@(x)1./((x.^2+1).*(x.^2+4).*(x.^2+9)),0, inf) = 0.0261799....*

Here's another example of the use of a partial fraction expansion to evaluate an integral. Here I'll do

$$\int_0^\infty \frac{dx}{x^4 + 2x^2 \cosh(2\alpha) + 1}$$

where α is an arbitrary constant. Writing the hyperbolic cosine in the denominator of the integrand out in exponential form, we have

$$x^4 + 2x^2 \cosh(2\alpha) + 1 = x^4 + 2x^2 \left[\frac{e^{2\alpha} + e^{-2\alpha}}{2}\right] + 1 = x^4 + x^2 e^{2\alpha} + x^2 e^{-2\alpha} + 1$$

$$= \left(x^2 + e^{2\alpha}\right)\left(x^2 + e^{-2\alpha}\right).$$

So, we can write the integrand in the following partial fraction form (with A and B as constants):

$$\frac{1}{x^4 + 2x^2 \cosh(2\alpha) + 1} = \frac{A}{(x^2 + e^{2\alpha})} + \frac{B}{(x^2 + e^{-2\alpha})}.$$

That is,

$$(A + B)x^2 + Ae^{-2\alpha} + Be^{2\alpha} = 1$$

which, since there is no x^2 term on the right, immediately tells us that $A = -B$. Thus,

$$-Be^{-2\alpha} + Be^{2\alpha} = 1$$

and so the constant B is given by

$$B = \frac{1}{e^{2\alpha} - e^{-2\alpha}}.$$

Therefore,

$$\int_0^\infty \frac{dx}{x^4 + 2x^2 \cosh(2\alpha) + 1} = \frac{1}{e^{2\alpha} - e^{-2\alpha}} \left[\int_0^\infty \frac{dx}{x^2 + e^{-2\alpha}} - \int_0^\infty \frac{dx}{x^2 + e^{2\alpha}}\right]$$

$$= \frac{1}{e^{2\alpha} - e^{-2\alpha}} \left[\frac{1}{e^{-\alpha}} \tan^{-1}\left(\frac{x}{e^{-\alpha}}\right) - \frac{1}{e^{\alpha}} \tan^{-1}\left(\frac{x}{e^{\alpha}}\right)\right]\Big|_0^\infty$$

$$= \frac{1}{e^{2\alpha} - e^{-2\alpha}} \left[e^\alpha \frac{\pi}{2} - e^{-\alpha} \frac{\pi}{2}\right] = \left(\frac{\pi}{2}\right) \frac{e^\alpha - e^{-\alpha}}{e^{2\alpha} - e^{-2\alpha}} = \left(\frac{\pi}{2}\right) \frac{e^\alpha - e^{-\alpha}}{(e^\alpha + e^{-\alpha})(e^\alpha - e^{-\alpha})}$$

$$= \left(\frac{\pi}{2}\right) \frac{1}{(e^\alpha + e^{-\alpha})} = \left(\frac{\pi}{2}\right) \frac{\frac{1}{2}}{\frac{(e^\alpha + e^{-\alpha})}{2}}$$

or, finally,

$$\int_0^\infty \frac{dx}{x^4 + 2x^2 \cosh(2\alpha) + 1} = \frac{\pi}{4 \cosh(\alpha)}. \tag{2.3.2}$$

For $\alpha = 1$, for example, this integral equals $0.5089806\ldots$ and MATLAB agrees, as $integral(@(x)1./(x.\wedge4+2*cosh(2)*x.\wedge2+1),0,inf) = 0.5089806\ldots$.

Back in Chap. 1 (Sect. 1.5) I showed you how the 'evenness' or 'oddness' of an integrand (if one of those two properties is present) can be of great help in transforming a 'hard' integral into an 'easy' one. As a more sophisticated example of this than was the example in Chap. 1, let's calculate the value of

$$\int_0^\infty \frac{dx}{x^4 + 2x^2 \cos(2\alpha) + 1}$$

where, as before, α is an arbitrary constant. This may superficially look a lot like the integral we just finished but, as you'll soon see, there is a really big difference in how we'll do this new one.

We start by making the change of variable $y = \frac{1}{x}$ (and so $\frac{dy}{dx} = -\frac{1}{x^2}$ or $dx = -x^2\,dy = -\frac{1}{y^2}\,dy$). Then,

$$I = \int_0^\infty \frac{dx}{x^4 + 2x^2 \cos(2\alpha) + 1} = \int_\infty^0 \frac{-\frac{1}{y^2}\,dy}{\frac{1}{y^4} + 2\frac{1}{y^2}\cos(2\alpha) + 1}$$

$$= \int_0^\infty \frac{y^2\,dy}{y^4 + 2y^2 \cos(2\alpha) + 1}.$$

If we then add our two versions of the integral (the left-most and the right-most integrals in the previous line, remembering that x and y are just dummy variables of integration) we have

$$2I = \int_0^\infty \frac{(1 + x^2)\,dx}{x^4 + 2x^2 \cos(2\alpha) + 1}$$

or,

$$I = \frac{1}{2}\int_0^\infty \frac{(1 + x^2)\,dx}{x^4 + 2x^2 \cos(2\alpha) + 1}.$$

And since the integrand is even, we can write

$$I = \frac{1}{4}\int_{-\infty}^\infty \frac{(1 + x^2)\,dx}{x^4 + 2x^2 \cos(2\alpha) + 1}.$$

Because $\cos(2\alpha) = 1 - 2\sin^2(\alpha)$ you can show by direct multiplication that

$$x^4 + 2x^2 \cos{(2\alpha)} + 1 = \left[x^2 - 2x \, \sin{(\alpha)} + 1\right]\left[x^2 + 2x \, \sin{(\alpha)} + 1\right]$$

and so

$$I = \frac{1}{4}\int_{-\infty}^{\infty} \frac{(1 + x^2) \, dx}{[x^2 - 2x \, \sin{(\alpha)} + 1][x^2 + 2x \, \sin{(\alpha)} + 1]}.$$

Now, since the integrand is even, then if we include in the numerator of the integrand an *odd* function like $2x \sin{(\alpha)}$ we do not change the value of the integral, and so

$$I = \frac{1}{4}\int_{-\infty}^{\infty} \frac{[x^2 - 2x \, \sin{(\alpha)} + 1] \, dx}{[x^2 - 2x \, \sin{(\alpha)} + 1][x^2 + 2x \, \sin{(\alpha)} + 1]}$$

or

$$I = \frac{1}{4}\int_{-\infty}^{\infty} \frac{dx}{x^2 + 2x \, \sin{(\alpha)} + 1}.$$

Or, since $\sin^2(\alpha) + \cos^2(\alpha) = 1$ then

$$I = \frac{1}{4}\int_{-\infty}^{\infty} \frac{dx}{x^2 + 2x \, \sin{(\alpha)} + \sin^2(\alpha) + \cos^2(\alpha)}$$

$$= \frac{1}{4}\int_{-\infty}^{\infty} \frac{dx}{[x + \sin{(\alpha)}]^2 + \cos^2(\alpha)}.$$

Let $u = x + \sin{(\alpha)}$ (and so $du = dx$), and then

$$I = \frac{1}{4}\int_{-\infty}^{\infty} \frac{du}{u^2 + \cos^2(\alpha)} = \frac{1}{4}\left(\frac{1}{\cos{(\alpha)}}\right)\left[\tan^{-1}\left\{\frac{u}{\cos{(\alpha)}}\right\}\right]\Big|_{-\infty}^{\infty}$$

$$= \frac{1}{4 \cos{(\alpha)}}\left[\tan^{-1}\{\infty\} - \tan^{-1}\{-\infty\}\right]$$

and so, at last,[4]

---

[4]A reader of the first edition of this book wrote to me to offer the observation that (2.3.3) follows almost immediately from (2.3.2) if one observes that $\cosh(iy) = \cos(y)$, $i = \sqrt{-1}$. That's true, too, but to see where that relation comes from requires Euler's famous identity which, while we *will* indeed make great use of, isn't what I'm trying to illustrate *here*. What I do *here* is with the specific intent of illustrating the 'evenness/oddness' trick, using *only* freshman calculus arguments.

$$\int_0^\infty \frac{dx}{x^4 + 2x^2 \cos(2\alpha) + 1} = \frac{\pi}{4\cos(\alpha)}. \tag{2.3.3}$$

Special, interesting cases occur for some obvious values of α. Specifically, for $\alpha = \frac{\pi}{4}$ we have

$$\int_0^\infty \frac{dx}{x^4 + 1} = \int_0^\infty \frac{x^2 dx}{x^4 + 1} = \frac{\pi\sqrt{2}}{4}. \tag{2.3.4}$$

These two integrals are therefore equal to 1.11072 ... and MATLAB agrees, as *integral(@(x)1./(x.^4+1),0,inf) = 1.11072 ...* and *integral(@(x)(x.^2)./(x.^4+1),0, inf) = 1.11072. ...* For $\alpha = 30°$ (that is, $\alpha = \frac{\pi}{6}$) we have

$$\int_0^\infty \frac{dx}{x^4 + x^2 + 1} = \frac{\pi}{2\sqrt{3}}. \tag{2.3.5}$$

This is equal to 0.906899 ..., and again MATLAB agrees as *integral(@(x)1./ (x.^4+x.^2+1),0,inf) = 0.906899 ....* For $\alpha = 60°$ (that is, $\alpha = \frac{\pi}{3}$) we have

$$\int_0^\infty \frac{dx}{x^4 - x^2 + 1} = \frac{\pi}{2}. \tag{2.3.6}$$

This is equal to 1.570796 ... and, indeed, *integral(@(x)1./(x.^4-x.^2+1),0, inf) = 1.570796.* And finally, for $\alpha = 0$ we have

$$\int_0^\infty \frac{dx}{x^4 + 2x^2 + 1} = \frac{\pi}{4}. \tag{2.3.7}$$

This is equal to 0.785398 ..., and *integral(@(x)1./(x.^4+2*x.^2+1),0, inf) = 0.785398 ....*

Here's a new trick, one using a difference equation to evaluate a class of definite integrals indexed on an integer-valued variable. Specifically,

$$I_n(\alpha) = \int_0^\pi \frac{\cos(n\theta) - \cos(n\alpha)}{\cos(\theta) - \cos(\alpha)} \, d\theta.$$

where α is a constant and n is a non-negative integer ($n = 0, 1, 2, 3, \ldots$). The first two integrals are easy to do by inspection: $I_0(\alpha) = 0$ and $I_1(\alpha) = \pi$. For $n > 1$, however, things get more difficult. What I'll do next is perfectly understandable as we go through the analysis step-by-step, but I have no idea what motivated the person who first did this. The mystery of mathematical genius!

If you recall the trigonometric identity

$$\cos\{(n+1)\theta\} + \cos\{(n-1)\theta\} = 2\cos(\theta)\cos(n\theta)$$

then perhaps you'd think of taking a look at the quantity $I_{n+1}(\alpha) + I_{n-1}(\alpha)$ to see if it is related in some 'nice' way to $I_n(\alpha)$. So, imagining that we have been so inspired, let's take a look at the quantity $Q = AI_{n+1}(\alpha) + BI_n(\alpha) + CI_{n-1}(\alpha)$, where A, B, and C are constants, to see what we get.

Thus,

$$Q = \int_0^\pi \frac{A[\cos\{(n+1)\theta\} - \cos\{(n+1)\alpha\}] + B[\cos(n\theta) - \cos(n\alpha)] + C[\cos\{(n-1)\theta\} - \cos\{(n-1)\alpha\}]}{\cos(\theta) - \cos(\alpha)} d\theta.$$

Suppose we now set $A = C = 1$ and $B = -2\cos(\alpha)$. Then,

$$Q = \int_0^\pi \frac{[\cos\{(n+1)\theta\} + \cos\{(n-1)\theta\} - 2\cos(\alpha)\cos(n\theta)] - [\cos\{(n+1)\alpha\} + \cos\{(n-1)\alpha\} - 2\cos(\alpha)\cos(n\alpha)]}{\cos(\theta) - \cos(\alpha)} d\theta.$$

From our trig identity the second term in the numerator vanishes and the first term reduces to

$$Q = \int_0^\pi \frac{2\cos(\theta)\cos(n\theta) - 2\cos(\alpha)\cos(n\theta)}{\cos(\theta) - \cos(\alpha)} d\theta$$

$$= \int_0^\pi \frac{2\cos(n\theta)[\cos(\theta) - \cos(\alpha)]}{\cos(\theta) - \cos(\alpha)} d\theta$$

$$= 2\int_0^\pi \cos(n\theta)d\theta = 2\left[\frac{\sin(n\theta)}{n}\right]\Big|_0^\pi = 0, n = 1, 2, \ldots$$

That is, we have the following second-order, linear difference equation:

$$I_{n+1}(\alpha) - 2\cos(\alpha)I_n(\alpha) + I_{n-1}(\alpha) = 0, n = 1, 2, 3, \ldots,$$

with the conditions $I_0(\alpha) = 0$ and $I_1(\alpha) = \pi$.

It is well-known that such a so-called *recursive equation* has solutions of the form $I_n = Ce^{sn}$ where C and s are constants. So,

$$Ce^{s(n+1)} - 2\cos(\alpha)Ce^{sn} + Ce^{s(n-1)} = 0$$

or, cancelling the common factor of Ce^{sn},

$$e^s - 2\cos(\alpha) + e^{-s} = 0$$

or,

$$e^{2s} - 2\cos(\alpha)e^s + 1 = 0.$$

This is a quadratic in e^s, and so

$$e^s = \frac{2\cos(\alpha) \pm \sqrt{4\cos^2(\alpha) - 4}}{2} = \frac{2\cos(\alpha) \pm 2i\sqrt{1 - \cos^2(\alpha)}}{2}$$

$$= \cos(\alpha) \pm i\sin(\alpha),$$

where $i = \sqrt{-1}$. Now, from Euler's fabulous formula we have $e^s = e^{\pm i\alpha}$ and thus $s = \pm i\alpha$. This means that the general solution for $I_n(\alpha)$ is

$$I_n(\alpha) = C_1 e^{in\alpha} + C_2 e^{-in\alpha}.$$

Since $I_0(\alpha) = 0$ then $C_1 + C_2 = 0$ or, $C_2 = -C_1$. Also, as $I_1(\alpha) = \pi$ we have

$$C_1 e^{i\alpha} - C_1 e^{-i\alpha} = \pi = C_1 i2\sin(\alpha)$$

which says that

$$C_1 = \frac{\pi}{i2\sin(\alpha)} \text{ and } C_2 = -\frac{\pi}{i2\sin(\alpha)}.$$

Thus,

$$I_n(\alpha) = \frac{\pi}{2\sin(\alpha)} \left[\frac{e^{in\alpha} - e^{-in\alpha}}{i} \right] = \frac{\pi}{2\sin(\alpha)} \left[\frac{i2\sin(n\alpha)}{i} \right]$$

or, at last, using x as the dummy variable of integration,

$$\int_0^\pi \frac{\cos(nx) - \cos(n\alpha)}{\cos(x) - \cos(\alpha)} dx = \pi \frac{\sin(n\alpha)}{\sin(\alpha)}. \tag{2.3.8}$$

For example, if $n = 6$ and $\alpha = \frac{\pi}{11}$ this result says our integral is equal to $\pi \frac{\sin\left(6\frac{\pi}{11}\right)}{\sin\left(\frac{\pi}{11}\right)}$ which is equal to $11.03747399\ldots$, and MATLAB agrees because $integral(@(x)(cos(6*x)-cos(6*pi/11))./(cos(x)-cos(pi/11)),0,pi) = 11.03747399\ldots$.

Here's a final, quick example of recursion used to solve an entire *class* of integrals: $I_n = \int_0^\infty x^{2n} e^{-x^2} dx$, $n = 0, 1, 2, 3, \ldots$.

We start by observing that

$$\frac{d}{dx} \left(x^{2n-1} e^{-x^2} \right) = (2n-1)x^{2n-2} e^{-x^2} - 2x^{2n} e^{-x^2}, n \geq 1.$$

So, integrating both sides,

$$\int_0^\infty \frac{d}{dx} \left(x^{2n-1} e^{-x^2} \right) dx = (2n-1) \int_0^\infty x^{2n-2} e^{-x^2} dx - 2 \int_0^\infty x^{2n} e^{-x^2} dx.$$

The right-most integral is I_n, and the middle integral is I_{n-1}. So,

$$\int_0^\infty \frac{d}{dx}\left(x^{2n-1}e^{-x^2}\right)dx = (2n-1)I_{n-1} - 2I_n.$$

Now, notice that the remaining integral (on the left-hand-side) is simply the integral of a derivative *and so is very easy to do*. In fact,

$$\int_0^\infty \frac{d}{dx}\left(x^{2n-1}e^{-x^2}\right)dx = \left(x^{2n-1}e^{-x^2}\right)\Big|_0^\infty = 0$$

since $x^{2n-1}e^{-x^2} = 0$ at $x = 0$ and as $x \to \infty$. So, we immediately have the recurrence

$$I_n = \frac{2n-1}{2}I_{n-1} = \frac{2n(2n-1)}{4n}I_{n-1}.$$

For the first few values n we have:

$$I_1 = \frac{2}{(4)(1)}I_0,$$

$$I_2 = \frac{(4)(3)}{(4)(2)}I_1 = \frac{(4)(3)(2)}{(4)(2)(4)(1)}I_0,$$

$$I_3 = \frac{(6)(5)}{(4)(3)}I_2 = \frac{(6)(5)(4)(3)(2)}{(4)(3)(4)(2)(4)(1)}I_0,$$

and by now you should see the pattern:

$$I_n = \frac{(2n)!}{4^n n!}I_0.$$

This is a nice result, but of course the next question is obvious: what is I_0? In fact,

$$I_0 = \int_0^\infty e^{-x^2}dx = \frac{1}{2}\sqrt{\pi},$$

which we have *not* shown (yet). In the next chapter, as result (3.1.4), I will show you (using a new trick) that

$$\int_{-\infty}^\infty e^{-x^2/2}dx = \sqrt{2\pi}.$$

If you make the change of variable $y = x\sqrt{2}$ (and remember that $\int_{-\infty}^\infty f(x)dx = 2\int_0^\infty f(x)dx$ if $f(x)$ is even) then the value of I_0 immediately follows. Thus,

$$\int_0^\infty x^{2n} e^{-x^2} dx = \frac{(2n)!}{4^n n!} \left(\frac{1}{2}\right) \sqrt{\pi}.$$ (2.3.9)

If $n = 5$ this says the integral is equal to 26.171388..., while.
integral(@(x)(x.^10).*exp(-(x.^2)),0,inf) = 26.171388 ....

2.4   Another Old Trick: Euler's Log-Sine Integral

In 1769 Euler computed (for $a = 1$) the value of (where $a \geq 0$)

$$I = \int_0^{\pi/2} \ln\{a \, \sin(x)\} dx,$$

which is equal to

$$\int_0^{\pi/2} \ln\{a \, \cos(x)\} dx.$$

The two integrals are equal because the integrands take-on the same values over the integration interval (sin(x) and cos(x) are mirror-images of each other over that interval). For many years it was commonly claimed in textbooks that these are quite difficult integrals to do, best tackled with the powerful techniques of contour integration. As you'll see with the following analysis, however, that is simply not the case.

So, to start we notice that

$$I = \frac{1}{2} \int_0^{\pi/2} [\ln\{a \, \sin(x)\} + \ln\{a \, \cos(x)\}] dx = \frac{1}{2} \int_0^{\pi/2} \ln\{a^2 \, \sin(x)\cos(x)\} dx.$$

Since $\sin(2x) = 2\sin(x)\cos(x)$, we have $\sin(x)\cos(x) = \frac{1}{2}\sin(2x)$ and therefore

$$I = \frac{1}{2} \int_0^{\pi/2} \ln\left\{a\frac{1}{2}a \, \sin(2x)\right\} dx = \frac{1}{2} \int_0^{\pi/2} \left[\ln(a) + \ln\left(\frac{1}{2}\right) + \ln\{a \, \sin(2x)\}\right] dx$$

$$= \frac{\pi}{4} \ln(a) - \frac{\pi}{4} \ln(2) + \frac{1}{2} \int_0^{\pi/2} \ln\{a\sin(2x)\} dx.$$

In the last integral, let $t = 2x$ (and so $dx = \frac{1}{2} dt$). Thus,

$$\frac{1}{2}\int_0^{\pi/2} \ln\{a\,\sin(2x)\}dx = \frac{1}{2}\int_0^{\pi} \ln\{a\,\sin(t)\}\frac{1}{2}dt = \frac{1}{2}I$$

where the last equality follows because (think of how sin(t) varies over the interval 0 to π)

$$\int_0^{\pi/2} \ln\{a\,\sin(t)\}dt = \frac{1}{2}\int_0^{\pi} \ln\{a\,\sin(t)\}dt.$$

So,

$$I = \frac{\pi}{4}\ln(a) - \frac{\pi}{4}\ln(2) + \frac{1}{2}I = \frac{\pi}{4}\ln\left(\frac{a}{2}\right) + \frac{1}{2}I$$

or,

$$\frac{1}{2}I = \frac{\pi}{4}\ln\left(\frac{a}{2}\right)$$

and so, at last, for a≥ 0,

$$\int_0^{\pi/2} \ln\{a\,\sin(x)\}\,dx = \int_0^{\pi/2} \ln\{a\,\cos(x)\}\,dx = \frac{\pi}{2}\ln\left(\frac{a}{2}\right). \tag{2.4.1}$$

Special cases of interest are a $= 1$ (Euler's integral) for which both integrals equal $-\frac{\pi}{2}\ln(2) = -1.088793\ldots$ and a $= 2$ for which both integrals are equal to zero. We can check both of these cases with MATLAB; *integral(@(x)log(sin(x)),0, pi/2) $= -1.0888035\ldots$* and *integral(@(x)log(cos(x)),0,pi/2) $= -1.0888043\ldots$*, while *integral(@(x)log(2∗sin(x)),0,pi/2) $= -1.0459 \times 10^{-5}$* and *integral(@(x)log (2∗cos(x)),0,pi/2) $= -1.1340 \times 10^{-5}$*. We'll see Euler's log-sine integral again, in Chap. 7.

With the result for a $= 1$, we can now calculate the interesting integral

$$\int_0^{\pi/2} \ln\left\{\frac{\sin(x)}{x}\right\}dx.$$

That's because this integral is

$$\int_0^{\pi/2} \ln\{\sin(x)\}dx - \int_0^{\frac{\pi}{2}} \ln\{x\}dx - \frac{\pi}{2}\ln(2) - [x\,\ln(x) - x]\Big|_0^{\pi/2}$$

$$= -\frac{\pi}{2}\ln(2) - \left[\frac{\pi}{2}\ln\left(\frac{\pi}{2}\right) - \frac{\pi}{2}\right] = -\frac{\pi}{2}\ln(2) - \left[\frac{\pi}{2}\ln(\pi) - \frac{\pi}{2}\ln(2) - \frac{\pi}{2}\right]$$

$$= -\frac{\pi}{2} \ln(2) - \frac{\pi}{2} \ln(\pi) + \frac{\pi}{2} \ln(2) + \frac{\pi}{2} = -\frac{\pi}{2} \ln(\pi) + \frac{\pi}{2}$$

and so

$$\int_0^{\pi/2} \ln\left\{\frac{\sin(x)}{x}\right\} dx = \frac{\pi}{2}[1 - \ln(\pi)]. \tag{2.4.2}$$

Our result says this integral is equal to $-0.22734\ldots$ and MATLAB agrees, as *integral(@(x)log(sin(x)./x),0,pi/2) = −0.22734 . . . .*

With a simple change of variable in Euler's log-sine integral we can get yet another pretty result. Since $\sin^2(\theta) = 1 - \cos^2(\theta)$ then

$$\frac{\sin^2(\theta)}{\cos^2(\theta)} = \tan^2(\theta) = \frac{1}{\cos^2(\theta)} - 1$$

and so

$$\tan^2(\theta) + 1 = \frac{1}{\cos^2(\theta)}$$

which says

$$\ln\left\{\frac{1}{\cos^2(\theta)}\right\} = -\ln\left\{\cos^2(\theta)\right\} = -2\ln\left\{\cos(\theta)\right\} = \ln\left\{\tan^2(\theta) + 1\right\}.$$

That is,

$$\ln\left\{\cos(\theta)\right\} = -\frac{1}{2}\ln\left\{\tan^2(\theta) + 1\right\}.$$

So, in the integral

$$\int_0^{\pi/2} \ln\left\{\cos(x)\right\} dx$$

replace the dummy variable of integration x with θ and write

$$\int_0^{\pi/2} \ln\left\{\cos(\theta)\right\} d\theta = \int_0^{\pi/2}\left[-\frac{1}{2}\ln\left\{\tan^2(\theta) + 1\right\}\right] d\theta = -\frac{\pi}{2}\ln(2)$$

or,

$$\int_0^{\pi/2} \ln\left\{\tan^2(\theta) + 1\right\}d\theta = \pi\ln(2).$$

Now, change variable to $x = \tan(\theta)$. Then

$$\frac{dx}{d\theta} = \frac{1}{\cos^2(\theta)}$$

and thus

$$d\theta = \cos^2(\theta)dx = \frac{1}{\tan^2(\theta) + 1}dx = \frac{1}{x^2 + 1}dx.$$

So, since $x = 0$ when $\theta = 0$ and $x = \infty$ when $\theta = \frac{\pi}{2}$, we have

$$\int_0^\infty \frac{\ln(x^2 + 1)}{x^2 + 1}dx = \pi\ln(2). \tag{2.4.3}$$

That is, this integral is equal to 2.177586 ..., and MATLAB agrees, as *integral* $(@(x)log(x.^2+1)./(x.^2+1),0,inf) = 2.177586....$ To end this discussion, here's a little calculation for *you* to play around with: writing $\int_0^\infty \frac{\ln(x^2+1)}{x^2+1}dx$ as $\int_0^1 + \int_1^\infty$, make the change of variable $u = \frac{1}{x}$ in the last integral and show that this leads to

$$\int_0^1 \frac{\ln\left(x + \frac{1}{x}\right)}{x^2 + 1}dx = \frac{\pi}{2}\ln(2). \tag{2.4.4}$$

This is equal to 1.088793 ..., and MATLAB agrees becauser *integral($@(x)log$ $(x+(1./x))./(x.^2+1),0,1) = 1.088799....$

The substitution $u = \frac{1}{x}$ is a trick well-worth keeping in mind. Here's another use of it to derive a result that almost surely would be much more difficult to get otherwise. Consider the integral

$$\int_0^\infty \frac{\ln(x^a + 1)}{x^2 - bx + 1}dx,$$

where $a \neq 0$ and b are constants. If we let $u = \frac{1}{x}$ (and so $dx = -\frac{1}{u^2}du$), we have

$$\int_0^\infty \frac{\ln(x^a + 1)}{x^2 - bx + 1}dx = \int_\infty^0 \frac{\ln\left(\frac{1}{u^a} + 1\right)}{\frac{1}{u^2} - b\frac{1}{u} + 1}\left(-\frac{1}{u^2}du\right) = \int_0^\infty \frac{\ln\left(\frac{1+u^a}{u^a}\right)}{1 - bu + u^2}du$$

$$= \int_0^\infty \frac{\ln(1 + u^a)}{1 - bu + u^2}du - a\int_0^\infty \frac{\ln(u)}{1 - bu + u^2}du.$$

That is,

$$\int_0^\infty \frac{\ln(x^a + 1)}{x^2 - bx + 1}\,dx = \int_0^\infty \frac{\ln(1 + x^a)}{1 - bx + x^2}\,dx - a\int_0^\infty \frac{\ln(x)}{1 - bx + x^2}\,dx$$

and so we immediately have

$$\int_0^\infty \frac{\ln(x)}{1 - bx + x^2}\,dx = 0. \qquad (2.4.5)$$

Notice that the case of $b = 0$ in (2.4.5) reduces this result to (1.5.1),

$$\int_0^\infty \frac{\ln(x)}{1 + x^2}\,dx = 0,$$

what we also get when we set $b = 1$ in (2.1.3). The value of b can't be just *anything*, however, and you'll be asked more on this point in the challenge problem section.

Now, let me remind you of a simple technique that you encountered way back in high school algebra—'completing the square.' This is a 'trick' that is well-worth keeping in mind when faced with an integral with a quadratic polynomial in the denominator of the integrand (but see also Challenge Problem (C2.2) for its use in a *cubic* denominator, and look back at the final integration of Sect. 1.6, too). As another example, let's calculate

$$\int_0^1 \frac{1 - x}{1 + x + x^2}\,dx.$$

We'll start by rewriting the denominator of the integrand by completing the square:

$$x^2 + x + 1 = x^2 + x + \frac{1}{4} + \left(1 - \frac{1}{4}\right) = \left(x + \frac{1}{2}\right)^2 + \frac{3}{4}.$$

Thus,

$$\int_0^1 \frac{1 - x}{1 + x + x^2}\,dx = \int_0^1 \frac{dx}{\left(x + \frac{1}{2}\right)^2 + \frac{3}{4}} - \int_0^1 \frac{x}{\left(x + \frac{1}{2}\right)^2 + \frac{3}{4}}\,dx.$$

Now, change variable to

$$u = x + \frac{1}{2}$$

(and so $dx = du$). Then, our integral becomes

$$\int_{\frac{1}{2}}^{\frac{3}{2}} \frac{du}{u^2 + \frac{3}{4}} - \int_{\frac{1}{2}}^{\frac{3}{2}} \frac{u - \frac{1}{2}}{u^2 + \frac{3}{4}} du = \frac{3}{2} \int_{\frac{1}{2}}^{\frac{3}{2}} \frac{du}{u^2 + \frac{3}{4}} - \int_{\frac{1}{2}}^{\frac{3}{2}} \frac{u}{u^2 + \frac{3}{4}} du.$$

The first integral on the right is

$$\int_{\frac{1}{2}}^{\frac{3}{2}} \frac{du}{u^2 + \left(\frac{\sqrt{3}}{2}\right)^2} = \frac{2}{\sqrt{3}} \tan^{-1}\left(\frac{u}{\sqrt{3}/2}\right)\Big|_{1/2}^{3/2} = \frac{2}{\sqrt{3}} \tan^{-1}\left(\frac{2u}{\sqrt{3}}\right)\Big|_{1/2}^{3/2}$$

$$= \frac{2}{\sqrt{3}}\left[\tan^{-1}\left(\frac{3}{\sqrt{3}}\right) - \tan^{-1}\left(\frac{1}{\sqrt{3}}\right)\right] = \frac{2}{\sqrt{3}}\left[\tan^{-1}\left(\sqrt{3}\right) - \tan^{-1}\left(\frac{1}{\sqrt{3}}\right)\right]$$

$$= \frac{2}{\sqrt{3}}\left[\frac{\pi}{3} - \frac{\pi}{6}\right] = \frac{\pi}{3\sqrt{3}}.$$

So,

$$\int_0^1 \frac{1 - x}{1 + x + x^2} dx = \frac{\pi}{2\sqrt{3}} - \int_{\frac{1}{2}}^{\frac{3}{2}} \frac{u}{u^2 + \frac{3}{4}} du.$$

In the integral on the right, change variable to

$$t = u^2 + \frac{3}{4}.$$

Then, $dt = 2u\, du$ and

$$\int_{\frac{1}{2}}^{\frac{3}{2}} \frac{u}{u^2 + \frac{3}{4}} du = \int_1^3 \frac{u}{t} \frac{dt}{2u} = \frac{1}{2} \int_1^3 \frac{dt}{t} = \frac{1}{2} \ln(t)\Big|_1^3 = \frac{1}{2} \ln(3).$$

So,

$$\int_0^1 \frac{1 - x}{1 + x + x^2} dx = \frac{\pi}{2\sqrt{3}} - \frac{1}{2} \ln(3)$$

or,

$$\int_0^1 \frac{1 - x}{1 + x + x^2} dx = \frac{1}{2}\left[\frac{\pi}{\sqrt{3}} - \ln(3)\right] \tag{2.4.6}$$

which equals 0.35759 ..., and MATLAB agrees because *integral(@(x)(1-x)./ (1+x+x.^2),0,1) = 0.357593....*

2.5 Trier's Double Integral

In this penultimate section of the chapter, let's return to an issue I touched on in the first chapter—how *physicists* encounter interesting integrals in their work. A classic problem in electrical physics imagines an infinitely large, two-dimensional network (or *grid*) of identical resistors, with node points at each of the lattice points (the points with integer coordinates). That is, each node is connected to each of its four nearest neighbor nodes through resistors of the same value, as shown in Fig. 2.5.1. For the analysis we'll do here, I'll take that value to be one-ohm (and for the general case, our results will simply scale in direct proportion to that value).

If we pick one of the nodes to be the *origin* node,[5] that is, to be node A at (0,0), the problem is to calculate the resistance R(m,n) that would be measured between the nodes (m,n) and (0,0). This problem has a long history, one that can be traced (at least in spirit) back to 1940, but probably the best (in the sense of being at the level of this book) mathematical development can be found in two short notes by the English electrical engineer Peter Trier (1919–2005). He shows[6] that the answer can be elegantly written in the form of a double integral:

$$R(m, n) = \frac{1}{(2\pi)^2} \int_{-\pi}^{\pi} \int_{-\pi}^{\pi} \frac{1 - \cos\{mx + ny\}}{2 - \cos(x) - \cos(y)} \, dx dy. \qquad (2.5.1)$$

Be sure to notice that R(m,n) = R(n,m), which makes sense given the physical symmetry of Fig. 2.5.1.

The integral in (2.5.1) says, by inspection, that R(0,0) = 0, and that is physically correct since the resistance between *any* node and itself is zero. Less immediately obvious is the value of R(1,0), but a simple, purely physical argument (that I'll show you in just a moment) says that $R(1,0) = \frac{1}{2}$. A somewhat more complicated physical argument (see the reference cited in note 27 in Chap. 1) shows that $R(1, 1) = \frac{2}{\pi} = 0.6366$. A question that now almost surely occurs to you is, does Trier's integral agree with these values for R(1,0) and R(1,1)? We can quickly answer that with MATLAB by typing

---

[5]Since the network is infinite in extent in all directions, which node we pick to be the origin node is arbitrary. In an *infinite* network, every node is 'at the center.'

[6]P. E. Trier, "An Electrical Resistance Network and Its Mathematical Undercurrents," *Bulletin of the Institute of Mathematics and Its Applications*, March/April 1985, pp. 58–60, and (in the same journal) "An Electrical Network—Some Further Undercurrents," January/February 1986, pp. 30–31. Trier's notes give some of the history of the problem, as well as a mathematical development of (2.5.1). Historical purists will no doubt object to me calling this 'Trier's integral,' since he was *not* the first to publish it. For example, the mathematician Harley Flanders (1925–2013) has it in his paper "Infinite Networks: II—Resistance in an Infinite Grid," *Journal of Mathematical Analysis and Applications*, October 1972, pp. 30–35, and even there he is simply quoting a result from an earlier (1964) work. Trier, however, *derives* (2.5.1) in a straightforward way from an infinite set of difference equations which I'll show you how to get later in this section.

Fig. 2.5.1 An infinitely
large, two-dimensional
network of identical
resistors

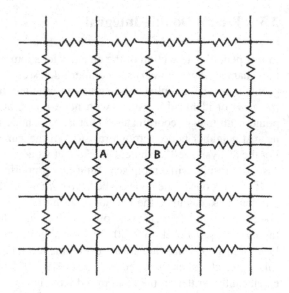

$R(1,0)$ = $integral2(@(x,y)(1\text{-}cos(x))./(2\text{-}cos(x)\text{-}cos(y)),\text{-}pi,pi,\text{-}pi,pi)/(2*pi)^2$
which produces the result 0.5000, and

$R(1,1)$ = $integral2(@(x,y)(1\text{-}cos(x+y))./(2\text{-}cos(x)\text{-}cos(y)),\text{-}pi,pi,\text{-}pi,pi)/(2*pi)^2$
which produces the result 0.6366. These computer results are, of course, nice,[7] but
a mathematician would like to see an analytical evaluation of (2.5.1).

Before I show you how to do that (using some of the integration tricks we've
already developed, plus one or two more), let me give you a little more *physical*
insight into what's going on in the infinite resistor network, insight that will help you
understand the central mathematical result our evaluation of (2.5.1) will produce. I'll
begin by describing the standard physical argument (often given in freshman physics
lectures) for why $R(1,0) = \frac{1}{2}$ ohm. This will help you understand the origin of (2.5.1).
We proceed in the following three simple, sequential steps.

Step 1 Imagine a 1-A current source that injects current into the resistor grid at node
A in Fig. 2.5.1. The current exits the grid along its 'edge at infinity' (use your
imagination with this!). By symmetry, the 1-A into A moves away from A by
splitting into four equal $\frac{1}{4}$ A currents, one of which flows from A to B (the node at

---

[7]Even more confidence in (2.5.1) is gained by using it to calculate the numerical value of R(1,2). In
1950 a theoretical calculation, using an approach totally different from (2.5.1), found the value R
$(1,2) = \frac{4}{\pi} - \frac{1}{2} = 0.7732$. (We'll calculate this theoretical result, too, at the end of this section.) Using
MATLAB, we simply type $integral2(@(x,y)(1\text{-}cos(x+2*y))./(2\text{-}cos(x)\text{-}cos(y)),\text{-}pi,pi,\text{-}pi,pi)/(2*pi)$
^{2} which gives the result 0.7732. You can't get better agreement (to four decimal places, anyway)
than that!

(1,0)). By *Ohm's law*[8] (voltage *drop* equals current times resistance), this results in a $\frac{1}{4}$ volt drop from A to B.

Step 2 Remove the 1-A current source of Step 1 and connect an identical source to the grid so that it injects 1-A into the grid *at infinity,* and removes it from the grid at B. By symmetry, the 1-A removed at B flows into B as four $\frac{1}{4}$ A currents, one of which is flowing from A to B. This results in a $\frac{1}{4}$ volt drop from A to B, with the same polarity as the $\frac{1}{4}$ volt drop from A to B that we got in Step 1.

Step 3 Finally, leaving the 1-A current source of Step 2 in place, reconnect the original 1-A current source of Step 1 that injects current into A. Then, by superposition, we can *add* the effects of the two currents, when individually applied, to get their combined effect when both are applied at the same time. That is, we have 1-A *into* node A, 1 A *out* of node B, zero current at infinity, and a total voltage drop of $\frac{1}{4} + \frac{1}{4} = \frac{1}{2}$ volt from A to B, that is, from (0,0) to (1,0). So a $\frac{1}{2}$ volt drop 'causes' 1-A of current, and therefore Ohm's law tells us that the resistance between (0,0) and (1,0) is $R(1,0) = \frac{1}{2}$ ohm.

This elegant, three-step solution only works because of the very high degree of symmetry in the problem. It does *not* work for two nodes that are not adjacent—try it and see. We *can*, however, use the basic ideas of current injection and extraction, along with superposition, to set-up the problem mathematically in such a way that (2.5.1) then follows. Here's how to do that. If we inject a current into the grid at infinity and remove that current at (0,0), then, clearly, there will be created at every node in the grid a voltage drop between each node and the origin node A at (0,0). For the node at (m,n), let's call that voltage drop v(m,n). What these voltage drops are depends, of course, on the magnitude of the injected current. The claim we make now is that if the injected current at infinity (and removed at (0,0)) is 2-A, then v(m,n) will *numerically* be equal to R(m,n). This is most easily seen, I think, with the aid of Fig. 2.5.2.

The left-hand-side of Fig. 2.5.2 shows 2-A in at infinity and out at (0,0), with a voltage drop of v(m,n) between the origin node and the node at (m,n). Then, as you can see, the right-hand-side of Fig. 2.5.2 shows two distinct situations that, *when added together*, give the left-hand-side. There are now just two crucial observations to make. To start, why in the first case on the right-hand-side is the voltage drop between (0,0) and (m,n) numerically equal to R(m,n)? That conclusion follows directly from Ohm's law because the current is 1-A. And second, why is the voltage drop equal to zero in the second case at the far-right of Fig. 2.5.2? That's because the nodes (0,0) and (m,n) are *indistinguishable*! That is, they each have 1-A leaving them, and they are in an *infinite* grid. In other words, there is nothing that makes them 'different' (their physical environments are identical) and so, in particular, they must have the same voltage (whatever it might be) which means a *zero* voltage *drop*. So, from superposition,

---

[8]Named after the German Georg Ohm (1787–1854), who published it in 1827.

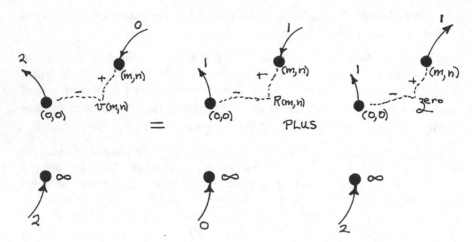

Fig. 2.5.2 Pictorial 'proof' that v(m,n) = R(m,n) + 0 = R(m,n)

$$v(m, n) = R(m, n) + 0 = R(m, n) \tag{2.5.2}$$

as claimed.

To solve for the $v(m,n)$, we write the following infinite set of difference equations, using conservation of electric charge that says the total instantaneous current (charge per unit time) into (or out of) a node is zero.[9] So, since all the resistors are one-ohm, we have (using Ohm's law),

$$[v(m + 1, n) - v(m, n)] + [v(m - 1, n) - v(m, n)] + [v(m, n + 1, -v(m, n)]$$

$$+ [v(m, n - 1) - v(m, n)] = \begin{cases} 2 & \text{if } m = n = 0 \\ 0 & \text{otherwise} \end{cases}.$$

Or, writing R instead of v,

$$R(m + 1, n) + R(m - 1, n) + R(m, n + 1) + R(m, n - 1) - 4R(m, n)$$

$$= \begin{cases} 2 & \text{if } m = n = 0 \\ 0 & \text{otherwise} \end{cases}. \tag{2.5.3}$$

This is the set of equations that Trier starts with,[10] a set that leads to the double integral of (2.5.1), the evaluation of which is our next task.

---

[9]Electrical engineers and physicists call this *Kirchhoff's current law*, after the German Gustav Robert Kirchhoff (1824–1887) who stated it in 1845.

[10]A discussion similar to Trier's, along with quite interesting commentary on some philosophical questions that a cautious mathematician might ask concerning the physical arguments I've made here, can be found in the paper by David Cameron, "The Square Grid of Unit Resistors," *The Mathematical Scientist*, September 1986, pp. 75–82.

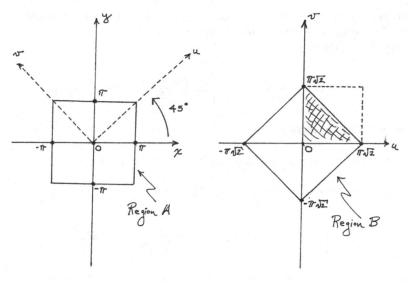

Fig. 2.5.3 Region of integration of (2.5.4) before (A) and after (B) a CCW axes rotation of 45°

In the following analysis I'll assume that m = n, and so we'll be calculating

$$R(n,n) = \frac{1}{(2\pi)^2} \int_{-\pi}^{\pi} \int_{-\pi}^{\pi} \frac{1 - \cos\{n(x+y)\}}{2 - \cos(x) - \cos(y)} dxdy, \qquad (2.5.4)$$

the so-called *diagonal resistances* of the infinite grid. This might seem to be a severe limitation but, as you'll see, when we combine R(n,n) with the difference equations of (2.5.3), and impose the physical symmetry I mentioned earlier, we'll actually be able to calculate *all* of the off-diagonal resistances as well. The region of integration for R(n,n) is the interior of the square (A) on the left-hand-side of Fig. 2.5.3.

It will prove to be convenient to rotate the coordinate axes 45° (you'll see why, soon), to give the new region of integration (B) on the right-hand-side of Fig. 2.5.3. I'll let you look in any good high school book on analytic geometry to confirm that if a point has coordinates x,y in the original coordinate system, then a CCW rotation of the axes through the angle α results in the following relationships with the coordinates u,v of the point in the new coordinate system:

$$x = u \cos(\alpha) - v \sin(\alpha), y = u \sin(\alpha) + v \cos(\alpha)$$

and so, for $\alpha = 45^\circ$, we have

$$x = \frac{u-v}{\sqrt{2}}, y = \frac{u+v}{\sqrt{2}}. \qquad (2.5.5)$$

This puts the new horizontal axis (u) along the diagonal of the original square region of integration (A) to give the region (B) of Fig. 2.5.3.

Using (2.5.5), the denominator of (2.5.4) is

$$2 - \cos(x) - \cos(y) = 2 - \left[\cos\left(\frac{u+v}{\sqrt{2}}\right) + \cos\left(\frac{u-v}{\sqrt{2}}\right)\right].$$

Remembering the identity

$$\cos(a+b) + \cos(a-b) = 2\cos(a)\cos(b),$$

the denominator becomes (with $a = \frac{u}{\sqrt{2}}$ and $b = \frac{v}{\sqrt{2}}$)

$$2 - \cos(x) - \cos(y) = 2 - 2\cos\left(\frac{u}{\sqrt{2}}\right)\cos\left(\frac{v}{\sqrt{2}}\right).$$

The numerator of (2.5.4) simplifies, once we notice from (2.5.5) that

$$x + y = \frac{2u}{\sqrt{2}} = u\sqrt{2},$$

to

$$1 - \cos\{n(x+y)\} = 1 - \cos\left(nu\sqrt{2}\right).$$

Thus, integrating *over region B* in Fig. 2.5.3, we have[11]

$$R(n,n) = \frac{1}{(2\pi)^2} \iint \frac{1 - \cos\left(nu\sqrt{2}\right)}{2\left[1 - \cos\left(\frac{u}{\sqrt{2}}\right)\cos\left(\frac{v}{\sqrt{2}}\right)\right]} du dv$$

or,

$$R(n,n) = \frac{1}{2(2\pi)^2} \int \left\{1 - \cos\left(nu\sqrt{2}\right)\right\} \left\{\int \frac{dv}{1 - \cos\left(\frac{u}{\sqrt{2}}\right)\cos\left(\frac{v}{\sqrt{2}}\right)}\right\} du. \quad (2.5.6)$$

---

[11]Notice, carefully, that the differential area in region B is dudv. To establish that, a mathematician would, probably, introduce the concept of the *Jacobian* of the coordinate transformation at this point and, in this particular case, show it is 1. I, instead, am using a physical argument, that a mere rotation of axes has no effect on the differential area, and so dxdy goes directly over to dudv.

Now, since you've surely noticed that I've left the integration limits off the integrals in (2.5.6), you are almost certainly wondering just what *are* those limits? Since the cosine function is unchanged with a change in the sign of its argument, we can do the v-integral over just the shaded triangular region of B *if* we multiply the result by 4. From Fig. 2.5.3 you can see that, for a given u, v varies from 0 to $-u + \pi\sqrt{2}$. To do the integral with those limits, however, leads to a very nasty calculation (that means I couldn't do it), but we are saved if we notice that the integrand of the v-integral is symmetrical in u and v which means we can integrate over the *entire* square (in dashed lines) defined by $0 \leq u \leq \pi\sqrt{2}, 0 \leq v \leq \pi\sqrt{2}$ *if* we divide by 2.[12]

So, our v-integral is most simply written as

$$\frac{1}{2}\left[4 \int_0^{\pi\sqrt{2}} \frac{dv}{1 - \cos\left(\frac{u}{\sqrt{2}}\right)\cos\left(\frac{v}{\sqrt{2}}\right)}\right] = 2 \int_0^{\pi\sqrt{2}} \frac{dv}{1 - \cos\left(\frac{u}{\sqrt{2}}\right)\cos\left(\frac{v}{\sqrt{2}}\right)}.$$

If we let $a - \frac{v}{\sqrt{2}}$ (and so $dv - \sqrt{2}da$) then our v integral becomes

$$2\sqrt{2} \int_0^{\pi} \frac{da}{1 - \cos\left(\frac{u}{\sqrt{2}}\right)\cos(a)}.$$

Now, recalling the change of variable trick we used in deriving (2.2.3), let $t = \tan\left(\frac{a}{2}\right)$. Then, as we showed earlier,

$$\cos(a) - \frac{1 - t^2}{1 + t^2}, da - \frac{2}{1 + t^2} dt,$$

and our v integral becomes

$$2\sqrt{2} \int_0^{\tan\left(\frac{\pi}{2}\right)} \frac{\frac{2}{1+t^2} dt}{1 - \cos\left(\frac{u}{\sqrt{2}}\right)\frac{1-t^2}{1+t^2}} = 4\sqrt{2} \int_0^{\infty} \frac{dt}{1 + t^2 - \cos\left(\frac{u}{\sqrt{2}}\right) + \cos\left(\frac{u}{\sqrt{2}}\right)t^2}$$

---

[12]Here's what I mean by all this (and this is *the* crucial observation in the analysis). For every point (u,v) in the shaded triangle of Fig. 2.5.3 there is a unique, matching point $\left(\pi\sqrt{2} - u, \pi\sqrt{2} - v\right)$ in the triangular region above the shaded one. (That is, these two points are in one-to-one correspondence.) These two triangles are two halves of the square region in dashed lines. Since the value of the v-integral integrand when evaluated at (u,v) is equal to the integrand evaluated at $\left(\pi\sqrt{2} - u, \pi\sqrt{2} - v\right)$—you should confirm this—then the integral over the shaded triangular region is one-half the integral over the dashed square (which is a *far* easier integral to do).

$$= 4\sqrt{2} \int_0^\infty \frac{dt}{1 - \cos\left(\frac{u}{\sqrt{2}}\right) + \left[1 + \cos\left(\frac{u}{\sqrt{2}}\right)\right] t^2} = \frac{4\sqrt{2}}{1 + \cos\left(\frac{u}{\sqrt{2}}\right)} \int_0^\infty \frac{dt}{\frac{1 - \cos\left(\frac{u}{\sqrt{2}}\right)}{1 + \cos\left(\frac{u}{\sqrt{2}}\right)} + t^2}$$

$$= \frac{4\sqrt{2}}{1 + \cos\left(\frac{u}{\sqrt{2}}\right)} \cdot \frac{1}{\sqrt{\frac{1 - \cos\left(\frac{u}{\sqrt{2}}\right)}{1 + \cos\left(\frac{u}{\sqrt{2}}\right)}}} \left\{ \tan^{-1}\left(\frac{t}{\sqrt{\frac{1 - \cos\left(\frac{u}{\sqrt{2}}\right)}{1 + \cos\left(\frac{u}{\sqrt{2}}\right)}}}\right)\right\}$$

$$= \frac{4\sqrt{2}}{\sqrt{1 + \cos\left(\frac{u}{\sqrt{2}}\right)}\sqrt{1 - \cos\left(\frac{u}{\sqrt{2}}\right)}} \tan^{-1}(\infty) = \frac{4\sqrt{2}}{\sqrt{1 - \cos^2\left(\frac{u}{\sqrt{2}}\right)}} \frac{\pi}{2} = \frac{2\sqrt{2}\pi}{\sin\left(\frac{u}{\sqrt{2}}\right)}.$$

Thus, (2.5.6) becomes

$$R(n, n) = \frac{1}{2(2\pi)^2} \int_0^{\pi\sqrt{2}} \frac{2\sqrt{2}\pi\{1 - \cos(nu\sqrt{2})\}}{\sin\left(\frac{u}{\sqrt{2}}\right)} du$$

or, if we let $b = \frac{u}{\sqrt{2}}$ (and so $du = \sqrt{2}db$), then

$$R(n, n) = \frac{2\sqrt{2}\pi}{2(2\pi)^2} \int_0^\pi \frac{1 - \cos(2nb)}{\sin(b)} \sqrt{2}db$$

or,

$$R(n, n) = \frac{2\pi}{(2\pi)^2} \int_0^\pi \frac{1 - \cos(2nb)}{\sin(b)} db = \frac{1}{2\pi} \int_0^\pi \frac{1 - \cos(2nb)}{\sin(b)} db. \qquad (2.5.7)$$

Now, concentrate on the numerator of the integrand in (2.5.7). Since

$$2\sin(\alpha)\sin(\beta) = \cos(\alpha - \beta) - \cos(\alpha + \beta)$$

then

$$2\sin(b)[\sin(b) + \sin(3b) + \sin(5b) + \ldots + \sin\{(2n - 1)b\}]$$
$$= [\cos(0) - \cos(2b)] + [\cos(2b) - \cos(4b)] + [\cos(4b) - \cos(6b)]$$
$$+ \ldots + [\cos\{(2n - 2)b\} - \cos(2nb)] = 1 - \cos(2nb)$$

because most of the terms cancel. That is,

$$1 - \cos(2nb) = 2\sin(b)[\sin(b) + \sin(3b) + \sin(5b) + \ldots + \sin\{(2n-1)b\}]$$

and so

$$R(n, n) = \frac{2}{2\pi} \int_0^\pi [\sin(b) + \sin(3b) + \sin(5b) + \ldots + \sin\{(2n-1)b\}]db$$

$$= \frac{1}{\pi}\left[-\cos(b) - \frac{1}{3}\cos(3b) - \ldots - \frac{1}{2n-1}\cos\{(2n-1)b\}\right]\Big|_0^\pi$$

$$= \frac{1}{\pi}\left[\left(1 + \frac{1}{3} + \ldots + \frac{1}{2n-1}\right) + \left(1 + \frac{1}{3} + \ldots + \frac{1}{2n-1}\right)\right]$$

or,

$$R(n, n) = \frac{2}{\pi}\left[\left(1 + \frac{1}{3} + \ldots + \frac{1}{2n-1}\right)\right] \qquad (2.5.8)$$

and we see, in particular, that $R(1, 1) = \frac{2}{\pi}$ as stated at the beginning of this section. One immediate, quite interesting result from (2.5.8) is that

$$\lim_{n\to\infty} R(n, n) = \infty$$

because, as Euler showed in 1737, the sum of the reciprocals of the primes, alone, diverges[13] and (2.5.8) contains all the prime reciprocals (with the exception of $\frac{1}{2}$, of course, as 2 is the lone even prime, but that exception has no effect on the divergence of $\lim_{n\to\infty} R(n, n)$).

The mathematical divergence of $R(n, n)$ as $n \to \infty$ has been interpreted as 'proof' that the physical current injection/extraction arguments used at the start of our analysis actually makes such an analysis invalid. That's because, so it has been argued, if the resistance between $(0, 0)$ and the 'edge' of the resistor grid at infinity is infinite, then it would be *impossible* to inject any current into $(0, 0)$ using anything less than infinite voltage! This argument, however, is itself, faulty, since the path from $(0, 0)$ to infinity *through the diagonal lattice points* is not the only path from $(0, 0)$ to infinity. There are infinitely many other paths from $(0, 0)$ to the remote edge of the grid. So, the question is, which infinity 'wins'—the infinity of the *number* of paths, or the infinity of the resistance of each individual path?

Something for you to think about when you need a break from doing integrals! (If you come-up with an answer, let me know.)

Now, to wrap this up, recall that I told you earlier that, even though we've found only the diagonal resistances, we can use them, together with the difference equations in (2.5.3), and symmetry, to find the off-diagonal resistances as well. As an example of how to do that, let's calculate $R(1, 2)$. We know from note 7 that the

---

[13]For a proof of this, see my *An Imaginary Tale*, Princeton 2016, pp. 150–152.

answer is $\frac{4}{\pi} - \frac{1}{2}$, but that's from somebody else's calculation. How can *we* calculate it? Just set m $= n = 1$ in (2.5.3) to get

$$R(2,1) + R(0,1) + R(1,2) + R(1,0) - 4R(1,1) = 0.$$

We know from symmetry that $R(2,1) = R(1,2)$, and that $R(0,1) = R(1,0)$. Thus,

$$2R(1,2) + 2R(1,0) - 4R(1,1) = 0$$

or,

$$R(1,2) = 2R(1,1) - R(1,0) = 2\left(\frac{2}{\pi}\right) - \frac{1}{2}$$

and so, just like that,

$$R(1,2) = \frac{4}{\pi} - \frac{1}{2}.$$

Using this approach, over and over, we can work our way outward from $(0,0)$ to eventually calculate the integrals $R(m,n)$ for any given values of m and n (see Challenge Problem (C2.7)).

2.6   A Final Example of a 'Hard' Integral Made 'Easy'

Back in Sect. 2.3 we discussed two simple examples of a so-called *quartic* integral (see (2.3.2) and (2.3.3)). What I'll show you here is how some of our earlier tricks can unravel the more general

$$\int_0^\infty \frac{dx}{bx^4 + 2ax^2 + 1} = ?$$

for a and b *any* positive constants. Our earlier results will then become special cases of the general solution. (In Challenge Problem (C2.10) you'll be asked to generalize this solution even a bit more.)

We start our analysis for the special case of b $= 1$, that is with

$$I = \int_0^\infty \frac{dx}{x^4 + 2ax^2 + 1},$$

and then make the change of variable u $= \frac{1}{x}$ (and so dx $= -\frac{du}{u^2}$), which gives

$$I = \int_\infty^0 \frac{-\frac{du}{u^2}}{\frac{1}{u^4} + 2a\frac{1}{u^2} + 1} = \int_0^\infty \frac{u^2}{1 + 2au^2 + u^4}\, du \qquad (2.6.1)$$

or,

$$I = \int_0^\infty \frac{x^2}{x^4 + 2ax^2 + 1}\, dx. \qquad (2.6.2)$$

Adding (2.6.1) and (2.6.2), we see that

$$I = \frac{1}{2}\int_0^\infty \frac{1 + x^2}{x^4 + 2ax^2 + 1}\, dx. \qquad (2.6.3)$$

Next, we make another change of variable, to $x = \tan(\theta)$ (and so $dx = \frac{d\theta}{\cos^2(\theta)}$). With that, (2.6.3) becomes

$$I = \frac{1}{2}\int_0^{\pi/2} \frac{1 + \tan^2(\theta)}{\tan^4(\theta) + 2a\tan^2(\theta) + 1} \left(\frac{d\theta}{\cos^2(\theta)}\right)$$

or, after just a bit of easy algebra,

$$I = \frac{1}{2}\int_0^{\pi/2} \frac{d\theta}{\sin^4(\theta) + 2a\sin^2(\theta)\cos^2(\theta) + \cos^4(\theta)}. \qquad (2.6.4)$$

Then making yet another change of variable to $u = 2\theta$ (and so $d\theta = \frac{1}{2}du$), (2.6.4) becomes (with the use of the double-angle identities of trigonometry)

$$I = \frac{1}{2}\int_0^\pi \frac{\frac{1}{2}\, du}{\sin^4\left(\frac{1}{2}u\right) + 2a\sin^2\left(\frac{1}{2}u\right)\cos^2\left(\frac{1}{2}u\right) + \cos^4\left(\frac{1}{2}u\right)}$$

$$= \frac{1}{2^2}\int_0^\pi \frac{du}{\left[\frac{1}{2} - \frac{1}{2}\cos(u)\right]^2 + 2a\left[\frac{1}{2} - \frac{1}{2}\cos(u)\right]\left[\frac{1}{2} + \frac{1}{2}\cos(u)\right] + \left[\frac{1}{2} + \frac{1}{2}\cos(u)\right]^2}$$

which reduces (if you're careful with the algebra) to

$$I = \frac{1}{2}\int_0^\pi \frac{du}{1 + a + (1 - a)\cos^2(u)}. \qquad (2.6.5)$$

Okay, now here's *another* change of variable, this time to $v = 2u$ (and so $du = \frac{1}{2}dv$). With that (2.6.5) becomes

$$I = \frac{1}{2} \int_0^{2\pi} \frac{\frac{1}{2} \, dv}{1 + a + (1-a)\cos^2\left(\frac{1}{2}v\right)} = \frac{1}{2^2} \int_0^{2\pi} \frac{dv}{1 + a + (1-a)\left[\frac{1}{2} + \frac{1}{2}\cos(v)\right]}$$

which (with just a bit more algebra) becomes

$$I = \frac{1}{2} \int_0^{2\pi} \frac{dv}{3 + a + (1-a)\cos(v)}$$

or, as $\int_0^{2\pi} = 2\int_0^{\pi}$ because of the symmetry of the cosine function about $v = \pi$,

$$I = \int_0^{\pi} \frac{dv}{3 + a + (1-a)\cos(v)} . \tag{2.6.6}$$

You almost certainly recognize the integral in (2.6.6) as being perfectly suited for the change of variable we used in Sect. 2.2, namely $z = \tan\left(\frac{v}{2}\right)$, where we showed

$$dv = \frac{2}{1 + z^2} \, dz$$

and

$$\cos(v) = \frac{1 - z^2}{1 + z^2} .$$

With this change of variable, (2.6.6) becomes

$$I = \int_0^{\infty} \frac{\frac{2}{1+z^2} \, dz}{3 + a + (1-a)\frac{1-z^2}{1+z^2}}$$

or, after a few simple algebraic manipulations,

$$I = \frac{1}{1+a} \int_0^{\infty} \frac{dz}{\frac{2}{1+a} + z^2} . \tag{2.6.7}$$

You of course now instantly see the integral in (2.6.7) as being the classic inverse-tangent result from freshman calculus, and so

$$I = \left(\frac{1}{1+a}\right) \frac{1}{\sqrt{\frac{2}{1+a}}} \tan^{-1}\left(\frac{z}{\sqrt{\frac{2}{1+a}}}\right)\Big|_0^{\infty} = \left(\frac{1}{1+a}\right) \frac{\sqrt{1+a}}{\sqrt{2}} \left(\frac{\pi}{2}\right)$$

or,

$$\int_0^\infty \frac{dx}{x^4 + 2ax^2 + 1} = \left(\frac{\pi}{2\sqrt{2}}\right)\frac{1}{\sqrt{1+a}}. \tag{2.6.8}$$

Now, the result in (2.6.8) isn't quite what I promised you at the start, with an *arbitrary* positive coefficient for x^4. So, here's how, with nothing but just a bit more algebra, to get that additional parameter into our result. Making the change of variable in (2.6.1) of $u = b^{-1/4}x$ (that is, $x = b^{1/4}u$ and so $dx = b^{1/4}du$) we have

$$I = \int_0^\infty \frac{b^{1/4}du}{bu^4 + 2ab^{1/2}u^2 + 1} = \left(\frac{\pi}{2\sqrt{2}}\right)\frac{1}{\sqrt{1+a}}$$

and so

$$\int_0^\infty \frac{du}{bu^4 + 2ab^{1/2}u^2 + 1} = \left(\frac{\pi}{2\sqrt{2}}\right)\frac{1}{b^{1/4}\sqrt{1+a}}.$$

Now, let's write $a' = ab^{1/2}$ (and so $a = \frac{a'}{b^{1/2}}$). Then,

$$\int_0^\infty \frac{du}{bu^4 + 2a'u^2 + 1} = \left(\frac{\pi}{2\sqrt{2}}\right)\frac{1}{b^{1/4}\sqrt{1+\frac{a'}{b^{1/2}}}} = \left(\frac{\pi}{2\sqrt{2}}\right)\frac{1}{\sqrt{b^{1/2} + b^{1/2}\frac{a'}{b^{1/2}}}}$$

or, writing a instead of a', and changing the dummy variable of integration back to x, we get a very pretty result called (for the obvious reason) the 'double-root' solution:

$$\int_0^\infty \frac{dx}{bx^4 + 2ax^2 + 1} = \left(\frac{\pi}{2\sqrt{2}}\right)\frac{1}{\sqrt{a + \sqrt{b}}}. \tag{2.6.9}$$

2.7 Challenge Problems

That was all pretty straightforward, but here are some problems that *will*, I think, give your brain a really good workout. And yet, like nearly everything else in this book, *if you see the trick* they will unfold for you like a butterfly at rest.

(C2.1) According to Edwards' *A Treatise on the Integral Calculus* (see the end of the original Preface), the following question appeared on an 1886 exam at the University of Cambridge: show that

$$\int_0^4 \frac{\ln(x)}{\sqrt{4x - x^2}}dx = 0.$$

Edwards included no solution and, given that it took me about 5 h spread over 3 days to do it (in my quiet office, under no pressure), my awe for the expected math level of some of the undergraduate students at nineteenth century Cambridge is unbounded. See if you can do it faster than I did (and if not, my solution is at the end of the book). This is the problem I mention in the new Preface (see note 5 there) that received even better solutions than mine from two readers of the first edition.

(C2.2) Calculate the value of $\int_0^1 \frac{dx}{x^3+1}$. Hint: first confirm the validity of the partial fraction expansion $\frac{1}{x^3+1} = \frac{1}{3}\left[\frac{1}{x+1} - \frac{x-2}{x^2-x+1}\right]$ and so $\int_0^1 \frac{dx}{x^3+1} = \frac{1}{3}\int_0^1 \frac{dx}{x+1} - \frac{1}{3}\int_0^1 \frac{x-2}{x^2-x+1}dx$. The first integral on the right will yield to an obvious change of variable, and the second integral on the right will yield to another change of variable, one just as obvious if you first complete the square in the denominator of the integrand. Your theoretical answer should have the numerical value 0.8356488....

(C2.3) Here's a pretty little recursive problem for you to work through. Suppose you know the value of

$$\int_0^\infty \frac{dx}{x^4 + 1}.$$

This integral is not particularly difficult to do—we did it in (2.3.4)—with a value of $\frac{\pi}{2\sqrt{2}}$. The point here is that with this knowledge you then also immediately know the values of

$$\int_0^\infty \frac{dx}{(x^4 + 1)^m}$$

for all integer m > 1 (and not just for m = 1). Show this is so by deriving the recursion

$$\int_0^\infty \frac{dx}{(x^4 + 1)^{m+1}} = \frac{4m - 1}{4m}\int_0^\infty \frac{dx}{(x^4 + 1)^m}.$$

For example,

$$\int_0^\infty \frac{dx}{(x^4 + 1)^3} = \frac{(4)(2) - 1}{(4)(2)}\int_0^\infty \frac{dx}{(x^4 + 1)^2} = \frac{7}{8}\int_0^\infty \frac{dx}{(x^4 + 1)^2}$$

and

$$\int_0^\infty \frac{dx}{(x^4 + 1)^2} = \frac{(4)(1) - 1}{(4)(1)}\int_0^\infty \frac{dx}{x^4 + 1} = \frac{3}{4}\int_0^\infty \frac{dx}{x^4 + 1}.$$

Thus,

$$\int_0^\infty \frac{dx}{(x^4+1)^3} = \left(\frac{7}{8}\right)\left(\frac{3}{4}\right)\int_0^\infty \frac{dx}{x^4+1} = \left(\frac{21}{32}\right)\frac{\pi}{2\sqrt{2}} = \frac{21\pi}{64\sqrt{2}} = 0.72891\ldots.$$

MATLAB agrees, as $integral(@(x)1./((x.^4+1).^3),0,inf) = 0.72891\ldots$.
Hint: Start with $\int_0^\infty \frac{dx}{(x^4+1)^m}$ and integrate by parts.

(C2.4) For what values of b does the integral in (2.4.5) make sense? Hint: think about where any singularities in the integrand are located.

(C2.5) Show that $\int_0^\infty \frac{\ln(1+x)}{x\sqrt{x}}dx = 2\pi$. We can 'check' this claim by noting that $2\pi = 6.283185\ldots$, and that $integral(@(x)log(1+x)./(x.*sqrt(x)),0,inf) = 6.283184\ldots$. Hint: integrate-by-parts.

(C2.6) Compute the value of $\int_0^{\pi/2} x\cot(x)dx = $? Hint: start with (2.4.1) and integrate-by-parts.

(C2.7) Calculate the exact value of the integral in (2.5.1) for $m = 0$ and $n = 2$. That is, what is $R(0,2)=$? What is the exact value of $R(3,1)=$? Hint: the numerical values of the integrals are
$$R(0,2) = integral2(@(x,y)(1-cos(2*y))./(2-cos(x)-cos(y)),-pi,pi,-pi,pi)/(2*pi)$$
$$^2 = 0.7267\ldots$$
$$R(3,1) = integral2(@(x,y)(1-cos(3*x+y))./(2-cos(x)-cos(y)),-pi,pi,-pi,pi)/(2*pi)$$
$$^2 = 0.8807\ldots.$$

(C2.8) In (2.3.4) we showed that the definite integral $\int_0^\infty \frac{x^2}{x^4+1}dx = \frac{\pi\sqrt{2}}{4}$. This integral can, however, be done *indefinitely* and then evaluated for *any* limits. If you look in math tables, you'll find a pretty complicated expression for the indefinite integral but, *if you see the trick*, it's actually pretty easy to do. Since this *is* a book of tricks, this is an appropriate exercise for you to tackle. Check your answer by confirming it agrees with the special case of (2.3.4), and then use it to calculate the *exact* value of $\int_0^1 \frac{x^2}{x^4+1}dx = $? The numerical value is given by.
$$integral(@(x)(x.^2)./(x.^4+1),0,1) = 0.2437477\ldots.$$

(C2.9) What is $\int_0^\infty \frac{x^2}{1-x^4}dx = $? Hint: Write $\int_0^\infty = \int_0^1 + \int_1^\infty$ and make the change of variable $y = \frac{1}{x}$ in the last integral.

(C2.10) What is $\int_0^\infty \frac{dx}{bx^4+2ax^2+c} = $? Hint: Factor c out of the denominator, and then use (2.6.9).

Chapter 3
Feynman's Favorite Trick

3.1 Leibniz's Formula

The starting point for Feynman's trick of 'differentiating under the integral sign,' mentioned at the end of Chap. 1, is Leibniz's formula. If we have the integral

$$I(\alpha) = \int_{a(\alpha)}^{b(\alpha)} f(x, \alpha) \, dx$$

where α is the so-called *parameter* of the integral (*not* the dummy variable of integration which is, of course, x), then we wish to calculate the derivative of I with respect to α. We do that in just the way you'd expect, from the very definition of the derivative:

$$\frac{dI}{d\alpha} = \lim_{\Delta\alpha \to 0} \frac{I(\alpha + \Delta\alpha) - I(\alpha)}{\Delta\alpha}.$$

Now, since the integration limits depend (in general) on α, then a $\Delta\alpha$ will cause a Δa and a Δb and so we have to write

$$I(\alpha + \Delta\alpha) - I(\alpha) = \int_{a+\Delta a}^{b+\Delta b} f(x, \alpha + \Delta\alpha) \, dx - \int_{a}^{b} f(x, \alpha) \, dx$$

$$= \left(\int_{a+\Delta a}^{a} + \int_{a}^{b} + \int_{b}^{b+\Delta b} \right) f(x, \alpha + \Delta\alpha) \, dx - \int_{a}^{b} f(x, \alpha) \, dx$$

$$= \int_{a}^{b} \{f(x, \alpha + \Delta\alpha) - f(x, \alpha)\} dx + \int_{b}^{b+\Delta b} f(x, \alpha + \Delta\alpha) \, dx - \int_{a}^{a+\Delta a} f(x, \alpha + \Delta\alpha) \, dx.$$

© Springer Nature Switzerland AG 2020
P. J. Nahin, *Inside Interesting Integrals*, Undergraduate Lecture Notes in Physics,
https://doi.org/10.1007/978-3-030-43788-6_3

As $\Delta\alpha \rightarrow 0$ we have $\Delta a \rightarrow 0$ and $\Delta b \rightarrow 0$, too, and so $\lim_{\Delta\alpha\rightarrow 0}$ $\{I(\alpha+\Delta\alpha) - I(\alpha)\} = \lim_{\Delta\alpha\rightarrow 0}\int_a^b \{f(x,\alpha+\Delta\alpha) - f(x,\alpha)\}dx + f(b,\alpha)\Delta b - f(a,\alpha)\Delta a$ where the last two terms follow because as $\Delta a \rightarrow 0$ and $\Delta b \rightarrow 0$ the value of x over the entire integration interval remains practically unchanged at $x = a$ or at $x = b$, respectively. Thus,

$$\frac{dI}{d\alpha} = \lim_{\Delta\alpha\rightarrow 0} \frac{I(\alpha + \Delta\alpha) - I(\alpha)}{\Delta\alpha}$$

$$= \lim_{\Delta\alpha\rightarrow 0} \frac{1}{\Delta\alpha} \int_a^b \{f(x,\alpha + \Delta\alpha) - f(x,\alpha)\}dx + \lim_{\Delta\alpha\rightarrow 0} f(b,\alpha)\frac{\Delta b}{\Delta\alpha}$$

$$- \lim_{\Delta\alpha\rightarrow 0} f(a,\alpha)\frac{\Delta a}{\Delta\alpha}$$

or, taking the $\frac{1}{\Delta\alpha}$ inside the integral (the Riemann integral itself is defined as a limit, so what we are doing is reversing the order of two limiting operations, something a pure mathematician would want to justify but, as usual in this book, we won't worry about it!),

$$\frac{dI}{d\alpha} = \int_a^b \frac{\partial f}{\partial\alpha} \, dx + f(b,\alpha)\frac{db}{d\alpha} - f(a,\alpha)\frac{da}{d\alpha}. \qquad (3.1.1)$$

This is the full-blown Leibniz formula for how to differentiate an integral, including the case where the integrand *and the limits* are functions of the parameter α. If the limits are not functions of the parameter (such as when the limits are constants) then the last two terms vanish and we simply (partially) differentiate the integrand under the integral sign with respect to the parameter α (*not* x).

Here's a quick example of the power of Leibniz's formula. We've used, numerous times, the elementary result

$$\int_0^\infty \frac{1}{x^2 + a^2} \, dx = \left\{\frac{1}{a} \tan^{-1}\left(\frac{x}{a}\right)\right\}\bigg|_0^\infty = \frac{\pi}{2a}.$$

Treating a as a parameter, differentiation then immediately gives us the new result that

$$\int_0^\infty \frac{-2a}{(x^2 + a^2)^2} \, dx = -\frac{2\pi}{4a^2}$$

or,

$$\int_0^\infty \frac{1}{(x^2 + a^2)^2} \, dx = \frac{\pi}{4a^3}. \qquad (3.1.2)$$

One could, of course, not stop here but continue to differentiate this new result again (and then again and again), each time producing a new result. For example, another differentiation gives

$$\int_0^\infty \frac{1}{(x^2 + a^2)^3}\, dx = \frac{3\pi}{16a^5}. \qquad (3.1.3)$$

For $a = 1$ this is $0.58904\ldots$ and *integral(@(x)1./((x.^2 + 1).^3),0,inf) = 0.58904\ldots*.

A more sophisticated example of the formula's use is the evaluation of

$$\int_{-\infty}^\infty e^{-\frac{x^2}{2}}\, dx.$$

This is the famous *probability integral*,[1] so-called because it appears in the theory of random quantities described by Gaussian (bell-shaped) probability density functions (about which you need know *nothing* for this book). Because this integral has an even integrand, we can instead study

$$\int_0^\infty e^{-\frac{x^2}{2}}\, dx$$

and then simply double the result.

To introduce a parameter (t) with which we can differentiate with respect to, let's define the function

$$g(t) = \left\{ \int_0^t e^{-\frac{x^2}{2}}\, dx \right\}^2,$$

and so what we are after is

$$\int_{-\infty}^\infty e^{-\frac{x^2}{2}}\, dx = 2\sqrt{g(\infty)}.$$

Note, *carefully*, that the parameter t and the dummy variable of integration x, are *independent* quantities.

---

[1]The probability integral is most commonly evaluated in textbooks with the trick of converting it to a double integral in polar coordinates (see, for example, my books *An Imaginary Tale: the story of* $\sqrt{-1}$, Princeton 2012, pp. 177–178, and *Mrs. Perkins's Electric Quilt*, Princeton 2009, pp. 282–283), and the use of Leibniz's formula that I'm going to show you here is uncommon. It is such an important integral that at the end of this chapter we will return to it with some additional analyses.

Differentiating g(t) using Leibniz's formula (notice that the upper limit on the integral is a function of the parameter t, but the integrand and the lower limit are not), we get

$$\frac{dg}{dt} = 2 \int_0^t e^{-\frac{x^2}{2}} \, dx \left\{ e^{-\frac{t^2}{2}} \right\} = 2 \int_0^t e^{-\frac{(t^2+x^2)}{2}} \, dx.$$

Next, change variable to $y = \frac{x}{t}$ (so that as x varies from 0 to t, y will vary from 0 to 1). That is, $x = yt$ and so $dx = t \, dy$ and we have

$$\frac{dg}{dt} = \int_0^1 2te^{-\frac{(t^2+y^2t^2)}{2}} \, dy = \int_0^1 2te^{-\frac{(1+y^2)t^2}{2}} \, dy.$$

Now, notice that the integrand can be written as a (partial) derivative as follows:

$$2te^{-\frac{(1+y^2)t^2}{2}} = \frac{\partial}{\partial t} \left\{ -\frac{2e^{-\frac{(1+y^2)t^2}{2}}}{1+y^2} \right\}.$$

That is,[2]

$$\frac{dg}{dt} = \int_0^1 \frac{\partial}{\partial t} \left\{ -\frac{2e^{-\frac{(1+y^2)t^2}{2}}}{1+y^2} \right\} dy = -2 \frac{d}{dt} \int_0^1 \frac{e^{-\frac{(1+y^2)t^2}{2}}}{1+y^2} \, dy.$$

And so, integrating,

$$g(t) = -2 \int_0^1 \frac{e^{-\frac{(1+y^2)t^2}{2}}}{1+y^2} \, dy + C$$

where C is the constant of integration. We can find C as follows. Let $t = 0$. Then

$$g(0) = \left\{ \int_0^0 e^{-\frac{x^2}{2}} \, dx \right\}^2 = 0$$

and, as $t \to 0$,

---

[2]The reason we write a *partial* derivative inside the integral and a total derivative outside the integral is that the integrand is a function of two variables (t and y) while the integral itself is a function only of t (we've 'integrated out' the y dependency).

$$\int_0^1 \frac{e^{-\frac{(1+y^2)t^2}{2}}}{1+y^2}\, dy \to \int_0^1 \frac{1}{1+y^2}\, dy = \tan^{-1}(y)\Big|_0^1 = \frac{\pi}{4}.$$

Thus, $0 = -2\left(\frac{\pi}{4}\right) + C$ and so $C = \frac{\pi}{2}$ and therefore

$$g(t) = \frac{\pi}{2} - 2\int_0^1 \frac{e^{-\frac{(1+y^2)t^2}{2}}}{1+y^2}\, dy.$$

Now, let $t \to \infty$. The integrand clearly vanishes over the entire interval of integration and we have $g(\infty) = \frac{\pi}{2}$. That is,

$$\int_{-\infty}^{\infty} e^{-\frac{x^2}{2}}\, dx = 2\sqrt{\frac{\pi}{2}} = \sqrt{2\pi}. \tag{3.1.4}$$

$\sqrt{2\pi} = 2.506628 \ldots$, and MATLAB agrees: *integral(@(x)exp(-(x.^2)/2),-inf, inf) = 2.506628 ....*

We can use differentiation under the integral sign, again, to generalize this result in the evaluation of

$$I(t) = \int_0^{\infty} \cos{(tx)}e^{-\frac{x^2}{2}}\, dx.$$

From our last result, we know that $I(0) = \sqrt{\frac{\pi}{2}}$. Then, differentiating with respect to t,

$$\frac{dI(t)}{dt} = \int_0^{\infty} -x \sin{(tx)}e^{-\frac{x^2}{2}}\, dx.$$

We can do this integral by-parts, let $u = \sin(tx)$ and $dv = xe^{-\frac{x^2}{2}}\, dx$. Then, $du = t\cos(tx)dx$ and $v = e^{-\frac{x^2}{2}}$. Thus,

$$\frac{dI(t)}{dt} = \left\{ \sin{(tx)}e^{-\frac{x^2}{2}} \right\}\Big|_0^{\infty} - \int_0^{\infty} t \cos{(tx)}e^{-\frac{x^2}{2}}\, dx.$$

The first term on the right vanishes at both the upper and lower limits, and so we have the elementary first-order differential equation

$$\frac{dI(t)}{dt} = -t\int_0^{\infty} \cos{(tx)}e^{-\frac{x^2}{2}}\, dx = -t\,I(t).$$

Rewriting, this is

$$\frac{dI(t)}{I(t)} = -t \, dt$$

and this is easily integrated to give

$$\ln \{I(t)\} = -\frac{t^2}{2} + C$$

where, as usual, C is a constant of integration. Since $I(0) = \sqrt{\frac{\pi}{2}}$, we have $C = \ln\left(\sqrt{\frac{\pi}{2}}\right)$ and so

$$\ln \{I(t)\} - \ln\left(\sqrt{\frac{\pi}{2}}\right) = \ln\left(\sqrt{\frac{2}{\pi}}I(t)\right) = -\frac{t^2}{2}$$

or, at last,

$$\int_0^\infty \cos(tx) e^{-\frac{x^2}{2}} \, dx = \sqrt{\frac{\pi}{2}} e^{-\frac{t^2}{2}}. \qquad (3.1.5)$$

For $t = 1$, for example, our result says the integral is equal to $\sqrt{\frac{\pi}{2e}} = 0.760173\ldots$ and *integral(@(x)cos(x).*exp(-(x.^2)/2),0,inf) = 0.760173 ....*

From this last result, if we return to the original limits of $-\infty$ to∞ we can write

$$\int_{-\infty}^\infty e^{-\frac{x^2}{2}} \cos(tx) \, dx = 2\sqrt{\frac{\pi}{2}} e^{-\frac{t^2}{2}} = \sqrt{2\pi} \, e^{-\frac{t^2}{2}}.$$

Then, if we write $\int_{-\infty}^\infty e^{-\frac{x^2}{2}} \cos(s + tx) \, dx = \int_{-\infty}^\infty e^{-\frac{x^2}{2}} \cos(s) \cos(tx) \, dx - \int_{-\infty}^\infty e^{-\frac{x^2}{2}} \sin(s) \sin(tx) \, dx$ where I've used the trig identity for the cosine of a sum, we have

$$\int_{-\infty}^\infty e^{-\frac{x^2}{2}} \cos(s + tx) \, dx = \cos(s) \int_{-\infty}^\infty e^{-\frac{x^2}{2}} \cos(tx) dx - \sin(s) \int_{-\infty}^\infty e^{-\frac{x^2}{2}} \sin(tx) dx.$$

But since the last integral on the right is zero because its integrand is odd, we have

$$\int_{-\infty}^\infty e^{-\frac{x^2}{2}} \cos(s + tx) \, dx = \cos(s) \int_{-\infty}^\infty e^{-\frac{x^2}{2}} \cos(tx) dx$$

and so, finally (using our result (3.1.5))

$$\int_{-\infty}^\infty e^{-\frac{x^2}{2}} \cos(s + tx) \, dx = \sqrt{2\pi} \, e^{-\frac{t^2}{2}} \cos(s). \qquad (3.1.6)$$

For t = s = 1 this is 0.82144... and *integral(@(x)exp(-(x.^2)/2).*cos(1+x),-inf, inf)* = 0.82144...

The trick of evaluating an integral by finding a differential equation for which the integral is the solution can be used to determine the value of

$$I(a) = \int_0^\infty \frac{\cos{(ax)}}{x^2 + b^2} \, dx$$

where a and b are each positive (a is the parameter and b is a constant). If we integrate by parts, writing

$$u = \frac{1}{x^2 + b^2}, dv = \cos{(ax)}dx$$

then

$$du = -\frac{2x}{\left(x^2 + b^2\right)^2} dx, v = \frac{\sin{(ax)}}{a}$$

and so

$$I(a) = \left\{ \frac{\sin{(ax)}}{a\left(x^2 + b^2\right)^2} \right\} \Bigg|_0^\infty + \frac{2}{a} \int_0^\infty \frac{x \, \sin{(ax)}}{\left(x^2 + b^2\right)^2} \, dx$$

or, as the first term on the right vanishes at both the upper and the lower limit, we have

$$I(a) = \frac{2}{a} \int_0^\infty \frac{x \, \sin{(ax)}}{\left(x^2 + b^2\right)^2} \, dx$$

and so it perhaps looks as though we are making things worse! As you'll soon see, we are not.

From our last result we multiply through by a and arrive at

$$aI(a) = 2 \int_0^\infty \frac{x \, \sin{(ax)}}{\left(x^2 + b^2\right)^2} \, dx$$

and then differentiate with respect to a to get

$$a\frac{dI(a)}{da} + I(a) = 2 \int_0^\infty \frac{x^2 \, \cos{(ax)}}{\left(x^2 + b^2\right)^2} \, dx.$$

The integrand can be re-written in a partial fraction expansion:

$$\frac{x^2 \cos(ax)}{\left(x^2+b^2\right)^2} = \frac{\cos(ax)}{x^2+b^2} - \frac{b^2 \cos(ax)}{\left(x^2+b^2\right)^2}.$$

Thus,

$$2\int_0^\infty \frac{x^2 \cos(ax)}{\left(x^2+b^2\right)^2}\, dx = 2\int_0^\infty \frac{\cos(ax)}{x^2+b^2}\, dx - 2b^2 \int_0^\infty \frac{\cos(ax)}{\left(x^2+b^2\right)^2}\, dx$$

and, since the first integral on right is I(a), we have

$$a\frac{dI(a)}{da} + I(a) = 2I(a) - 2b^2 \int_0^\infty \frac{\cos(ax)}{\left(x^2+b^2\right)^2}\, dx$$

or,

$$a\frac{dI(a)}{da} - I(a) = -2b^2 \int_0^\infty \frac{\cos(ax)}{\left(x^2+b^2\right)^2}\, dx.$$

Differentiating this with respect to a, we get

$$a\frac{d^2I(a)}{da^2} + \frac{dI(a)}{da} - \frac{dI(a)}{da} = 2b^2 \int_0^\infty \frac{x \sin(ax)}{\left(x^2+b^2\right)^2}\, dx$$

or,

$$a\frac{d^2I(a)}{da^2} = 2b^2 \int_0^\infty \frac{x \sin(ax)}{\left(x^2+b^2\right)^2}\, dx.$$

If you look back at the start of the last paragraph, you'll see that we found the integral on the right to be

$$\int_0^\infty \frac{x \sin(ax)}{\left(x^2+b^2\right)^2}\, dx = \frac{a}{2}\, I(a),$$

and so

$$a\frac{d^2I(a)}{da^2} = 2b^2 \frac{a}{2}\, I(a)$$

or, rearranging, we have the following *second*-order, linear differential equation for I(a):

$$\frac{d^2 I(a)}{da^2} - b^2 I(a) = 0.$$

Such equations are well-known to have exponential solutions—$I(a) = Ce^{ka}$, where C and k are constants—and substitution into the differential equation gives

$$Ck^2 e^{ka} - b^2 Ce^{ka} = 0$$

and so $k^2 - b^2 = 0$ or $k = \pm b$. Thus, the general solution to the differential equation is the sum of these two particular solutions:

$$I(a) = C_1 e^{ab} + C_2 e^{-ab}.$$

We need two conditions on I(a) to determine the constants C_1 and C_2, and we can get them from our two different expressions for I(a): the original

$$I(a) = \int_0^\infty \frac{\cos(ax)}{x^2 + b^2} \, dx$$

and the expression we got by integrating by parts

$$I(a) = \frac{2}{a} \int_0^\infty \frac{x \sin(ax)}{\left(x^2 + b^2\right)^2} \, dx.$$

From the first we see that

$$I(0) = \int_0^\infty \frac{1}{x^2 + b^2} \, dx - \frac{1}{b} \left\{ \tan^{-1}\left(\frac{x}{b}\right) \right\} \Big|_0^\infty = \frac{\pi}{2b},$$

and from the second we see that $\lim_{a \to \infty} I(a) = 0$.
Thus,

$$I(0) = \frac{\pi}{2b} = C_1 + C_2$$

and

$$I(\infty) = 0$$

which says that $C_1 = 0$. Thus, $C_2 = \frac{\pi}{2b}$ and we have this beautiful result:

$$\int_0^\infty \frac{\cos(ax)}{x^2 + b^2} \, dx = \frac{\pi}{2b} e^{-ab}, \tag{3.1.7}$$

discovered in 1810 by Laplace. If $b = 1$ and $a = \pi$ then the integral is equal to $\frac{\pi}{2}e^{-\pi}$ $= 0.06788 \ldots$, and *integral(@(x)cos(pi*x)./(x.^2+1),0,inf) = 0.06788 ....*

Before moving on to new tricks, let me observe that with a simple change of variable we can often get some spectacular results from previously derived ones. For example, since (as we showed earlier in (3.1.4))

$$\int_{-\infty}^{\infty} e^{-\frac{x^2}{2}}dx = \sqrt{2\pi}$$

it follows (with $u = x\sqrt{2}$) that

$$\int_{0}^{\infty} e^{-u^2}\,du = \frac{1}{2}\sqrt{\pi}\,.$$

Then, letting $t = e^{-x^2}$, we have $x = \sqrt{-\ln(t)}$ (and so $dx = -\frac{dt}{2xe^{-x^2}} = \frac{dt}{-2xt}$) and thus

$$\int_{0}^{\infty} e^{-x^2}dx = \int_{1}^{0} t\,\frac{dt}{-2xt} = \frac{1}{2}\int_{0}^{1}\frac{dt}{\sqrt{-\ln(t)}} = \frac{1}{2}\sqrt{\pi}$$

which says

$$\int_{0}^{1}\frac{dx}{\sqrt{-\ln(x)}} = \sqrt{\pi} \qquad (3.1.8)$$

which is $1.77245 \ldots$, and agreeing is *integral(@(x)1./sqrt(-log(x)),0,1) = 1.77245* ....

3.2 An Amazing Integral

We start with the function g(y), defined as

$$g(y) = \int_{0}^{\infty} e^{-xy}\,\frac{\sin(ax)}{x}\,dx, \quad y > 0,$$

where a is some constant (more about a, soon). Differentiating with respect to the parameter y (notice, carefully, that x is the dummy variable of integration),

$$\frac{dg}{dy} = \int_{0}^{\infty}\frac{\partial}{\partial y}\left\{e^{-xy}\,\frac{\sin(ax)}{x}\right\}dx = \int_{0}^{\infty} -xe^{-xy}\,\frac{\sin(ax)}{x}\,dx$$

or,

$$\frac{dg}{dy} = -\int_0^\infty e^{-xy} \sin (ax) \, dx.$$

If this integral is then integrated-by-parts *twice*, it is easy to show that

$$\frac{dg}{dy} = -\frac{a}{a^2 + y^2}$$

which is easily integrated to give

$$g(y) = C - \tan^{-1}\left(\frac{y}{a}\right)$$

where C is an arbitrary constant of integration. We can calculate C by noticing, in the original integral definition of g(y), that $g(\infty) = 0$ because the e^{-xy} factor in the integrand goes to zero *everywhere* as $y \to \infty$ (because, over the entire interval of integration, $x \geq 0$).

Thus,

$$0 = C - \tan^{-1}(\pm\infty),$$

where we use the $+$ sign if $a > 0$ and the $-$ sign if $a < 0$. So, $C = \pm\frac{\pi}{2}$ and we have

$$g(y) = \pm\frac{\pi}{2} \quad \tan^{-1}\left(\frac{y}{a}\right).$$

The special case of $y = 0$ (and so $\tan^{-1}\left(\frac{y}{a}\right) = 0$) gives us the following wonderful result, called *Dirichlet's discontinuous* integral (after the German mathematician Gustav Dirichlet (1805–1859)):

$$\int_0^\infty \frac{\sin (ax)}{x} \, dx = \begin{cases} \frac{\pi}{2} & \text{if } a > 0 \\ 0 & \text{if } a = 0 \, . \\ -\frac{\pi}{2} & \text{if } a < 0 \end{cases} \qquad (3.2.1)$$

The value of the integral *at* $a = 0$ is zero, as the numerator of the integrand is then zero over the entire interval of integration. Euler derived the $a = 1$ special case sometime between 1776 and his death in 1783. The plot in Fig. 3.2.1, generated by using *integral* to evaluate the integral for 1000 values of a over the interval $-0.05 < a < 0.05$, hints at the correctness of our calculations, with the all-important sudden jump as a goes from negative to positive dramatically

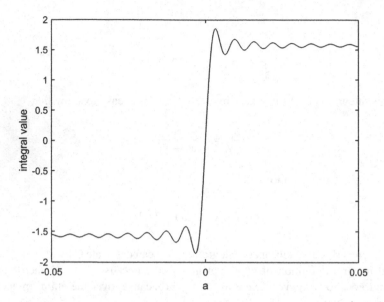

Fig. 3.2.1 Dirichlet's discontinuous integral

illustrated (the wiggles are due to the *Gibbs phenomenon*,[3] which is inherent in any attempt to represent a discontinuous function as a finite sum of sinusoids and, after all, integrating with *integral is* such a sum—see any book on Fourier series).

3.3   Frullani's Integral

As yet another example of differentiating under the integral sign, let's calculate

$$I(a, b) = \int_0^\infty \frac{\tan^{-1}(ax) - \tan^{-1}(bx)}{x}\, dx.$$

Notice that $I(a, a) = 0$. Differentiating with respect to a (using a *partial* derivative as I is a function of the *two* parameters a and b):

---

[3] After the American physicist J. W. Gibbs (1839–1903), who was the *wrong* person to have his name attached to the wiggles. For the history of this, see my book *Dr. Euler's Fabulous Formula*, Princeton 2011, pp. 163–173.

$$\frac{\partial I}{\partial a} = \int_0^\infty \left(\frac{1}{x}\right) \frac{\partial}{\partial a} \tan^{-1}(ax)\, dx = \int_0^\infty \left(\frac{1}{x}\right) \left\{\frac{x}{1+a^2x^2}\right\} dx = \int_0^\infty \frac{1}{1+a^2x^2}\, dx$$

$$= \frac{1}{a} \left\{ \tan^{-1}\left(\frac{x}{a}\right)\right\}\Big|_0^\infty = \frac{1}{a} \tan^{-1}(\infty) = \frac{\pi}{2a}.$$

So, integrating with respect to a,

$$I(a, b) = \frac{\pi}{2} \ln(a) + C(b),$$

where C(b) is an arbitrary *function* (of b) of integration. Writing C(b) in the alternative form of $\frac{\pi}{2} \ln(C(b))$—a more convenient constant!—we have

$$I(a, b) = \frac{\pi}{2} \ln(a) + \frac{\pi}{2} \ln(C(b)) = \frac{\pi}{2} \ln(aC(b)).$$

Since I(a, a) = 0, we have aC(a) = 1 or, $C(a) = \frac{1}{a}$. So, $C(b) = \frac{1}{b}$ and thus

$$\int_0^\infty \frac{\tan^{-1}(ax) - \tan^{-1}(bx)}{x}\, dx = \frac{\pi}{2} \ln\left(\frac{a}{b}\right). \qquad (3.3.1)$$

If $a = \pi$ and $b = 1$, for example, then

$$\int_0^\infty \frac{\tan^{-1}(\pi x) - \tan^{-1}(x)}{x}\, dx = \frac{\pi \ln(\pi)}{2} = 1.798137\ldots$$

and, in agreement, *integral(@(x)(atan(pi*x)-atan(x))./x,0,inf) = 1.798137 ....*
This result is a special case of what is called *Frullani's integral*, after the Italian mathematician Giuliano Frullani (1795–1834), who first wrote of it in an 1821 letter to a friend. We'll use Frullani's integral later in the book, and so it is worth doing a general derivation, which is usually *not* given in textbooks. Here's one way to do it, where we assume we have a function f(x) such that both f(0) and f(∞) exist. Our derivation starts with the definition of U as (where a is some positive constant)

$$U = \int_0^{h/a} \frac{f(ax) - f(0)}{x}\, dx$$

where h is some positive constant (for *now* h is a finite constant, but in just a bit we are going to let h→∞). Replacing a with b, we could just as well write

$$U = \int_0^{h/b} \frac{f(bx) - f(0)}{x}\, dx.$$

(To see that these two expressions for U are indeed equal, remember that a is an *arbitrary* constant and so it doesn't matter if we write a or if we write b.) So,

$$\int_0^{h/a} \frac{f(ax)}{x}\, dx = U + f(0)\int_0^{h/a} \frac{dx}{x}$$

and

$$\int_0^{h/b} \frac{f(bx)}{x}\, dx = U + f(0)\int_0^{h/b} \frac{dx}{x}.$$

Subtracting the last equation from the preceding one,

$$\int_0^{h/a} \frac{f(ax)}{x}\, dx - \int_0^{h/b} \frac{f(bx)}{x}\, dx = f(0)\left[\int_0^{h/a} \frac{dx}{x} - \int_0^{h/b} \frac{dx}{x}\right] = f(0)\int_{h/b}^{h/a} \frac{dx}{x}$$

$$= f(0)\{\ln(x)\}\big|_{h/b}^{h/a} = f(0)\ln\left(\frac{h/a}{h/b}\right) = f(0)\ln\left(\frac{b}{a}\right).$$

Then, subtracting and adding $\int_{h/b}^{h/a} \frac{f(bx)}{x}\, dx$ to the initial left-hand-side of this last equation, we have

$$\int_0^{h/a} \frac{f(ax)}{x}\, dx - \int_0^{h/b} \frac{f(bx)}{x}\, dx - \int_{h/b}^{h/a} \frac{f(bx)}{x}\, dx + \int_{h/b}^{h/a} \frac{f(bx)}{x}\, dx = f(0)\ln\left(\frac{b}{a}\right)$$

or, combining the second and third terms on the left,

$$\int_0^{h/a} \frac{f(ax)}{x}\, dx - \int_0^{h/a} \frac{f(bx)}{x}\, dx + \int_{h/b}^{h/a} \frac{f(bx)}{x}\, dx = f(0)\ln\left(\frac{b}{a}\right)$$

or,

$$\int_0^{h/a} \frac{f(ax) - f(bx)}{x}\, dx + \int_{h/b}^{h/a} \frac{f(bx)}{x}\, dx = f(0)\ln\left(\frac{b}{a}\right).$$

Now, imagine that $h \to \infty$. Then the upper limit on the left-most integral $\to \infty$, and *both* limits on the second integral $\to \infty$. That means $f(bx) \to f(\infty)$ over the entire interval of integration in the second integral and so

$$\lim_{h\to\infty}\int_{h/b}^{h/a} \frac{f(bx)}{x}\, dx = \lim_{h\to\infty} f(\infty)\int_{h/b}^{h/a} \frac{dx}{x} = f(\infty)\lim_{h\to\infty}\ln\left(\frac{b}{a}\right)$$

$$= f(\infty)\ln\left(\frac{b}{a}\right).$$

This all means that our last equation becomes, as $h \to \infty$,

$$\int_0^\infty \frac{f(ax) - f(bx)}{x}\, dx + f(\infty)\ln\left(\frac{b}{a}\right) = f(0)\ln\left(\frac{b}{a}\right)$$

or, writing $\ln\left(\frac{b}{a}\right) = -\ln\left(\frac{a}{b}\right)$, we have our result (Frullani's integral):

$$\int_0^\infty \frac{f(ax) - f(bx)}{x}\, dx = \{f(\infty) - f(0)\}\ln\left(\frac{a}{b}\right). \qquad (3.3.2)$$

This result assumes, as I said before, that $f(x)$ is such that *both* $f(\infty)$ and $f(0)$ exist. This is the case in the first example of $f(x) = \tan^{-1}(x)$, and $f(x) = e^{-x}$ works, too, as then $f(0) = 1$ and $f(\infty) = 0$ and so

$$\int_0^\infty \frac{e^{-ax} - e^{-bx}}{x}\, dx = -\ln\left(\frac{a}{b}\right) = \ln\left(\frac{b}{a}\right), a, b > 0. \qquad (3.3.3)$$

But not every $f(x)$ meets this requirement.

For example, can we calculate

$$I(a, b) = \int_0^\infty \frac{\cos(ax) - \cos(bx)}{x}\, dx$$

using Frullani's integral? No, because here we have $f(x) = \cos(x)$ and so while f$(0) = 1$ exists, $f(\infty)$ does not (the cosine function oscillates endlessly between -1 and $+1$ and has no limiting value). What we *can* do, however, is calculate

$$I(t) = \int_0^\infty e^{-tx}\left\{\frac{\cos(ax) - \cos(bx)}{x}\right\} dx$$

and see what happens as $t \to 0$. I'll defer doing this calculation until the next section, where I'll first show you yet another neat trick for doing integrals and then we'll use it to do $I(t)$. Interestingly, it turns out to be sort of the 'inverse' of Feynman's trick of differentiating under the integral sign.

3.4   The Flip-Side of Feynman's Trick

I'll demonstrate this new trick by using it to calculate the value of

$$I(a, b) = \int_0^\infty \frac{\cos(ax) - \cos(bx)}{x^2}\, dx.$$

We can write the integrand of this integral as an integral, itself, because

$$\frac{\cos{(ax)} - \cos{(bx)}}{x^2} = -\int_b^a \frac{\sin{(xy)}}{x} dy.$$

This works because the integral, when evaluated (remember, the integration is with respect to y and so the x can be treated as a constant), is

$$-\int_b^a \frac{\sin{(xy)}}{x} dy = -\frac{1}{x}\int_b^a \sin{(xy)}dy = \frac{1}{x}\left\{\frac{\cos{(xy)}}{x}\right\}\Big|_b^a = \frac{\cos{(ax)} - \cos{(bx)}}{x^2}.$$

So,

$$I(a,b) = -\int_0^\infty \left\{\int_b^a \frac{\sin{(xy)}}{x} dy \right\} dx.$$

This may look like a (huge!) step backward, as we now have a *double* integral to do. One thing you can do with double integrals that you can't do with single integrals, however, is *reverse the order* of integration. Of course we might run into objections that we haven't proven we are justified in doing that but, remember the philosophical approach we are taking: *we will just go ahead and do anything we please and not worry about it*, and only when we are done will we 'check' our formal answer with *integral*. So, reversing,

$$I(a,b) = -\int_b^a \left\{\int_0^\infty \frac{\sin{(xy)}}{x} dx \right\} dy.$$

Now, recall our earlier result, the discontinuous integral of Dirichlet in (3.2.1):

$$\int_0^\infty \frac{\sin{(cx)}}{x} dx = \begin{cases} \frac{\pi}{2} & \text{if } c > 0 \\ 0 & \text{if } c = 0 \\ -\frac{\pi}{2} & \text{if } c < 0 \end{cases}.$$

In the inner integral of our double integral, y (taken as positive) plays the role of c and so

$$I(a,b) = -\int_b^a \frac{\pi}{2} dy$$

and so, *just like that*, we have our answer:

$$\int_0^\infty \frac{\cos{(ax)} - \cos{(bx)}}{x^2} dx = \frac{\pi}{2}(b - a). \tag{3.4.1}$$

If a = 0 and b = 1, for example, then

$$\int_0^\infty \frac{1 - \cos(x)}{x^2}\, dx = \frac{\pi}{2} = 1.57079\ldots$$

and *integral* agrees: *integral(@(x)(1-cos(x))./x.^2,0,inf) = 1.57079 ....* 'More' dramatically, if $a = \sqrt{2}$ and $b = \sqrt{3}$ then our result says the integral is equal to $\frac{\pi}{2} \times$ $(\sqrt{3} - \sqrt{2}) = 0.499257\ldots$, and again *integral* agrees as *integral(@(x)(cos(sqrt(2)* x)-cos(sqrt(3)*x))./x.^2,0,inf) = 0.499257....*

Flipping Feynman's differentiating trick on its head and integrating an integral is a useful new trick, applicable in many interesting ways. For example, recall from earlier the result

$$\int_0^\infty e^{-x^2}\, dx = \frac{1}{2}\sqrt{\pi}\,.$$

Change variable to $x = t\sqrt{a}$ (and so $dx = \sqrt{a}\, dt$) and thus

$$\int_0^\infty e^{-a\,t^2} \sqrt{a}\, dt = \frac{1}{2}\sqrt{\pi}$$

or,

$$\int_0^\infty e^{-a\,t^2}\, dt = \frac{\sqrt{\pi}}{2\sqrt{a}}\,.$$

Now integrate both sides of this with respect to a, from p to q:

$$\int_p^q \left\{ \int_0^\infty e^{-a\,t^2}\, dt \right\} da = \int_p^q \frac{\sqrt{\pi}}{2\sqrt{a}}\, da = \frac{1}{2}\sqrt{\pi} \int_p^q \frac{da}{\sqrt{a}} = \frac{1}{2}\sqrt{\pi}\,\{2\sqrt{a}\}\Big|_p^q = \sqrt{\pi}(\sqrt{q} - \sqrt{p}).$$

Reversing the order of integration in the double integral,

$$\int_0^\infty \left\{ \int_p^q e^{-a\,t^2}\, da \right\} dt = \int_0^\infty \left\{ -\frac{e^{-a\,t^2}}{t^2} \right\}\Big|_p^q dt = \int_0^\infty \frac{e^{-p\,t^2} - e^{-q\,t^2}}{t^2}\, dt$$

or, finally,

$$\int_0^\infty \frac{e^{-p\,t^2} - e^{-q\,t^2}}{t^2}\, dt = \sqrt{\pi}\,(\sqrt{q} - \sqrt{p}). \tag{3.4.2}$$

With $p = 1$ and $q = 2$ the integral is equal to $\sqrt{\pi}\,(\sqrt{2} - 1) = 0.73417\ldots$, and *integral(@(x)(exp(-(x.^2))-exp(-2*(x.^2)))./x.^2,0,inf) = 0.73417 ....*

As another example of the technique, suppose you were faced with evaluating

$$\int_0^1 \frac{x^a - 1}{\ln(x)} \, dx, \quad a \geq 0.$$

We can do this by first observing that the integrand can be written as an integral as follows:

$$\int_0^a x^y \, dy = \int_0^a e^{\ln(x^y)} \, dy = \int_0^a e^{y \ln(x)} \, dy = \left\{ \frac{e^{y \ln(x)}}{\ln(x)} \right\}\Big|_0^a = \frac{e^{a \ln(x)} - 1}{\ln(x)} = \frac{x^a - 1}{\ln(x)}.$$

So,

$$\int_0^1 \frac{x^a - 1}{\ln(x)} \, dx = \int_0^1 \left\{ \int_0^a x^y \, dy \right\} dx$$

or, upon reversing the order of integration,

$$\int_0^1 \frac{x^a - 1}{\ln(x)} \, dx = \int_0^a \left\{ \int_0^1 x^y \, dx \right\} dy = \int_0^a \left\{ \frac{x^{y+1}}{y+1} \right\}\Big|_0^1 dy = \int_0^a \frac{1}{y+1} \, dy.$$

Making the obvious change of variable $u = y + 1$, we have

$$\int_0^1 \frac{x^a - 1}{\ln(x)} \, dx = \ln(u)\Big|_1^{a+1}.$$

or, at last, we arrive at the very pretty

$$\int_0^1 \frac{x^a - 1}{\ln(x)} \, dx = \ln(a+1), \quad a \geq 0. \tag{3.4.3}$$

For example,

$$\int_0^1 \frac{x - 1}{\ln(x)} \, dx = \ln(2) = 0.693147\ldots$$

and

$$\int_0^1 \frac{x^2 - 1}{\ln(x)} \, dx = \ln(3) = 1.098612\ldots$$

In agreement, *integral(@(x)(x-1)./log(x),0,1) = 0.693147* ... and *integral(@(x) (x.^2-1)./log(x),0,1) = 1.098612* ....
From this result we can write

$$\int_0^1 \frac{x^a - x^b}{\ln(x)} \, dx = \int_0^1 \frac{x^a - 1}{\ln(x)} \, dx - \int_0^1 \frac{x^b - 1}{\ln(x)} \, dx = \ln(a+1) - \ln(b+1)$$

and so

$$\int_0^1 \frac{x^a - x^b}{\ln(x)} \, dx = \ln\left(\frac{a+1}{b+1}\right). \tag{3.4.4}$$

Now that we've seen how integrating under the integral sign works, let's return to the integral I didn't do at the end of the previous section,

$$I(t) = \int_0^\infty e^{-tx} \left\{ \frac{\cos(ax) - \cos(bx)}{x} \right\} dx.$$

Notice that we can write the portion of the integrand that is in the braces as an integral:

$$\frac{\cos(ax) - \cos(bx)}{x} = \int_a^b \sin(xs) ds.$$

Thus,

$$I(t) = \int_0^\infty e^{-tx} \left\{ \int_a^b \sin(xs) ds \right\} dx$$

or, upon reversing the order of integration (note, *carefully*, that since we are going to the x-integration first we have to bring the exponential factor into the inner integral, too) we have

$$I(t) = \int_a^b \left\{ \int_0^\infty e^{-tx} \sin(xs) dx \right\} ds.$$

From standard integration tables we find

$$\int e^{-tx} \sin(xs) dx = \frac{e^{-tx} \{ s\sin(xs) - s \cos(xs) \}}{t^2 + s^2}$$

and so

$$I(t) = \int_a^b \left[\frac{e^{-tx} \{ s\sin(xs) - s\cos(xs) \}}{t^2 + s^2} \right] \Bigg|_0^\infty ds = \int_a^b \frac{s}{t^2 + s^2} ds.$$

Making the change of variable $u = t^2 + s^2$ (and so $ds = \frac{du}{2s}$), gives

$$I(t) = \int_{t^2+a^2}^{t^2+b^2} \frac{s}{u} \left(\frac{du}{2s}\right) = \frac{1}{2}\int_{t^2+a^2}^{t^2+b^2} \frac{du}{u} = \frac{1}{2}\ln\left(\frac{t^2+b^2}{t^2+a^2}\right)$$

or, at last

$$I(t) = \int_0^\infty e^{-tx}\left\{\frac{\cos(ax) - \cos(bx)}{x}\right\} dx = \ln\sqrt{\left(\frac{t^2+b^2}{t^2+a^2}\right)}. \tag{3.4.5}$$

And so, as $t \to 0$, we find

$$\int_0^\infty \frac{\cos(ax) - \cos(bx)}{x} dx = \ln\left(\frac{b}{a}\right). \tag{3.4.6}$$

Notice that if either a or b is zero the integral blows-up. That is,

$$\int_0^\infty \frac{1 - \cos(x)}{x} dx$$

does not exist.

A twist on the use of integral differentiation is illustrated with the problem of calculating integrals like

$$\int_0^1 \{\ln(x)\}^2 dx$$

or

$$\int_0^1 \sqrt{x}\{\ln(x)\}^2 dx.$$

A trick that does the job is to introduce a parameter (I'll call it a) that we can differentiate with respect to and then, if we set that parameter equal to a specific value, reduces the integral to one of the above. For example, consider

$$I(a) = \int_0^1 x^a\{\ln(x)\}^2 dx,$$

which reduces to the above first integral if $a = 0$ and to the second integral if $a = \frac{1}{2}$.

So, differentiating,

$$\frac{dI}{da} = \frac{d}{da}\int_0^1 e^{\ln(x^a)}\{\ln(x)\}^2 dx = \frac{d}{da}\int_0^1 e^{a\ln(x)}\{\ln(x)\}^2 dx$$

$$= \int_0^1 e^{a\ln(x)}\ln(x)\{\ln(x)\}^2 dx = \int_0^1 x^a\{\ln(x)\}^3 dx.$$

Now this may look as though we've just made things worse—and we have! *But—* it also should give you a hint on how to turn the situation around to going our way. Look at what happened: differentiating I(a) gave us the same integral back except that the power of ln(x) *increased* by 1. This should nudge you into seeing that our original integral can be thought of as coming from differentiating $\int_0^1 x^a \ln(x)dx$, with that integral in turn coming from differentiating $\int_0^1 x^a dx$, an integral that is *easy* to do. That is,

$$\int_0^1 x^a \ln(x)dx = \frac{d}{da} \int_0^1 x^a dx = \frac{d}{da}\left[\left\{\frac{x^{a+1}}{a+1}\right\}\Big|_0^1\right] = \frac{d}{da}\left(\frac{1}{a+1}\right) = -\frac{1}{(a+1)^2}.$$

Thus,

$$\int_0^1 x^a\{\ln(x)\}^2 dx = \frac{d}{da}\int_0^1 x^a \ln(x)dx = \frac{d}{da}\left\{-\frac{1}{(a+1)^2}\right\} = \frac{2(a+1)}{(a+1)^4}$$

or,

$$\int_0^1 x^a\{\ln(x)\}^2 dx = \frac{2}{(a+1)^3}. \tag{3.4.7}$$

So, for $a = 0$ we have

$$\int_0^1 \{\ln(x)\}^2 dx = 2$$

and, in agreement, *integral(@(x)log(x).^2,0,1)* = 2.00000003.... And for $a = \frac{1}{2}$ we have the integral equal to $\frac{2}{\left(\frac{3}{2}\right)^3} = \frac{16}{27} = 0.5925925\ldots$; in agreement, *integral(@(x)sqrt (x).*log(x).^2,0,1)* = 0.5925925....

As the final examples of this section, let's see how to do the integrals

$$I(a) = \int_0^\pi \ln\{a + b\,\cos(x)\}\,dx$$

where $a > b \geq 0$, and

$$I(b) = \int_0^\pi \frac{\ln\{1 + b\,\cos(x)\}}{\cos(x)}dx$$

where $0 \leq b < 1$. The analysis will be a two-step process, with the first step being the evaluation of the derivative with respect to a of I(a), that is, the integral

$$\frac{dI}{da} = \int_0^\pi \frac{1}{a + b\,\cos\,(x)}\,dx$$

where $a > b \geq 0$. For this integral, change variable to $z = \tan\left(\frac{x}{2}\right)$. Then, as shown back in Chap. 2 (in the analysis leading up to (2.2.3)), we established that

$$\cos\,(x) = \frac{1 - z^2}{1 + z^2}$$

and

$$dx = \frac{2}{1 + z^2}\,dz,$$

and so

$$\int_0^\pi \frac{1}{a + b\,\cos\,(x)}\,dx = \int_0^\infty \frac{1}{a + b\frac{1 - z^2}{1 + z^2}}\left(\frac{2}{1 + z^2}\right)\,dz = 2\int_0^\infty \frac{dz}{a(1 + z^2) + b(1 - z^2)}$$

$$= 2\int_0^\infty \frac{dz}{(a + b) + z^2(a - b)} = \frac{2}{a - b}\int_0^\infty \frac{dz}{\frac{a + b}{a - b} + z^2} = \left(\frac{2}{a - b}\right)\frac{1}{\sqrt{\frac{a + b}{a - b}}}\tan^{-1}\left(\frac{z}{\sqrt{\frac{a + b}{a - b}}}\right)\Bigg|_0^\infty$$

$$= \frac{2}{\sqrt{a^2 - b^2}}\tan^{-1}\left(z\sqrt{\frac{a - b}{a + b}}\right)\Bigg|_0^\infty .$$

Thus,[4]

$$\int_0^\pi \frac{1}{a + b\,\cos\,(x)}\,dx = \frac{\pi}{\sqrt{a^2 - b^2}},\,a > b. \qquad (3.4.8)$$

Now, for the evaluation of our two integrals, starting with I(a). Differentiating with respect to a, we have from (3.4.8) that

---

[4]The integral in (3.4.8) occurs in a paper by R. M. Dimeo, "Fourier Transform Solution to the Semi-Infinite Resistance Ladder," *American Journal of Physics*, July 2000, pp. 669–670, where it is done by contour integration (see the comments immediately following (8.8.11)). Our derivation here shows contour integration is actually *not* necessary, although Professor Dimeo's comment that contour integration is "well within the abilities of the undergraduate physics major," is in agreement with the philosophical position I take in the original Preface. Notice that we can use (3.4.8) to derive all sorts of new integrals by differentiation. For example, suppose $b = -1$, and so we have $\int_0^\pi \frac{1}{a - \cos\,(x)}\,dx = \frac{\pi}{\sqrt{a^2 - 1}}$. Then, differentiating both sides with respect to a, we get $\int_0^\pi \frac{1}{[a - \cos\,(x)]^2}\,dx = \frac{\pi a}{(a^2 - 1)^{3/2}}$. If, for example, $a = 5$, this new integral is equal to $\frac{5\pi}{24^{3/2}} = \frac{5\pi}{48\sqrt{6}} = 0.1335989\ldots$. To check, we see that $integral(@(x)1./((5-\cos(x)).^2),0,pi) = 0.1335989\ldots$.

$$\frac{dI}{da} = \int_0^\pi \frac{1}{a+b \cos(x)}\, dx = \frac{\pi}{\sqrt{a^2 - b^2}}.$$

So, integrating indefinitely with respect to a,

$$I(a) = \int \frac{\pi}{\sqrt{a^2 - b^2}}\, da$$

or, from standard integration tables,

$$I(a) = \pi \ln\left\{a + \sqrt{a^2 - b^2}\right\} + C$$

where C is the constant of integration. We find C as follows: when $b = 0$ *in the integral* we have

$$I(a) = \int_0^\pi \ln(a)\, da = \pi\ln(a),$$

and so setting this equal to our integration result (with $b = 0$) we have

$$\pi \ln\{2a\} + C = \pi\ln(a)$$

which tells us that $C = -\pi\ln(2)$. Thus,

$$I(a) = \pi \ln\left\{a + \sqrt{a^2 - b^2}\right\} - \pi\ln(2)$$

Or

$$\int_0^\pi \ln\{a + b \cos(x)\}\, dx = \pi \ln\left\{\frac{a + \sqrt{a^2 - b^2}}{2}\right\}, a > b. \qquad (3.4.9)$$

If $a = 2$ and $b = 1$, for example, our integral is equal to π $\ln\left\{\frac{2+\sqrt{3}}{2}\right\} = 1.959759\ldots$, and checking: *integral(@(x)log(2+cos(x)),0,pi)* = 1.959759. . . .

Finally, turning to the I(b) integral, differentiation with respect to b gives

$$\frac{dI}{db} = \int_0^\pi \frac{\frac{\cos(x)}{1+b \cos(x)}}{\cos(x)}\, dx = \int_0^\pi \frac{1}{1+b \cos(x)}\, dx.$$

But from (3.4.8) we have (with $a = 1$)

$$\frac{dI}{db} = \frac{\pi}{\sqrt{1-b^2}}.$$

So, integrating indefinitely with respect to b, we have

$$I(b) = \pi \ \sin^{-1}(b) + C$$

where C is the constant of integration. Since $I(0) = 0$ by inspection of the original integral, we finally have our answer:

$$\int_0^\pi \frac{\ln\{1+b\ \cos(x)\}}{\cos(x)}\,dx = \pi\ \sin^{-1}(b). \qquad (3.4.10)$$

If, for example, $b = \frac{1}{3}$, this integral has a value of $\pi \ \sin^{-1}\left(\frac{1}{3}\right) = 1.067629\ldots$ and checking, we see that $integral(@(x)log(1+(cos(x)/3))./cos(x),0,pi) = 1.067629\ldots$

3.5   Combining Two Tricks and Hardy's Integral

In this section we'll combine the differentiation of an integral with the recursion trick that we used back in Chap. 2 (Sect. 2.3) to solve a whole class of integrals. Here we'll tackle

$$I_n = \int_0^{\pi/2} \frac{1}{\{a\ \cos^2(x) + b\ \sin^2(x)\}^n}\,dx, n = 1,2,3,\ldots$$

where we'll take both a and b as parameters we can differentiate with respect to. If you calculate $\frac{\partial I_n}{\partial a}$ and $\frac{\partial I_n}{\partial b}$ you should be able to see, by inspection, that

$$\frac{\partial I_n}{\partial a} = -n\int_0^{\pi/2} \frac{\cos^2(x)}{\{a\ \cos^2(x) + b\ \sin^2(x)\}^{n+1}}\,dx$$

and

$$\frac{\partial I_n}{\partial b} = -n\int_0^{\frac{\pi}{2}} \frac{\sin^2(x)}{\{a\ \cos^2(x) + b\ \sin^2(x)\}^{n+1}}\,dx.$$

From these two results we then immediately have

$$\frac{\partial I_n}{\partial a} + \frac{\partial I_n}{\partial b} = -n\int_0^{\frac{\pi}{2}} \frac{1}{\{a\ \cos^2(x) + b\ \sin^2(x)\}^{n+1}}\,dx = -nI_{n+1}.$$

Or, if we replace n with $n-1$, we have the recursion

$$I_n = -\frac{1}{n-1}\left[\frac{\partial I_{n-1}}{\partial a} + \frac{\partial I_{n-1}}{\partial b}\right].$$

From the recursion we see that the first value of n we can use with it is $n = 2$,[5] and even then only if we already know I_1. Then, once we have I_2 we can use it to find I_3, and so on. But first, we need to find

$$I_1 = \int_0^{\pi/2} \frac{1}{a\cos^2(x) + b\sin^2(x)}\,dx = \int_0^{\pi/2} \frac{\frac{1}{\cos^2(x)}}{a + b\tan^2(x)}\,dx.$$

Changing variable to $y = \tan(x)$, we have $\frac{dy}{dx} = \frac{1}{\cos^2(x)}$ or, $dx = \cos^2(x)\,dy$. Thus,

$$I_1 = \int_0^\infty \frac{\frac{1}{\cos^2(x)}}{a + b\,y^2}\cos^2(x)\,dy = \frac{1}{b}\int_0^\infty \frac{1}{\frac{a}{b} + y^2}\,dy = \frac{1}{b}\left\{\sqrt{\frac{b}{a}}\tan^{-1}\left(\frac{y}{\sqrt{\frac{a}{b}}}\right)\right\}\Big|_0^\infty$$

$$= \frac{1}{\sqrt{ab}}\tan^{-1}(\infty)$$

or,

$$I_1 = \int_0^{\pi/2} \frac{1}{a\cos^2(x) + b\sin^2(x)}\,dx = \frac{\pi}{2\sqrt{ab}}. \tag{3.5.1}$$

To find I_2, we first calculate

$$\frac{\partial I_1}{\partial a} = \frac{-2\pi\sqrt{b}\,\frac{1}{2}a^{-\frac{1}{2}}}{4ab} = -\left(\frac{\pi}{4}\right)\frac{\sqrt{\frac{b}{a}}}{ab}$$

and

$$\frac{\partial I_1}{\partial b} = \frac{-2\pi\sqrt{a}\,\frac{1}{2}b^{-\frac{1}{2}}}{4ab} = -\left(\frac{\pi}{4}\right)\frac{\sqrt{\frac{a}{b}}}{ab}.$$

Thus, using the recursion with $n = 2$,

$$I_2 = \frac{\pi}{4ab}\left[\sqrt{\frac{b}{a}} + \sqrt{\frac{a}{b}}\right] - \left(\frac{\pi}{4ab}\right)\frac{b+a}{ab}$$

or,

---

[5] For $n = 1$, the recursion gives I_1 in terms of I_0, where $I_0 = \int_0^{\pi/2} dx = \frac{\pi}{2}$ with no dependency on either a or b. That is, $\frac{\partial I_0}{\partial a} = \frac{\partial I_0}{\partial b} = 0$ and the recursion becomes the useless, *indeterminate* $I_1 = \frac{0}{0}I_0$.

$$I_2 = \int_0^{\pi/2} \frac{1}{\{a\,\cos^2(x) + b\,\sin^2(x)\}^2}\,dx = \left(\frac{\pi}{4\sqrt{ab}}\right)\left(\frac{1}{a}+\frac{1}{b}\right). \tag{3.5.2}$$

We can repeat this process endlessly (although the algebra rapidly gets ever grubbier!), and I'll let you fill-in the details to show that the recursion (with $n = 3$) gives

$$I_3 = \int_0^{\pi/2} \frac{1}{\{a\,\cos^2(x) + b\,\sin^2(x)\}^3}\,dx = \left(\frac{\pi}{16\sqrt{ab}}\right)\left(\frac{3}{a^2}+\frac{3}{b^2}+\frac{2}{ab}\right). \tag{3.5.3}$$

To check this (as well as the results we got earlier, since a mistake earlier would show-up here, too), if $a = 1$ and $b = 2$ then $I_3 = \frac{4.75\pi}{16\sqrt{2}} = 0.65949....$ Using MATLAB directly on the integral, *integral(@(x)1./((cos(x).^2+2*sin(x).^2).^3),0, pi/2) = 0.65949....*

As a final example of recursion, consider a generalized version of the integral that appeared in the first section of this chapter, in (3.1.2) and (3.1.3):

$$I_n(y) = \int_0^y \frac{dx}{(x^2+a^2)^n}.$$

Earlier, we had the special case of $y = \infty$. We can use recursion, combined with integration by parts, to evaluate this more general integral. In the integration by parts formula let

$$u = \frac{1}{(x^2+a^2)^n}$$

and $dv = 1$. Then obviously $v = x$ and you can quickly confirm that

$$du = -n\frac{2x}{(x^2+a^2)^{n+1}}\,dx.$$

Thus,

$$I_n(y) = \left\{\frac{x}{(x^2+a^2)^n}\right\}\Bigg|_0^y + n\int_0^y \frac{2x^2\,dx}{(x^2+a^2)^{n+1}},$$

or, as $2x^2 = 2(x^2+a^2) - 2a^2$, we have

$$I_n(y) = \frac{y}{(y^2+a^2)^n} + n\left[\int_0^y \frac{2(x^2+a^2)\,dx}{(x^2+a^2)^{n+1}} - 2a^2\int_0^y \frac{dx}{(x^2+a^2)^{n+1}}\right]$$

or,

$$I_n(y) = \frac{y}{(y^2 + a^2)^n} + 2nI_n(y) - 2na^2I_{n+1}(y)$$

and so, finally, we arrive at the recursion

$$I_{n+1}(y) = \frac{y}{2na^2(y^2 + a^2)^n} + \frac{2n-1}{2na^2}I_n(y). \tag{3.5.4}$$

We start the recursion in (3.5.4) with

$$I_1(y) = \int_0^y \frac{dx}{x^2 + a^2} = \frac{1}{a}\tan^{-1}\left(\frac{x}{a}\right)\Big|_0^y = \frac{1}{a}\tan^{-1}\left(\frac{y}{a}\right),$$

and so for the case of $y = \infty$ we have $I_1(y) = \frac{\pi}{2a}$. Then, as

$$I_{n+1}(\infty) = \frac{2n-1}{2na^2}I_n(\infty),$$

we have

$$I_2(\infty) = \frac{1}{2a^2}I_1(\infty) = \frac{\pi}{4a^3}$$

and

$$I_3(\infty) = \frac{3}{4a^2}I_2(\infty) = \frac{3\pi}{16a^5}$$

which are exactly what we calculated in (3.1.2) and (3.1.3), respectively. And a new result that we didn't have before is

$$I_4(\infty) = \int_0^\infty \frac{dx}{(x^2 + a^2)^4} = \frac{5}{6a^2}I_3(\infty) = \frac{15\pi}{96a^7}.$$

As a check, if $a = 1$ the value of $I_4(\infty)$ is 0.490873..., and in agreement MATLAB says *integral(@(x)1./((x.^2+1).^4),0,inf)* = 0.490873....

To end this section on doubling-up on tricks, we'll combine the idea of differentiating an integral with the idea of a power series expansion to evaluate an integral first done by the great G. H. Hardy. Appearing as a challenge problem in an 1899 issue of *The Educational Times*, Hardy showed impressive ingenuity in computing the value of

$$\int_0^\infty \frac{\ln(x)}{1 + x^3}dx.$$

At a crucial point in his analysis, Hardy needed to use

$$\int_0^1 x^n \ln(x)\,dx = -\frac{1}{(n+1)^2}, \quad n > -1. \tag{3.5.5}$$

You did this earlier, as a Challenge Problem (C1.6), using integration-by-parts, and I'll do it again here using Feynman's trick, to have it fresh in your mind when we need it. We start by writing

$$I(n) = \int_0^1 x^n\,dx = \left(\frac{x^{n+1}}{n+1}\right)\Big|_0^1 = \frac{1}{n+1}, \quad n > -1. \tag{3.5.6}$$

The reason this is useful is because we can write I(n) as

$$I(n) = \int_0^1 e^{\ln(x^n)}\,dx = \int_0^1 e^{n\,\ln(x)}\,dx$$

and so, differentiating with respect to n,

$$\frac{dI}{dn} = \int_0^1 \ln(x)e^{n\,\ln(x)}\,dx = \int_0^1 x^n \ln(x)\,dx.$$

But since from (3.5.6) we see that

$$\frac{dI}{dn} = -\frac{1}{(n+1)^2},$$

we then immediately have (3.5.5). *Now* we are ready for Hardy's integral.
   We start the analysis by writing

$$\int_0^\infty \frac{\ln(x)}{1+x^3}\,dx = \int_0^1 \frac{\ln(x)}{1+x^3}\,dx + \int_1^\infty \frac{\ln(x)}{1+x^3}\,dx.$$

In the right-most integral, change variable to $x = \frac{1}{y}$ and so $\frac{dx}{dy} = -\frac{1}{y^2}$ or $dx = -\frac{dy}{y^2}$. Thus,

$$\int_1^\infty \frac{\ln(x)}{1+x^3}\,dx = \int_1^0 \frac{\ln\left(\frac{1}{y}\right)}{1+\frac{1}{y^3}}\left(-\frac{dy}{y^2}\right) = \int_0^1 \frac{-\ln(y)}{1+y^3}y^3\left(\frac{dy}{y^2}\right) = \int_0^1 \frac{-y\,\ln(y)}{1+y^3}\,dy,$$

and so

$$\int_0^\infty \frac{\ln(x)}{1+x^3}\,dx = \int_0^1 \frac{(1-x)\ln(x)}{1+x^3}\,dx.$$

Now,

$$\frac{1-x}{1+x^3} = (1-x)(1-x^3+x^6-x^9+x^{12}-x^{15}+\ldots)$$

$$= (1-x^3+x^6-x^9+x^{12}-x^{15}+\ldots) + (-x+x^4-x^7+x^{10}-x^{13}+x^{16}-\ldots)$$

$$= 1-x-x^3+x^4+x^6-x^7-x^9+x^{10}+x^{12}-x^{13}-x^{15}+x^{16}-\ldots$$

and so, integrating,

$$\int_0^\infty \frac{\ln(x)}{1+x^3}\,dx = \int_0^1 \left(\begin{array}{l}1-x-x^3+x^4+x^6-x^7-x^9+x^{10}\\ +x^{12}-x^{13}-x^{15}+x^{16}-\ldots\end{array}\right)\ln(x)\,dx.$$

Thinking of $1 = x^0$, we have from (3.5.5) that

$$\int_0^\infty \frac{\ln(x)}{1+x^3}\,dx = -\frac{1}{1^2}+\frac{1}{2^2}+\frac{1}{4^2}-\frac{1}{5^2}-\frac{1}{7^2}+\frac{1}{8^2}+\frac{1}{10^2}-\frac{1}{11^2}-\frac{1}{13^2}+\frac{1}{14^2}+\frac{1}{16^2}$$

$$-\frac{1}{17^2}-\ldots$$

$$= -1+\frac{1}{4}+\frac{1}{16}-\frac{1}{25}-\frac{1}{49}+\frac{1}{64}+\frac{1}{100}-\frac{1}{121}-\frac{1}{169}+\frac{1}{196}+\frac{1}{256}-\ldots$$

Put this perhaps odd-looking infinite series temporarily to the side, and let's now write down a couple of preliminary results that will remove the 'odd-looking' nature of the series solution to Hardy's integral. First, recall Euler's famous result (mentioned in Sect. 1.3 and which we'll derive later, in Chap. 7):

$$\frac{1}{1^2}+\frac{1}{2^2}+\frac{1}{3^2}+\frac{1}{4^2}+\frac{1}{5^2}+\frac{1}{6^2}+\ldots = \frac{\pi^2}{6}.$$

Notice that if we use just the *even* integers on the left,

$$\frac{1}{2^2}+\frac{1}{4^2}+\frac{1}{6^2}+\ldots = \frac{1}{(2\cdot1)^2}+\frac{1}{(2\cdot2)^2}+\frac{1}{(2\cdot3)^2}+\ldots = \frac{1}{4}\left(\frac{1}{1^2}+\frac{1}{2^2}+\frac{1}{3^2}+\ldots\right)$$

$$= \left(\frac{1}{4}\right)\left(\frac{\pi^2}{6}\right) = \frac{\pi^2}{24}.$$

This means that if we use just the *odd* integers we must get

$$\frac{1}{1^2}+\frac{1}{3^2}+\frac{1}{5^2}+\ldots = \frac{\pi^2}{6}-\frac{\pi^2}{24} = \frac{3\pi^2}{24} = \frac{\pi^2}{8}.$$

These two results then tells us that the *alternating* series

$$\frac{1}{1^2} - \frac{1}{2^2} + \frac{1}{3^2} - \frac{1}{4^2} + \frac{1}{5^2} - \frac{1}{6^2} + \cdots = \left(\frac{1}{1^2} + \frac{1}{3^2} + \frac{1}{5^2} + \cdots \right) - \left(\frac{1}{2^2} + \frac{1}{4^2} + \frac{1}{6^2} + \cdots \right) = \frac{\pi^2}{8} - \frac{\pi^2}{24}$$

which says

$$1 - \frac{1}{4} + \frac{1}{9} - \frac{1}{16} + \frac{1}{25} - \frac{1}{36} + \cdots = \frac{\pi^2}{12}$$

or, multiplying through by -1,

$$\mathbf{A} \qquad -1 + \frac{1}{4} - \frac{1}{9} + \frac{1}{16} - \frac{1}{25} + \frac{1}{36} - \frac{1}{49} + \frac{1}{64} - \frac{1}{81} + \frac{1}{100} - \frac{1}{121} + \cdots = -\frac{\pi^2}{12}.$$

Next, write the alternating series using just the integers that are multiples of 3:

$$\frac{1}{3^2} - \frac{1}{6^2} + \frac{1}{9^2} - \frac{1}{12^2} + \frac{1}{15^2} - \cdots = \frac{1}{(3 \cdot 1)^2} - \frac{1}{(3 \cdot 2)^2} + \frac{1}{(3 \cdot 3)^2} - \frac{1}{(3 \cdot 4)^2}$$

$$+ \frac{1}{(3 \cdot 5)^2} - \cdots$$

$$= \left(\frac{1}{9} \right) \left(\frac{1}{1^2} - \frac{1}{2^2} + \frac{1}{3^2} - \frac{1}{4^2} + \frac{1}{5^2} - \frac{1}{6^2} + \cdots \right) = \left(\frac{1}{9} \right) \left(\frac{\pi^2}{12} \right)$$

or,

$$\mathbf{B} \qquad \frac{1}{9} - \frac{1}{36} + \frac{1}{81} - \frac{1}{144} + \frac{1}{225} - \cdots = \frac{\pi^2}{108}.$$

Finally, add **A** and **B** to get

$$-1 + \frac{1}{4} + \frac{1}{16} - \frac{1}{25} - \frac{1}{49} + \frac{1}{64} + \frac{1}{100} - \frac{1}{121} - \cdots$$

which, to our infinite pleasure, is Hardy's 'odd-looking' infinite series! That is, $\int_0^\infty \frac{\ln(x)}{1+x^3} dx = -\frac{\pi^2}{12} + \frac{\pi^2}{108} = -\frac{8\pi^2}{108} = -\frac{2\pi^2}{27} = -0.7310818\ldots$. MATLAB agrees, as *integral(@(x)log(x)./(1+x.^3),0,inf) = −0.7310818....*

How, you may wonder, did Hardy manage to see through all the 'fraction noise' to the particular **A** and **B** series that solved the problem? Only God knows! (I certainly don't.) Take this as a specific example of why Hardy is considered to have been a genius.

3.6 Uhler's Integral and Symbolic Integration

This chapter celebrates Feynman's association with 'differentiation under the integral sign,' but of course the technique greatly pre-dates him. As his own words at the end of Chap. 1 make clear, he learned the method from a well-known textbook (the 1926 *Advanced Calculus* by MIT math professor Frederick Woods (1864–1950)), and the method was popular among mathematicians and physicists *long* before Woods wrote his book and Feynman entered the picture. For example, in the October 1914 issue of *The American Mathematical Monthly* a little challenge problem appeared, posed by a Yale physics professor named Horace Scudder Uhler[6] (1872–1956). There he asked readers how they would evaluate the double integral

$$I = \int_0^a \left(a^2 - x^2\right) x \, dx \int_{a-x}^{a+x} \frac{e^{-cy}}{y} \, dy, \qquad (3.6.1)$$

which had occurred while "in the process of solving a certain physical problem." The values of a and c are positive.

Uhler actually already knew how to do it—he even included the final answer when he made his challenge—but he was curious about how *others* might approach this double integral. A solution from a reader was published in the December issue, and finally Uhler showed (in the January 1915 issue, more than 3 years before Feynman was born)) how *he* had done it, by differentiation with respect to the parameter c. Here's what he wrote: "Differentiate I with respect to the parameter c, then

$$\frac{dI}{dc} = -\int_0^a \left(a^2 - x^2\right) x \, dx \int_{a-x}^{a+x} e^{-cy} \, dy,$$

which can be evaluated at once [I think this "at once" is just a bit misleading as, even using standard integration tables, I found the routine algebra pretty grubby] and gives

$$\frac{dI}{dc} = \frac{2a^2}{c^3} - \frac{6a}{c^4} + \frac{6}{c^5} - \left(\frac{6a}{c^4} + \frac{6}{c^5} + \frac{2a^2}{c^3}\right) e^{-2ac}.$$

---

[6]Uhler was a pioneer in heroic numerical calculation, made all the more impressive in that he worked in the pre-electronic computer days. His major tool was a good set of log tables. I describe his 1921 calculation of, to well over 100 decimal digits, the value of $\left(\sqrt{-1}\right)^{\sqrt{-1}} = e^{-\frac{\pi}{2}}$ in my book *An Imaginary Tale: the story $\sqrt{-1}$*, Princeton 2010, pp. 235–237. *Why* did Uhler perform such a monstrous calculation? For the same reason kids play with mud—he thought it was fun!

We can now integrate both sides of this equation [for a condition to determine the constant of integration, notice that $I(\infty) = 0$ because the integrand of the inner y-integral goes to zero as c $\rightarrow\infty$], thus [and then Uhler gives the answer]

$I = \frac{1}{c^2}\left[a^2 - \frac{3a}{c} + \frac{1}{c}\left(a + \frac{3}{2c}\right)(1 - e^{-2ac})\right]$."

We can make a partial computer check of this as follows. For the case of a $= 1$ and c $= 1$, for example, Uhler's formula becomes

$$I = \left[1 - 3 + \left(1 + \frac{3}{2}\right)(1 - e^{-2})\right] = -2 + \frac{5}{2}(1 - e^{-2}) = 0.16166179\ldots.$$

MATLAB (actually, the Symbolic Math Toolbox which is part of MATLAB) can numerically evaluate (with double precision) Uhler's double integral *directly* as follows for these particular values of a and c, where the symbols x and y are first declared to be *symbolic* quantities:

syms x y
double(int(int((1-x^2)*x*exp(-y)/y,y,1-x,1+x),x,0,1))

The computer result is in *outstanding* agreement with Uhler's formula: 0.16166179 .... The nested *int* (for 'integrate') commands actually perform *symbolic* integrations, resulting in the *exact* answer of −5/2 * exp(−2) + 1/2, and the final command of *double* converts that exact symbolic answer into double precision numerical form.

The syntax for the use of nested *int* commands to symbolically evaluate multiple integrals should now be obvious but, *just to be sure*, here's a teaser. You may recall that at the end of the original Preface I mentioned something called the 'Hardy-Schuster optical integral.' We'll treat this integral analytically later in the book where, a little ways into the analysis, we'll get it into the form of a *triple* definite integral:

$$\int_0^\infty \left\{\int_x^\infty \left\{\int_x^\infty \cos\left(t^2 - u^2\right) dt\right\} du\right\} dx.$$

Our analysis will show that this perhaps intimidating expression is equal to $\frac{1}{2}\sqrt{\frac{\pi}{2}}$ $= 0.62665706$ .... To have the Symbolic Math Toolbox evaluate it, however, all that is required are the following two lines:

syms x t u
int(int(int(cos(t^2-u^2),t,x,inf),u,x,inf),x,0,inf)

When MATLAB returns the answer, it begins by telling you that no explicit integral could be found, but then it prints out a somewhat cryptic expression that it calls *ans* (for 'answer'). If you then convert *ans* to numeric form by typing *double (ans)* you get ...(really BIG drum roll): 0.62665706 ....

So far I have used *integral* to check one dimensional integrations, and now *int* to check multi-dimensional integrations. But of course *int* can easily do one-dimensional integrals, too. To see *int* used on a one-dimensional integral, recall our result (2.1.4):

$$\int_0^\infty \frac{dx}{1 + e^{ax}} = \frac{\ln(2)}{a}.$$

Differentiating with respect to the parameter a, we get

$$\int_0^\infty \frac{xe^{ax}}{(1 + e^{ax})^2}\, dx = \frac{\ln(2)}{a^2}.$$

Now, change variable to $t = e^{ax}$ and so $x = \frac{1}{a}\ln(t)$. This means $dx = \frac{dt}{at}$ and thus

$$\int_1^\infty \frac{\frac{1}{a}\ln(t)t}{(1 + t)^2}\left(\frac{dt}{at}\right) = \frac{\ln(2)}{a^2}$$

or, after cancelling the a's and t's and changing the dummy variable of integration back to x,

$$\int_1^\infty \frac{\ln(x)}{(1 + x)^2}\, dx = \ln(2). \qquad\qquad (3.6.2)$$

Using the Symbolic Math Toolbox, we can check this result by writing

syms x
int(log(x)/((1+x)^2),x,1,inf)

which returns the exact *symbolic* result of:

$$ans - \log(2).$$

Or, for another one-dimensional example, recall the aerodynamic integral from Chap. 1 (the opening paragraph of Sect. 1.4),

$$\int_{-1}^1 \sqrt{\frac{1 + x}{1 - x}}\, dx = \pi.$$

The Symbolic Math Toolbox agrees, as

syms x
int(sqrt((1+x)/(1-x)),x,-1,1)

produces the exact *symbolic* result of:

$$ans = pi$$

while *integral(@(x)sqrt((1+x)./(1-x)),-1,1)* = 3.14159265 ... (the actual *numerical* value of pi is, of course, 3.14159265 ...).

The Symbolic Math Toolbox is a powerful application, but I mention it here mostly for completeness, and we'll use *integral* (and its variants) as our main workhorse for numerical integration. As two additional examples of that here, recall the second challenge problem from Chap. 1, where you were asked to show that

$$\int_1^\infty \frac{dx}{\sqrt{x^3-1}} < 4.$$

With MATLAB's *integral* we can get a far more precise value for the integral: *integral(@(x)(1./sqrt(x.^3-1)),1,inf)* = 2.42865. . . .

And finally, returning to Uhler's integral in (3.6.1), *integral2* does the job with these three lines:

ymin=@(x)1-x;
ymax=@(x)1+x;
integral2(@(x,y)(1-x.^2).*x.*exp(-y)./y,0,1,ymin,ymax)

which produces the result 0.16166179. . . .

3.7   The Probability Integral Revisited

In footnote 1 of this chapter I promised you some more discussion of the probability integral

$$\int_{-\infty}^\infty e^{-\frac{x^2}{2}} dx.$$

To start, I'll derive it again but in a different way (but still using differentiation under the integral sign). Let

$$I = \int_0^\infty e^{-x^2} dx$$

and further define

$$f(x) = \left(\int_0^x e^{-t^2} dt\right)^2 \text{ and } g(x) = \int_0^1 \frac{e^{-x^2(1+t^2)}}{1+t^2} dt.$$

Then,

$$\frac{df}{dx} = 2\left\{\int_0^x e^{-t^2} dt\right\}e^{-x^2} = 2e^{-x^2}\int_0^x e^{-t^2} dt$$

and

$$\frac{dg}{dx} = \int_0^1 \frac{-2x(1+t^2)e^{-x^2(1+t^2)}}{1+t^2} dt = -2\int_0^1 xe^{-x^2(1+t^2)} dt$$

$$= -2x \int_0^1 e^{-x^2} e^{-x^2 t^2} dt = -2xe^{-x^2} \int_0^1 e^{-x^2 t^2} dt.$$

Let $u = tx$ (which means $du = x\,dt$) and so $dt = \frac{du}{x}$. Thus,

$$\frac{dg}{dx} = -2xe^{-x^2} \int_0^x e^{-u^2} \frac{du}{x} = -2e^{-x^2} \int_0^x e^{-u^2} du = -\frac{df}{dx}.$$

So,

$$\frac{df}{dx} + \frac{dg}{dx} = 0$$

or, with C a constant, we have

$$f(x) + g(x) = C.$$

In particular $f(0) + g(0) = C$, but since $f(0) = 0$ and since

$$g(0) = \int_0^1 \frac{1}{1+t^2} dt = \tan^{-1}(t) \Big|_0^1 = \frac{\pi}{4}$$

we have $C = \frac{\pi}{4}$. That is, $f(x) + g(x) = \frac{\pi}{4}$.

But that means $f(\infty) + g(\infty) = \frac{\pi}{4}$ and since $f(\infty) = I^2$ and $g(\infty) = 0$ (because the integrand of the $g(x)$ integral $\to 0$ as $x \to \infty$) then

$$I^2 = \frac{\pi}{4}.$$

That is,

$$I = \int_0^\infty e^{-x^2} dx = \frac{1}{2}\sqrt{\pi}$$

or, doubling the interval of integration,

$$\int_{-\infty}^\infty e^{-x^2} dx = \sqrt{\pi}. \qquad (3.7.1)$$

Now, let $x = \frac{u}{\sqrt{2}}$ (and so $dx = \frac{du}{\sqrt{2}}$). Then,

$$\int_{-\infty}^\infty e^{-x^2} dx = \int_{-\infty}^\infty e^{-\frac{u^2}{2}} \frac{du}{\sqrt{2}} = \sqrt{\pi}$$

or, as we derived in (3.1.4), with the dummy variable of integration changed back to x,

$$\int_{-\infty}^{\infty} e^{-\frac{x^2}{2}} dx = \sqrt{2\pi}.$$

What I'll show you now is a variation on the probability integral, using the following integral identity: if f(x) is continuous, and if the integrals exist, then

$$\int_{-\infty}^{\infty} f\left(x - \frac{1}{x}\right) dx = \int_{-\infty}^{\infty} f(x) dx. \tag{3.7.2}$$

This is a form of the *Cauchy-Schlömilch transformation* (after Cauchy who knew it by 1823, and the German mathematician Oskar Schlömilch (1823–1901) who popularized it in an 1848 textbook), and (3.7.2) is remarkably easy to establish. We start by writing

$$\int_{-\infty}^{\infty} f\left(x - \frac{1}{x}\right) dx = \int_{-\infty}^{0} f\left(x - \frac{1}{x}\right) dx + \int_{0}^{\infty} f\left(x - \frac{1}{x}\right) dx.$$

In the first integral on the right, let $x = -e^{-u}$, and so $\frac{dx}{du} = e^{-u}$ which says $dx = e^{-u} du$. We see that we have $u = -\infty$ when $x = -\infty$ and $u = \infty$ when $x = 0$. So,

$$\int_{-\infty}^{0} f\left(x - \frac{1}{x}\right) dx = \int_{-\infty}^{\infty} f(-e^{-u} + e^{u}) e^{-u} du.$$

In the second integral on the right, let $x = e^u$, and so $\frac{dx}{du} = e^u$ which says $dx = e^u du$. We see that we have $u = -\infty$ when $x = 0$ and $u = \infty$ when $x = \infty$. So,

$$\int_{0}^{\infty} f\left(x - \frac{1}{x}\right) dx = \int_{-\infty}^{\infty} f(e^{u} - e^{-u}) e^{u} du$$

and therefore

$$\int_{-\infty}^{\infty} f\left(x - \frac{1}{x}\right) dx = \int_{-\infty}^{\infty} f(e^{u} - e^{-u}) e^{-u} du + \int_{-\infty}^{\infty} f(e^{u} - e^{-u}) e^{u} du$$

$$= \int_{-\infty}^{\infty} f(e^{u} - e^{-u})(e^{u} + e^{-u}) du.$$

Now, make the change of variable $y = e^u - e^{-u}$ (and so $\frac{dy}{du} = e^u + e^{-u}$ which means that $dy = (e^u + e^{-u}) du$). Thus, we have $y = -\infty$ when $u = -\infty$, and we have $y = \infty$ when $u = \infty$, and so, just like that, we have

$$\int_{-\infty}^{\infty} f\left(x - \frac{1}{x}\right) dx = \int_{-\infty}^{\infty} f(y)\,dy = \int_{-\infty}^{\infty} f(x)\,dx.$$

Now, in our earlier result of (3.7.1),

$$\int_{-\infty}^{\infty} e^{-x^2} dx = \sqrt{\pi},$$

change variable to $y = \frac{x}{\sqrt{a}}$ (dx $= \sqrt{a}\,dy$), where a is an arbitrary positive constant. Then,

$$\int_{-\infty}^{\infty} e^{-ay^2} \sqrt{a}\,dy = \sqrt{\pi}$$

or,

$$\int_{-\infty}^{\infty} e^{-ax^2} dx = \sqrt{\frac{\pi}{a}}.$$

Using (3.7.2), this tells us that

$$\int_{-\infty}^{\infty} e^{-a\left(x-\frac{1}{x}\right)^2} dx = \int_{-\infty}^{\infty} e^{-a\left(x^2-2+\frac{1}{x^2}\right)^2} dx = e^{2a} \int_{-\infty}^{\infty} e^{-ax^2-\frac{a}{x^2}} dx = \sqrt{\frac{\pi}{a}}.$$

or,

$$\int_{-\infty}^{\infty} e^{-ax^2-\frac{a}{x^2}} dx = e^{-2a} \sqrt{\frac{\pi}{a}}$$

and so, as the integrand is even, we have the very pretty

$$\int_{0}^{\infty} e^{-ax^2-\frac{a}{x^2}} dx = \frac{1}{2} e^{-2a} \sqrt{\frac{\pi}{a}}. \qquad (3.7.3)$$

To numerically check this result, for $a = 1$ (3.7.3) says the integral is equal to $\frac{\sqrt{\pi}}{2e^2} = 0.11993777\ldots$ and, in agreement, MATLAB computes
integral(@(x)exp(-((x.^2)+1./(x.^2))),0,inf) = 0.11993777....
We can further generalize (3.7.3) a bit more as follows, as was first done by the French mathematician Pierre-Simon Laplace (1749–1827) 200 years ago. What we'll calculate is the integral

$$I = \int_0^\infty e^{-ax^2 - \frac{b}{x^2}} dx.$$

To start, let $t = x\sqrt{a}$. Then, $dx = \frac{dt}{\sqrt{a}}$ and so

$$I = \int_0^\infty e^{-t^2 - \frac{ab}{t^2}} \frac{dt}{\sqrt{a}} = \frac{1}{\sqrt{a}} \int_0^\infty e^{-t^2 - \frac{ab}{t^2}} dt = \frac{1}{\sqrt{a}} I_2$$

where

$$I_2 = \int_0^\infty e^{-t^2 - \frac{ab}{t^2}} dt.$$

Next, define

$$y = \frac{\sqrt{ab}}{t}.$$

Then

$$\frac{dy}{dt} = -\frac{\sqrt{ab}}{t^2} = -\frac{\sqrt{ab}}{\frac{ab}{y^2}} = -\frac{y^2}{\sqrt{ab}}$$

or,

$$dt = -\sqrt{ab}\, \frac{dy}{y^2}.$$

So, as $ab = y^2 t^2$ we have

$$I_2 = \int_\infty^0 e^{-\frac{ab}{y^2} - y^2} \left(-\sqrt{ab}\, \frac{dy}{y^2} \right) = \sqrt{ab} \int_0^\infty \frac{e^{-y^2 - \frac{ab}{y^2}}}{y^2} dy.$$

Thus,

$$2I_2 = \int_0^\infty e^{-t^2 - \frac{ab}{t^2}} dt + \sqrt{ab} \int_0^\infty \frac{e^{-t^2 - \frac{ab}{t^2}}}{t^2} dt$$

or,

$$2I_2 = \int_0^\infty e^{-\left(t^2 + \frac{ab}{t^2}\right)} \left\{ 1 + \frac{\sqrt{ab}}{t^2} \right\} dt.$$

Now, change variable again, this time to

$$s = t - \frac{\sqrt{ab}}{t}.$$

So,

$$\frac{ds}{dt} = 1 + \frac{\sqrt{ab}}{t^2}$$

or,

$$dt = \frac{ds}{1 + \frac{\sqrt{ab}}{t^2}}.$$

Then, since $s^2 = t^2 - 2\sqrt{ab} + \frac{ab}{t^2}$ we have (notice that $s = -\infty$ when $t = 0$)

$$2I_2 = \int_{-\infty}^{\infty} e^{-s^2 - 2\sqrt{ab}} \left\{ 1 + \frac{\sqrt{ab}}{t^2} \right\} \frac{ds}{1 + \frac{\sqrt{ab}}{t^2}}.$$

or,

$$I_2 = \frac{1}{2} \int_{-\infty}^{\infty} e^{-s^2 - 2\sqrt{ab}} ds = \frac{e^{-2\sqrt{ab}}}{2} \int_{-\infty}^{\infty} e^{-s^2} ds.$$

and so

$$I_2 = e^{-2\sqrt{ab}} \int_{0}^{\infty} e^{-s^2} ds.$$

Now, since $I = \frac{1}{\sqrt{a}} I_2$ then

$$I = \int_{0}^{\infty} e^{-ax^2 - \frac{b}{x^2}} dx = \frac{1}{\sqrt{a}} e^{-2\sqrt{ab}} \int_{0}^{\infty} e^{-s^2} ds,$$

and since (3.7.1) tells us that

$$\int_{0}^{\infty} e^{-s^2} ds = \frac{\sqrt{\pi}}{2}.$$

we then have the beautiful

$$\int_{0}^{\infty} e^{-ax^2 - \frac{b}{x^2}} dx = \frac{1}{2} \sqrt{\frac{\pi}{a}} e^{-2\sqrt{ab}}. \tag{3.7.4}$$

Notice that if $b = a$, (3.7.4) reduces to (3.7.3). To numerically check (3.7.4), for $a = \frac{1}{2}$ and $b = 2$ (for example) the integral is equal to $\frac{1}{e^2} \sqrt{\frac{\pi}{2}} = 0.169617623758\ldots$

and, in agreement, MATLAB computes *integral(@(x)exp(-((0.5*x.^2)+2./(x.^2))),0,inf) = 0.169617623758. . . .*

3.8 Dini's Integral

To wrap this chapter up, let's work our way through a logarithmic integral with important applications in mathematical physics and engineering, one first evaluated in 1878 by the Italian mathematician Ulisse Dini (1845–1918). It is

$$I(\alpha) = \int_0^\pi \ln\left\{1 - 2\alpha\cos(x) + \alpha^2\right\}dx$$

where the parameter α is a real number. Notice that since $\cos(x)$ runs through all its values from -1 to $+1$ as x varies from 0 to π, we see that the sign of α is immaterial, that is, $I(\alpha) = I(|\alpha|)$. So, from now on I'll discuss just the two cases of $0 \le \alpha < 1$ and $\alpha > 1$ (for $\alpha = \pm1$, Dini's integral reduces to (2.4.1)).

Differentiating, we have

$$\frac{dI}{d\alpha} = \int_0^\pi \frac{-2\cos(x) + 2\alpha}{1 - 2\alpha\cos(x) + \alpha^2}dx = \frac{1}{\alpha}\int_0^\pi\left\{1 - \frac{1 - \alpha^2}{1 - 2\alpha\cos(x) + \alpha^2}\right\}dx$$

$$= \frac{\pi}{\alpha} - \frac{1}{\alpha}\int_0^\pi \frac{1 - \alpha^2}{1 - 2\alpha\cos(x) + \alpha^2}dx.$$

Next, make the change of variable $z = \tan\left(\frac{x}{2}\right)$ which means, as we showed back in Chap. 2 (see the analysis leading up to (2.2.3)), that

$$\cos(x) = \frac{1 - z^2}{1 + z^2}$$

and

$$dx = \frac{2}{1 + z^2}dz.$$

Thus,

$$\frac{dI}{d\alpha} = \frac{\pi}{\alpha} - \frac{1}{\alpha}\int_0^\infty \frac{1 - \alpha^2}{1 - 2\alpha\frac{1-z^2}{1+z^2} + \alpha^2}\left(\frac{2}{1 + z^2}dz\right)$$

$$= \frac{\pi}{\alpha} - \frac{2(1 - \alpha^2)}{\alpha}\int_0^\infty \frac{dz}{1 + z^2 - 2\alpha(1 - z^2) + \alpha^2(1 + z^2)}$$

which becomes, with just a little algebra,

$$\frac{dI}{d\alpha} = \frac{\pi}{\alpha} - \left(\frac{2}{\alpha}\right)\left(\frac{1-\alpha}{1+\alpha}\right)\int_0^\infty \frac{dz}{\left(\frac{1-\alpha}{1+\alpha}\right)^2 + z^2}.$$

This integrates as follows:

$$\frac{dI}{d\alpha} = \frac{\pi}{\alpha} - \frac{2}{\alpha}\left(\frac{1-\alpha}{1+\alpha}\right)\frac{1}{\frac{1-\alpha}{1+\alpha}}\left\{\tan^{-1}\left(\frac{z}{\frac{1-\alpha}{1+\alpha}}\right)\right\}\Big|_0^\infty$$

$$= \frac{1}{\alpha}\left[\pi - 2\tan^{-1}\left(\frac{1+\alpha}{1-\alpha}z\right)\Big|_0^\infty\right].$$

If $\alpha > 1$, then $1 + \alpha > 0$ and $1 - \alpha < 0$ which means $\frac{1+\alpha}{1-\alpha} < 0$ and therefore $\tan^{-1}\left(\frac{1+\alpha}{1-\alpha}z\right)\Big|_0^\infty = -\frac{\pi}{2}$ and so $\frac{dI}{d\alpha} = \frac{1}{\alpha}\left[\pi - 2\left(-\frac{\pi}{2}\right)\right] = \frac{2\pi}{\alpha}$, $\alpha > 1$.

If $0 \leq \alpha < 1$, however, then $1 + \alpha > 0$ and $1 - \alpha > 0$ and so $\frac{1+\alpha}{1-\alpha} > 0$ and therefore $\tan^{-1}\left(\frac{1+\alpha}{1-\alpha}z\right)\Big|_0^\infty = +\frac{\pi}{2}$ and so $\frac{dI}{d\alpha} = \frac{1}{\alpha}\left[\pi - 2\left(\frac{\pi}{2}\right)\right] = 0$, $0 \leq \alpha < 1$.

Now, since

$$I(0) = \int_0^\pi \ln(1)d\alpha = 0,$$

then

$$\frac{dI}{d\alpha} = 0, 0 \leq \alpha < 1$$

integrates *by inspection* to

$$I(\alpha) = 0, 0 \leq \alpha < 1.$$

To integrate

$$\frac{dI}{d\alpha} = \frac{2\pi}{\alpha} \text{ for } \alpha > 1$$

we need to get a specific value of $I(\alpha)$ for some $\alpha > 1$ in order to evaluate the constant of integration. Alas, there is no obvious value of α that leads to a 'nice' integral to evaluate (like we had for $I(0)$)! Fortunately there is a clever trick that will get us around this difficulty. What we'll do is first pick a value (call it β) such that $0 \leq \beta < 1$, which means

$$I(\beta) = \int_0^\pi \ln\left\{1 - 2\beta\cos(x) + \beta^2\right\}dx = 0$$

by our previous argument. Then, we'll pick the specific α value we're going to use to be $\alpha = \frac{1}{\beta}$ which will, of course, mean that $\alpha > 1$. That is,

$$I(\beta) = 0 = I\left(\frac{1}{\alpha}\right) = \int_0^\pi \ln\left\{1 - 2\frac{1}{\alpha}\cos(x) + \frac{1}{\alpha^2}\right\}dx$$

$$= \int_0^\pi \ln\left\{\frac{\alpha^2 - 2\alpha\cos(x) + 1}{\alpha^2}\right\}dx$$

$$= \int_0^\pi \ln\left\{\alpha^2 - 2\alpha\cos(x) + 1\right\}dx - \int_0^\pi \ln\left\{\alpha^2\right\}dx$$

$$= \int_0^\pi \ln\left\{\alpha^2 - 2\alpha\cos(x) + 1\right\}dx - \int_0^\pi 2\ln\left\{\alpha\right\}dx$$

which, you'll recall, equals zero (look back to where this line of math starts: $I(\beta) = 0 = I(\frac{1}{\alpha})$). That is, when $\alpha > 1$ we have

$$\int_0^\pi \ln\left\{\alpha^2 - 2\alpha\cos(x) + 1\right\}dx = \int_0^\pi 2\ln\left\{\alpha\right\}dx = 2\ln\left\{\alpha\right\}\int_0^\pi dx = 2\pi \ln\left\{\alpha\right\}$$

and we see, in fact, that this is indeed in agreement with $\frac{dI}{d\alpha} = \frac{2\pi}{\alpha}$ for $\alpha > 1$.
   So,

$$I(\alpha) = \begin{cases} 0, 0 \leq \alpha < 1 \\ 2\pi \ln\left\{\alpha\right\}, \alpha > 1 \end{cases}$$

as well as $I(\alpha) = I(|\alpha|)$. The easiest way to write all this compactly is

$$\int_0^\pi \ln\left\{1 - 2\alpha\cos(x) + \alpha^2\right\}dx = \begin{cases} 0, \alpha^2 < 1 \\ \pi \ln\left\{\alpha^2\right\}, \alpha^2 > 1 \end{cases}. \qquad (3.8.1)$$

To check, for $\alpha = \frac{1}{2}$ for example, we have *integral(@(x)log((5/4)-cos(x)),0, pi)* $= 2.08 \times 10^{-16}$ (pretty close to zero), and for $\alpha = 2$ we have *integral(@(x)log (5-4∗cos(x)),0,pi)* $= 4.35517218\ldots$ while $\pi \ln(2^2) = 2\pi \ln(2) = 4.35517218\ldots.$

3.9   Feynman's Favorite Trick Solves a Physics Equation

In a physics paper published[7] several years ago, a classic problem in mechanics was given an interesting new treatment. The details of that analysis are not important here, just that once all the dust had settled after the *physics* had been done, the author arrived at the following equation to be solved:

---

[7]Waldemar Klobus, "Motion On a Vertical Loop with Friction," *American Journal of Physics*, September 2011, pp. 913–918.

$$V^2(\phi) = V^2(0) + 2\{\cos(\phi) - \mu\sin(\phi)\} - 2\mu\int_0^\phi V^2(x)dx. \qquad (3.9.1)$$

In (3.9.1) μ is a non-negative constant, and $V^2(0)$ is also a constant (it is, of course, the value of $V^2(\phi)$ when $\phi = 0$). The goal is to solve (3.9.1) for $V^2(\phi)$ as a function of ϕ, μ, and $V^2(0)$.

Once he had arrived at (3.9.1) the author wrote "[This] is a Volterra [after the Italian mathematician Vito Volterra (1860–1940)] integral equation of the second kind, for which the solution is straightforward." He didn't provide any solution details; after referring readers to a brief appendix, where he cited the general formula for the solution to such an equation (in general, integral equations can be difficult because the unknown quantity to be found appears both inside *and* outside an integral), he simply wrote down the answer. Using this formula allowed "the solution [to be] found straightforwardly, albeit laboriously." Now, in fact he was perfectly correct in declaring the solution to be "straightforward," but there really isn't anything that has to be "laborious" about solving (3.9.1). That is, there doesn't *if* you know how to differentiate an integral.

So, differentiating (3.9.1) with respect to ϕ using (3.1.1), we have

$$\frac{dV^2(\phi)}{d\phi} = 2\{-\sin(\phi) - \mu\cos(\phi)\} - 2\mu V^2(\phi)$$

or, rearranging, we have the following first order differential equation for $V^2(\phi)$:

$$\frac{dV^2(\phi)}{d\phi} + 2\mu V^2(\phi) = -2\{\sin(\phi) + \mu\cos(\phi)\}. \qquad (3.9.2)$$

Next, we multiply through (3.9.2) by $e^{2\mu\phi}$ to get

$$\frac{dV^2(\phi)}{d\phi}e^{2\mu\phi} + 2\mu V^2(\phi)e^{2\mu\phi} = 2\{\sin(\phi) + \mu\cos(\phi)\}e^{2\mu\phi}.$$

This is useful because the left-hand-side is now expressible as a derivative, as follows:

$$\frac{d}{d\phi}\{V^2(\phi)e^{2\mu\phi}\} = -2\{\sin(\phi) + \mu\cos(\phi)\}e^{2\mu\phi}. \qquad (3.9.3)$$

This last step is, in fact, an application of a technique routinely taught in a first course in differential equations, called an *integrating factor* (the $e^{2\mu\phi}$ is the factor). And that is our next step, to *integrate* (3.9.3) from 0 to ϕ:

$$\int_0^\phi \frac{d}{dx}\{V^2(x)e^{2\mu x}\}dx = -2\int_0^\phi \{\sin(x) + \mu\cos(x)\}e^{2\mu x}dx.$$

The left-hand-side is

$$\int_0^\phi \frac{d}{dx}\{V^2(x)e^{2\mu x}\}dx = \int_0^\phi d\{V^2(x)e^{2\mu x}\} = \{V^2(x)e^{2\mu x}\}\Big|_0^\phi = V^2(\phi)e^{2\mu\phi} - V^2(0),$$

and so

$$V^2(\phi)e^{2\mu\phi} = V^2(0) - 2\int_0^\phi \{\sin(x) + \mu\cos(x)\}e^{2\mu x}dx$$

or,

$$V^2(\phi) = e^{-2\mu\phi}V^2(0) - 2e^{-2\mu\phi}\int_0^\phi \{\sin(x) + \mu\cos(x)\}e^{2\mu x}dx. \tag{3.9.4}$$

Both of the integrals on the right-hand-side of (3.9.4) are easy to do (either by-parts, or just look them up in a table of integrals) and, if you work through the details (which I'll leave to you), you'll arrive at

$$V^2(\phi) = e^{-2\mu\phi}\left[V^2(0) + \frac{2(2\mu^2 - 1)}{1 + 4\mu^2}\right] - \frac{2}{1 + 4\mu^2}$$
$$\times \left[3\mu\sin(\phi) + (2\mu^2 - 1)\cos(\phi)\right] \tag{3.9.5}$$

which is, indeed, the solution given by the author of the physics paper, and so we have yet another success story for 'Feynman's favorite trick.'

3.10 Challenge Problems

(C3.1) Treat a as a parameter and calculate

$$\int_0^\infty \frac{\ln(1 + a^2x^2)}{b^2 + x^2}dx,$$

where a and b are positive. If you look back at (2.4.3) you'll see that this is a generalization of that integral, for the special case of $a = b = 1$. That is, your answer for the above integral should reduce to $\pi\ln(2)$ if $a = b = 1$.

(C3.2) Calculate the Cauchy Principal Value of

$$\int_{-\infty}^\infty \frac{\cos(ax)}{b^2 - x^2}dx, \ a > 0, \ b > 0.$$

Hint: make a partial fraction expansion, follow that with the appropriate change of variable (it should be obvious), and then finally recall Dirichlet's integral.

(C3.3) In (3.1.7) we found that

$$\int_0^\infty \frac{\cos(ax)}{x^2 + b^2}\, dx = \frac{\pi}{2b} e^{-ab}.$$

Combine that result with your result from the previous challenge problem to calculate

$$\int_{-\infty}^\infty \frac{\cos(ax)}{b^4 - x^4}\, dx.$$

(C3.4) Show that the Cauchy Principal Value of

$$\int_0^\infty \frac{x \sin(ax)}{x^2 - b^2}\, dx = \frac{\pi}{2} \cos(ab).$$

Again, don't forget Dirichlet's integral.

(C3.5) Since we are using Dirichlet's integral so much, here's another look at it. Suppose a and b are both real numbers, with $b > 0$ but a can have either sign. Show that:

$$\int_0^\infty \cos(ax) \frac{\sin(bx)}{x}\, dx = \begin{cases} \frac{\pi}{2}, & |a| < b \\ 0, & |a| > b \\ \frac{\pi}{4}, & |a| - b \end{cases}.$$

(C3.6) I've mentioned the "lifting theory" integral several times in the book, and told you its value is π. MATLAB agrees with that and, indeed, I showed you the *indefinite* integration result in Chap. 1 (Sect. 1.7) that you could confirm by differentiation. See if you can *directly* derive the specific case $\int_{-1}^1 \sqrt{\frac{1+x}{1-x}}\, dx = \pi$. Hint: try the change of variable $x = \cos(2u)$, and remember the double-angle identities.

(C3.7) Here's a classic puzzler involving double integrals for you to think about (since we touched on double integrals in Sect. 3.6). A standard trick to try on double integrals, as I discussed in Sect. 3.4, is to *reverse* the order of integration, with the idea being the order of integration shouldn't matter but maybe one order is easier to do than the other. That is, the assumption is

$$\int_a^b \left\{ \int_c^d f(x,y)\, dx \right\} dy = \int_c^d \left\{ \int_a^b f(x,y)\, dy \right\} dx.$$

'Usually' this is okay *if* f(x,y) is what mathematicians call 'well-behaved.' *But it's not always true.* If, for example, $a = c = 0$ and $b = d = 1$, and $f(x, y) = \frac{x-y}{(x+y)^3}$, then the equality fails. Each of the two double integrals does indeed exist, but they are not equal. Show that this is so by direct calculation of each double integral. Any idea on *why* they aren't equal? Historical note: Curious calculations like this one can be traced back to 1814, when Cauchy began his studies of complex function theory, studies that eventually led to contour integration.

(C3.8) Show that

$$\int_{-\infty}^{\infty} xe^{-x^2-x} \, dx = -\frac{1}{2}\sqrt{\pi\sqrt{e}}$$

and that

$$\int_{-\infty}^{\infty} x^2 e^{-x^2-x} \, dx = \frac{3}{4}\sqrt{\pi\sqrt{e}}.$$

Hint: Consider $I(a, b) = \int_{-\infty}^{\infty} e^{-ax^2+bx} \, dx$, complete the square in the exponent, make the obvious change of variable, and recall (3.7.4). Then, differentiate $I(a, b)$ with respect to a and with respect to b, and in each case set $a = 1$ and $b = -1$.

(C3.9) Given that $\int_0^{\infty} \frac{\sin(mx)}{x(x^2+a^2)} dx = \frac{\pi}{2}\left(\frac{1-e^{-am}}{a^2}\right)$ for $a > 0$, $m > 0$—later, in Chap. 8 (in Challenge Problem C8.2), you'll be asked to derive this result using contour integration—differentiate with respect to the parameter a and thereby evaluate $\int_0^{\infty} \frac{\sin(mx)}{x(x^2+a^2)^2} dx$. Hint: If $m = a = 1$ then your result should reduce to $\frac{\pi}{2} \times \left(1 - \frac{3}{2e}\right) = 0.70400\ldots$, and MATLAB agrees: *integral(@(x)sin(x)./(x.*((x.^2+1).^2)),0,inf)* $= 0.70400.\ldots$

(C3.10) Despite Feynman's enthusiasm for the technique of differentiating integrals, he certainly did use other techniques as well; his famous 1949 paper (see note 9 in Chap. 1), for example, is full of contour integrations. And yet, even there, he returned to his favorite trick. Recall his enthusiasm for the integral $\frac{1}{ab} = \int_0^1 \frac{dx}{[ax+b(1-x)]^2}$. In a note at the end of the 1949 paper Feynman observed how additional identities could be easily derived from this integral, such as $\frac{1}{2a^2b} = \int_0^1 \frac{x \, dx}{[ax+b(1-x)]^3}$. Differentiate his original integral with respect to a, and so confirm this second integral. (While a good exercise—and you *should* do this—it is so easy I've not bothered to include a solution at the end of the book.)

(C3.11) In one of his Hughes/Malibu talks that I mentioned in the new Preface, Feynman stated that he was once asked to evaluate $\int_0^{\pi/2} \cos\{m \tan(\theta)\} d\theta$ where m is an arbitrary real constant. Show that this can be done with (3.1.7). Hint: make the change of variable

$$x = \tan(\theta).$$

(C3.12) The two integrals $\int_1^\infty \frac{dx}{x^2}$ and $\int_1^\infty \frac{dy}{y^2}$ clearly exist (both are equal to 1). Does this imply that the double integral $\int_1^\infty \int_1^\infty \frac{1}{x^2+y^2} dx dy$ also exists? We might try to use MATLAB to experimentally explore this question as follows. If we use MATLAB's *integral2*, we typeintegral2(@(x,y)1./(x.^2+y.^2),1,inf,1,inf) which produces 28.78445, which seems to say the double integral exists. But if we use the Symbolic Math Toolbox, we type

syms x y
f = 1/(x^2 + y^2);
double(int(int(f,x,1,inf),y,1,inf))

which produces the result *Inf* which says the double integral *doesn't* exist. See if you can resolve this conflict analytically, and comment on what this experience suggests about the risk of accepting computer calculations without at least *some* cautious skepticism.

(C3.13) Show that $\int_0^\infty \left(e^{-ax} - e^{-bx}\right) \frac{\cos(x)}{x} dx = \ln\left(\sqrt{b^2+1}\right) - \ln\left(\sqrt{a^2+1}\right)$.
Hint: Start with (3.4.5), and set $a = 0$ and $b = 1$ to get $\int_0^\infty e^{-tx}\left\{\frac{1-\cos(x)}{x}\right\} dx = \ln\left(\frac{\sqrt{t^2+1}}{t}\right)$. Then, first set $t = a$, then set $t = b$, then do a little algebra, and don't forget (3.3.3).

(C3.14) Consider the following seemingly slight variant of the integral in the previous problem: $\int_0^\infty \left(e^{-ax} - e^{-bx}\right) \frac{\sin(x)}{x} dx$. Show that the result of this 'slight' change is very different from the result in the previous problem. Hint: Start with $\frac{e^{-ax}-e^{-bx}}{x} = \int_a^b e^{-xy} dy$, and recall this freshman calculus result (easily derived with integration-by-parts):

$$\int_0^\infty e^{-xy} \sin(x) dx = \frac{1}{1+y^2}.$$

(C3.15) Show that $I = \int_0^\infty \frac{\left(1-e^{-ax}\right)\left(1-e^{-bx}\right)}{x^2} dx = (a+b)\ln(a+b) - a\ln(a) - b\ln(b)$.

Hint: Notice that $\frac{1-e^{-ax}}{x} = \int_0^a e^{-xy} dy$ and that $\frac{1-e^{-bx}}{x} = \int_0^b e^{-xz} dz$. Thus, substitution of these expressions into the original integral gives the *triple* integral $I = \int_0^\infty \left[\int_0^a e^{-xy} dy\right] \left[\int_0^b e^{-xz} dz\right] dx$. To finish, simply reverse the order of integration to $I = \int_0^a \int_0^b \int_0^\infty$ and evaluate (you should find the resulting calculations to be quite easy).

(C3.16) What is $\int_{-\infty}^{\infty} e^{-\frac{(x^2-ax-1)^2}{bx^2}} dx =?$ Hint: use (3.7.2).

(C3.17) Suppose $f(x)$ is any function such that $f\left(\frac{1}{x}\right) = x^2 f(x)$. Show that $\int_0^\infty \frac{f(x)}{x^b+1} dx$ is independent of b. Hint: Write the integral as $I(b)$, compute $\frac{dI}{db}$, and explain why the result equals zero.

(C3.18) The Cauchy-Schlömilch transformation can be generalized far beyond its form in (3.7.2). Here's another version: if $f(x)$ is continuous, and if the integrals exist, then $I = \int_0^\infty f\left\{\left(x - \frac{1}{x}\right)^2\right\} dx = \int_0^\infty f(x^2) dx$. Can you prove this? Hint: Start with the integral on the left and make the change of variable $y = \frac{1}{x}$.

Chapter 4
Gamma and Beta Function Integrals

4.1 Euler's Gamma Function

In two letters written as 1729 turned into 1730, the great Euler created what is today called the *gamma function*, $\Gamma(n)$, defined today in textbooks by the integral

$$\Gamma(n) = \int_0^\infty e^{-x} x^{n-1} \, dx, \quad n > 0. \tag{4.1.1}$$

This definition is the modern one (equivalent to Euler's), dating from 1809, and due to the French mathematician Adrien-Marie Legendre (1752–1833). For $n = 1$ it is an easy integration to see that

$$\Gamma(1) = \int_0^\infty e^{-x} \, dx = \{-e^{-x}\} \Big|_0^\infty = 1.$$

Then, using integration by parts on

$$\Gamma(n + 1) = \int_0^\infty e^{-x} x^n \, dx,$$

where we define $u = x^n$ and $dv = e^{-x} \, dx$ (and so $du = nx^{n-1} \, dx$ and $v = -e^{-x}$), we get

$$\Gamma(n + 1) = \{-x^n e^{-x}\} \Big|_0^\infty + \int_0^\infty e^{-x} nx^{n-1} \, dx = n \int_0^\infty e^{-x} x^{n-1} \, dx.$$

That is, we have the so-called *functional equation* for the gamma function:

© Springer Nature Switzerland AG 2020
P. J. Nahin, *Inside Interesting Integrals*, Undergraduate Lecture Notes in Physics,
https://doi.org/10.1007/978-3-030-43788-6_4

$$\Gamma(n+1) = n\Gamma(n). \tag{4.1.2}$$

So, in particular, for n a positive integer,

$$\Gamma(2) = 1 \cdot \Gamma(1) = 1 \cdot 1 = 1,$$

$$\Gamma(3) = 2 \cdot \Gamma(2) = 2 \cdot 1 = 2,$$

$$\Gamma(4) = 3 \cdot \Gamma(3) = 3 \cdot 2 = 6,$$

$$\Gamma(5) = 4 \cdot \Gamma(4) = 4 \cdot 6 = 24,$$

and so on. In general, you can see that for positive integer values of n

$$\Gamma(n+1) = n!$$

or, equivalently,

$$\Gamma(n) = (n-1)!, \quad n \geq 1. \tag{4.1.3}$$

The gamma function is intimately related to the factorial function (and this connection was, in fact, the original motivation for Euler's interest in $\Gamma(n)$).

The importance of the functional equation (4.1.2) is that it allows the extension of the gamma function to *all* of the real numbers, not just to the positive integers. For example, if we put $n = -\frac{1}{2}$ into (4.1.2) we get $\Gamma\left(\frac{1}{2}\right) = -\frac{1}{2}\Gamma\left(-\frac{1}{2}\right)$ and so $\Gamma\left(-\frac{1}{2}\right) = -2\Gamma\left(\frac{1}{2}\right)$. We'll calculate $\Gamma\left(\frac{1}{2}\right)$ to be $\sqrt{\pi}$ later in this chapter, and thus $\Gamma\left(-\frac{1}{2}\right) = -2\sqrt{\pi}$. We can use this same technique to extend the values of $\Gamma(n)$ for positive values of n (as calculated from the integral in (4.1.1)) to all the negative numbers. The gamma function can be extended to handle complex arguments, as well. When we get to Chap. 8 I'll show you how Riemann did the same sort of thing for the *zeta function* (to be defined in the next chapter). Riemann's functional equation for the zeta function extends (or *continues*) the zeta function's definition from all complex numbers with a real part >1, to the *entire* complex plane. That, in turn, has resulted in the greatest unsolved problem in mathematics today. Exciting stuff! But that's for later.

(4.1.3) is particularly interesting because it shows that, contrary to the initial reaction of most students when first encountering the factorial function, $0! \neq 0$. Rather, setting $n = 1$ in (4.1.3) gives

$$\Gamma(1) = (1-1)! = 0! = 1.$$

The gamma function occurs in many applications, ranging from pure mathematics to esoteric physics to quite practical engineering problems. As with other such useful functions (like the trigonometric functions), the values of $\Gamma(n)$ have been extensively computed and tabulated. MATLAB even has a special command for it named (you shouldn't be surprised) *gamma*.

As an example of its appearance in mathematics, let's evaluate

$$\int_0^\infty e^{-x^3}dx.$$

Making the change of variable $y = x^3$ (which means $x = y^{\frac{1}{3}}$), and so

$$dx = \frac{dy}{3x^2} = \frac{dy}{3y^{\frac{2}{3}}}.$$

Thus,

$$\int_0^\infty e^{-x^3}dx = \int_0^\infty e^{-y}\frac{dy}{3y^{\frac{2}{3}}} = \frac{1}{3}\int_0^\infty e^{-y}y^{-\frac{2}{3}}.$$

This looks just like (4.1.1) with $n - 1 = -\frac{2}{3}$, and so $n = \frac{1}{3}$ and we have

$$\int_0^\infty e^{-x^3}dx = \frac{1}{3}\Gamma\left(\frac{1}{3}\right).$$

Or, using (4.1.2),

$$\frac{1}{3}\Gamma\left(\frac{1}{3}\right) = \Gamma\left(\frac{4}{3}\right).$$

and so

$$\int_0^\infty e^{-x^3}dx = \Gamma\left(\frac{4}{3}\right). \tag{4.1.4}$$

Using MATLAB's gamma function, *gamma(4/3)* = 0.8929795... and *integral* agrees; *integral(@(x)exp(-x.^3),0,inf)* = 0.8929795....

4.2   Wallis' Integral and the Beta Function

Around 1650 the English mathematician John Wallis (1616–1703) studied the integral

$$I(n) = \int_0^1 (x - x^2)^n \, dx$$

which he could directly evaluate for small integer values of n. For example,

$$I(0) = \int_0^1 dx = 1,$$

$$I(1) = \int_0^1 (x - x^2)\, dx = \left\{ \frac{1}{2}x^2 - \frac{1}{3}x^3 \right\}\bigg|_0^1 = \frac{1}{2} - \frac{1}{3} = \frac{1}{6},$$

$$I(2) = \int_0^1 (x - x^2)^2\, dx = \int_0^1 (x^2 - 2x^3 + x^4)\, dx = \left\{ \frac{1}{3}x^3 - \frac{1}{2}x^4 + \frac{1}{5}x^5 \right\}\bigg|_0^1$$

$$= \frac{1}{3} - \frac{1}{2} + \frac{1}{5} = \frac{10}{30} - \frac{15}{30} + \frac{6}{30} = \frac{1}{30},$$

and so on. From this short list of numbers, Wallis somehow then *guessed* (!) the general result for I(n). Can you? In this section we'll derive I(n), but see if you can duplicate Wallis' inspired feat before we finish the formal derivation.

We start by defining the *beta function*:

$$B(m, n) = \int_0^1 x^{m-1}(1 - x)^{n-1}\, dx, \quad m > 0, n > 0. \tag{4.2.1}$$

The beta function is intimately related to the gamma function, as I'll now show you. Changing variable in (4.1.1) to $x = y^2$ (and so $dx = 2y\, dy$) we have

$$\Gamma(n) = \int_0^\infty e^{-y^2} y^{2n-2} 2y\, dy = 2 \int_0^\infty e^{-y^2} y^{2n-1}\, dy.$$

We get another true equation if we replace n with m, and the dummy integration variable y with the dummy integration variable x, and so

$$\Gamma(m) = 2 \int_0^\infty e^{-x^2} x^{2m-1}\, dx. \tag{4.2.2}$$

Thus,

$$\Gamma(m)\Gamma(n) = 4 \int_0^\infty e^{-x^2} x^{2m-1}\, dx \int_0^\infty e^{-y^2} y^{2n-1}\, dy$$

$$= 4 \int_0^\infty \int_0^\infty e^{-(x^2+y^2)} x^{2m-1} y^{2n-1} dx\, dy.$$

This double integral looks pretty awful, but the trick that brings it to its knees is to switch from Cartesian coordinates to polar coordinates. That is, we'll write $r^2 = x^2 + y^2$ where $x = r \cos(\theta)$ and $y = r \sin(\theta)$, and the differential area $dx\, dy$ transforms to $r\, dr\, d\theta$. When we integrate the double integral over the region $0 \leq x$, $y < \infty$ we are integrating over the entire first quadrant of the plane, which is equivalent to integrating over the region $0 \leq r < \infty$ and $0 \leq \theta \leq \frac{\pi}{2}$. So,

$$\Gamma(m)\Gamma(n) = 4\int_0^{\frac{\pi}{2}}\int_0^{\infty} e^{-r^2}\{r\,\cos(\theta)\}^{2m-1}\{r\,\sin(\theta)\}^{2n-1}\, r\, dr\, d\theta$$

or,

$$\Gamma(m)\Gamma(n) = \left[2\int_0^{\infty} e^{-r^2} r^{2(m+n)-1} dr\right]\left[2\int_0^{\frac{\pi}{2}} \cos^{2m-1}(\theta)\sin^{2n-1}(\theta)d\theta\right]. \quad (4.2.3)$$

Let's now examine, in turn, each of the integrals in square brackets on the right in (4.2.3). First, if you compare

$$2\int_0^{\infty} e^{-r^2} r^{2(m+n)-1} dr$$

to (4.2.2), you see that they are the same if we associate $x \leftrightarrow r$ and $m \leftrightarrow (m+n)$. Making those replacements, the first square-bracket term in (4.2.3) becomes

$$2\int_0^{\infty} e^{-r^2} r^{2(m+n)-1} dr = \Gamma(m+n).$$

Thus,

$$\Gamma(m)\Gamma(n) = \Gamma(m+n)\left[2\int_0^{\frac{\pi}{2}} \cos^{2m-1}(\theta)\sin^{2n-1}(\theta)d\theta\right]. \quad (4.2.4)$$

Next, returning to (4.2.1), the definition of the beta function, make the change of variable $x = \cos^2(\theta)$ (and so $dx = -2\sin(\theta)\cos(\theta)d\theta$), which says that $1 - x = \sin^2(\theta)$. So,

$$B(m,n) = \int_0^1 x^{m-1}(1-x)^{n-1}\, dx = -2\int_{\frac{\pi}{2}}^0 \cos^{2m-2}(\theta)\sin^{2n-2}(\theta)\sin(\theta)\cos(\theta)d\theta$$

or

$$B(m,n) = 2\int_0^{\frac{\pi}{2}} \cos^{2m-1}(\theta)\sin^{2n-1}(\theta)d\theta$$

which is the integral in the square brackets of (4.2.4). Therefore,

$$\Gamma(m)\Gamma(n) = \Gamma(m+n)B(m,n)$$

or, rewriting, we have a very important result, one that ties the gamma and beta functions together:

$$B(m, n) = \frac{\Gamma(m)\Gamma(n)}{\Gamma(m+n)}. \tag{4.2.5}$$

Now we can write down the answer to Wallis' integral almost immediately. Just observe that

$$I(k) = \int_0^1 \left(x - x^2\right)^k dx = \int_0^1 x^k (1-x)^k dx = B(k+1, k+1) = \frac{\Gamma(k+1)\Gamma(k+1)}{\Gamma(2k+2)}.$$

or, using (4.1.3),

$$I(k) = \frac{k!k!}{(2k+1)!} = \frac{(k!)^2}{(2k+1)!}.$$

So,

$$I(n) = \int_0^1 \left(x - x^2\right)^n dx = \frac{(n!)^2}{(2n+1)!}. \tag{4.2.6}$$

If you look back at the start of this section, at the first few values for $I(n)$ that we (Wallis) calculated by direct integration, you'll see that (4.2.6) gives those same results. For 'large' values of n, however, direct evaluation of the integral becomes pretty grubby, while the right-hand side of (4.2.6) is easy to do. For example, if $n = 7$ we have

$$\int_0^1 \left(x - x^2\right)^7 dx = \frac{(7!)^2}{15!} = 1.94\ldots \times 10^{-5}$$

and MATLAB agrees, as $integral(@(x)(x-x.\wedge 2).\wedge 7, 0, 1) = 1.94 \ldots \times 10^{-5}$.

One quite useful result comes directly from (4.2.6), for the case of $n = \frac{1}{2}$. This is, of course, a *non*-integer value, while we have so far (at least implicitly) taken n to be a positive integer. In keeping with the spirit of this book, however, let's just blissfully ignore that point and see where it takes us. For $n = \frac{1}{2}$, (4.2.6) becomes the claim

$$\int_0^1 \sqrt{x - x^2}\, dx = \frac{\left(\frac{1}{2}!\right)^2}{2!} = \frac{1}{2}\left(\frac{1}{2}!\right)^2.$$

Alternatively, we can *directly* evaluate the integral on the left using the area interpretation of the Riemann integral, as follows. The integral is the area under the curve (and above the x-axis) described by $y(x) = \sqrt{x - x^2}$ from $x = 0$ to $x = 1$. That curve is probably more easily recognized if we write it as $y^2 = x - x^2$ or, $x^2 - x + y^2 = 0$ or, by completing the square, as

$$\left(x - \frac{1}{2}\right)^2 + y^2 = \left(\frac{1}{2}\right)^2.$$

That is, the curve is simply the circle with radius $\frac{1}{2}$ centered on the x-axis at $x = \frac{1}{2}$. The area associated with our integral is, then, the area of the *upper-half* of the circle, which is $\frac{\pi}{8}$. So,

$$\frac{1}{2}\left(\frac{1}{2}!\right)^2 = \frac{\pi}{8}$$

or,

$$\left(\frac{1}{2}\right)! = \frac{1}{2}\sqrt{\pi}. \tag{4.2.7}$$

Does (4.2.7) 'make sense'? Yes, and here's one check of it. Recall (4.1.3),

$$\Gamma(n) = (n-1)!$$

that, for $n = \frac{3}{2}$ gives $\left(\frac{1}{2}\right)!$ and so, putting $n = \frac{3}{2}$ in (4.1.1) and using (4.2.7), we arrive at

$$\int_0^\infty e^{-x}\sqrt{x}\,dx = \frac{1}{2}\sqrt{\pi}. \tag{4.2.8}$$

This is $0.886226,\ldots$ and MATLAB agrees, as *integral(@(x)exp(-x).*sqrt(x),0, inf) = 0.886226*.

We can use (4.2.8) to establish yet another interesting integral, that of

$$\int_0^1 \sqrt{-\ln(x)}\,dx.$$

Make the change of variable $y = -\ln(x)$. Then,

$$e^y = e^{-\ln(x)} = e^{\ln(x^{-1})} = e^{\ln\left(\frac{1}{x}\right)} = \frac{1}{x},$$

and since

$$\frac{dy}{dx} = -\frac{1}{x} = -e^y,$$

we have

$$dx = -\frac{dy}{e^y}.$$

Thus,

$$\int_0^1 \sqrt{-\ln(x)}\, dx = \int_\infty^0 \sqrt{y}\left(-\frac{dy}{e^y}\right) = \int_0^\infty \sqrt{y}e^{-y}\, dy.$$

But this last integral is (4.2.8) and so

$$\int_0^1 \sqrt{-\ln(x)}\, dx = \frac{1}{2}\sqrt{\pi}. \tag{4.2.9}$$

MATLAB agrees, as $integral(@(x)sqrt(-log(x)),0,1) = 0.886226....$ (It's interesting to compare (4.2.9) with (3.1.8).)

With our work in this chapter, we can even talk of *negative* factorials; suppose, for example, $n = \frac{1}{2}$ in (4.1.3), which gives

$$\Gamma\left(\frac{1}{2}\right) = \left(-\frac{1}{2}\right)!.$$

From (4.1.1), setting $n = \frac{1}{2}$ we have

$$\Gamma\left(\frac{1}{2}\right) = \int_0^\infty \frac{e^{-x}}{\sqrt{x}}dx,$$

an integral we can find using (3.7.1). There, if we set $a = 1$ and $b = 0$ we have

$$\int_0^\infty e^{-x^2}dx = \frac{1}{2}\sqrt{\pi}.$$

If we now change variable to $s = x^2$ then $\frac{ds}{dx} = 2x = 2\sqrt{s}$ and so $dx = \frac{ds}{2\sqrt{s}}$. That gives us

$$\int_0^\infty e^{-x^2}dx = \int_0^\infty e^{-s}\frac{ds}{2\sqrt{s}} = \frac{1}{2}\sqrt{\pi}$$

or, replacing the dummy variable of integration s back to x,

$$\int_0^\infty \frac{e^{-x}}{\sqrt{x}}dx = \sqrt{\pi}. \tag{4.2.10}$$

This is $1.77245 ...$, and MATLAB agrees: $integral(@(x)exp(-x)./sqrt(x),0, inf) = 1.77245....$

Thus,

$$\Gamma\left(\frac{1}{2}\right) = \left(-\frac{1}{2}\right)! = \sqrt{\pi}. \qquad (4.2.11)$$

Another way to see (4.2.11) is to use (4.2.4). There, with $m = n = \frac{1}{2}$, we have

$$\Gamma\left(\frac{1}{2}\right)\Gamma\left(\frac{1}{2}\right) = \Gamma(1)\left[2\int_0^{\frac{\pi}{2}}d\theta\right] = \Gamma(1)\left[2\left(\frac{\pi}{2}\right)\right] = \Gamma(1)\pi$$

or, as $\Gamma(1) = 1$,

$$\Gamma^2\left(\frac{1}{2}\right) = \pi$$

and so, again,

$$\Gamma\left(\frac{1}{2}\right) = \sqrt{\pi}.$$

Together, the gamma and beta functions let us do some downright nasty-looking integrals. For example, even though the integrand itself is pretty simple looking, the integral

$$\int_0^{\frac{\pi}{2}}\sqrt{\sin(x)}\,dx$$

seems to be invulnerable to attack by any of the tricks we've developed up to now—but look at what happens when we make the change of variable $u = \sin^2(x)$. Then

$$dx = \frac{du}{2\sin(x)\cos(x)}$$

and, as

$$\sin(x) = u^{\frac{1}{2}}$$

and

$$\cos(x) = \sqrt{1 - \sin^2(x)} = (1 - u)^{\frac{1}{2}},$$

we have

$$\int_0^{\frac{\pi}{2}} \sqrt{\sin(x)}\, dx = \int_0^1 \frac{u^{\frac{1}{4}}}{2u^{\frac{1}{2}}(1-u)^{\frac{1}{2}}}\, du = \frac{1}{2}\int_0^1 u^{-\frac{1}{4}}(1-u)^{-\frac{1}{2}}\, du = \frac{1}{2}B\left(\frac{3}{4},\frac{1}{2}\right).$$

From (4.2.5) this becomes (the integral on the far-right follows because, over the integration interval, the sine and cosine take-on the same values),

$$\int_0^{\frac{\pi}{2}} \sqrt{\sin(x)}\, dx = \frac{\Gamma\left(\frac{3}{4}\right)\Gamma\left(\frac{1}{2}\right)}{2\Gamma\left(\frac{5}{4}\right)} = \int_0^{\frac{\pi}{2}} \sqrt{\cos(x)}\, dx. \qquad (4.2.12)$$

MATLAB's gamma function says this is equal to *gamma(3/4)*gamma(1/2)/ (2*gamma(5/4))* = 1.19814..., and MATLAB agrees as *integral(@(x)sqrt(sin (x)),0,pi/2)* = 1.19814....

The same substitution works for

$$\int_0^{\frac{\pi}{2}} \frac{dx}{\sqrt{\sin(x)\cos(x)}},$$

which transforms into

$$\frac{1}{2}\int_0^1 \frac{du}{\left\{u^{\frac{1}{2}}(1-u)^{\frac{1}{2}}\right\}^{\frac{3}{2}}} = \frac{1}{2}\int_0^1 u^{-\frac{3}{4}}(1-u)^{-\frac{3}{4}}du = \frac{1}{2}B\left(\frac{1}{4},\frac{1}{4}\right) = \frac{\Gamma^2\left(\frac{1}{4}\right)}{2\Gamma\left(\frac{1}{2}\right)}$$

or, using (4.2.11),

$$\int_0^{\frac{\pi}{2}} \frac{dx}{\sqrt{\sin(x)\cos(x)}} = \frac{\Gamma^2\left(\frac{1}{4}\right)}{2\sqrt{\pi}}. \qquad (4.2.13)$$

MATLAB's gamma function says this is equal to *(gamma(1/4)^2)/(2*sqrt (pi))* = 3.708149... and *integral* agrees: *integral(@(x)1./sqrt(sin(x).*cos(x)),0,pi/ 2)* = 3.708149....

As another example, we can use (4.2.13) and the double-angle formula from trigonometry to write

$$\int_0^{\frac{\pi}{2}} \frac{dx}{\sqrt{\sin(x)\cos(x)}} = \int_0^{\frac{\pi}{2}} \frac{dx}{\sqrt{\frac{1}{2}\sin(2x)}} = \frac{\Gamma^2\left(\frac{1}{4}\right)}{2\sqrt{\pi}}.$$

Make the change of variable u = 2x and so dx = $\frac{1}{2}$du and thus

$$\int_0^{\pi} \frac{du}{2\sqrt{\frac{1}{2}\sin(u)}} = \frac{\sqrt{2}}{2}(2)\int_0^{\frac{\pi}{2}} \frac{du}{\sqrt{\sin(u)}} = \frac{\Gamma^2\left(\frac{1}{4}\right)}{2\sqrt{\pi}}$$

or,

$$\int_0^{\frac{\pi}{2}} \frac{dx}{\sqrt{\sin{(x)}}} = \left(\frac{1}{\sqrt{2}}\right) \frac{\Gamma^2\left(\frac{1}{4}\right)}{2\sqrt{\pi}}.$$

Or, finally (the integral on the far-right follows because, over the integration interval, the sine and cosine take-on the same values),

$$\int_0^{\frac{\pi}{2}} \frac{dx}{\sqrt{\sin{(x)}}} = \frac{\Gamma^2\left(\frac{1}{4}\right)}{2\sqrt{2\pi}} = \int_0^{\frac{\pi}{2}} \frac{dx}{\sqrt{\cos{(x)}}}. \qquad (4.2.14)$$

MATLAB's gamma function says this is equal to *(gamma(1/4)^2)/(2*sqrt (2*pi))* = 2.62205..., and *integral* agrees: *integral(@(x)1./sqrt(sin(x)),0,pi/ 2)* = 2.62205....

Finally, in the definition (4.2.1) of the beta function make the change of variable

$$x = \frac{y}{1+y}$$

and so

$$1 - x = 1 - \frac{y}{1+y} = \frac{1}{1+y}$$

and

$$-\frac{dx}{dy} = \frac{1}{(1+y)^2}$$

and so

$$dx = \frac{dy}{(1+y)^2}.$$

Thus,

$$B(m, n) = \int_0^\infty \left(\frac{y}{1+y}\right)^{m-1} \left(\frac{1}{1+y}\right)^{n-1} \frac{dy}{(1+y)^2} = \int_0^\infty \frac{y^{m-1}}{(1+y)^{m+n}} dy.$$

Then, setting n = 1 − m, we have

$$\int_0^\infty \frac{y^{m-1}}{1+y} dy = B(m, 1-m) = \Gamma(m)\Gamma(1-m) \qquad (4.2.15)$$

where the second equality follows by (4.2.5) because $\Gamma(1) = 1$. Later, in Chap. 8 (in (8.7.9)), we'll show (using contour integration) that

$$\int_0^\infty \frac{y^{m-1}}{1+y}\,dy = \frac{\pi}{\sin(m\pi)}$$

and so

$$\Gamma(m)\Gamma(1-m) = \frac{\pi}{\sin(m\pi)} \qquad (4.2.16)$$

which is the famous *reflection formula* for the gamma function, discovered by Euler in 1771.

The reflection formula will be of great use to us when we get to Chap. 9 (the Epilogue). There I'll show you Riemann's crowning achievement, his derivation of the functional equation for the zeta function (the zeta function will be introduced in the next chapter) in which (4.2.16) will play a central role. There is one additional result we'll need there, as well, one that we can derive right now with the aid of the beta function. We start with (4.2.5),

$$B(m,n) = \frac{\Gamma(m)\Gamma(n)}{\Gamma(m+n)} = \frac{(m-1)!(n-1)!}{(m+n-1)!},$$

from which it follows that

$$B(m+1,n+1) = \frac{m!n!}{(m+n+1)!}.$$

So, writing $m = n = z$, we have

$$B(z+1,z+1) = \frac{z!z!}{(2z+1)!}.$$

From the definition of the beta function in (4.2.1),

$$B(z+1,z+1) = \int_0^1 x^z(1-x)^z dx$$

and so

$$\frac{z!z!}{(2z+1)!} = \int_0^1 x^z(1-x)^z dx.$$

Next, make the change of variable $x = \frac{1+s}{2}$ (and so $1 - x = \frac{1-s}{2}$) to get

$$\frac{z!z!}{(2z+1)!} = \int_{-1}^{1} \left(\frac{1+s}{2}\right)^z \left(\frac{1-s}{2}\right)^z \frac{1}{2} ds = 2^{-2z-1} \int_{-1}^{1} (1-s^2)^z \, ds$$

or, since the integrand is even,

$$\frac{z!z!}{(2z+1)!} = 2^{-2z} \int_{0}^{1} (1-s^2)^z \, ds.$$

Make a second change of variable now, to $u = s^2$ and (so $ds = \frac{du}{2\sqrt{u}}$), to arrive at

$$\frac{z!z!}{(2z+1)!} = 2^{-2z} \int_{0}^{1} (1-u)^z \frac{du}{2\sqrt{u}} = 2^{-2z-1} \int_{0}^{1} (1-u)^z u^{-\frac{1}{2}} \, du.$$

The last integral is, from (4.2.1),

$$B\left(\frac{1}{2}, z+1\right) = \frac{\Gamma(\frac{1}{2})\Gamma(z+1)}{\Gamma(z+\frac{3}{2})} = \frac{(-\frac{1}{2})! z!}{(z+\frac{1}{2})!}$$

and so, recalling from (4.2.11) that $\left(-\frac{1}{2}\right)! = \sqrt{\pi}$, we have

$$\frac{z!z!}{(2z+1)!} = 2^{-2z-1} \frac{z!\sqrt{\pi}}{(z+\frac{1}{2})!}.$$

Cancelling a z! on each side, and then cross-multiplying, gives us

$$z!\left(z+\frac{1}{2}\right)! = 2^{-2z-1}\sqrt{\pi}\,(2z+1)! \tag{4.2.17}$$

and since

$$\left(z+\frac{1}{2}\right)! = \left(z+\frac{1}{2}\right)\left(z-\frac{1}{2}\right)! = \left(\frac{2z+1}{2}\right)\left(z-\frac{1}{2}\right)!$$

and also

$$(2z+1)! = (2z+1)(2z)!,$$

then we can alternatively write

$$z!\left(z-\frac{1}{2}\right)! = 2^{-2z}\sqrt{\pi}\,(2z)! \tag{4.2.18}$$

(4.2.17) and (4.2.18) are variations on what mathematicians commonly call the *Legendre duplication formula*. We'll use (4.2.18), in particular, in Chap. 9.

4.3 Double Integration Reversal

In Chap. 3 (Sect. 3.4) I discussed how reversing the order of integration in a double integral can be a useful technique to have tucked away in your bag of tricks. In this section I'll show you some more examples of that trick in which the gamma function will appear. Our starting point is the double integral

$$\int_0^\infty \int_0^\infty \sin{(bx)}y^{p-1}e^{-xy}dx\, dy.$$

If we argue that the value of this integral is independent of the order of integration, then we can write

$$\int_0^\infty \sin{(bx)}\left\{\int_0^\infty y^{p-1}e^{-xy}dy\right\}dx = \int_0^\infty y^{p-1}\left\{\int_0^\infty \sin{(bx)}e^{-xy}dx\right\}dy. \quad (4.3.1)$$

Concentrate your attention for now on the right-hand-side of (4.3.1). We have, for the inner x-integral,

$$\int_0^\infty \sin{(bx)}e^{-xy}dx = \frac{b}{b^2 + y^2},$$

a result you can get either by integrating-by-parts, or find worked-out using yet another trick that I'll show you in Chap. 7, in the result (7.1.2). For now, however, you can just accept it. Then, the right-hand-side of (4.3.1) is calculated as

$$\int_0^\infty y^{p-1}\frac{b}{b^2 + y^2}dy = b\int_0^\infty \frac{y^{p-1}}{b^2\left(1 + \frac{y^2}{b^2}\right)}dy = \frac{1}{b}\int_0^\infty \frac{y^{p-1}}{1 + \left(\frac{y}{b}\right)^2}dy.$$

Now, make the change of variable

$$t = \frac{y}{b},$$

and so $dy = b\, dt$, which means the right-hand-side of (4.3.1) is

$$\frac{1}{b}\int_0^\infty \frac{(tb)^{p-1}}{1 + t^2}b\, dt = b^{p-1}\int_0^\infty \frac{t^{p-1}}{1 + t^2}dt.$$

As I'll show you in Chap. 8, when we get to contour integration, we'll find in (8.7.8) that

$$\int_0^\infty \frac{x^m}{1 + x^n}dx = \frac{\frac{\pi}{n}}{\sin{\left\{(m + 1)\frac{\pi}{n}\right\}}}.$$

So, if we let n = 2 and m = p − 1 we have

$$\int_0^\infty \frac{t^{p-1}}{1+t^2}\, dt = \frac{\frac{\pi}{2}}{\sin\left\{\frac{p\pi}{2}\right\}}.$$

Thus, the right-hand-side of (4.3.1) is

$$\frac{b^{p-1}\pi}{2\sin\left\{\frac{p\pi}{2}\right\}}.$$

Next, shift your attention to the left-hand-side of (4.3.1). For the inner, y-integral, make the change of variable u = xy (where, of course, at this point we treat x as a constant). Then, du = x dy and so

$$\int_0^\infty y^{p-1}e^{-xy}dy = \int_0^\infty \left(\frac{u}{x}\right)^{p-1}e^{-u}\frac{du}{x} = \frac{1}{x^p}\int_0^\infty u^{p-1}e^{-u}\, du.$$

The last integral is just the definition of Γ(p) in (4.1.1), and so

$$\int_0^\infty y^{p-1}e^{-xy}dy = \frac{1}{x^p}\Gamma(p).$$

Thus, we immediately see that the left-hand-side of (4.3.1) is

$$\Gamma(p)\int_0^\infty \frac{\sin(bx)}{x^p}\, dx.$$

Equating our two results for each side of (4.3.1), we have (for 0 < p < 2),

$$\int_0^\infty \frac{\sin(bx)}{x^p}\, dx = \frac{b^{p-1}\pi}{2\Gamma(p)\sin\left\{\frac{p\pi}{2}\right\}}. \qquad (4.3.2)$$

This result reproduces Dirichlet's integral if b = 1 and p = 1 (that's *good*, of course!), but the real value of (4.3.2) comes from its use in doing an integral that, until now, would have stopped us cold:

$$\int_0^\infty \frac{\sin(x^q)}{x^q}\, dx = ?$$

For example, if q = 3, what is

$$\int_0^\infty \frac{\sin(x^3)}{x^3}\, dx = ?$$

Note, carefully, that this is *not*

$$\int_0^\infty \left\{ \frac{\sin(x)}{x} \right\}^3 dx$$

which we'll do later in Chap. 7, in (7.6.2).

We start with the obvious step of changing variable to $u = x^q$. Then

$$\frac{du}{dx} = qx^{q-1} = q\frac{x^q}{x} = \frac{qu}{u^{1/q}}$$

and so

$$dx = \frac{u^{1/q}}{qu} du.$$

Thus,

$$\int_0^\infty \frac{\sin(x^q)}{x^q} dx = \int_0^\infty \frac{\sin(u)}{u} \left(\frac{u^{1/q}}{qu} du \right) = \frac{1}{q} \int_0^\infty \frac{\sin(u)}{u^{2-1/q}} du.$$

The last integral is of the form (4.3.2), with $b = 1$ and $p = 2 - \frac{1}{q}$, and so

$$\int_0^\infty \frac{\sin(x^q)}{x^q} dx = \frac{\pi}{2q\Gamma\left(2 - \frac{1}{q}\right) \sin\left\{ \left(2 - \frac{1}{q}\right)\frac{\pi}{2} \right\}}.$$

Since

$$\sin\left\{ \left(2 - \frac{1}{q}\right)\frac{\pi}{2} \right\} = \sin\left(\pi - \frac{\pi}{2q}\right) = \sin(\pi)\cos\left(\frac{\pi}{2q}\right) - \cos(\pi)\sin\left(\frac{\pi}{2q}\right)$$

$$= \sin\left(\frac{\pi}{2q}\right)$$

we have

$$\int_0^\infty \frac{\sin(x^q)}{x^q} dx = \frac{\pi}{2q\Gamma\left(2 - \frac{1}{q}\right) \sin\left(\frac{\pi}{2q}\right)}. \tag{4.3.3}$$

The result in (4.3.3) is formally an answer to our question, but we can simplify it a bit. The following is a summary of the properties of the gamma function where, because of the functional equation (4.1.2), z is now *any* real number and not just a positive integer.

(a) $\Gamma(z) = (z - 1)!$;

(b) $\Gamma(z + 1) = z!$;

(c) $(z - 1)! = \frac{z!}{z}$ (alternatively, $z! = z(z - 1)!$;

(d) $\Gamma(z)\Gamma(1 - z) = \frac{\pi}{\sin(\pi z)}$.

From (b) we can write $\Gamma\left(2 - \frac{1}{q}\right) = \left(1 - \frac{1}{q}\right)!$ and then, from (c),

$$\left(1 - \frac{1}{q}\right)! = \left(1 - \frac{1}{q}\right)\left(-\frac{1}{q}\right)!$$

So, (4.3.3) becomes

$$\int_0^\infty \frac{\sin(x^q)}{x^q}\, dx = \frac{\pi}{2q\left(1 - \frac{1}{q}\right)\left(-\frac{1}{q}\right)! \sin\left\{\frac{\pi}{2q}\right\}}$$

or,

$$\int_0^\infty \frac{\sin(x^q)}{x^q}\, dx = \frac{\pi}{2(q - 1)\left(\frac{1}{q}\right)! \sin\left\{\frac{\pi}{2q}\right\}}. \tag{4.3.4}$$

Next, from (a), (b), and (d) we have

$$\Gamma(z)\Gamma(1 - z) = \frac{\pi}{\sin(\pi z)} = (z - 1)!(-z)! = \frac{z(z-1)!(-z)!}{z} = \frac{z!(-z)!}{z}$$

and so

$$(-z)! = \frac{\pi z}{z! \sin(\pi z)}.$$

Writing $\frac{1}{q}$ for z, this becomes

$$\left(-\frac{1}{q}\right)! = \frac{\pi \frac{1}{q}}{\left(\frac{1}{q}\right)! \sin\left(\frac{\pi}{q}\right)}. \tag{4.3.5}$$

Substituting (4.3.5) into (4.3.4), we get

$$\int_0^\infty \frac{\sin(x^q)}{x^q}\, dx = \frac{\pi}{2(q - 1)\frac{\pi \frac{1}{q}}{\left(\frac{1}{q}\right)! \sin\left(\frac{\pi}{q}\right)} \sin\left\{\frac{\pi}{2q}\right\}} = \frac{\left(\frac{1}{q}\right)! \sin\left(\frac{\pi}{q}\right)}{2(q - 1)\left(\frac{1}{q}\right) \sin\left(\frac{\pi}{2q}\right)}$$

$$= \frac{\left(\frac{1}{q}\right)! 2 \sin\left(\frac{\pi}{2q}\right) \cos\left(\frac{\pi}{2q}\right)}{2(q - 1)\left(\frac{1}{q}\right) \sin\left(\frac{\pi}{2q}\right)}$$

or,

$$\int_0^\infty \frac{\sin(x^q)}{x^q} dx = \frac{\left(\frac{1}{q}\right)!}{\frac{1}{q}} \left(\frac{1}{q-1}\right) \cos\left(\frac{\pi}{2q}\right). \tag{4.3.6}$$

From (c), which says $z! = z(z-1)!$, we can write

$$\frac{\left(\frac{1}{q}\right)!}{\frac{1}{q}} = \frac{\frac{1}{q}\left(\frac{1}{q}-1\right)!}{\frac{1}{q}} = \left(\frac{1}{q}-1\right)! = \Gamma\left(\frac{1}{q}\right)$$

where the last equality follows from (a). Using this in (4.3.6), we at last have our answer:

$$\int_0^\infty \frac{\sin(x^q)}{x^q} dx = \frac{\Gamma\left(\frac{1}{q}\right)}{q-1} \cos\left(\frac{\pi}{2q}\right), \quad q > 1. \tag{4.3.7}$$

If $q = 3$, for example, (4.3.7) says

$$\int_0^\infty \frac{\sin(x^3)}{x^3} dx = \frac{\Gamma\left(\frac{1}{3}\right)}{2} \cos\left(\frac{\pi}{6}\right) = \frac{\Gamma\left(\frac{1}{3}\right)}{2}\left(\frac{\sqrt{3}}{2}\right) = \frac{\sqrt{3}}{4}\Gamma\left(\frac{1}{3}\right)$$

and MATLAB agrees: *sqrt(3)*gamma(1/3)/4 = 1.16001...* and *integral(@(x)sin(x. ^3)./x.^3,0,inf) = 1.16001....*

The trick of double integration reversal is so useful, let's use it again to evaluate

$$\int_0^\infty \frac{\cos(bx)}{x^p} dx,$$

which is an obvious variation on (4.3.2). We start with

$$\int_0^\infty \int_0^\infty \cos(bx) y^{p-1} e^{-xy} dx\, dy$$

and then, as before, assume that the value of the double integral is independent of the order of integration. That is,

$$\int_0^\infty \cos(bx) \left\{\int_0^\infty y^{p-1} e^{-xy} dy\right\} dx = \int_0^\infty y^{p-1} \left\{\int_0^\infty \cos(bx) e^{-xy} dx\right\} dy. \tag{4.3.8}$$

Concentrate your attention for now on the right-hand side of (4.3.8). We have, for the inner, x-integral (from integration-by-parts)

$$\int_0^\infty \cos(bx)e^{-xy}dx = \frac{y}{b^2+y^2}.$$

Thus, the right-hand side of (4.3.8) is

$$\int_0^\infty y^{p-1}\frac{y}{b^2+y^2}\,dy = \int_0^\infty \frac{y^p}{b^2\left(1+\frac{y^2}{b^2}\right)}dy = \frac{1}{b^2}\int_0^\infty \frac{y^p}{1+\frac{y^2}{b^2}}dy.$$

Now, make the change-of-variable

$$t = \frac{y}{b}$$

and so $dy = b\,dt$, which means that the right-hand side of (4.3.8) is

$$\frac{1}{b^2}\int_0^\infty \frac{(tb)^p}{1+t^2}b\,dt = b^{p-1}\int_0^\infty \frac{t^p}{1+t^2}\,dt.$$

Again, as before, we next use a result to be established later as (8.7.8):

$$\int_0^\infty \frac{x^m}{1+x^n}dx = \frac{\frac{\pi}{n}}{\sin\left\{(m+1)\frac{\pi}{n}\right\}}.$$

So, if we let $n = 2$ and $m = p$, we have

$$\int_0^\infty \frac{t^p}{1+t^2}\,dt = \frac{\frac{\pi}{2}}{\sin\left\{\frac{(p+1)\pi}{2}\right\}} = \frac{\frac{\pi}{2}}{\cos\left\{\frac{p\pi}{2}\right\}}.$$

Thus, the right-hand side of (4.3.8) is

$$\frac{b^{p-1}\pi}{2\cos\left\{\frac{p\pi}{2}\right\}}.$$

Next, shift your attention to the left-hand side of (4.3.8). For the inner, y-integral, you'll recall that we've already worked it out to be

$$\int_0^\infty y^{p-1}e^{-xy}dy = \frac{1}{x^p}\Gamma(p).$$

So, the left-hand side of (4.3.8) is

$$\Gamma(p) \int_0^\infty \frac{\cos{(bx)}}{x^p} \, dx.$$

Equating our two results for each side of (4.3.8), we have

$$\int_0^\infty \frac{\cos{(bx)}}{x^p} \, dx = \frac{b^{p-1}\pi}{2\Gamma(p)\cos\left\{\frac{p\pi}{2}\right\}}, \quad 0 < p < 1. \tag{4.3.9}$$

There is still a *lot* more we can do with these results. For example, if we write (4.3.2) with u rather than x as the dummy variable of integration, then

$$\int_0^\infty \frac{\sin{(bu)}}{u^p} \, du = \frac{b^{p-1}\pi}{2\Gamma(p)\sin\left\{\frac{p\pi}{2}\right\}},$$

and if we then let $p = 1 - \frac{1}{k}$ this becomes

$$\int_0^\infty \frac{\sin{(bu)}}{u^{1-\frac{1}{k}}} \, du = \frac{b^{-\frac{1}{k}}\pi}{2\Gamma\left(1 - \frac{1}{k}\right)\sin\left\{\frac{\left(1-\frac{1}{k}\right)\pi}{2}\right\}}$$

or,

$$\int_0^\infty u^{\frac{1}{k}} \frac{\sin{(bu)}}{u} \, du = \frac{\pi}{2b^{\frac{1}{k}}\Gamma\left(1 - \frac{1}{k}\right)\sin\left\{\frac{\pi}{2} - \frac{\pi}{2k}\right\}}.$$

Now, let $u = x^k$ and so

$$\frac{du}{dx} = k\,x^{k-1}$$

and so

$$du = k\frac{x^k}{x} = k\frac{u}{u^{\frac{1}{k}}} \, dx.$$

Thus,

$$\int_0^\infty u^{\frac{1}{k}} \frac{\sin{(bu)}}{u} \, du = \int_0^\infty u^{\frac{1}{k}} \frac{\sin{(bx^k)}}{u} k\frac{u}{u^{\frac{1}{k}}} \, dx = k\int_0^\infty \sin{(bx^k)} \, dx$$

and so

$$\int_0^\infty \sin\left(bx^k\right)dx = \frac{\pi}{2kb^{\frac{1}{k}}\Gamma\left(1 - \frac{1}{k}\right)\sin\left\{\frac{\pi}{2} - \frac{\pi}{2k}\right\}}.$$

From the reflection formula (4.2.16) for the gamma function

$$\Gamma\left(\frac{1}{k}\right)\Gamma\left(1 - \frac{1}{k}\right) = \frac{\pi}{\sin\left\{\frac{\pi}{k}\right\}}$$

and so

$$\Gamma\left(1 - \frac{1}{k}\right) = \frac{\pi}{\Gamma\left(\frac{1}{k}\right)\sin\left\{\frac{\pi}{k}\right\}}$$

or,

$$\int_0^\infty \sin\left(bx^k\right)dx = \frac{\Gamma\left(\frac{1}{k}\right)\sin\left\{\frac{\pi}{k}\right\}}{2kb^{\frac{1}{k}}\sin\left\{\frac{\pi}{2} - \frac{\pi}{2k}\right\}} = \frac{2\Gamma\left(\frac{1}{k}\right)\sin\left\{\frac{\pi}{2k}\right\}\cos\left\{\frac{\pi}{2k}\right\}}{2kb^{\frac{1}{k}}\sin\left\{\frac{\pi}{2} - \frac{\pi}{2k}\right\}}.$$

Thus, as

$$\sin\left\{\frac{\pi}{2} - \frac{\pi}{2k}\right\} = \sin\left\{\frac{\pi}{2}\right\}\cos\left\{\frac{\pi}{2k}\right\} - \cos\left\{\frac{\pi}{2}\right\}\sin\left\{\frac{\pi}{2k}\right\} = \cos\left\{\frac{\pi}{2k}\right\},$$

we arrive at

$$\int_0^\infty \sin\left(bx^k\right)dx = \frac{\Gamma\left(\frac{1}{k}\right)\sin\left\{\frac{\pi}{2k}\right\}}{kb^{\frac{1}{k}}}, b > 0, k > 1. \qquad (4.3.10)$$

For example, if $b = 1$ and $k = 3$ then (4.3.10) says that

$$\int_0^\infty \sin\left(x^3\right)dx = \frac{\Gamma\left(\frac{1}{3}\right)\sin\left\{\frac{\pi}{6}\right\}}{3} = \frac{\frac{1}{2}\Gamma\left(\frac{1}{3}\right)}{3} = \frac{1}{6}\Gamma\left(\frac{1}{3}\right) = 0.446489\ldots.$$

If, in (4.3.9), we use u rather than x as the dummy variable of integration, then

$$\int_0^\infty \frac{\cos\left(bu\right)}{u^p}du = \frac{b^{p-1}\pi}{2\Gamma(p)\cos\left\{\frac{p\pi}{2}\right\}}.$$

As before, let $p = 1 - \frac{1}{k}$ (and then $u = x^k$) and so (skipping a few steps)

$$k\int_0^\infty \cos\left(bx^k\right)dx = \frac{b^{-\frac{1}{k}}\pi}{2\Gamma\left(1 - \frac{1}{k}\right)\cos\left\{\frac{\pi}{2} - \frac{\pi}{2k}\right\}}$$

and thus

$$\int_0^\infty \cos\left(bx^k\right)dx = \frac{\pi}{2kb^{\frac{1}{k}}\Gamma\left(1-\frac{1}{k}\right)\cos\left\{\frac{\pi}{2}-\frac{\pi}{2k}\right\}}.$$

Since

$$\cos\left\{\frac{\pi}{2}-\frac{\pi}{2k}\right\} = \cos\left\{\frac{\pi}{2}\right\}\cos\left\{\frac{\pi}{2k}\right\} + \sin\left\{\frac{\pi}{2}\right\}\sin\left\{\frac{\pi}{2k}\right\} = \sin\left\{\frac{\pi}{2k}\right\},$$

and, since as before

$$\Gamma\left(1-\frac{1}{k}\right) = \frac{\pi}{\Gamma\left(\frac{1}{k}\right)\sin\left\{\frac{\pi}{k}\right\}} = \frac{\pi}{2\Gamma\left(\frac{1}{k}\right)\sin\left\{\frac{\pi}{2k}\right\}\cos\left\{\frac{\pi}{2k}\right\}},$$

we have

$$\int_0^\infty \cos\left(bx^k\right)dx = \frac{\Gamma\left(\frac{1}{k}\right)\cos\left\{\frac{\pi}{2k}\right\}}{kb^{\frac{1}{k}}}, \quad b > 0, \ k > 1. \tag{4.3.11}$$

If $b = 1$ and $k = 3$ then (4.3.11) says that

$$\int_0^\infty \cos\left(x^3\right)dx = \frac{\Gamma\left(\frac{1}{3}\right)\cos\left\{\frac{\pi}{6}\right\}}{3} = \frac{\frac{\sqrt{3}}{2}\Gamma\left(\frac{1}{3}\right)}{3} = \frac{1}{2\sqrt{3}}\Gamma\left(\frac{1}{3}\right) = 0.77334\ldots.$$

Notice that $\int_0^\infty \cos\left(x^3\right)dx \neq \int_0^\infty \sin\left(x^3\right)dx$.

Finally, here's one more example of the reversal of double integration order trick. Recall our earlier starting point in deriving (4.3.9):

$$\int_0^\infty \cos\left(bx\right)e^{-xy}dx = \frac{y}{b^2+y^2}.$$

If we integrate both sides *with respect to y* from 0 to $c \geq 0$ we have

$$\int_0^c\left\{\int_0^\infty \cos\left(bx\right)e^{-xy}dx\right\}dy = \int_0^c\frac{y}{b^2+y^2}dy,$$

or, reversing the order of integration on the left,

$$\int_0^\infty \cos\left(bx\right)\left\{\int_0^c e^{-xy}dy\right\}dx = \int_0^c\frac{y}{b^2+y^2}dy. \tag{4.3.12}$$

Clearly,

$$\int_0^c e^{-xy} dy = \left(-\frac{e^{-xy}}{x} \right)\Big|_0^c = \frac{1 - e^{-cx}}{x}.$$

Now, on the right-hand side of (4.3.12) let $u = b^2 + y^2$, and so

$$\frac{du}{dy} = 2y$$

or,

$$dy = \frac{du}{2y}.$$

Thus,

$$\int_0^c \frac{y}{b^2 + y^2} dy = \int_{b^2}^{b^2+c^2} \frac{y}{u}\left(\frac{du}{2y} \right) = \frac{1}{2}\int_{b^2}^{b^2+c^2} \frac{du}{u} = \frac{1}{2}\ln(u)\Big|_{b^2}^{b^2+c^2} = \frac{1}{2}\ln\left(\frac{b^2 + c^2}{b^2} \right).$$

So, (4.3.12) becomes

$$\int_0^\infty \frac{1 - e^{-cx}}{x}\cos(bx)dx = \frac{1}{2}\ln\left(\frac{b^2 + c^2}{b^2} \right). \qquad (4.3.13)$$

This is obviously correct for $c = 0$ (giving zero on both sides of (4.3.13)). If $b - c = 1$, then (4.3.13) says that

$$\int_0^\infty \frac{1 - e^{-x}}{x}\cos(x)dx = \frac{1}{2}\ln(2) = 0.3465\ldots$$

and MATLAB 'agrees' as *integral(@(x)(1-exp(-x)).*cos(x)./x,0,600)* $= 0.3466\ldots$. If we write (4.3.13) twice, first with $c = r$ and $b = p$, we have

$$\int_0^\infty \frac{1 - e^{-rx}}{x}\cos(px)dx = \frac{1}{2}\ln\left(\frac{p^2 + r^2}{p^2} \right),$$

and then again with $c = s$ and $b = q$, we have

$$\int_0^\infty \frac{1 - e^{-sx}}{x}\cos(qx)dx = \frac{1}{2}\ln\left(\frac{q^2 + s^2}{q^2} \right).$$

Thus,

$$\int_0^\infty \frac{1 - e^{-sx}}{x} \cos(qx)dx - \int_0^\infty \frac{1 - e^{-rx}}{x} \cos(px)dx$$

$$= \frac{1}{2} \ln\left(\frac{q^2 + s^2}{q^2}\right) - \frac{1}{2} \ln\left(\frac{p^2 + r^2}{p^2}\right)$$

or,

$$\int_0^\infty \frac{\cos(qx) - \cos(px)}{x}dx + \int_0^\infty \frac{e^{-rx}\cos(px) - e^{-sx}\cos(qx)}{x}dx = \frac{1}{2} \ln\left(\frac{\frac{q^2+s^2}{q^2}}{\frac{p^2+r^2}{p^2}}\right)$$

$$= \frac{1}{2} \ln\left(\frac{q^2+s^2}{p^2+r^2} \cdot \frac{p^2}{q^2}\right) = \frac{1}{2} \ln\left(\frac{q^2+s^2}{p^2+r^2}\right) + \frac{1}{2} \ln\left(\frac{p^2}{q^2}\right) = \frac{1}{2} \ln\left(\frac{q^2+s^2}{p^2+r^2}\right) + \ln\left(\frac{p}{q}\right).$$

But since the first integral on the left in the last equation is $\ln\left(\frac{p}{q}\right)$ by (3.4.6), we have the result

$$\int_0^\infty \frac{e^{-rx}\cos(px) - e^{-sx}\cos(qx)}{x}dx = \frac{1}{2} \ln\left(\frac{q^2 + s^2}{p^2 + r^2}\right) \qquad (4.3.14)$$

For example, if $r = q = 0$ and $p = s = 1$, then (4.3.14) reduces to

$$\int_0^\infty \frac{\cos(x) - e^{-x}}{x}dx = 0,$$

a result which we'll derive again in Chap. 8, in (8.6.4), in an entirely different way using contour integration.

4.4 The Gamma Function Meets Physics

Back in Chap. 1 (Sect. 1.8) I showed you some examples of integrals occurring in *physical* problems and, as we are now approximately half-way through the book, I think the time is right for another such illustration. In the following analysis you'll see how gamma functions appear in a problem involving the motion of a point mass under the influence of an inverse power force field (*not* necessarily restricted to the inverse-*square* law of gravity). Specifically, imagine a point mass m held *at rest* at $y = a$, in a force field of magnitude $\frac{k}{y^{p+1}}$ directed straight down the y-axis towards the origin, where k and p are both positive constants. (An inverse-square gravitational force field is the special case of $p = 1$, but our analysis will hold for *any* $p > 0$. For the case of $p = 0$, see Challenge Problem C4.7.) Our problem is simple to state: The mass is allowed to move at time $t = 0$ and so, of course, it begins accelerating down the y-axis towards the origin. At what time $t = T$ does the mass arrive at the origin?

From Mr. Newton we have the observation 'force equals mass times acceleration' or, in math,[1]

$$F = m\frac{d^2y}{dt^2}.$$

Now, if we denote the speed of the mass by v then

$$v = \frac{dy}{dt}$$

and so

$$\frac{dv}{dt} = \frac{d^2y}{dt^2}$$

which says (invoking the amazingly useful chain-rule from calculus) that

$$F = m\frac{dv}{dt} = m\left(\frac{dv}{dy}\right)\left(\frac{dy}{dt}\right) = mv\frac{dv}{dy}.$$

So,

$$-\frac{k}{y^{p+1}} = mv\frac{dv}{dy}, \tag{4.4.1}$$

where we use the minus sign because the force field is operating in the negative direction, down the y-axis towards the origin. This differential equation is easily integrated, as the variables are separable:

$$-\frac{k\,dy}{m\,y^{p+1}} = v\,dv.$$

Now, integrating indefinitely,

$$\frac{1}{2}v^2 = -\frac{k}{m}\int\frac{dy}{y^{p+1}} = -\frac{k}{m}\int y^{-p-1}dy = \left(-\frac{k}{m}\right)\left(\frac{y^{-p}}{-p}\right) + C = \frac{k}{mpy^p} + C$$

---

[1]Newton didn't actually say this (what is known as the second law of motion), but rather the much more profound 'force is the rate of change of momentum.' If the mass doesn't change with time (as is the case in our problem here) then the above is okay, but if you want to study rocket physics (the mass of a rocket decreases as it burns fuel and ejects the combustion products) then you have to use what Newton really said.

or,

$$v^2 = \frac{2k}{mpy^p} + 2C$$

where C is the arbitrary constant of integration. Using the initial condition $v = 0$ when $y = a$, we see that

$$C = -\frac{k}{mpa^p}.$$

So,

$$v^2 = \frac{2k}{mpy^p} - \frac{2k}{mpa^p} = \frac{2k}{mp}\left(\frac{1}{y^p} - \frac{1}{a^p}\right).$$

Thus,

$$\left(\frac{dy}{dt}\right)^2 = \frac{2k}{mp}\left(\frac{1}{y^p} - \frac{1}{a^p}\right)$$

or, solving[2] for the differential dt,

$$dt = -\sqrt{\frac{mp}{2k}}\left(\frac{dy}{\sqrt{\frac{1}{y^p} - \frac{1}{a^p}}}\right) = -\sqrt{\frac{mp}{2k}}\frac{dy}{\sqrt{\frac{a^p - y^p}{a^p\,y^p}}} = -\sqrt{\frac{mp}{2k}}\frac{a^{\frac{p}{2}}}{\sqrt{\left(\frac{a}{y}\right)^p - 1}}dy.$$

Integrating t from 0 to T on the left (which means we integrate y from a to 0 on the right),

$$\int_0^T dt = T = -a^{\frac{p}{2}}\sqrt{\frac{mp}{2k}}\int_a^0 \frac{dy}{\sqrt{\left(\frac{a}{y}\right)^p - 1}}.$$

Changing variable to $u = \frac{y}{a}$ (and so $dy = a\,du$), this becomes

$$T = -a^{\frac{p}{2}}\sqrt{\frac{mp}{2k}}\int_1^0 \frac{a}{\sqrt{\frac{1}{u^p} - 1}}du = -a^{\left(\frac{p}{2}\right)+1}\sqrt{\frac{mp}{2k}}\int_1^0 \frac{u^{p/2}}{\sqrt{1 - u^p}}du$$

---

[2]Notice that dt is written with a *minus* sign. We do that because we *physically* know $\frac{dy}{dt} < 0$ (the point mass is moving downward towards the origin) and so when we solve for $\frac{dy}{dt}$ we use the *negative* square root.

or,

$$T = a^{\left(\frac{p}{2}\right)+1} \sqrt{\frac{mp}{2k}} \int_0^1 \frac{u^{p/2}}{\sqrt{1-u^p}} du.$$

Now, let

$$x = u^p$$

and so

$$\frac{dx}{du} = pu^{p-1} = \frac{pu^p}{u} = \frac{px}{x^{\frac{1}{p}}}$$

or,

$$du = \frac{x^{\frac{1}{p}}}{px} dx.$$

Thus,

$$T = a^{\left(\frac{p}{2}\right)+1} \sqrt{\frac{mp}{2k}} \int_0^1 \frac{x^{\frac{1}{2}}}{\sqrt{1-x}} \left(\frac{x^{\frac{1}{p}}}{px}\right) dx = a^{\left(\frac{p}{2}\right)+1} \sqrt{\frac{mp}{2k}} \left(\frac{1}{p}\right) \int_0^1 \frac{x^{\frac{1}{p}-\frac{1}{2}}}{\sqrt{1-x}} dx$$

$$= a^{\left(\frac{p}{2}\right)+1} \sqrt{\frac{m}{2kp}} \int_0^1 x^{\frac{1}{p}-\frac{1}{2}}(1-x)^{-\frac{1}{2}} dx.$$

This last integral has the form of the beta function integral in (4.2.1), with

$$m - 1 = \frac{1}{p} - \frac{1}{2}$$

and

$$n - 1 = -\frac{1}{2}.$$

That is,

$$m = \frac{1}{p} + \frac{1}{2}$$

and

$$n = \frac{1}{2}.$$

So,

$$T = a^{\left(\frac{p}{2}\right)+1}\sqrt{\frac{m}{2kp}}B\left(\frac{1}{p}+\frac{1}{2},\frac{1}{2}\right)$$

and since, from (4.2.5),

$$B\left(\frac{1}{p}+\frac{1}{2},\frac{1}{2}\right) = \frac{\Gamma\left(\frac{1}{p}+\frac{1}{2}\right)\Gamma\left(\frac{1}{2}\right)}{\Gamma\left(\frac{1}{p}+1\right)} = \frac{\Gamma\left(\frac{1}{p}+\frac{1}{2}\right)\sqrt{\pi}}{\Gamma\left(\frac{1}{p}+1\right)}$$

we at last have our answer:

$$T = a^{\left(\frac{p}{2}\right)+1}\sqrt{\frac{m\pi}{2kp}}\left\{\frac{\Gamma\left(\frac{1}{p}+\frac{1}{2}\right)}{\Gamma\left(\frac{1}{p}+1\right)}\right\}, \quad p > 0. \tag{4.4.2}$$

A final observation: the mass arrives at the origin with infinite speed because the force field becomes arbitrarily large as the mass approaches y = 0. That means there is some point on the positive y-axis where the speed of the mass exceeds that of light which, it has been known since Einstein's special theory of relativity, is impossible. This is a nice freshman math/physics textbook analysis in *classical* physics, but it is not relativistically correct.

4.5   Gauss 'Meets' Euler

The title of this brief, final section of the chapter is, of course, a 'teaser,' as the year Euler died (1783) in Russia the genius Carl Friedrich Gauss (1777–1855) was finishing-up his kindergarten year in Germany. Gauss was a child prodigy, yes, but it seems unlikely that Euler would have had much to say (on mathematics, anyway) to a 5 year old. Matters were quite different 14 years later, however, when a still teenage Gauss read a posthumously published (1786) paper by Euler, describing his discovery that if $A = \int_0^1 \frac{dx}{\sqrt{1-x^4}}$ and if $B = \int_0^1 \frac{x^2 \, dx}{\sqrt{1-x^4}}$ then $AB = \frac{\pi}{4}$. Gauss thought this to be a remarkable result, and it fascinated him all his life.[3]

---

[3]Gauss kept a mathematical diary during his younger years, and from it we read that he learned $AB = \frac{\pi}{4}$ in January 1797, a few months before his 20th birthday.

Before Euler and his development of the gamma function, the A and B integrals would have been impossible to evaluate other than numerically, but *with* the gamma function, its functional equation, and the reflection formula, the derivation of $AB = \frac{\pi}{4}$ reduces to being a 'routine' exercise. So, as the final example of this chapter to tie these concepts together, here's how to show $AB = \frac{\pi}{4}$.

In the A integral make the change of variable $x^2 = \sin(\theta)$. Then

$$2x\frac{dx}{d\theta} = \cos(\theta)$$

and so

$$dx = \frac{\cos(\theta)}{2x}d\theta = \frac{\cos(\theta)}{2\sqrt{\sin(\theta)}}d\theta.$$

Thus,

$$A = \int_0^{\pi/2} \frac{\frac{\cos(\theta)}{2\sqrt{\sin(\theta)}}d\theta}{\sqrt{1-\sin^2(\theta)}} = \frac{1}{2}\int_0^{\pi/2}\frac{\frac{\cos(\theta)}{\sqrt{\sin(\theta)}}}{\cos(\theta)}d\theta$$

or,

$$A = \frac{1}{2}\int_0^{\pi/2}\frac{d\theta}{\sqrt{\sin(\theta)}}.$$

This integral is one we've already done—see (4.2.14)—and we have

$$A = \left(\frac{1}{2}\right)\frac{\Gamma^2\left(\frac{1}{4}\right)}{2\sqrt{2\pi}}. \tag{4.5.1}$$

In the B integral make the change of variable $x^2 = \cos(\theta)$. Then

$$2x\frac{dx}{d\theta} = -\sin(\theta)$$

and so

$$dx = -\frac{\sin(\theta)}{2x}d\theta = -\frac{\sin(\theta)}{2\sqrt{\cos(\theta)}}d\theta.$$

Thus,

$$B = \int_{\pi/2}^{0} \frac{\cos(\theta)\left[-\frac{\sin(\theta)}{2\sqrt{\cos(\theta)}} \, d\theta\right]}{\sqrt{1-\cos^2(\theta)}} = \frac{1}{2}\int_{0}^{\pi/2} \frac{\sqrt{\cos(\theta)}\sin(\theta)}{\sin(\theta)} \, d\theta$$

or,

$$B = \frac{1}{2}\int_{0}^{\pi/2} \sqrt{\cos(\theta)} \, d\theta.$$

This integral is one we've already done—see (4.2.12)—and we have

$$B = \left(\frac{1}{2}\right)\frac{\Gamma\left(\frac{3}{4}\right)\Gamma\left(\frac{1}{2}\right)}{2\Gamma\left(\frac{5}{4}\right)}. \tag{4.5.2}$$

Combining (4.5.1) and (4.5.2), we arrive at

$$AB = \left(\frac{1}{4}\right)\left\{\frac{\Gamma^2\left(\frac{1}{4}\right)\Gamma\left(\frac{3}{4}\right)\Gamma\left(\frac{1}{2}\right)}{2\sqrt{2\pi}2\Gamma\left(\frac{5}{4}\right)}\right\}, \tag{4.5.3}$$

a rather complicated-looking expression! For (4.5.3) to reduce to Euler's result, all the gamma functions in the curly brackets have to combine with the $2\sqrt{2\pi}$ factor to give π. Can we show that is, indeed, what happens? Yes, and it's not at all hard to do with the aid of the functional equation of the gamma function, and the reflection formula. Here's how it all goes.

Setting m $= \frac{1}{4}$ in the reflection formula of (4.2.16), we have

$$\Gamma\left(\frac{1}{4}\right)\Gamma\left(\frac{3}{4}\right) = \frac{\pi}{\sin\left(\frac{\pi}{4}\right)} = \frac{\pi}{1/\sqrt{2}} = \pi\sqrt{2}$$

and so

$$\Gamma\left(\frac{1}{4}\right) = \frac{\pi\sqrt{2}}{\Gamma\left(\frac{3}{4}\right)}. \tag{4.5.4}$$

Also, from the functional equation of (4.1.2) we can write

$$\Gamma\left(\frac{5}{4}\right) = \Gamma\left(\frac{1}{4}+1\right) = \frac{1}{4}\Gamma\left(\frac{1}{4}\right). \tag{4.5.5}$$

So, putting (4.5.4) and (4.5.5) into (4.5.3), and recalling that $\Gamma\left(\frac{1}{2}\right) = \sqrt{\pi}$, we have

$$\frac{\Gamma^2\left(\frac{1}{4}\right)\Gamma\left(\frac{3}{4}\right)\Gamma\left(\frac{1}{2}\right)}{2\sqrt{2\pi}2\Gamma\left(\frac{5}{4}\right)} = \frac{\Gamma^2\left(\frac{1}{4}\right)\Gamma\left(\frac{3}{4}\right)\sqrt{\pi}}{4\sqrt{2\pi}\frac{1}{4}\Gamma\left(\frac{1}{4}\right)} = \frac{\Gamma\left(\frac{1}{4}\right)\Gamma\left(\frac{3}{4}\right)}{\sqrt{2}} = \left(\frac{1}{\sqrt{2}}\right)\left(\frac{\pi\sqrt{2}}{\Gamma\left(\frac{3}{4}\right)}\right)\Gamma\left(\frac{3}{4}\right) = \pi,$$

and Euler is vindicated.

4.6 Challenge Problems

(C4.1) Find

$$I(n) = \int_0^1 \left(1 - \sqrt{x}\right)^n dx$$

and, in particular, use your general result to show that

$$I(9) = \int_0^1 \left(1 - \sqrt{x}\right)^9 dx = \frac{1}{55}.$$

(C4.2) Prove that

$$\int_0^1 x^m \ln^n(x) dx = (-1)^n \frac{n!}{(m+1)^{n+1}}.$$

We'll use this result later, in Chap. 6, to do some really impressive, historically important integrals. You'll recognize this as a generalization of (3.5.5), which is for the special case of $n = 1$ in our result here.

(C4.3) Show that the integral of $x^a y^b$ over the triangular region defined by the x and y axes, and the line $x + y = 1$, is $\frac{a!b!}{(a+b+2)!}$, where a and b are both non-negative constants.

(C4.4) Use (4.3.2) to evaluate $\int_0^\infty \frac{\sin(x)}{\sqrt{x}} dx$ and (4.3.9) to evaluate $\int_0^\infty \frac{\cos(x)}{\sqrt{x}} dx$. These are famous integrals, which we'll later re-derive in an entirely different way in (7.5.2). Also, use (4.3.10) and (4.3.11) to evaluate $\int_0^\infty \sin(x^2) dx$ and $\int_0^\infty \cos(x^2) dx$. These are also famous integrals, and we'll re-derive them later in an entirely different way in (7.2.1) and (7.2.2). (Notice that while we showed $\int_0^\infty \cos(x^3) dx \neq \int_0^\infty \sin(x^3) dx$, $\int_0^\infty \cos(x^2) dx$ *does* equal $\int_0^\infty \sin(x^2) dx$.)

(C4.5) Use the double-integration reversal trick we used to derive (4.3.13) to evaluate (for b > 0, c > 0) $\int_0^\infty \frac{\sin(bx)}{x} e^{-cx} dx$. Hint: start with the integral we used to begin the derivation of (4.3.2), $\int_0^\infty \sin(bx) e^{-xy} dx = \frac{b}{b^2+y^2}$, and then integrate both sides with respect to y from c to infinity. Notice that the result is a generalization of Dirichlet's integral in (3.2.1), to which it reduces as we let $c \to 0$.

(C4.6) Evaluate the integral $\int_0^\infty \frac{x^a}{(1+x^b)^c} dx$ in terms of gamma functions, where a, b, and c are constants such that $a > -1$, $b > 0$, and $bc > a + 1$, and use your formula to calculate the value of $\int_0^\infty \frac{x\sqrt{x}}{(1+x)^3} dx$. Hint: make the change-of-variable $y = x^b$ and then recall the zero-to-infinity integral form of the beta function that we derived just before (4.2.15): $B(m, n) = \int_0^\infty \frac{y^{m-1}}{(1+y)^{m+n}} dy$.

(C4.7) Notice that T in (4.4.2) is undefined for the $p = 0$ case (the case of an inverse *first* power force field). See if you can find T for the $p = 0$ case. Hint: Set $p = 0$ in (4.4.1) and *then* integrate. You'll find (3.1.8) to be quite useful in evaluating the integral you'll encounter.

(C4.8) How does the gamma function behave at the negative integers? Hint: use the reflection formula in (4.2.16).

(C4.9) Show that $\int_0^\infty \frac{\sin(x^2)}{\sqrt{x}} dx = \frac{\pi}{4\Gamma(\frac{3}{4})\sin(\frac{3\pi}{8})}$ and that $\int_0^\infty \frac{\cos(x^2)}{\sqrt{x}} dx = \frac{\pi}{4\Gamma(\frac{3}{4})\cos(\frac{3\pi}{8})}$. To 'check' these results, the evaluations of the expressions on the right-hand-sides of the equality signs are *pi/(4*gamma(3/4)*sin(3*pi/8))* = 0.69373... and *pi/ (4*gamma(3/4)*cos(3*pi/8))* = 1.67481..., respectively, while the integrals themselves evaluate as (using the Symbolic Math Toolbox, after defining x as a symbolic variable)) *double(int(sin(x^2)/sqrt(x),0,inf))* = 0.69373... and *double(int(cos(x^2)/ sqrt(x),0,inf))* = 1.67481.... Hint: In (4.3.2) and in (4.3.9) set $b = 1$ and make the substitution $x = y^2$.

(C4.10) Show that $\int_0^\infty e^{-x^n} dx = \Gamma\left(\frac{n+1}{n}\right)$. Hint: Review the derivation of (4.1.4).

(C4.11) Show that $\int_0^\infty \left(e^{-ax} - e^{-bx}\right) x^{n-1} dx = \left(\frac{1}{a^n} - \frac{1}{b^n}\right)\Gamma(n)$. Notice that the left-hand-side reduces to the left-hand-side of (3.3.3) for $n = 0$, but the right-hand-side becomes (shall we say) 'difficult.' The right-hand-side of (3.3.3), fortunately, tells what happens at $n = 0$. Hint: Start with (4.1.1) and make the obvious change of variables.

(C4.12) Consider the class of integrals $I_{2k} = \int_{-\infty}^\infty x^{2k} e^{-x^2} dx$. By making the change of variable $y = x^2$, show that $I_{2k} = \Gamma\left(k + \frac{1}{2}\right)$. Hint: A little algebra, and (4.1.1), are all you'll need. Compare with (2.3.9)—does this give you a formula for $\Gamma\left(k + \frac{1}{2}\right)$?

Chapter 5
Using Power Series to Evaluate Integrals

5.1 Catalan's Constant

To start this chapter, here's a very simple illustration of the power series technique for the calculation of integrals, in this case giving us what is called *Catalan's constant* (mentioned in the original Preface) and written as $G = 0.9159655 \ldots$. The power series expansion of $\tan^{-1}(x)$, for $|x| \leq 1$, is our starting point, and it can be found as follows. The key idea is to write

$$\int_0^x \frac{dy}{1+y^2} = \tan^{-1}(y)\Big|_0^x = \tan^{-1}(x).$$

Then, by long division of the integrand, we have

$$\tan^{-1}(x) = \int_0^x \left(1 - y^2 + y^4 - y^6 + y^8 - \ldots\right) dy$$

$$= \left(y - \frac{y^3}{3} + \frac{y^5}{5} - \frac{y^7}{7} + \frac{y^9}{9} - \ldots\right)\Big|_0^x = x - \frac{x^3}{3} + \frac{x^5}{5} - \frac{x^7}{7} + \ldots$$

and so, for $|x| \leq 1$

$$\frac{\tan^{-1}(x)}{x} = 1 - \frac{x^2}{3} + \frac{x^4}{5} - \frac{x^6}{7} + \ldots$$

and therefore, integrating term-by-term, we have

© Springer Nature Switzerland AG 2020
P. J. Nahin, *Inside Interesting Integrals*, Undergraduate Lecture Notes in Physics,
https://doi.org/10.1007/978-3-030-43788-6_5

$$\int_0^1 \frac{\tan^{-1}(x)}{x} dx = \left[x - \frac{x^3}{3^2} + \frac{x^5}{5^2} - \frac{x^7}{7^2} + \cdots \right]\Big|_0^1$$

$$= \frac{1}{1^2} - \frac{1}{3^2} + \frac{1}{5^2} - \frac{1}{7^2} + \cdots = 0.9159655\ldots..$$

That is,

$$\int_0^1 \frac{\tan^{-1}(x)}{x} dx = G. \tag{5.1.1}$$

We check our calculation with *integral(@(x)atan(x)./x,0,1)* = 0.9159655 . . . .

We can use power series to show that an entirely different integral is also equal to G. Specifically, let's take a look at the following integral (mentioned in the new Preface):

$$\int_1^\infty \frac{\ln(x)}{x^2 + 1} dx.$$

We can write the 'bottom-half' of the integrand as

$$\frac{1}{x^2 + 1} = \frac{1}{x^2 \left(1 + \frac{1}{x^2} \right)} = \frac{1}{x^2} \left[1 - \frac{1}{x^2} + \frac{1}{x^4} - \frac{1}{x^6} + \cdots \right]$$

where the endless factor in the brackets follows from long division. Thus,

$$\int_1^\infty \frac{\ln(x)}{x^2 + 1} dx = \int_1^\infty \left\{ \frac{\ln(x)}{x^2} - \frac{\ln(x)}{x^4} + \frac{\ln(x)}{x^6} - \cdots \right\} dx.$$

We see that all the integrals on the right are of the general form

$$\int_1^\infty \frac{\ln(x)}{x^k} dx, \quad k \text{ an even integer} \geq 2,$$

an integral that is easily done by parts. Letting u = ln(x)—and so $\frac{du}{dx} = \frac{1}{x}$ which says du = $\frac{dx}{x}$—and letting dv = $\frac{1}{x^k} dx$ and so v = $\frac{x^{-k+1}}{-k+1} = -\frac{1}{k-1} \left(\frac{1}{x^{k-1}} \right)$, we then have

$$\int_1^\infty \frac{\ln(x)}{x^k} dx = -\left\{ \frac{\ln(x)}{k-1} \left(\frac{1}{x^{k-1}} \right) \right\}\Big|_1^\infty + \frac{1}{k-1} \int_1^\infty \left(\frac{1}{x^{k-1}} \right) \frac{dx}{x} = \frac{1}{k-1} \int_1^\infty \frac{dx}{x^k}$$

$$= \frac{1}{k-1} \left\{ \frac{x^{-k+1}}{-k+1} \right\}\Big|_1^\infty = -\left\{ \frac{1}{(k-1)^2 x^{k-1}} \right\}\Big|_1^\infty = \frac{1}{(k-1)^2}.$$

So,

$$\int_1^\infty \frac{\ln(x)}{x^2+1}\,dx = \frac{1}{(2-1)^2} - \frac{1}{(4-1)^2} + \frac{1}{(6-1)^2} - \cdots = \frac{1}{1^2} - \frac{1}{3^2} + \frac{1}{5^2} - \cdots$$

or,

$$\int_1^\infty \frac{\ln(x)}{x^2+1}\,dx = G. \tag{5.1.2}$$

And so it is, as *integral(@(x)log(x)./(x.^2+1),1,inf)* =0.9159656.... .
We can combine this result with an earlier one to now calculate

$$\int_0^\infty \frac{\ln(x+1)}{x^2+1}\,dx,$$

which we can write as

$$\int_0^1 \frac{\ln(x+1)}{x^2+1}\,dx + \int_1^\infty \frac{\ln(x+1)}{x^2+1}\,dx.$$

We've already done the first integral on the right, in our result (2.2.4) that found it equal to $\frac{\pi}{8}\ln(2)$, and so we have

$$\int_0^\infty \frac{\ln(x+1)}{x^4+1}\,dx = \frac{\pi}{8}\ln(2) + \int_1^\infty \frac{\ln(x+1)}{x^2+1}\,dx = \frac{\pi}{8}\ln(2) + \int_1^\infty \frac{\ln\left\{x\left(1+\frac{1}{x}\right)\right\}}{x^2+1}\,dx$$

$$= \frac{\pi}{8}\ln(2) + \int_1^\infty \frac{\ln(x)}{x^2+1}\,dx + \int_1^\infty \frac{\ln\left(1+\frac{1}{x}\right)}{x^2+1}\,dx.$$

The first integral on the right is (5.1.2) and so

$$\int_0^\infty \frac{\ln(x+1)}{x^2+1}\,dx = \frac{\pi}{8}\ln(2) + G + \int_1^\infty \frac{\ln\left(1+\frac{1}{x}\right)}{x^2+1}\,dx,$$

and in the remaining integral on right make the change of variable $u = \frac{1}{x}$ (and so $\frac{du}{dx} = -\frac{1}{x^2}$ or, $dx = -x^2 du = -\frac{du}{u^2}$). Thus,

$$\int_1^\infty \frac{\ln\left(1+\frac{1}{x}\right)}{x^2+1}\,dx = \int_1^0 \frac{\ln(1+u)}{\frac{1}{u^2}+1}\left(-\frac{du}{u^2}\right) = \int_0^1 \frac{\ln(1+u)}{u^2+1}\,du,$$

but this is just (2.2.4) again! So,

$$\int_0^\infty \frac{\ln(x+1)}{x^2+1}\,dx = \frac{\pi}{8}\ln(2) + G + \frac{\pi}{8}\ln(2)$$

or, finally,

$$\int_0^\infty \frac{\ln(x+1)}{x^2+1}\,dx = \frac{\pi}{4}\ln(2) + G. \tag{5.1.3}$$

This is equal to 1.46036 ... while *integral(@(x)log(x+1)./(x.^2+1),0,inf)* = 1.46036. . . . .

Returning to the main theme of this chapter, as our last example of the use of power series in this section, consider now the integral

$$I = \int_0^\pi \frac{\theta\,\sin(\theta)}{a + b\,\cos^2(\theta)}\,d\theta$$

where a > b > 0. Then, expanding the integrand as a power series,

$$I = \frac{1}{a}\int_0^\pi \theta\sin(\theta)\left\{ 1 - \left(\frac{b}{a}\right)\cos^2(\theta) + \left(\frac{b}{a}\right)^2\cos^4(\theta) - \left(\frac{b}{a}\right)^3\cos^6(\theta) + \ldots \right\}d\theta$$

$$= \frac{1}{a}\left[\int_0^\pi \theta\sin(\theta)d\theta - \left(\frac{b}{a}\right)\int_0^\pi \theta\sin(\theta)\cos^2(\theta)d\theta + \left(\frac{b}{a}\right)^2\int_0^\pi \theta\sin(\theta)\cos^4(\theta)d\theta - \ldots \right].$$

Looking at the general form of the integrals on the right, we see that for n = 0, 1, 2, ... we have

$$\int_0^\pi \theta\sin(\theta)\cos^{2n}(\theta)d\theta.$$

This is easily integrated by parts to give (use $u = \theta$ and $dv = \sin(\theta)\cos^{2n}(\theta)d\theta$)

$$\int_0^\pi \theta\sin(\theta)\cos^{2n}(\theta)d\theta = \left\{ -\frac{\theta\cos^{2n+1}(\theta)}{2n+1} \right\}\Big|_0^\pi + \frac{1}{2n+1}\int_0^\pi \cos^{2n+1}(\theta)d\theta = \frac{\pi}{2n+1}$$

because it is clear from symmetry (notice that 2n + 1 is *odd* for all n) that

$$\int_0^\pi \cos^{2n+1}(\theta)d\theta = 0.$$

So,

$$I = \frac{\pi}{a}\left[1 - \frac{1}{3}\left(\frac{b}{a}\right) + \frac{1}{5}\left(\frac{b}{a}\right)^2 - \frac{1}{7}\left(\frac{b}{a}\right)^3 + \cdots\right]$$

$$= \frac{\pi}{a}\frac{\sqrt{a}}{\sqrt{b}}\left[\frac{\sqrt{b}}{\sqrt{a}} - \frac{1}{3}\left(\frac{b\sqrt{b}}{a\sqrt{a}}\right) + \frac{1}{5}\left(\frac{b^2\sqrt{b}}{a^2\sqrt{a}}\right) - \frac{1}{7}\left(\frac{b^3\sqrt{b}}{a^3\sqrt{a}}\right) + \cdots\right]$$

$$= \frac{\pi}{\sqrt{ab}}\left[\left(\frac{b}{a}\right)^{1/2} - \frac{1}{3}\left(\frac{b}{a}\right)^{\frac{3}{2}} + \frac{1}{5}\left(\frac{b}{a}\right)^{\frac{5}{2}} - \frac{1}{7}\left(\frac{b}{a}\right)^{\frac{7}{2}} + \cdots\right]$$

or, if you recall the power series expansion for $\tan^{-1}(x)$ from the start of this section and set $x = \left(\frac{b}{a}\right)^{1/2}$, we see that

$$\int_0^\pi \frac{\theta \, \sin(\theta)}{a + b \, \cos^2(\theta)} \, d\theta = \frac{\pi}{\sqrt{ab}} \tan^{-1}\left(\sqrt{\frac{b}{a}}\right), \quad a > b. \qquad (5.1.4)$$

For example, if $a = 3$ and $b = 1$ then

$$\int_0^\pi \frac{x \, \sin(x)}{a + b \, \cos^2(x)} \, dx = \frac{\pi}{\sqrt{3}} \tan^{-1}\left(\frac{1}{\sqrt{3}}\right) = 0.94970312\ldots$$

while *integral(@(x)x.*sin(x)./((3+cos(x).^2)),0,pi)* = 0.94970312. . . .

5.2   Power Series for the Log Function

Using essentially the same approach that we just used to study Catalan's constant, we can evaluate

$$\int_0^1 \frac{\ln(1+x)}{x} \, dx \, .$$

We first get a power series expansion for $\ln(x+1)$, for $|x| < 1$, by writing

$$\int_0^x \frac{dy}{1+y} = \ln(1+y)\Big|_0^x = \ln(1+x) = \int_0^x (1 - y + y^2 - y^3 + \ldots)dy$$

$$= \left(y - \frac{y^2}{2} + \frac{y^3}{3} - \frac{y^4}{4} + \ldots\right)\Big|_0^x$$

or,

$$\ln(1+x) = x - \frac{x^2}{2} + \frac{x^3}{3} - \frac{x^4}{4} + \dots, \ |x| < 1.$$

(You may recall that I previously cited this series (without derivation) in a footnote in Chap. 2.)

Thus,

$$\int_0^1 \frac{\ln(1+x)}{x} dx = \int_0^1 \left(1 - \frac{x}{2} + \frac{x^2}{3} - \frac{x^3}{4} + \dots\right) dx$$

$$= \left(x - \frac{x^2}{2^2} + \frac{x^3}{3^2} - \frac{x^4}{4^2} + \dots\right)\Big|_0^1 = 1 - \frac{1}{2^2} + \frac{1}{3^2} - \frac{1}{4^2} + \dots$$

In 1734 Euler showed that[1]

$$\sum_{k=1}^{\infty} \frac{1}{k^2} = 1 + \frac{1}{2^2} + \frac{1}{3^2} + \frac{1}{4^2} + \dots = \frac{\pi^2}{6},$$

which is the sum of the reciprocals squared of *all* the positive integers. If we write just the sum of the reciprocals squared of all the *even* positive integers we have

$$\sum_{k=1}^{\infty} \frac{1}{(2k)^2} = \frac{1}{4} \sum_{k=1}^{\infty} \frac{1}{k^2} = \frac{\pi^2}{24}.$$

So, the sum of the reciprocals squared of all the *odd* positive integers must be

$$\frac{\pi^2}{6} - \frac{\pi^2}{24} = \frac{\pi^2}{8}.$$

If you look at the result we got for our integral, you'll see it is the sum of the reciprocals squared of all the odd integers *minus* the sum of the reciprocals squared of all the even integers (that is, $\frac{\pi^2}{8} - \frac{\pi^2}{24} = \frac{\pi^2}{12}$) and so

$$\int_0^1 \frac{\ln(1+x)}{x} dx = \frac{\pi^2}{12}. \tag{5.2.1}$$

---

[1] I mentioned this sum earlier, in the Introduction (Sect. 1.3). Later in this book, in Chap. 7, I'll show you a beautiful way that Euler—using integrals, of course!—derived his famous and very important result. What I'll show you there is *not* the way Euler originally did it, but it has the distinct virtue of being perfectly correct while Euler's original approach (while that of a genius—for details, see my book *An Imaginary Tale*, Princeton 2010, pp. 148–149) is open to some serious mathematical concerns.

This is equal to 0.822467..., and MATLAB agrees: *integral(@(x)log(1+x)./x,0,1) = 0.822467....* It is a trivial modification to show (I'll let *you* do this!—see also Challenge Problem 7) that

$$\int_0^1 \frac{\ln(1-x)}{x}\,dx = -\frac{\pi^2}{6}. \tag{5.2.2}$$

This is $-1.644934\ldots$, and MATLAB again agrees: *integral(@(x)log(1-x)./x,0,1) = -1.644934....*

Okay, what can be done with *this* integral:

$$\int_0^1 \frac{1}{x}\ln\left\{\left(\frac{1+x}{1-x}\right)^2\right\}dx = ?$$

You, of course, suspect that the answer is: *use power series*! To start, write

$$\int_0^1 \frac{1}{x}\ln\left\{\left(\frac{1+x}{1-x}\right)^2\right\}dx = 2\int_0^1 \frac{1}{x}\{\ln(1+x) - \ln(1-x)\}dx.$$

Earlier we wrote out the power series expansion for $\ln(1 + x)$, and so we can immediately get the expansion for $\ln(1 - x)$ by simply replacing x with –x. Our integral becomes

$$2\int_0^1 \frac{1}{x}\left\{\left(x - \frac{x^2}{2} + \frac{x^3}{3} - \frac{x^4}{4} + \ldots\right) - \left(-x - \frac{x^2}{2} - \frac{x^3}{3} - \frac{x^4}{4} - \ldots\right)\right\}dx$$

$$= 2\int_0^1 \frac{1}{x}\left(2x + 2\frac{1}{3}x^3 + 2\frac{1}{5}x^5 + \ldots\right)dx = 4\int_0^1\left(1 + \frac{1}{3}x^2 + \frac{1}{5}x^4 + \ldots\right)dx$$

$$= 4\left(x + \frac{1}{9}x^3 + \frac{1}{25}x^5 + \ldots\right)\Big|_0^1 = 4\left(\frac{1}{1^2} + \frac{1}{3^2} + \frac{1}{5^2} + \ldots\right) = 4\left(\frac{\pi^2}{8}\right)$$

or, finally,

$$\int_0^1 \frac{1}{x}\ln\left\{\left(\frac{1+x}{1-x}\right)^2\right\}dx = \frac{\pi^2}{2}. \tag{5.2.3}$$

This is equal to 4.934802..., and MATLAB agrees: *integral(@(x)log(((1+x)./(1-x)).^2)./x,0,1) = 4.934802....*

We can mix trig and log functions using the power series idea to evaluate an integral as nasty-looking as

$$\int_0^{\pi/2} \cot(x)\ln\{\sec(x)\}dx.$$

Let $t = \cos(x)$. Then $\frac{dt}{dx} = -\sin(x)$ or, $dx = -\frac{dt}{\sin(x)}$. So,

$$\int_0^{\pi/2} \cot(x) \ln\{\sec(x)\}dx = \int_0^{\pi/2} \frac{1}{\tan(x)} \ln\left\{\frac{1}{\cos(x)}\right\}dx$$

$$= -\int_1^0 \frac{\cos(x)}{\sin(x)} \ln\{\cos(x)\}\left\{-\frac{dt}{\sin(x)}\right\}$$

$$= -\int_0^1 \frac{\cos(x)}{\sin^2(x)} \ln\{\cos(x)\}\,dt = -\int_0^1 \frac{t}{1-t^2} \ln(t)dt.$$

Over the entire interval of integration we can write

$$\frac{t}{1-t^2} = t\{1 + t^2 + t^4 + t^6 + \ldots\}$$

which is easily verified by cross-multiplying. That is,

$$\frac{t}{1-t^2} = t + t^3 + t^5 + t^7 + \ldots = \sum_{n=0}^{\infty} t^{2n+1}.$$

Thus, our integral is

$$-\int_0^1 \left\{\sum_{n=0}^{\infty} t^{2n+1}\right\} \ln(t)dt$$

or, assuming that we can reverse the order of integration and summation,[2] we have

$$\int_0^{\pi/2} \cot(x) \ln\{\sec(x)\}dx = -\sum_{n=0}^{\infty}\left\{\int_0^1 t^{2n+1} \ln(t)dt\right\}.$$

We can do this last integral by parts (or simply look back at (3.5.5)): let $u = \ln(t)$ and $dv = t^{2n+1}\,dt$. Thus $\frac{du}{dt} = \frac{1}{t}$ or, $du = \frac{1}{t}dt$ and $v = \frac{t^{2n+2}}{2n+2}$. So,

$$\int_0^1 t^{2n+1} \ln(t)dt = \left\{\frac{t^{2n+2}}{2n+2} \ln(t)\right\}\Big|_0^1 - \int_0^1 \frac{t^{2n+2}}{2n+2}\left(\frac{1}{t}\right)dt = -\frac{1}{2n+2}\int_0^1 t^{2n+1}\,dt$$

$$= -\frac{1}{2n+2}\left\{\frac{t^{2n+2}}{2n+2}\right\}\Big|_0^1 = -\frac{1}{(2n+2)^2}.$$

So,

---

[2]This reversal is an example of a step where a mathematician would feel obligated to first show uniform convergence before continuing. I, on the other hand, with a complete lack of shame, will just plow ahead and *do* the reversal and then, once I have the 'answer,' will ask *integral* what it 'thinks.'

$$\int_0^{\pi/2} \cot(x) \ln\{\sec(x)\}dx = \sum_{n=0}^{\infty} \frac{1}{(2n+2)^2} = \frac{1}{4}\sum_{n=0}^{\infty}\frac{1}{(n+1)^2} = \frac{1}{4}\sum_{k=1}^{\infty}\frac{1}{k^2}$$

$$= \frac{1}{4}\left(\frac{\pi^2}{6}\right)$$

or,

$$\int_0^{\pi/2} \cot(x)\ln\{\sec(x)\}dx = \frac{\pi^2}{24}. \qquad (5.2.4)$$

This is $0.4112335\ldots$, and MATLAB agrees: *integral(@(x)cot(x).*log(sec(x)),0, pi/2) = 0. 4112335....* .

For one more example of a power series integration involving log functions, let's see what we can do with

$$\int_0^1 \ln(1+x)\ln(1-x)dx.$$

If you look at this for just a bit then you should be able to convince yourself that, by symmetry of the integrand around $x = 0$, we can write

$$\int_0^1 \ln(1+x)\ln(1-x)dx = \frac{1}{2}\int_{-1}^1 \ln(1+x)\ln(1-x)dx.$$

(Or, if you prefer, make the change of variable $u = -x$ and observe that the integral becomes $\int_{-1}^0 \ln(1-u)\ln(1+u)du$. *The integrand is unchanged*. So, $\int_0^1 = \int_{-1}^0$ and thus $\int_{-1}^0 + \int_0^1 = \int_{-1}^1 = 2\int_0^1$, as claimed.) Now, make the change of variable $y = \frac{x+1}{2}$ (and so $dx = 2dy$). Since $x = 2y - 1$ then $1 + x = 2y$ and $1 - x = 1 - (2y - 1) = 2 - 2y$, all of which says

$$\int_0^1 \ln(1+x)\ln(1-x)dx = \frac{1}{2}\int_0^1 \ln(2y)\ln(2-2y)2dy = \int_0^1 \ln(2y)\ln\{2(1-y)\}dy$$

$$= \int_0^1 [\ln(2) + \ln(y)][\ln(2) + \ln(1-y)]dy$$

$$= \int_0^1 \left[\{\ln(2)\}^2 + \ln(2)\{\ln(y) + \ln(1-y)\} + \ln(y)\ln(1-y)\right]dy$$

$$= \{\ln(2)\}^2 + \ln(2)\left\{\int_0^1 \ln(y)dy + \int_0^1 \ln(1-y)dy\right\} + \int_0^1 \ln(y)\ln(1-y)dy.$$

Since

$$\int_0^1 \ln(y)dy = \int_0^1 \ln(1-y)dy,$$

either 'by inspection' or by making the change of variable $u = 1 - y$ in one of the integrals, we have (changing the dummy variable of integration back to x)

$$\int_0^1 \ln(1+x)\ln(1-x)dx = \{\ln(2)\}^2 + 2\ln(2)\int_0^1 \ln(x)dx + \int_0^1 \ln(x)\ln(1-x)dx.$$

Since

$$\int_0^1 \ln(x)dx = \{x\ln(x) - x\}\Big|_0^1 = -1,$$

we have

$$\int_0^1 \ln(1+x)\ln(1-x)dx = \{\ln(2)\}^2 - 2\ln(2) + \int_0^1 \ln(x)\ln(1-x)dx$$

and so all we have left to do is the integral on the right. To do

$$\int_0^1 \ln(x)\ln(1-x)dx$$

we'll use the power series expansion of $\ln(1-x)$. As shown earlier,

$$\ln(1-x) = -x - \frac{x^2}{2} - \frac{x^3}{3} - \frac{x^4}{4} - \dots, \quad |x| < 1,$$

and so we have

$$\int_0^1 \ln(x)\ln(1-x)dx = -\sum_{k=1}^\infty \frac{1}{k}\int_0^1 x^k \ln(x)dx.$$

If you look back at our analysis that that resulted in (5.2.4), you'll see that we've already done the integral on the right. There we showed that

$$\int_0^1 t^{2n+1}\ln(t)dt = -\frac{1}{(2n+2)^2}$$

and so we immediately have (replacing $2n + 1$ with k) the result

$$\int_0^1 x^k \ln (x)dx = -\frac{1}{(k+1)^2}.$$

Thus,

$$\int_0^1 \ln (x) \ln (1-x)dx = \sum_{k=1}^{\infty} \frac{1}{k(k+1)^2} = \sum_{k=1}^{\infty} \left\{ \frac{1}{k(k+1)} - \frac{1}{(k+1)^2} \right\}$$

$$= \sum_{k=1}^{\infty} \left[\left\{ \left(\tfrac{1}{k}\right) - \left(\tfrac{1}{k+1}\right) \right\} - \frac{1}{(k+1)^2} \right]$$

$$= \left(1 + \frac{1}{2} + \frac{1}{3} + \frac{1}{4} + \ldots\right) - \left(\frac{1}{2} + \frac{1}{3} + \frac{1}{4} + \ldots\right) - \sum_{k=1}^{\infty} \frac{1}{(k+1)^2}$$

$$= 1 - \left(\frac{1}{2^2} + \frac{1}{3^2} + \frac{1}{4^2} + \ldots\right) = 1 - \left(\frac{\pi^2}{6} - 1\right) = 2 - \frac{\pi^2}{6}.$$

So, finally,

$$\int_0^1 \ln (1+x) \ln (1-x)dx = \{\ln (2)\}^2 - 2\ln (2) + 2 - \frac{\pi^2}{6}. \tag{5.2.5}$$

This is equal to $-0.550775\ldots$, and in agreement is

$$integral(@(x) \log (1+x). * \log (1-x), 0, 1) = -0.550775\ldots.$$

I'll end this section with the calculation of

$$\int_0^1 \frac{\{\ln (x)\}^2}{1+x^2} dx,$$

an integral that will find an important use later in the book. The trick here is to make the change of variable $x = e^{-y}$, which is equivalent to $y = -\ln (x)$. Thus, $dy = -\frac{1}{x}dx$ and so $dx = -x\, dy = -e^{-y}dy$. So,

$$\int_0^1 \frac{\{\ln (x)\}^2}{1+x^2} dx = \int_{\infty}^0 \frac{y^2}{1+e^{-2y}} \{-e^{-y}dy\}$$

$$= \int_0^{\infty} y^2 (1 - e^{-2y} + e^{-4y} - e^{-6y} + \ldots)e^{-y}dy$$

$$= \int_0^{\infty} y^2 (e^{-y} - e^{-3y} + e^{-5y} - e^{-7y} + \ldots)dy.$$

From standard integration tables (or integration by parts)

$$\int_0^\infty y^2 e^{ay} dy = \frac{e^{ay}}{a}\left(y^2 - \frac{2y}{a} + \frac{2}{a^2}\right)\Big|_0^\infty = -\frac{2}{a^3} > 0$$

because $a < 0$ in every term in our above infinite series integrand. Thus,

$$\int_0^1 \frac{\{\ln(x)\}^2}{1+x^2} dx = 2\left[\frac{1}{1^3} - \frac{1}{3^3} + \frac{1}{5^3} - \frac{1}{7^3} + \cdots\right],$$

but it is known[3] that the alternating series is equal to $\frac{\pi^3}{32}$. So,

$$\int_0^1 \frac{\{\ln(x)\}^2}{1+x^2} dx = \frac{\pi^3}{16}. \tag{5.2.6}$$

This value is 1.937892... and, using *integral* to check: *integral(@(x)log(x).^2./(1+x.^2),0,1)* = 1.937892....

There are two additional comments we can make about this integral. First,

$$\int_0^\infty \frac{\{\ln(x)\}^2}{1+x^2} dx = 2\int_0^1 \frac{\{\ln(x)\}^2}{1+x^2} dx = \frac{\pi^3}{8}.$$

I'll let you fill-in the details (simply notice that $\int_0^\infty = \int_0^1 + \int_1^\infty$ and then make the change of variable $y = \frac{1}{x}$ in the $\int_1^\infty$ integral). And second, look at what we get when we consider the *seemingly* unrelated integral

$$\int_0^{\pi/2} [\ln\{\tan(\theta)\}]^2 d\theta$$

if we make the change of variable $x = \tan(\theta)$. Because

$$\frac{dx}{d\theta} = \frac{\cos^2(\theta) + \sin^2(\theta)}{\cos^2(\theta)} = 1 + \tan^2(\theta)$$

we have

$$d\theta = \frac{1}{1+\tan^2(\theta)} dx = \frac{1}{1+x^2} dx$$

and so

---

[3]See my *Dr. Euler's Fabulous Formula*, Princeton 2011, p. 149 for the derivation of this result using Fourier series.

$$\int_0^{\pi/2} [\ln\{\tan(\theta)\}]^2 d\theta = \int_0^\infty \frac{\{\ln(x)\}^2}{1+x^2} dx.$$

Thus,

$$\int_0^{\pi/2} [\ln\{\tan(\theta)\}]^2 d\theta = \frac{\pi^3}{8}. \tag{5.2.7}$$

This is 3.875784..., and *integral* confirms it: *integral(@(x)log(tan(x)).^2,0,pi/2)* = 3.875784..... (The appearance of pi-*cubed* is worthy of special note.)

5.3   Zeta Function Integrals

In the previous section I mentioned Euler's discovery of

$$\sum_{k=1}^\infty \frac{1}{k^2} = \frac{\pi^2}{6}.$$

This is actually a special case of what is called the *Riemann zeta function*:

$$\zeta(s) = \sum_{k=1}^\infty \frac{1}{k^s},$$

where s is a complex number with a real part greater than 1 to insure that the sum converges.[4] Euler's sum is for s = 2 but, in fact, his method for finding $\zeta(2)$—spoken as *zeta-two*—works for *all* positive, *even* integer values of s; $\zeta(4) = \frac{\pi^4}{90}$, $\zeta(6) = \frac{\pi^6}{945}$, $\zeta(8) = \frac{\pi^8}{9,450}$, and $\zeta(10) = \frac{1,280\pi^{10}}{119,750,400}$. In general, $\zeta(s) = \frac{m}{n}\pi^s$, where m and n are integers, *if* s is an even positive integer. Euler's method fails for all positive *odd* integer values of s, however, and nobody has succeeded in the nearly 300 years since Euler's discovery of $\zeta(2)$ in finding a formula that gives $\zeta(s)$ for even a single *odd* value of s.

Mathematicians can, of course, calculate the *numerical* value of $\zeta(s)$, for any s, as accurately as desired. For example, $\zeta(3) = 1.20205...$, and the values of those dots are now known out to *at least* a hundred *billion* (!) decimal places. But it is a *formula* that mathematicians want, and the discovery of one for $\zeta(3)$ would be an event of supernova-magnitude in the world of mathematics.

---

[4]For s = 1, $\zeta(1)$ is just the harmonic series, which has been known *for centuries* before Euler's day to diverge.

Ironically, it is easy to write $\zeta(s)$, for *any* integer value of s, even or odd, as a double integral. To see this, write the power series expansion of $\frac{x^a y^a}{1-xy}$, where a is (for now) a constant:

$$\frac{x^a y^a}{1-xy} = x^a y^a \left\{ 1 + xy + (xy)^2 + (xy)^3 + (xy)^4 + (xy)^5 + \dots \right\}, \quad |xy| < 1.$$

Then, using this as the integrand of a double integral, we have

$$\int_0^1 \int_0^1 \frac{x^a y^a}{1-xy} \, dx \, dy = \int_0^1 x^a \left\{ \int_0^1 \left\{ y^a + xy^{a+1} + x^2 y^{a+2} + x^3 y^{a+3} + \dots \right\} dy \right\} dx$$

$$= \int_0^1 x^a \left\{ \frac{y^{a+1}}{a+1} + x\frac{y^{a+2}}{a+2} + x^2 \frac{y^{a+3}}{a+3} + x^3 \frac{y^{a+4}}{a+4} + \dots \right\} \Bigg|_0^1 dx$$

$$= \int_0^1 \left\{ \frac{x^a}{a+1} + \frac{x^{a+1}}{a+2} + \frac{x^{a+2}}{a+3} + \frac{x^{a+3}}{a+4} + \dots \right\} dx$$

$$= \left\{ \frac{x^{a+1}}{(a+1)^2} + \frac{x^{a+2}}{(a+2)^2} + \frac{x^{a+3}}{(a+3)^2} + \frac{x^{a+4}}{(a+4)^2} + \dots \right\} \Bigg|_0^1$$

$$= \frac{1}{(a+1)^2} + \frac{1}{(a+2)^2} + \frac{1}{(a+3)^2} + \dots$$

and so

$$\int_0^1 \int_0^1 \frac{x^a y^a}{1-xy} \, dx \, dy = \sum_{n=1}^{\infty} \frac{1}{(n+a)^2}. \tag{5.3.1}$$

Now, if we set $a = 0$ then (5.3.1) obviously reduces to

$$\int_0^1 \int_0^1 \frac{1}{1-xy} \, dx \, dy = \sum_{n=1}^{\infty} \frac{1}{n^2} = \zeta(2),$$

but we can actually do *much* more with (5.3.1) than just $\zeta(2)$. Here's how. Remembering Feynman's favorite trick of differentiating under the integral sign, let's differentiate (5.3.1) with respect to a (which started off as a constant but which we'll now treat as a parameter). Then, on the right-hand side we get

$$-2\sum_{n=1}^{\infty} \frac{1}{(n+a)^3}.$$

On the left-hand side we first write

$$x^a y^a = (xy)^a = e^{\ln\{(xy)^a\}} = e^{a \ln(xy)}$$

and so differentiation of the integral gives

$$\frac{d}{da} \int_0^1 \int_0^1 \frac{x^a y^a}{1-xy} \, dx \, dy = \int_0^1 \int_0^1 \frac{\ln(xy)e^{a \ln(xy)}}{1-xy} \, dx \, dy.$$

Thus,

$$\int_0^1 \int_0^1 \frac{\ln(xy)e^{a \ln(xy)}}{1-xy} \, dx \, dy = -2 \sum_{n=1}^{\infty} \frac{1}{(n+a)^3}.$$

Now, differentiate again, and get

$$\int_0^1 \int_0^1 \frac{\{\ln(xy)\}^2 e^{a \ln(xy)}}{1-xy} \, dx \, dy = (-2)(-3) \sum_{n=1}^{\infty} \frac{1}{(n+a)^4}.$$

Indeed, if we differentiate over and over you should be able to see the pattern:

$$\int_0^1 \int_0^1 \frac{\{\ln(xy)\}^{s-2} e^{a \ln(xy)}}{1-xy} \, dx \, dy = (-2)(-3) \ldots (-\{s-1\}) \sum_{n=1}^{\infty} \frac{1}{(n+a)^s}.$$

So, returning $e^{a \ln(xy)}$ back to $(xy)^a$ we have, for $s \geq 2$,

$$\int_0^1 \int_0^1 \frac{(xy)^a \{\ln(xy)\}^{s-2}}{1-xy} \, dx \, dy = (-1)^s (s-1)! \sum_{n=1}^{\infty} \frac{1}{(n+a)^s}. \tag{5.3.2}$$

Or, for the particularly interesting case of $a = 0$,

$$\zeta(s) = \frac{(-1)^s}{(s-1)!} \int_0^1 \int_0^1 \frac{\{\ln(xy)\}^{s-2}}{1-xy} \, dx \, dy. \tag{5.3.3}$$

For $s = 5$, for example, it is easy to calculate $\zeta(5)$ directly from the sum-definition of the zeta function to get $1.036927\ldots$, while the right-hand-side of (5.3.3) is found with the following MATLAB code (*factor* $= -\frac{1}{24}$ is the value of $\frac{(-1)^5}{(4)!}$):

integral2 $(@(x,y)(\log(x*y)).^3./(1-x.*y),0,1,0,1) * factor = 1.036927\ldots.$

To end this section, I'll now show you a beautiful result that connects the gamma function from Chap. 4 with the zeta function. You'll recall from (4.1.1) that

$$\Gamma(s) = \int_0^\infty e^{-x} x^{s-1} \, dx.$$

So, starting with the integral

$$\int_0^\infty e^{-kx} x^{s-1} \, dx, k > 0,$$

make the change of variable $u = kx$ which gives

$$\int_0^\infty e^{-kx} x^{s-1} \, dx = \int_0^\infty e^{-u} \left(\frac{u}{k}\right)^{s-1} \frac{du}{k} = \frac{1}{k^s} \int_0^\infty e^{-u} u^{s-1} \, du = \frac{\Gamma(s)}{k^s}.$$

Then, summing over both sides, we have

$$\sum_{k=1}^\infty \int_0^\infty e^{-kx} x^{s-1} \, dx = \sum_{k=1}^\infty \frac{\Gamma(s)}{k^s} = \Gamma(s) \sum_{k=1}^\infty \frac{1}{k^s} = \Gamma(s)\zeta(s).$$

Next, assuming that we can interchange the order of summation and integration, we have

$$\Gamma(s)\zeta(s) = \int_0^\infty x^{s-1} \sum_{k=1}^\infty e^{-kx} \, dx.$$

The summation in the integrand is a geometric series, easily calculated to be

$$\sum_{k=1}^\infty e^{-kx} = \frac{1}{e^x - 1},$$

and so we immediately have the amazing

$$\int_0^\infty \frac{x^{s-1}}{e^x - 1} \, dx = \Gamma(s)\zeta(s) \tag{5.3.4}$$

that was discovered in 1859 by none-other than Riemann, himself, the hero of this book. For $s = 4$, for example, (5.3.4) says that (recalling Euler's result $\zeta(4) = \frac{\pi^4}{90}$)

$$\int_0^\infty \frac{x^3}{e^x - 1} \, dx = \Gamma(4)\zeta(4) = (3!)\left(\frac{\pi^4}{90}\right) = \frac{\pi^4}{15}$$

and you'll recall both this integral and its value from the original Preface (see note 7 there). This theoretical value is 6.493939... and MATLAB agrees, as

$$integral\left(@(x)\left(x.^3\right)./\left(exp\left(x\right) - 1\right), 0, inf\right) = 6.493939\ldots.$$

With a little preliminary work we can use the same approach to do an interesting variant of (5.3.4), namely

$$\int_0^\infty \frac{x^{s-1}}{e^x + 1} dx.$$

To set things up, define the two functions

$$u_s = \frac{1}{1^s} + \frac{1}{3^s} + \frac{1}{5^s} + \frac{1}{7^s} + \cdots$$

and

$$v_s = \frac{1}{1^s} - \frac{1}{2^s} + \frac{1}{3^s} - \frac{1}{4^s} + \cdots$$

and so

$$\zeta(s) = u_s + \frac{1}{2^s} + \frac{1}{4^s} + \frac{1}{6^s} + \cdots = u_s + \frac{1}{2^s}\left[\frac{1}{1^s} + \frac{1}{2^s} + \frac{1}{3^s} + \cdots\right] = u_s + \frac{1}{2^s}\zeta(s).$$

Thus,

$$u_s = \zeta(s)\left[1 - \frac{1}{2^s}\right]. \tag{5.3.5}$$

Also,

$$v_s = \left[\frac{1}{1^s} + \frac{1}{3^s} + \frac{1}{5^s} + \cdots\right] - \left[\frac{1}{2^s} + \frac{1}{4^s} + \cdots\right] = u_s - \frac{1}{2^s}\left[\frac{1}{1^s} + \frac{1}{2^s} + \frac{1}{3^s} + \cdots\right]$$

$$= u_s - \frac{1}{2^s}\zeta(s).$$

So, using (5.3.5),

$$v_s = \zeta(s)\left[1 - \frac{1}{2^s}\right] - \frac{1}{2^s}\zeta(s) = \zeta(s)\left[1 - \frac{2}{2^s}\right].$$

Thus,

$$v_s = \zeta(s)\left[1 - 2^{1-s}\right]. \tag{5.3.6}$$

But, noticing that

$$V_s = \sum_{k=1}^{\infty} \frac{(-1)^{k-1}}{k^s},$$

we have

$$\sum_{k=1}^{\infty} \frac{(-1)^{k-1}}{k^s} = \zeta(s)\left[1 - 2^{1-s}\right]. \tag{5.3.7}$$

Now we can do our integral! We start with

$$\int_0^{\infty} (-1)^{k-1} e^{-kx} x^{s-1} \, dx,$$

again make the change of variable $u = kx$, and then repeat what we did for (5.3.4). Everything will go through as before, except that you'll get one sum that will yield to (5.3.7), and another that will be $\sum_{k=1}^{\infty} (-1)^{k-1} e^{-kx}$ instead of $\sum_{k=1}^{\infty} e^{-kx}$. That will still give a geometric series, one still easy to do, with a sum of

$$\frac{e^{-x}}{1 + e^{-x}} = \frac{1}{e^x + 1}.$$

The end-result is

$$\int_0^{\infty} \frac{x^{s-1}}{e^x + 1} \, dx = \left[1 - 2^{1-s}\right] \Gamma(s)\zeta(s), \quad s > 1. \tag{5.3.8}$$

For $s = 4$, for example, this says

$$\int_0^{\infty} \frac{x^3}{e^x + 1} \, dx = \left[1 - 2^{-3}\right] \Gamma(4)\zeta(4) = \left(\frac{7}{8}\right) \frac{\pi^4}{15} = 5.68219\ldots$$

and MATLAB agrees, as *integral(@(x)(x.^3)./(exp(x)+1),0,inf) = 5.68219....*

By the way, notice that for $s = 1$ (5.3.8) gives an *indeterminate* value for the integral;

$$\int_0^{\infty} \frac{1}{e^x + 1} \, dx = \left[1 - 2^{1-1}\right] \Gamma(1)\zeta(1) = (0)(1)(\infty) = ?$$

because $\zeta(1)$ is the divergent harmonic series. This, despite the fact that the integral clearly exists since the integrand is always finite and goes to zero *very quickly* as x goes to infinity. This indeterminacy is, of course, precisely why the restriction $s > 1$ is there. So, what *is* the value of the integral for $s = 1$? We've already answered this, back in our result (2.1.4), where we showed that

$$\int_0^\infty \frac{1}{e^{ax}+1}\,dx = \frac{\ln(2)}{a}.$$

For $s = 1$ in (5.3.8) we set $a = 1$ in (2.1.4) and our integral is equal to $\ln(2)$.

To end this section, we'll do several interesting integrals that each lead directly to the zeta function. We begin with

$$\int_0^1 \frac{\ln^2(1+x)}{x}\,dx$$

and, curiously, it will be an *algebraic* trick that cracks the calculation of this integral wide-open for us. As our starting point, you should have no difficulty in convincing yourself of the truth of the identity

$$B^2 = \frac{(A+B)^2 + (A-B)^2 - 2A^2}{2}.$$

So, if we define $A = \ln(1-x)$ and $B = \ln(1+x)$ it then immediately follows that

$$\ln^2(1+x) = \frac{\{\ln(1-x)+\ln(1+x)\}^2 + \{\ln(1-x)-\ln(1+x)\}^2 - 2\ln^2(1-x)}{2}$$

or, if we divide through by x and integrate from 0 to 1,

$$\int_0^1 \frac{\ln^2(1+x)}{x}\,dx - \frac{1}{2}\int_0^1 \frac{\ln^2(1-x^2)}{x}\,dx + \frac{1}{2}\int_0^1 \frac{\ln^2\left(\frac{1-x}{1+x}\right)}{x}\,dx$$

$$- \int_0^1 \frac{\ln^2(1-x)}{x}\,dx. \tag{5.3.9}$$

Now, consider each of the integrals on the right in (5.3.9), in turn.

For the first integral, make the change of variable $y = 1 - x^2$ (and so $dx = -\frac{dy}{2x}$). Then,

$$\int_0^1 \frac{\ln^2(1-x^2)}{x}\,dx = \int_1^0 \int_0^1 \frac{\ln^2(y)}{x}\left(-\frac{dy}{2x}\right) = \frac{1}{2}\int_0^1 \frac{\ln^2(y)}{x^2}\,dy = \frac{1}{2}\int_0^1 \frac{\ln^2(y)}{1-y}\,dy$$

$$= \frac{1}{2}\int_0^1 \{1+y+y^2+y^3+\ldots\}\ln^2(y)\,dy = \frac{1}{2}\int_0^1 \sum_{k=1}^\infty y^{k-1}\ln^2(y)\,dy$$

$$= \frac{1}{2}\sum_{k=1}^\infty \int_0^1 x^{k-1}\ln^2(x)\,dx.$$

Recalling (3.4.7), where we showed

$$\int_0^1 x^a \ln^2(x)dx = \frac{2}{(a+1)^3},$$

we have upon setting $a = k - 1$ that

$$\int_0^1 \frac{\ln^2(1-x^2)}{x}dx = \frac{1}{2}\sum_{k=1}^{\infty}\frac{2}{k^3}$$

or,

$$\int_0^1 \frac{\ln^2(1-x^2)}{x}dx = \zeta(3). \qquad (5.3.10)$$

MATLAB agrees, as *zeta(3) = 1.2020569...* and

$$integral\left(@(x)\left(\left(log\left(1-x.^2\right)\right).^2\right)./x, 0, 1\right) = 1.2020569....$$

For the second integral on the right in (5.3.9), make the change of variable $y = \frac{1-x}{1+x}$ (and so $\frac{dy}{dx} = -\frac{2}{(1+x)^2}$ or, $dx = -\frac{(1+x)^2}{2}dy$). Now, notice that

$$1 - y^2 = 1 - \left(\frac{1-x}{1+x}\right)^2 = \frac{4x}{(1+x)^2}$$

and so

$$(1+x)^2 = \frac{4x}{1-y^2}.$$

Thus,

$$dx = -\frac{4x}{2(1-y^2)}dy = -\frac{2x}{(1-y^2)}dy$$

and therefore

$$\int_0^1 \frac{\ln^2\left(\frac{1-x}{1+x}\right)}{x}dx = \int_1^0 \frac{\ln^2(y)}{x}\left(-\frac{2x}{(1-y^2)}dy\right) = 2\int_0^1 \frac{\ln^2(y)}{1-y^2}dy$$

$$= 2\int_0^1 \{1 + y^2 + y^4 + y^6 + \ldots\}\ln^2(y)dy = 2\int_0^1 \sum_{k=1}^{\infty}y^{2(k-1)}\ln^2(y)dy$$

$$= 2\sum_{k=1}^{\infty}\int_0^1 x^{2(k-1)}\ln^2(x)dx.$$

Again recalling (3.4.7), with a now set to 2k − 1, we have

$$\int_0^1 \frac{\ln^2\left(\frac{1-x}{1+x}\right)}{x} dx = 2\sum_{k=1}^{\infty} \frac{2}{(2k-1)^3} = 4\sum_{k=1}^{\infty} \frac{1}{(2k-1)^3}.$$

This last sum, that of the reciprocals-cubed of the odd positive integers, is just u_s in (5.3.5) for s = 3. That is,

$$\sum_{k=1}^{\infty} \frac{1}{(2k-1)^3} = \frac{1}{1^3} + \frac{1}{3^3} + \frac{1}{5^3} + \ldots = \zeta(3)\left[1 - \frac{1}{2^3}\right] = \frac{7}{8}\zeta(3)$$

and so

$$\int_0^1 \frac{\ln^2\left(\frac{1-x}{1+x}\right)}{x} dx = \frac{7}{2}\zeta(3). \tag{5.3.11}$$

MATLAB again agrees, as *(7/2)*zeta(3)= 4.2071991...* and

$$integral\left(@(x)\left((log\,((1-x)./(1+x))).^2\right)./x, 0, 1\right) = 4.2071991\,\ldots.$$

Finally, for the third, right-most integral in (5.3.9), I'll let you verify that making the change of variable $x = y^2$ quickly leads to the result

$$\int_0^1 \frac{\ln^2(1-x)}{x} dx = 2\zeta(3). \tag{5.3.12}$$

Alternatively, an equally fast derivation of (5.3.12) is 'assigned' in Challenge Problem 5.6, using (5.3.4). In any case, MATLAB agrees, as *2*zeta(3)= 2.4041138...* and

$$integral\left(@(x)\left((log\,((1-x))).^2\right)./x, 0, 1\right) = 2.4041138\,\ldots.$$

With our results of (5.3.10), (5.3.11), and (5.3.12) in hand, we can now plug them into (5.3.9) to compute

$$\int_0^1 \frac{\ln^2(1+x)}{x} dx = \frac{1}{2}\{\zeta(3)\} + \frac{1}{2}\left\{\frac{7}{2}\zeta(3)\right\} - 2\zeta(3)$$

and arrive at

$$\int_0^1 \frac{\ln^2(1+x)}{x} dx = \frac{1}{4}\zeta(3). \tag{5.3.13}$$

A MATLAB check 'confirms' (5.3.13), as *zeta(3)/4=0.3005142257...* while

$$integral\big(@(x)\big((\,log\,((1+x)))^2\big)./x,0,1\big)=0.3005142257\ldots.$$

To end this section, consider the integral

$$\int_0^\infty x^{s-1}\{1-\tanh{(x)}\}dx.$$

We start by writing

$$1-\tanh{(x)}=1-\frac{\sinh{(x)}}{\cosh{(x)}}=1-\frac{e^x-e^{-x}}{e^x+e^{-x}}=\frac{2e^{-x}}{e^x+e^{-x}}=\frac{2e^{-2x}}{1+e^{-2x}}.$$

Thus,

$$1-\tanh{(x)}=2e^{-2x}\big(1-e^{-2x}+e^{-4x}-e^{-6x}+\ldots\big)$$
$$=2\big(e^{-2x}-e^{-4x}+e^{-6x}-e^{-8x}+\ldots\big)$$
$$=2\sum_{n=1}^\infty(-1)^{n-1}e^{-2nx}.$$

So,

$$\int_0^\infty x^{s-1}\{1-\tanh{(x)}\}dx=2\int_0^\infty x^{s-1}\sum_{n=1}^\infty(-1)^{n-1}e^{-2nx}dx$$
$$=2\sum_{n=1}^\infty(-1)^{n-1}\int_0^\infty x^{s-1}e^{-2nx}dx.$$

Changing variable to $y=2nx$ ($dx=\frac{dy}{2n}$), we have

$$\int_0^\infty x^{s-1}\{1-\tanh{(x)}\}dx=2\sum_{n=1}^\infty(-1)^{n-1}\int_0^\infty\Big(\frac{y}{2n}\Big)^{s-1}e^{-y}\frac{dy}{2n}$$
$$=2\sum_{n=1}^\infty(-1)^{n-1}\frac{1}{(2n)^s}\int_0^\infty y^{s-1}e^{-y}dy.$$

Remembering (4.1.1), the integral definition of the gamma function, we have

$$\int_0^\infty x^{s-1}\{1-\tanh{(x)}\}dx=2^{1-s}\Gamma(s)\sum_{n=1}^\infty\frac{(-1)^{n-1}}{n^s}.$$

But, recalling (5.3.6)

$$\sum_{n=1}^{\infty} \frac{(-1)^{n-1}}{n^s} = \frac{1}{1^s} - \frac{1}{2^s} + \frac{1}{3^s} - \frac{1}{4^s} + \ldots = \zeta(s)\left[1 - 2^{1-s}\right]$$

and so

$$\int_0^{\infty} x^{s-1}\{1 - \tanh(x)\}dx = 2^{1-s}\Gamma(s)\zeta(s)\left[1 - 2^{1-s}\right]. \tag{5.3.14}$$

We can use Feynman's integral differentiation trick to get another interesting integral from (5.3.14), as follows. First, change variable to $x = ay$, where a is positive ($dx = a\,dy$). Then,

$$\int_0^{\infty} (ay)^{s-1}\{1 - \tanh(ay)\}a\,dy = a^s \int_0^{\infty} y^{s-1}\{1 - \tanh(ay)\}dy$$

and so

$$\int_0^{\infty} y^{s-1}\{1 - \tanh(ay)\}dy = \frac{2^{1-s}\Gamma(s)\zeta(s)\left[1 - 2^{1-s}\right]}{a^s}.$$

Then, differentiating with respect to a,

$$\int_0^{\infty} y^{s-1}\{-y\,\text{sech}^2(ay)\}dy = 2^{1-s}\Gamma(s)\zeta(s)\left[1 - 2^{1-s}\right]\left(-sa^{-s-1}\right)$$

or, setting $a = 1$ (and returning to x as the dummy variable of integration) and remembering from (4.1.2) that $s\Gamma(s) = \Gamma(s + 1)$, we have

$$\int_0^{\infty} \frac{x^s}{\cosh^2(x)}dx = 2^{1-s}\Gamma(s + 1)\zeta(s)\left[1 - 2^{1-s}\right] \tag{5.3.15}$$

I won't go through the details here, but if you repeat what we've just done in getting (5.3.14) and (5.3.15), then starting with the integral

$$\int_0^{\infty} x^{s-1}\{\coth(x) - 1\}dx,$$

you should be able to derive (in fact, consider the following an unassigned challenge problem)

$$\int_0^{\infty} x^{s-1}\{\coth(x) - 1\}dx = 2^{1-s}\Gamma(s)\zeta(s) \tag{5.3.16}$$

and then, from (5.3.16) via Feynman's trick,

$$\int_0^\infty \frac{x^s}{\sinh^2(x)} \, dx = 2^{1-s} \Gamma(s+1) \zeta(s). \tag{5.3.17}$$

5.4   Euler's Constant and Related Integrals

Since the thirteenth century it has been known that the harmonic series diverges. That is,

$$\lim_{n \to \infty} \sum_{k=1}^n \frac{1}{k} = \lim_{n \to \infty} H(n) = \infty.$$

The divergence is quite slow, growing only as the logarithm of n, and so it seems quite reasonable to suspect that the *difference* between H(n) and $\log(n) = \ln(n)$ might *not* diverge. In fact, this difference, in the limit, is famous in mathematics as *Euler's constant* (or simply as *gamma*, written as γ).[5] That is, $\gamma = \lim_{n \to \infty} \gamma(n)$ where

$$\gamma(n) = \sum_{k=1}^n \frac{1}{k} - \ln(n).$$

It is not, I think, at all obvious that this expression does approach a limit as $n \to \infty$ but, using the area interpretation of the Riemann integral, it is easy to establish both that the limit exists and that it is somewhere in the interval 0 to 1. Here's how to do that.
Since

$$\gamma(n) = 1 + \frac{1}{2} + \frac{1}{3} + \ldots + \frac{1}{n} - \ln(n)$$

then

---

[5]A technically sophisticated yet quite readable treatment of 'all about γ,' at the level of this book, is Julian Havil's *Gamma*, Princeton University Press 2003. The constant is also sometimes called the *Euler-Mascheroni constant*, to give some recognition to the Italian mathematician Lorenzo Mascheroni (1750–1800) who, in 1790, calculated γ to 32 decimal places (but, alas, not without error). As I write, γ has been machine-calculated to literally *billions* of decimal places, with the first few digits being 0.5772156649 . . . . . Unlike π or e which are known to be irrational (transcendental, in fact), the rationality (or not) of γ is unknown. There isn't a mathematician on the planet who doesn't believe γ is irrational, but there is no known *proof* of that belief.

$$\gamma(n+1) - \gamma(n) = \left\{1 + \frac{1}{2} + \frac{1}{3} + \ldots + \frac{1}{n} + \frac{1}{n+1} - \ln(n+1)\right\}$$

$$- \left\{1 + \frac{1}{2} + \frac{1}{3} + \ldots + \frac{1}{n} - \ln(n)\right\}$$

$$= \frac{1}{n+1} + \ln(n) - \ln(n+1) = \frac{1}{n+1} + \ln\left(\frac{n}{n+1}\right)$$

or,

$$\gamma(n+1) - \gamma(n) = \frac{1}{n+1} + \ln\left(1 - \frac{1}{n+1}\right).$$

Clearly, $\ln\left(1 - \frac{1}{n+1}\right) < 0$ as it is the logarithm of a number less than 1. But we can actually say much more than just that.

Recalling the power series for $\ln(1+x)$ that we established at the beginning of Sect. 5.2, setting $x = -\frac{1}{n+1}$ gives us

$$\ln\left(1 - \frac{1}{n+1}\right) = -\frac{1}{n+1} - \frac{1}{2(n+1)^2} - \frac{1}{3(n+1)^3} - \ldots$$

This says that $\ln\left(1 - \frac{1}{n+1}\right)$ is *more negative* than $-\frac{1}{n+1}$ and so

$$\gamma(n+1) - \gamma(n) < 0.$$

That is, $\gamma(n)$ steadily *decreases* as n *increases* and, in fact, since $\gamma(1) - 1 = \ln(1) = 1$, we have established that $\gamma(n)$ steadily decreases from 1 as n increases.

Next, observe that

$$\int_1^n \frac{dt}{t} = \ln(t)\Big|_1^n = \ln(n) = \int_1^2 \frac{dt}{t} + \int_2^3 \frac{dt}{t} + \ldots + \int_{n-1}^n \frac{dt}{t}.$$

Now,

$$\int_j^{j+1} \frac{dt}{t} < \frac{1}{j}$$

because $\frac{1}{t}$ steadily decreases over the integration interval and, taking the integrand as a constant equal to its *greatest* value in that interval, *over*estimates the integral. Thus,

$$\ln(n) < 1 + \frac{1}{2} + \frac{1}{3} + \ldots + \frac{1}{n-1}$$

and so

$$0 < 1 + \frac{1}{2} + \frac{1}{3} + \ldots + \frac{1}{n-1} - \ln(n).$$

Adding $\frac{1}{n}$ to both sides of this inequality,

$$\frac{1}{n} < 1 + \frac{1}{2} + \frac{1}{3} + \ldots + \frac{1}{n-1} + \frac{1}{n} - \ln(n) = \gamma(n)$$

which, combined with our first result, says

$$0 < \frac{1}{n} < \gamma(n) \le 1.$$

So, what we have shown is that $\gamma(n)$ is (for all $n \ge 1$) in the interval 0 to 1 and that, as n increases, $\gamma(n)$ steadily decreases from 1 *without ever reaching* 0. Thus, $\gamma(n)$ must approach a limiting values as $n \to \infty$. But what *is* that limiting value?

Figure 5.4.1 shows a semi-log plot of $\gamma(n) = H(n) - \ln(n)$ as n varies over the interval $1 \le n \le 10,000$, and γ does appear to both exist (with an approximate value of 0.57) and to be approached fairly quickly. Such a plot is, of course, not a *proof* that γ exists (we've already established that), but it is fully in the spirit of this book. By adding γ to our catalog of constants, joining such workhorses as e, π, ln(2), and G, we can now 'do' some new, very interesting integrals.

I'll start by showing you how to write H(n) as an integral, and then we'll manipulate this integral into a form that will express γ as an integral, too. In the integral

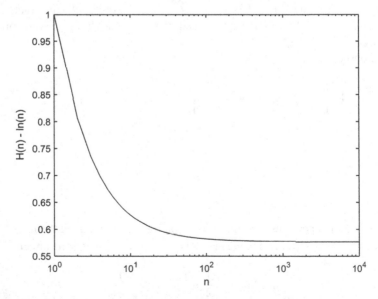

Fig. 5.4.1 Euler's constant as a limit change variable to u = 1 − x (and so dx=−du) to get

$$\int_0^1 \frac{1 - (1 - x)^n}{x}\, dx$$

$$\int_0^1 \frac{1 - (1 - x)^n}{x}\, dx = \int_1^0 \frac{1 - u^n}{1 - u}(-du)$$

$$= \int_0^1 \frac{(1 - u)(1 + u + u^2 + u^3 + \ldots + u^{n-1})}{1 - u}\, du$$

$$= \int_0^1 \{1 + u + u^2 + u^3 + \ldots + u^{n-1}\}\, du$$

$$= \left\{ u + \frac{1}{2} u^2 + \frac{1}{3} u^3 + \frac{1}{4} u^4 + \ldots + \frac{1}{n} u^n \right\} \Big|_0^1$$

$$= 1 + \frac{1}{2} + \frac{1}{3} + \frac{1}{4} + \ldots + \frac{1}{n} = H(n).$$

Next, change variable in our original integral (an integral we now know is H(n)) to $u = nx$ (and so $dx = \frac{1}{n} du$) and write

$$H(n) = \int_0^n \frac{1 - \left(1 - \frac{u}{n}\right)^n}{\frac{u}{n}} \left(\frac{1}{n} du\right) = \int_0^n \frac{1 - \left(1 - \frac{u}{n}\right)^n}{u}\, du$$

$$= \int_0^1 \frac{1 - \left(1 - \frac{u}{n}\right)^n}{u}\, du + \int_1^n \frac{1 - \left(1 - \frac{u}{n}\right)^n}{u}\, du$$

$$= \int_0^1 \frac{1 - \left(1 - \frac{u}{n}\right)^n}{u}\, du + \int_1^n \frac{du}{u} - \int_1^n \frac{\left(1 - \frac{u}{n}\right)^n}{u}\, du$$

$$= \int_0^1 \frac{1 - \left(1 - \frac{u}{n}\right)^n}{u}\, du + \ln(n) - \int_1^n \frac{\left(1 - \frac{u}{n}\right)^n}{u}\, du.$$

We thus have

$$\gamma(n) = H(n) - \ln(n) = \int_0^1 \frac{1 - \left(1 - \frac{u}{n}\right)^n}{u}\, du - \int_1^n \frac{\left(1 - \frac{u}{n}\right)^n}{u}\, du.$$

Letting $n \to \infty$, and recalling the definition of e^u as

$$e^u = \lim_{n \to \infty} \left(1 + \frac{u}{n}\right)^n,$$

we arrive at

$$\gamma = \lim_{n \to \infty} \gamma(n) = \int_0^1 \frac{1 - e^{-u}}{u} du - \int_1^\infty \frac{e^{-u}}{u} du. \qquad (5.4.1)$$

The curious expression in (5.4.1) for γ is *extremely* important and I'll show you some applications of it in the calculations that follow. Checking it with MATLAB, *integral(@(x)(1-exp(-x))./x,0,1)-integral(@(x)exp(-x)./x,1,inf)* $= 0.577215\ldots$
To start, consider the integral

$$\int_0^\infty e^{-x} \ln(x) dx$$

which we can write as

$$\int_0^\infty e^{-x} \ln(x) dx = \int_0^1 e^{-x} \ln(x) dx + \int_1^\infty e^{-x} \ln(x) dx. \qquad (5.4.2)$$

Alert! Pay *careful* attention at our next step, where we'll observe that

$$e^{-x} = -\frac{d}{dx}(e^{-x} - 1).$$

This is of course true, but *why* (you are surely wondering) are we bothering to write the simple expression on the left in the more complicated form on the right? The answer is that we are going to do the first integral on the right in (5.4.2) by parts and, without this trick in notation, things will not go well (when we get done, you should try to do it all over again without this notational trick).
So, continuing, we have for the first integral on the right

$$-\int_0^1 \frac{d}{dx}(e^{-x} - 1) \ln(x) dx$$

which becomes, with

$$u = \ln(x)$$

and

$$dv = \frac{d}{dx}(e^{-x} - 1) dx$$

(and so $du = \frac{1}{x} dx$ and $v = e^{-x} - 1$),

$$\int_0^1 e^{-x} \ln(x)dx = -\int_0^1 \frac{d}{dx}(e^{-x}-1)\ln(x)dx$$

$$= -\left(\{\ln(x)[e^{-x}-1]\}\big|_0^1 - \int_0^1 \frac{e^{-x}-1}{x}dx\right) = -\int_0^1 \frac{1-e^{-x}}{x}dx.$$

For the second integral on the right in (5.4.2) we also integrate by parts, now with $u = \ln(x)$ and $dv = e^{-x}$, and so

$$\int_1^\infty e^{-x}\ln(x)dx = \{-e^{-x}\ln(x)\}\Big|_1^\infty + \int_1^\infty e^{-x}\frac{1}{x}dx = \int_1^\infty e^{-x}\frac{1}{x}dx.$$

So,

$$\int_0^\infty e^{-x}\ln(x)dx = -\int_0^1 \frac{1-e^{-x}}{x}dx + \int_1^\infty e^{-x}\frac{1}{x}dx.$$

By (5.4.1) the right-hand-side is $-\gamma$ (remember, I *told* you (5.4.1) would be important!) and so we see that

$$\int_0^\infty e^{-x}\ln(x)dx = -\gamma. \tag{5.4.3}$$

Checking this with MATLAB, *integral(@(x)exp(-x).*log(x),0,inf)* $=$ $-0.577215....$

Sometimes, particularly when reading an advanced math book, you'll run across a comment that says (5.4.3) can be established by differentiating the integral definition of the gamma function $\Gamma(z)$ and then setting $z = 1$. If you write

$$x^{z-1} = e^{\ln(x^{z-1})} = e^{(z-1)\ln(x)} = e^{z\ln(x)}e^{-\ln(x)}$$

it isn't difficult to differentiate

$$\Gamma(z) = \int_0^\infty e^{-x}x^{z-1}\,dx$$

to get

$$\frac{d\Gamma(z)}{dz} = \Gamma'(z) = \int_0^\infty e^{-x}\ln(x)x^{z-1}\,dx$$

and so

$$\Gamma'(1) = \int_0^\infty e^{-x} \ln(x) \, dx.$$

But what tells us that $\Gamma'(1) = -\gamma$?????

The authors of such books are assuming their readers are familiar with what mathematicians call the *digamma function*, which as an infinite series is[6]

$$\frac{d}{dz} \ln\{\Gamma(z)\} = \frac{\Gamma'(z)}{\Gamma(z)} = -\frac{1}{z} - \gamma + \sum_{r=1}^{\infty} \left(\frac{1}{r} - \frac{1}{r+z}\right).$$

Setting $z = 1$ gives (because $\Gamma(1) = 0\,! = 1$)

$$\Gamma'(1) = -1 - \gamma + \sum_{r=1}^{\infty} \left(\frac{1}{r} - \frac{1}{r+1}\right) = -1 - \gamma + \left\{\left(1 - \frac{1}{2}\right) + \left(\frac{1}{2} - \frac{1}{3}\right) + \ldots\right\}$$

or, because the fractions in the curly brackets telescope,

$$\Gamma'(1) = -\gamma$$

as claimed and we get (5.4.3).

Okay, let's next consider the interesting integral

$$\int_0^1 \ln\{-\ln(x)\} dx.$$

Make the obvious change of variable $u = -\ln(x)$ to arrive at

$$\int_0^1 \ln\{-\ln(x)\} \, dx = \int_0^\infty e^{-u} \ln(u) du$$

which is (5.4.3). Thus,

$$\int_0^1 \ln\{-\ln(x)\} = -\gamma. \tag{5.4.4}$$

Checking with MATLAB, *integral(@(x)log(-log(x)),0,1) = −0.577215....*

---

[6] I won't pursue the derivation of $\frac{\Gamma'(z)}{\Gamma(z)}$, but you can find a discussion in Havil's book (see note 5). We'll use the digamma function in the final calculation of this chapter, in a derivation of (5.4.11). See also Challenge Problem 5.10. An integral form of the digamma function will be derived in Section 8.10 (and see Challenge Problem 8.9).

Another dramatic illustration of the power of (5.4.1) comes from the early twentieth century mathematician Ramanujan (who will be discussed in more length in Chap. 9), who evaluated the exponentially-stuffed integral

$$I = \int_0^\infty \{e^{-\alpha e^x} + e^{-\alpha e^{-x}} - 1\}dx$$

for α any positive constant. Observing that the integrand is an even function, he started by doubling the integration interval and writing

$$2I = \int_{-\infty}^\infty \{e^{-\alpha e^x} + e^{-\alpha e^{-x}} - 1\}dx$$

and then changing variable to $u = e^x$ (and so $dx = \frac{du}{u}$). Thus,

$$2I = \int_0^\infty \frac{e^{-\alpha u} + e^{-\frac{\alpha}{u}} - 1}{u}du.$$

Next, he broke this integral into a sum of two integrals:

$$2I = \int_0^{\frac{1}{\alpha}} \frac{e^{-\alpha u} + e^{-\frac{\alpha}{u}} - 1}{u}du + \int_{\frac{1}{\alpha}}^\infty \frac{e^{-\alpha u} + e^{-\frac{\alpha}{u}} - 1}{u}du.$$

He then broke *each* of these two integrals into two more:

$$2I = \int_0^{\frac{1}{\alpha}} \frac{e^{-\alpha u} - 1}{u}du + \int_0^{\frac{1}{\alpha}} \frac{e^{-\frac{\alpha}{u}}}{u}du + \int_{\frac{1}{\alpha}}^\infty \frac{e^{-\alpha u}}{u}du + \int_{\frac{1}{\alpha}}^\infty \frac{e^{-\frac{\alpha}{u}} - 1}{u}du$$

$$= \left\{-\int_0^{\frac{1}{\alpha}} \frac{1 - e^{-\alpha u}}{u}du + \int_{\frac{1}{\alpha}}^\infty \frac{e^{-\alpha u}}{u}du\right\} + \left\{\int_0^{\frac{1}{\alpha}} \frac{e^{-\frac{\alpha}{u}}}{u}du - \int_{\frac{1}{\alpha}}^\infty \frac{1 - e^{-\frac{\alpha}{u}}}{u}du\right\}.$$

Now, for the two integrals in the first pair of curly brackets make the change of variable $x = \alpha u$ (and so $du = \frac{1}{\alpha}dx$), and for the two integrals in the second pair of curly brackets make the change of variable $x = \frac{\alpha}{u}$ (and so $du = -\frac{\alpha}{x^2}dx$). Then,

$$2I = -\int_0^1 \frac{1 - e^{-x}}{\frac{x}{\alpha}}\left(\frac{1}{\alpha}dx\right) + \int_1^\infty \frac{e^{-x}}{\frac{x}{\alpha}}\left(\frac{1}{\alpha}dx\right) + \int_\infty^{\alpha^2} \frac{e^{-x}}{\frac{\alpha}{x}}\left(-\frac{\alpha}{x^2}dx\right) - \int_{\alpha^2}^0 \frac{1 - e^{-x}}{\frac{\alpha}{x}}\left(-\frac{\alpha}{x^2}dx\right)$$

$$= -\int_0^1 \frac{1 - e^{-x}}{x}dx + \int_1^\infty \frac{e^{-x}}{x}dx + \int_{\alpha^2}^\infty \frac{e^{-x}}{x}dx - \int_0^{\alpha^2} \frac{1 - e^{-x}}{x}dx.$$

Since the first two integrals on the right are $-\gamma$ by (5.4.1), then

$$2I = -\gamma + \int_{\alpha^2}^{\infty} \frac{e^{-x}}{x} dx - \int_0^{\alpha^2} \frac{dx}{x} + \int_0^{\alpha^2} \frac{e^{-x}}{x} dx$$

or, combining the first and last integrals on the right,

$$2I = -\gamma + \int_0^{\infty} \frac{e^{-x}}{x} dx - \int_0^{\alpha^2} \frac{dx}{x}.$$

Continuing,

$$2I = -\gamma + \left\{ \int_0^1 \frac{e^{-x}}{x} dx + \int_1^{\infty} \frac{e^{-x}}{x} dx \right\} - \left\{ \int_0^1 \frac{dx}{x} - \int_{\alpha^2}^1 \frac{dx}{x} \right\}$$

or, combining the first and third integrals on the right,

$$2I = -\gamma + \left\{ \int_1^{\infty} \frac{e^{-x}}{x} dx + \int_0^1 \frac{e^{-x} - 1}{x} dx \right\} + \int_{\alpha^2}^1 \frac{dx}{x}$$

$$= -\gamma + \left\{ \int_1^{\infty} \frac{e^{-x}}{x} dx - \int_0^1 \frac{1 - e^{-x}}{x} dx \right\} + \ln(x) \Big|_{\alpha^2}^1 .$$

The two integrals in the final pair of curly brackets are, again by (5.4.1), $-\gamma$, and so

$$2I = -2\gamma - \ln(\alpha^2) = -2\gamma - 2\ln(\alpha).$$

Thus, finally, we have Ramanujan's integral

$$\int_0^{\infty} \left\{ e^{-\alpha e^x} + e^{-\alpha e^{-x}} - 1 \right\} dx = -\gamma - \ln(\alpha). \tag{5.4.5}$$

To check this, for $\alpha = 1$ the integral is $-\gamma$, while for $\alpha = 2$ the integral is $-\gamma - \ln(2) = -1.270362\ldots$, and MATLAB agrees as *integral(@(x)(exp(-exp (x))+exp(-exp(-x))-1),0,inf)* $= -0.577215\ldots$ and *integral(@(x)(exp(-2*exp(x))+ exp(-2*exp(-x))-1),0,inf)* $= -1.270362\ldots$

As yet another example of (5.4.1), consider the following mysterious-looking integral:

$$\int_0^{\infty} \frac{e^{-x^a} - e^{-x^b}}{x} dx$$

where a and b are both positive constants. This integral 'looks like' a Frullani integral (see Chap. 3 again), but it is not. It is *much* deeper. Without our equally mysterious (5.4.1) I think it would be impossible to make any headway with this integral. Here's how to do it.

$$\int_0^\infty \frac{e^{-x^a} - e^{-x^b}}{x}\,dx = \int_0^\infty \frac{e^{-x^a} - e^{-(x^a)^{\frac{b}{a}}}}{x}\,dx = \int_0^\infty \frac{e^{-x^a} - e^{-(x^a)^p}}{x}\,dx, p = \frac{b}{a}.$$

Let

$$u = x^a$$

and so

$$dx = \frac{du}{ax^{a-1}}.$$

Thus,

$$\int_0^\infty \frac{e^{-x^a} - e^{-x^b}}{x}\,dx = \int_0^\infty \frac{e^{-u} - e^{-u^p}}{x}\left(\frac{du}{ax^{a-1}}\right) = \frac{1}{a}\int_0^\infty \frac{e^{-u} - e^{-u^p}}{u}\,du$$

$$= \frac{1}{a}\left[\int_0^\infty \frac{e^{-u} - 1}{u}\,du - \int_0^\infty \frac{e^{-u^p} - 1}{u}\,du\right]$$

$$= \frac{1}{a}\left[\int_0^1 \frac{e^{-u} - 1}{u}\,du + \int_1^\infty \frac{e^{-u} - 1}{u}\,du - \int_0^1 \frac{e^{-u^p} - 1}{u}\,du - \int_1^\infty \frac{e^{-u^p} - 1}{u}\,du\right]$$

$$= \frac{1}{a}\left[\left\{-\int_0^1 \frac{1 - e^{-u}}{u}\,du + \int_1^\infty \frac{e^{-u}}{u}\,du\right\} - \left\{\int_0^1 \frac{e^{-u^p} - 1}{u}\,du + \int_1^\infty \frac{e^{-u^p}}{u}\,du\right\}\right]$$

$$= \frac{1}{a}\left[-\gamma - \left\{\int_0^1 \frac{e^{-u^p} - 1}{u}\,du + \int_1^\infty \frac{e^{-u^p}}{u}\,du\right\}\right],$$

where once again (5.4.1) comes into play.

Next, in the two integrals on the right in the previous line, make the change of variable

$$y = u^p$$

and so

$$du = \frac{dy}{pu^{p-1}}.$$

Thus,

$$\left\{ \int_0^1 \frac{e^{-u^p} - 1}{u}\, du + \int_1^\infty \frac{e^{-u^p}}{u}\, du \right\} = \frac{1}{p}\left\{ \int_0^1 \frac{e^{-y} - 1}{y}\, dy + \int_1^\infty \frac{e^{-y}}{y}\, dy \right\}$$

and we have (where (5.4.1) is used yet again)

$$\int_0^\infty \frac{e^{-x^a} - e^{-x^b}}{x}\, dx = \frac{1}{a}\left[-\gamma - \frac{1}{p}\left\{ \int_0^1 \frac{e^{-u} - 1}{u}\, du + \int_1^\infty \frac{e^{-u}}{u}\, du \right\} \right]$$

$$= \frac{1}{a}\left[-\gamma - \frac{1}{p}\left\{ -\int_0^1 \frac{1 - e^{-u}}{u}\, du + \int_1^\infty \frac{e^{-u}}{u}\, du \right\} \right] = \frac{1}{a}\left[-\gamma - \frac{1}{p}\{-\gamma\} \right]$$

$$= \gamma\left[-\frac{1}{a} + \frac{1}{ap} \right] = \gamma\left[\frac{1}{b} - \frac{1}{a} \right]$$

or, at last,

$$\int_0^\infty \frac{e^{-x^a} - e^{-x^b}}{x}\, dx = \gamma \frac{a - b}{ab}. \tag{5.4.6}$$

To check this, suppose $a = 2$ and $b = 1$. Then (5.4.6) says

$$\int_0^\infty \frac{e^{-x^2} - e^{-x}}{x}\, dx = \frac{1}{2}\gamma$$

or, equivalently,

$$2\int_0^\infty \frac{e^{-x^2} - e^{-x}}{x}\, dx = \gamma.$$

And, in fact, MATLAB agrees, as $2*integral(@(x)(exp(-(x.\char`\^2))-exp(-x))./x,0,inf)$ $= 0.5772153\ldots$.

Our result in (5.4.1) is so useful that I think it helpful to see it developed in an alternative, quite different way. We start with a result from earlier in the book, (3.3.3), where we showed that

$$\ln\left(\frac{b}{a}\right) = \int_0^\infty \frac{e^{-ax} - e^{-bx}}{x}\, dx.$$

In particular, if we set $a = 1$ and $b = t$, we have

$$\ln(t) = \int_0^\infty \frac{e^{-x} - e^{-tx}}{x}\, dx. \tag{5.4.7}$$

Next, for n any positive integer, we have

$$\int_0^\infty e^{-nx}\,dx = \left(-\frac{1}{n}e^{-nx}\right)\Big|_0^\infty = \frac{1}{n}.$$

So, summing over all n from 1 to N, we have

$$\sum_{n=1}^N \frac{1}{n} = \sum_{n=1}^N \int_0^\infty e^{-nx}\,dx = \int_0^\infty \left\{\sum_{n=1}^N e^{-nx}\right\}dx.$$

The sum is a geometric series, easily evaluated to give

$$\sum_{n=1}^N e^{-nx} = \frac{e^{-x} - e^{-(N+1)x}}{1 - e^{-x}}.$$

Thus,

$$\sum_{n=1}^N \frac{1}{n} = \int_0^\infty \frac{e^{-x} - e^{-(N+1)x}}{1 - e^{-x}}\,dx. \qquad (5.4.8)$$

Now, recall the definition of γ from the beginning of this section:

$$\gamma = \lim_{N\to\infty}\left\{\sum_{n-1}^N \frac{1}{n} - \ln(N)\right\}.$$

Putting (5.4.7) and (5.4.8) into the expression on the right, we have

$$\gamma = \lim_{N\to\infty}\int_0^\infty \left[\left\{\frac{e^{-x} - e^{-(N+1)x}}{1 - e^{-x}}\right\} - \left\{\frac{e^{-x} - e^{-Nx}}{x}\right\}\right]dx$$

and so

$$\gamma = \int_0^\infty \frac{e^{-x}}{1-e^{-x}}\,dx - \int_0^\infty \frac{e^{-x}}{x}\,dx = \int_0^\infty \frac{e^{-x}}{1-e^{-x}}\,dx - \left\{\int_0^1 \frac{e^{-x}}{x}\,dx + \int_1^\infty \frac{e^{-x}}{x}\,dx\right\}.$$

In the first integral on the right, let $s = 1 - e^{-x}$. Then, $\frac{ds}{dx} = e^{-x}$ and so $dx = \frac{ds}{e^{-x}}$.
Thus,

$$\gamma = \int_0^1 \frac{e^{-x}}{s}\left(\frac{ds}{e^{-x}}\right) - \int_0^1 \frac{e^{-x}}{x}\,dx - \int_1^\infty \frac{e^{-x}}{x}\,dx$$

$$= \int_0^1 \frac{1}{s}\,ds - \int_0^1 \frac{e^{-x}}{x}\,dx - \int_1^\infty \frac{e^{-x}}{x}\,dx$$

or,

$$\gamma = \int_0^1 \frac{1 - e^{-x}}{x} dx - \int_1^\infty \frac{e^{-x}}{x} dx$$

which is just (5.4.1).

(5.4.1) isn't the only tool we have for working with γ, and to illustrate that let's do the exotic integral

$$\int_0^\infty e^{-x^2} \ln(x) dx$$

as the final calculation for this section. This will be a calculation that will require us to recall Feynman's favorite trick from Chap. 3 of differentiating an integral, as well as the power series for the log function that we used earlier in this chapter. The answer will feature a (perhaps) surprising appearance of γ. We start with a class of integrals indexed on the parameter m:

$$I(m) = \int_0^\infty x^m e^{-x^2} dx. \tag{5.4.9}$$

Differentiating with respect to m,

$$\frac{dI}{dm} = \frac{d}{dm} \int_0^\infty e^{\ln(x^m)} e^{-x^2} dx = \frac{d}{dm} \int_0^\infty e^{m \ln(x)} e^{-x^2} dx = \int_0^\infty \ln(x) e^{m \ln(x)} e^{-x^2} dx$$

and so

$$\frac{dI}{dm} = \int_0^\infty x^m \ln(x) e^{-x^2} dx.$$

This tells us that the integral we are after is the $m = 0$ case, that is,

$$\int_0^\infty e^{-x^2} \ln(x) dx = \frac{dI}{dm}\bigg|_{m=0}. \tag{5.4.10}$$

Next, returning to (5.4.9), make the change of variable $t = x^2$ (and so $dx = \frac{dt}{2x} = \frac{dt}{2\sqrt{t}}$). Then,

$$I(m) = \int_0^\infty t^{\frac{m}{2}} e^{-t} \frac{dt}{2\sqrt{t}} = \frac{1}{2} \int_0^\infty e^{-t} t^{\frac{m-1}{2}} dt.$$

Recalling (4.1.1), this last integral is the gamma function $\Gamma(n)$ for the case of $n - 1 = \frac{m}{2} - \frac{1}{2}$ (for $n = \frac{m+1}{2}$). Thus,

$$I(m) = \frac{1}{2}\Gamma\left(\frac{m+1}{2}\right)$$

and so, from (5.4.10),

$$\int_0^\infty e^{-x^2}\ln(x)dx = \frac{1}{2}\left\{\frac{d}{dm}\Gamma\left(\frac{m+1}{2}\right)\right\}\Bigg|_{m=0}. \qquad (5.4.11)$$

To do the differentiation of the gamma function, recall the digamma function from earlier in this section, where we have

$$\Gamma'(z) = \frac{d\Gamma(z)}{dz} = \Gamma(z)\left[-\frac{1}{z} - \gamma + \sum_{r=1}^\infty\left(\frac{1}{r} - \frac{1}{r+z}\right)\right].$$

So, with $z = \frac{m+1}{2}$, we have

$$\frac{d\Gamma\left(\frac{m+1}{2}\right)}{d\left(\frac{m+1}{2}\right)} = \frac{d\Gamma\left(\frac{m+1}{2}\right)}{\frac{1}{2}\,dm} = 2\frac{d\Gamma\left(\frac{m+1}{2}\right)}{dm}$$

$$= \Gamma\left(\frac{m+1}{2}\right)\left[-\frac{1}{\frac{m+1}{2}} - \gamma + \sum_{r=1}^\infty\left(\frac{1}{r} - \frac{1}{r+\frac{m+1}{2}}\right)\right].$$

That is, for the case of $m = 0$,

$$\left\{\frac{d}{dm}\Gamma\left(\frac{m+1}{2}\right)\right\}\Bigg|_{m=0} = \frac{1}{2}\Gamma\left(\frac{1}{2}\right)\left[-2 - \gamma + \sum_{r=1}^\infty\left(\frac{1}{r} - \frac{1}{r+\frac{1}{2}}\right)\right]$$

$$= \frac{1}{2}\sqrt{\pi}\left[-2 - \gamma + \sum_{r=1}^\infty\left(\frac{2}{2r} - \frac{2}{2r+1}\right)\right]$$

and so, putting this into (5.4.11), we have

$$\int_0^\infty e^{-x^2}\ln(x)dx = \frac{1}{4}\sqrt{\pi}\left[-2 - \gamma + 2\sum_{r=1}^\infty\left(\frac{1}{2r} - \frac{1}{2r+1}\right)\right] \qquad (5.4.12)$$

Now, concentrate on the summation

$$\sum_{r=1}^\infty\left(\frac{1}{2r} - \frac{1}{2r+1}\right) = \left(\frac{1}{2} - \frac{1}{3}\right) + \left(\frac{1}{4} - \frac{1}{5}\right) + \left(\frac{1}{6} - \frac{1}{7}\right) + \ldots.$$

Then, recalling the power for $\ln(1 + x)$ that we derived at the start of Sect. 5.2, we see that

$$\ln(2) = 1 - \frac{1}{2} + \frac{1}{3} - \frac{1}{4} + \frac{1}{5} - \cdots$$

and so

$$\frac{1}{2} - \frac{1}{3} + \frac{1}{4} - \frac{1}{5} + \frac{1}{6} - \frac{1}{7} + \ldots = 1 - \ln(2),$$

which means

$$\sum_{r=1}^{\infty} \left(\frac{1}{2r} - \frac{1}{2r+1} \right) = 1 - \ln(2).$$

Putting this into (5.4.12) we have

$$\int_0^\infty e^{-x^2} \ln(x)dx = \frac{1}{4}\sqrt{\pi}[-2 - \gamma + 2\{1 - \ln(2)\}]$$

or, at last, we have the beautiful

$$\int_0^\infty e^{-x^2} \ln(x)dx = -\frac{1}{4}\sqrt{\pi}\,[\gamma + 2\ln(2)]. \tag{5.4.13}$$

This is equal to $-0.8700577\ldots$ and MATLAB agrees as *integral(@(x)exp(-(x.^2)).*log(x),0,inf)* $= -0.8700577\ldots$

5.5 Generalizing Catalan's Constant

A natural generalization of (5.1.3) is to write

$$I(t) = \int_0^\infty \frac{\ln(1 + tx)}{1 + x^2}dx, t \geq 0 \tag{5.5.1}$$

which we can evaluate using Feynman's trick. (The special case of $t = 1$, of course, *is* (5.1.3). Also obvious, by inspection, is that $I(0) = 0$.) Differentiating (5.5.1) with respect to t, we have

$$\frac{dI}{dt} = \int_0^\infty \frac{x}{(1+tx)(1+x^2)}dx = \int_0^\infty \frac{1}{t^2+1}\left(\frac{x+t}{1+x^2} - \frac{t}{1+tx}\right)dx$$

or,

$$\frac{dI}{dt} = \frac{1}{t^2+1}\int_0^\infty \frac{x}{1+x^2}dx + \frac{t}{t^2+1}\int_0^\infty \frac{dx}{1+x^2} - \frac{t}{t^2+1}\int_0^\infty \frac{dx}{1+tx}. \tag{5.5.2}$$

Our goal is probably now clear. The idea is to evaluate the three integrals in (5.5.2), thus obtaining $\frac{dI}{dt}$ as some function of t which can then (we hope!) be integrated *indefinitely*, followed finally with the use of the value of $I(0)$ or $I(1)$ (our result in (5.1.3)) to evaluate the constant of indefinite integration. In this way we would arrive at a formula for $I(t)$, a formula valid for *all $t \geq 0$*.

The immediate issue we run into is that the first and third integrals in (5.5.2) are each divergent (both blow-up logarithmically), but that doesn't necessarily mean all is lost. That's because (5.5.2) shows that all that matters is their *difference*, which could be finite even as each integral individually goes to infinity. Thus, let's write (5.5.2) as

$$\frac{dI}{dt} = \frac{1}{t^2+1} \lim_{\epsilon \to \infty} \left\{ \int_0^\epsilon \frac{x}{1+x^2} dx - t \int_0^\epsilon \frac{dx}{1+tx} \right\} + \frac{t}{t^2+1} \int_0^\infty \frac{dx}{1+x^2}. \quad (5.5.3)$$

For the first integral in the curly brackets write $y = x^2$ (and so $dx = \frac{dy}{2x}$), and for the second integral in the curly brackets write $y = 1 + tx$ (and so $dx = \frac{dy}{t}$). Then,

$$\lim_{\epsilon \to \infty} \left\{ \int_0^\epsilon \frac{x}{1+x^2} dx - t \int_0^\epsilon \frac{dx}{1+tx} \right\} = \lim_{\epsilon \to \infty} \left\{ \int_1^{1+\epsilon^2} \left(\frac{x}{y}\right) \frac{dy}{2x} - t \int_1^{1+t\epsilon} \left(\frac{1}{y}\right) \frac{dy}{t} \right\}$$

$$= \lim_{\epsilon \to \infty} \left\{ \frac{1}{2} \ln(y) \Big|_1^{1+\epsilon^2} - \ln(y)\Big|_1^{1+t\epsilon} \right\} = \lim_{\epsilon \to \infty} \left\{ \frac{1}{2} \ln(1+\epsilon^2) \quad \ln(1+t\epsilon) \right\}$$

$$= \lim_{\epsilon \to \infty} \left\{ \frac{1}{2} \ln(\epsilon^2) - \ln(t\epsilon) \right\} = \lim_{\epsilon \to \infty} \left\{ \ln(\epsilon) - \ln(t) - \ln(\epsilon) \right\} = -\ln(t).$$

Thus, (5.5.3) becomes

$$\frac{dI}{dt} = -\frac{\ln(t)}{t^2+1} + \frac{t}{t^2+1} \int_0^\infty \frac{dx}{1+x^2} - \frac{\ln(t)}{t^2+1} + \frac{t}{t^2+1} \tan^{-1}(x)\Big|_0^\infty$$

or,

$$\frac{dI}{dt} = -\frac{\ln(t)}{t^2+1} + \frac{\pi t}{2(t^2+1)}, \quad (5.5.4)$$

from which we immediately write

$$I(t) = -\int \frac{\ln(t)}{t^2+1} dt + \frac{\pi}{2} \int \frac{t}{t^2+1} dt + C \quad (5.5.5)$$

where C is the constant of indefinite integration.

The second integral in (5.5.5) is easy; just let $y = t^2 + 1$ (and so $dt = \frac{dy}{2t}$). Then,

$$\frac{\pi}{2} \int \frac{t}{t^2+1} dt = \frac{\pi}{2} \int \left(\frac{t}{y}\right) \frac{dy}{2t} = \frac{\pi}{4} \int \frac{dy}{y} = \frac{\pi}{4} \ln(y) = \frac{\pi}{4} \ln(t^2+1)$$

and so (5.5.5) becomes

$$I(t) = -\int \frac{\ln (t)}{t^2 + 1} dt + \frac{\pi}{4} \ln (t^2 + 1) + C. \tag{5.5.6}$$

The remaining integral in (5.5.6) is just a bit more challenging, but not by much and we can do it by-parts. That is, in

$$\int u \, dv = uv - \int v \, du$$

let $u = \ln (t)$ and $dv = \frac{dt}{t^2+1}$. Then $du = \frac{dt}{t}$ and $v = \tan^{-1}(t)$. Thus,

$$\int \frac{\ln (t)}{t^2 + 1} dt = \ln (t) \tan^{-1}(t) - \int \frac{\tan^{-1}(t)}{t} dt. \tag{5.5.7}$$

As we showed at the start of Sect. 5.1,

$$\tan^{-1}(t) = t - \frac{t^3}{3} + \frac{t^5}{5} - \frac{t^7}{7} + \dots, 0 \leq t \leq 1 \tag{5.5.8}$$

but since we are allowing the possibility of $t > 1$ in (5.5.1) we need to also write[7]

$$\tan^{-1}(t) = \frac{\pi}{2} - \frac{1}{t} + \frac{1}{3t^3} - \frac{1}{5t^5} + \frac{1}{7t^7} - \dots, t \geq 1. \tag{5.5.9}$$

Thus,

$$\frac{\tan^{-1}(t)}{t} = \begin{cases} 1 - \dfrac{t^2}{3} + \dfrac{t^4}{5} - \dfrac{t^6}{7} + \dots, 0 \leq t \leq 1 \\[2mm] \dfrac{\pi}{2t} - \dfrac{1}{t^2} + \dfrac{1}{3t^4} - \dfrac{1}{5t^6} + \dfrac{1}{7t^8} - \dots, t \geq 1 \end{cases}$$

and so

$$\int \frac{\tan^{-1}(t)}{t} dt = \begin{cases} t - \dfrac{t^3}{3^2} + \dfrac{t^5}{5^2} - \dfrac{t^7}{7^2} + \dots, 0 \leq t \leq 1 \\[2mm] \dfrac{\pi}{2} \ln (t) + \dfrac{1}{t} - \dfrac{1}{3^2 t^3} + \dfrac{1}{5^2 t^5} - \dfrac{1}{7^2 t^7} + \dots, t \geq 1 \end{cases}.$$

Putting this into (5.5.7), we have

---

[7] If you draw a right triangle with perpendicular sides 1 and x, it is clear from high school geometry that $\tan^{-1}(x) + \tan^{-1}\left(\frac{1}{x}\right) = \frac{\pi}{2}$, from which (5.5.9) immediately follows (in conjunction with (5.5.8)).

$$\int \frac{\ln(t)}{t^2+1}\,dt = \ln(t)\tan^{-1}(t) - \begin{cases} t - \frac{t^3}{3^2} + \frac{t^5}{5^2} - \frac{t^7}{7^2} + \ldots, 0 \le t \le 1 \\ \frac{\pi}{2}\ln(t) + \frac{1}{t} - \frac{1}{3^2 t^3} + \frac{1}{5^2 t^5} - \frac{1}{7^2 t^7} + \ldots, t \ge 1 \end{cases}$$

and then putting this into (5.5.6), we arrive at

$$I(t) = C + \frac{\pi}{4}\ln(t^2+1) - \ln(t)\tan^{-1}(t)$$

$$+ \begin{cases} t - \frac{t^3}{3^2} + \frac{t^5}{5^2} - \frac{t^7}{7^2} + \ldots, 0 \le t \le 1 \\ \frac{\pi}{2}\ln(t) + \frac{1}{t} - \frac{1}{3^2 t^3} + \frac{1}{5^2 t^5} - \frac{1}{7^2 t^7} + \ldots, t \ge 1 \end{cases}.$$

The value of the indefinite constant of integration is seen to be $C = 0$ by imposing either of the known values of $I(0)$ or $I(1)$, and so

$$I(t) = \frac{\pi}{4}\ln(t^2+1) - \ln(t)\tan^{-1}(t)$$

$$+ \begin{cases} t - \frac{t^3}{3^2} + \frac{t^5}{5^2} - \frac{t^7}{7^2} + \ldots, 0 \le t \le 1 \\ \frac{\pi}{2}\ln(t) + \frac{1}{t} - \frac{1}{3^2 t^3} + \frac{1}{5^2 t^5} - \frac{1}{7^2 t^7} + \ldots, t \ge 1 \end{cases}. \qquad (5.5.10)$$

The result in (5.5.10) may look complicated, but in fact it is no more so than is the result in (5.1.3). That's because we can simply define a generalized Catalan's *function* as

$$G(t) = \begin{cases} t - \frac{t^3}{3^2} + \frac{t^5}{5^2} - \frac{t^7}{7^2} + \ldots, 0 \le t \le 1 \\ \frac{\pi}{2}\ln(t) + \frac{1}{t} - \frac{1}{3^2 t^3} + \frac{1}{5^2 t^5} - \frac{1}{7^2 t^7} + \ldots, t \ge 1 \end{cases} \qquad (5.5.11)$$

and write

$$\int_0^\infty \frac{\ln(1+tx)}{1+x^2}\,dx = \frac{\pi}{4}\ln(t^2+1) - \ln(t)\tan^{-1}(t) + G(t), t \ge 0 \qquad (5.5.12)$$

where $G(1) = G$ (Catalan's *constant*) which is, of course, itself computed by numerically evaluating the power series for $t = 1$ (which of the two expressions we use in (5.5.11) is irrelevant for that particular value of t). For $t = 1$, (5.5.12) reduces to (5.1.3).

As we've done before, let's do a quick numerical check of (5.5.12) and (5.5.11). If $t = \frac{1}{2}$ a direct evaluation of the integral gives

$$integral\left(@(x)\, log\,(1 + 0.5 * x)./(1 + x.^2), 0,\, inf\,\right) = 0.9838549\ldots$$

while evaluation of (5.5.10) produces[8] $0.9838549.\ldots$ And for t $= 2$, the results are

$$integral\left(@(x)\, log\,(1 + 2 * x)./(1 + x.^2), 0,\, inf\,\right) = 2.0726479\ldots$$

while evaluation of (5.5.10) produces $2.0726479.\ldots$. As always, such excellent numerical agreement *proves* nothing, but who would deny that it certainly does calm the nerves!

5.6   Challenge Problems

(C5.1) Consider the class of definite integrals defined by

$$I(m, n) = \int_0^1 \frac{1 - x^m}{1 - x^n}\, dx,$$

where m and n are positive integers. I(m, n) exists for all such m and n since the integrand is everywhere finite over the interval of integration.[9] Make a power series expansion of the integrand and then integrate term-by-term to arrive at an infinite sum that can be easily evaluated with a simple computer program (no use of MATLAB's *integral* or any other similar high-powered software allowed!). If you have access to a computer, write a code to evaluate I(m, n) for given values of m and n. You can check your code by seeing if it gives the correct answers for those cases easy to do by hand (obviously, I(n, n) $= 1$ for all n). For example,

$$I(2, 1) = \int_0^1 \frac{1 - x^2}{1 - x}\, dx = \int_0^1 (1 + x)dx = \left(x + \frac{1}{2}x^2\right)\Big|_0^1 = 1.5$$

and

$$I(1, 2) = \int_0^1 \frac{1 - x}{1 - x^2}\, dx = \int_0^1 \frac{1}{1 + x}\, dx = \int_1^2 \frac{du}{u} = \ln\,(2) = 0.6931.$$

Then, use your code to compute, *with at least the first six decimal digits correct,* the values of I(5, 7) and I(7, 5).

---

[8]The MATLAB code used to evaluate (5.5.10) summed the first 30 terms.

[9]At the upper limit the integrand does become the indeterminate $\frac{0}{0}$, but we can use L'Hospital's rule to compute the perfectly respectable $\lim_{x\to 1}\frac{1-x^m}{1-x^n} = \lim_{x\to 1}\frac{-m\,x^{m-1}}{-n\,x^{n-1}} = \frac{m}{n}$.

(C5.2) Show that

$$\int_1^\infty \frac{\{x\}}{x^2}\,dx = 1 - \gamma$$

where $\{x\}$ is the fractional part of x. To give you confidence that this is correct, MATLAB calculates *integral(@(x)(x-floor(x))./x.^2,1,inf)* $= 0.42278\ldots$, while $1 - \gamma = 0.42278\ldots.$. Hint: you might find it helpful to look back at Chap. 1 and review the derivation of (1.8.1).

(C5.3) If you *really* understand the trick involved in doing integrals with fractional-part integrands then you should now be able to show that the great nemesis of the great Euler, $\zeta(3)$, can be written as the quite interesting integral

$$\zeta(3) = \frac{3}{2} - 3\int_1^\infty \frac{\{x\}}{x^4}\,dx.$$

To give you confidence that this is correct, recall that the value of $\zeta(3)$ is $1.20205\ldots$, while MATLAB calculates *1.5-3*integral(@(x)(x-floor(x))./x.^4,1,inf)* $=$ $1.20205\ldots$.

(C5.4) It can be shown (using the contour integration technique we'll discuss in Chap. 8) that $\int_0^\infty \frac{dx}{(x+a)\{\ln^2(x)+\pi^2\}} = \frac{1}{1-a} + \frac{1}{\ln(a)}$, $a > 0$. Each term on the right blows-up when $a = 1$, and so it isn't immediately apparent what the value of the integral is when $a = 1$. The integrand, itself, doesn't do anything 'weird' at $a = 1$ however, and MATLAB encounters no problem at $a = 1$: *integral(@(x)1./((x+1).*(log(x).^2 +pi^2)),0,inf)* $= 0.486\ldots$. Using the power series expansion for $\ln(1 + x)$ for $-1 < x < 1$, show the actual value of the integral at $a = 1$ is $\frac{1}{2}$. (Notice that $\frac{1}{1-a}$ and $\frac{1}{\ln(a)}$ individually blow-up in *opposite* directions as $a \to 1$, either from below or from above, and so it is *a priori* plausible that their sum could be finite.)

(C5.5) For the case of $s = 2$, (5.3.7) says that

$$\zeta(2) = \frac{1}{1 - 2^{-1}}\sum_{k=1}^\infty \frac{(-1)^{k-1}}{k^2} = \frac{1}{1-\frac{1}{2}}\left[\frac{1}{1^2} - \frac{1}{2^2} + \frac{1}{3^2} - \frac{1}{4^2} + \cdots\right]$$

$$= 2\left[\frac{1}{1^2} - \frac{1}{2^2} + \frac{1}{3^2} - \frac{1}{4^2} + \cdots\right].$$

But by Euler's original definition we know that

$$\zeta(2) = \frac{1}{1^2} + \frac{1}{2^2} + \frac{1}{3^2} + \frac{1}{4^2} + \cdots.$$

Thus, it must be true that

$$2\left[\frac{1}{1^2} - \frac{1}{2^2} + \frac{1}{3^2} - \frac{1}{4^2} + \cdots\right] = \frac{1}{1^2} + \frac{1}{2^2} + \frac{1}{3^2} + \frac{1}{4^2} + \cdots.$$

Starting with the expression on the left, show that it is, *indeed*, equal to the expression on the right.

(C5.6) Show, in a way different from the derivation of (5.3.12) in the text, that

$$\int_0^1 \frac{\ln^2(1-x)}{x} dx = 2\zeta(3).$$

Hint: try the change of variable $1 - x = e^{-t}$, and then remember (5.3.4).

(C5.7) Show that $\int_0^1 \frac{\{-\ln(x)\}^p}{1-x} dx = \Gamma(p+1)\zeta(p+1), p > 0$. Notice that the case of $p = 1$ says $\zeta(2) = \int_0^1 \frac{-\ln(x)}{1-x} dx = \frac{\pi^2}{6} = 1.644934\ldots$ and MATLAB agrees: *integral (@(x)-log(x)./(1-x),0,1)* = 1.644934.... Hint: Make the appropriate change of variable in (5.3.4). Notice, too, that if you make the change of variable $u = 1 - x$ then the result for $p = 1$ gives another derivation of (5.2.2). You should confirm this.

(C5.8) Right after deriving (5.3.1) we showed that $\int_0^1 \int_0^1 \frac{1}{1-xy} dx\, dy = \zeta(2)$. Show that this is the $n = 2$ special case of a more general integration over an n-dimensional unit hypercube. That is, show that $\int_0^1 \int_0^1 \cdots \int_0^1 \frac{1}{1-x_1 x_2 \ldots x_n} dx_1\, dx_2 \ldots dx_n = \zeta(n)$.

(C5.9) Show that $\int_0^\infty \ln\left(\frac{e^x+1}{e^x-1}\right) dx = \frac{\pi^2}{4}$.

(C5.10) Show that $\int_0^\infty e^{-x} \ln^2(x) dx = \gamma^2 + \frac{\pi^2}{6}$. MATLAB agrees with this, as $\gamma^2 + \frac{\pi^2}{6}$ equals 1.978111..., and *integral(@(x)exp(-x).*(log(x).^2),0,inf)* = 1.978112.... Hint: Start with $I(m) = \int_0^\infty x^m e^{-x} dx$ and then think about how to express $I(m)$ as the *second* derivative of a gamma function. To do the differentiations involved, you'll need to apply a *double* dose of the digamma function.

(C5.11) Starting with (5.4.1), show that $\gamma = \int_0^1 \frac{1-e^{-x}-e^{-\frac{1}{x}}}{x} dx$. This is a particularly useful expression for γ because, over the entire *finite* interval of integration, the integrand is *finite*. MATLAB computes the integral as *integral(@(x)(1-exp(-x)-exp(-1./x))./x,0,1)* = 0.577215...

(C5.12) During one of his Hughes/Malibu lectures that I mentioned in the new Preface, Feynman stated that one integral he thought requires "special ingenuity" is $\int_0^\infty \frac{x}{\sinh(x)} dx = \frac{\pi^2}{4}$. Show that Feynman's value is correct, but that to get it is actually a pretty straightforward calculation *if* one uses a power series expansion. Do this by deriving the more general result $\int_0^\infty \frac{x^{s-1}}{\sinh(x)} dx = 2(1 - 2^{-s})\Gamma(s)\zeta(s)$, for which Feynman's integral is the special case $s = 2$. Hint: Start by writing $\sinh(x) = \frac{e^x - e^{-x}}{2}$, and so $\frac{1}{\sinh(x)} = \frac{2}{e^x - e^{-x}} = 2\frac{e^{-x}}{1-e^{-2x}}$. Finally, keep (4.1.1) and (5.3.5) in mind.

(C5.13) Show that $\int_0^1 \int_0^1 \frac{dxdy}{1+x^2y^2} = G$ (Catalan's constant). We can numerically check this claim by typing *integral2(@(x,y)1./(1+(x.*y).^2),0,1,0,1) = 0.9159655....*

(C5.14) Show that $\int_0^\infty \frac{\ln^2(x)}{1+x^2} dx = \frac{\pi^3}{8}$. MATLAB agrees, as $\frac{\pi^3}{8} = 3.875784\ldots$ and *integral(@(x)(log(x).^2)./(1+x.^2),0,inf) = 3.875784....* Hint: Transform $\int_0^\infty$ into $\int_0^1$, make the obvious power series expansion, and recall the result from C4.2. You may also find it useful to know that Euler proved that $\frac{1}{1^3} - \frac{1}{3^3} + \frac{1}{5^3} - \frac{1}{7^3} + \ldots = \frac{\pi^3}{32}$ (for how he did that, at the level of this book, see my *Dr. Euler's Fabulous Formula*, Princeton 2017, p. 149).

(C5.15) Show that $\int_0^1 \frac{\ln(1-x)\ln(1+x)}{x} dx = -\frac{5}{8}\zeta(3)$. MATLAB agrees as

$$-(5/8)*zeta(3) = -0.7512855\ldots \text{ while } integral(@(x)log(1-x).*log(1+x)./x,0,1)$$
$$= -0.7512855\ldots$$

Hint: In the algebraic identity $(A + B)^2 = A^2 + 2AB + B^2$ set $A = \ln(1 - x)$, $B = \ln(1 + x)$, do the obvious division by x and integration, and then recall (5.3.10), (5.3.12), and (5.3.13).

(C5.16) Show that $\int_0^1 \frac{\ln^3(1-x)}{x} dx = -6\zeta(4) = -\frac{\pi^4}{15}$. MATLAB agrees, as

$$-(pi^4)/15 = -6.4939394\ldots \text{ while } integral(@(x)((log(1-x)).^3)./x,0,1)$$
$$= -6.4939394\ldots.$$

Hint: Change variable to $y = 1 - x$, write $\frac{1}{1-y}$ as a power series, and recall the result in (C1.2).

(C5.17) What is $\int_0^1 \int_0^1 \frac{(xy)^2 \ln(xy)}{1-xy} dxdy = ?$ Hint: Use (5.3.2). Check your answer against MATLAB's numerical evaluation of the integral:

$$integral2(@(x,y)((x*y).^2).*log(x*y)./(1-x*y),0,1,0,1)$$
$$= -0.154113806\ldots.$$

(C5.18) In the text we showed that $\int_0^1 \int_0^1 \frac{dxdy}{1-xy} = \zeta(2)$. Here we up-the-ante just a bit. Calculate the values of $\int_0^1 \int_0^1 \frac{dxdy}{1-x^2y^2}$ and $\int_0^1 \int_0^1 \frac{dxdy}{1+xy}$. Hint: Start by defining

$$D = \int_0^1 \int_0^1 \frac{dxdy}{1-xy} - \int_0^1 \int_0^1 \frac{dxdy}{1+xy} = \int_0^1 \int_0^1 \frac{2xy\,dxdy}{1-x^2y^2}$$

and

$$S = \int_0^1 \int_0^1 \frac{dxdy}{1-xy} + \int_0^1 \int_0^1 \frac{dxdy}{1+xy} = 2 \int_0^1 \int_0^1 \frac{dxdy}{1-x^2y^2}$$

and then (in D)change variables to $u = x^2$ and $v = y^2$. Then, add the two expressions and do a little algebra.

(**C5.19**) In 1882 the German mathematician Adolf Hurwitz (1859–1919) generalized the Riemann zeta function to what is called the *Hurwitz zeta function*: $\zeta_h(s, a) = \sum_{n=0}^{\infty} \frac{1}{(n+a)^s}$, $a > 0$. Notice that $\zeta_h(s, 1) = \sum_{n=0}^{\infty} \frac{1}{(n+1)^s} = \frac{1}{1^s} + \frac{1}{2^s} + \ldots = \zeta(s)$. Following the same approach we used to get (5.3.4), show that $\Gamma(s)\zeta_h(s, a) = \int_0^{\infty} \frac{e^{(1-a)x}}{e^x - 1} x^{s-1} dx$. Notice that this reduces to (5.3.4) if $a = 1$.

Chapter 6
Some Not-So-Easy Integrals

6.1 Bernoulli's Integral

As I mentioned in the original Preface, in 1697 John Bernoulli evaluated the exotic-looking integral

$$\int_0^1 x^x dx.$$

How can this be done? And what about other similar integrals, ones that Bernoulli's integral might inspire, like

$$\int_0^1 x^{-x} dx$$

and

$$\int_0^1 x^{x^2} dx$$

and

$$\int_0^1 x^{\sqrt{x}} dx$$

and ... well, you get the picture! In this opening section I'll show you a unified way to do all of these calculations.

We start with the identity

© Springer Nature Switzerland AG 2020   227
P. J. Nahin, *Inside Interesting Integrals*, Undergraduate Lecture Notes in Physics,
https://doi.org/10.1007/978-3-030-43788-6_6

$$x^{cx^a} = e^{\ln\left(x^{cx^a}\right)} = e^{cx^a \ln(x)}$$

where a and c are constants. Then, since the power series expansion of the exponential is

$$e^y = 1 + y + \frac{y^2}{2!} + \frac{y^3}{3!} + \cdots$$

then, with $y = cx^a \ln(x)$, we have

$$x^{cx^a} = 1 + cx^a \ln(x) + \frac{1}{2!}c^2 x^{2a} \ln^2(x) + \frac{1}{3!}c^3 x^{3a} \ln^3(x) + \cdots$$

and so

$$\int_0^1 x^{cx^a} dx = \int_0^1 dx + c\int_0^1 x^a \ln(x)dx + \frac{c^2}{2!}\int_0^1 x^{2a} \ln^2(x)dx$$
$$+ \frac{c^3}{3!}\int_0^1 x^{3a} \ln^3(x)dx + \cdots. \qquad (6.1.1)$$

You can do all of the integrals in (6.1.1) if you did the challenge problem I gave you at the end of Chap. 4: the derivation of

$$\int_0^1 x^m \ln^n(x)dx = (-1)^n \frac{n!}{(m+1)^{n+1}}.$$

If you didn't do this (or got stuck) now is a good time to take a look at the Challenge Problem solutions at the end of the book. All of the integrals in (6.1.1) are of this form, with different values for m and n.[1]

Using this general result on each of the integrals on the right-hand side of (6.1.1), we arrive at

$$\int_0^1 x^{cx^a} dx = 1 - \frac{c}{(a+1)^2} + \frac{c^2}{2!}\left\{\frac{2!}{(2a+1)^3}\right\} - \frac{c^3}{3!}\left\{\frac{3!}{(3a+1)^4}\right\} + \frac{c^4}{4!}\left\{\frac{4!}{(4a+1)^5}\right\}$$
$$- \cdots$$

or,

---

[1]This is *not* the way Bernoulli did his original evaluation, but rather is the modern way. The evaluation of $\int_0^1 x^m \ln^n(x)dx$ that I give in the solutions uses the gamma function, which was still in the future in Bernoulli's day. Bernoulli used repeated integration by parts, which in fact is perfectly fine. The lack of a specialized tool doesn't stop a genius!

$$\int_0^1 x^{cx^a} dx = 1 - \frac{c}{(a+1)^2} + \frac{c^2}{(2a+1)^3} - \frac{c^3}{(3a+1)^4} + \frac{c^4}{(4a+1)^5} - \cdots. \qquad (6.1.2)$$

So, with $a = c = 1$ we have Bernoulli's integral:

$$\int_0^1 x^x dx = 1 - \frac{1}{2^2} + \frac{1}{3^3} - \frac{1}{4^4} + \frac{1}{5^5} - \cdots. \qquad (6.1.3)$$

It is easy to write a little program to sum the right-hand side; using the first ten terms, we get $0.78343\ldots$, and in agreement is $integral(@(x)x.^x,0,1) = 0.78343\ldots$. If $c = -1$ and $a = 1$ then (6.1.2) becomes

$$\int_0^1 x^{-x} dx = 1 + \frac{1}{2^2} + \frac{1}{3^3} + \frac{1}{4^4} + \frac{1}{5^5} + \cdots. \qquad (6.1.4)$$

Summing the first ten terms on the right gives $1.29128\ldots$, and in agreement is $integral(@(x)x.^(-x),0,1) = 1.29128\ldots$. By the way, since (6.1.4) can be written as

$$\int_0^1 \frac{1}{x^x} dx = \sum_{k=1}^{\infty} \frac{1}{k^k}$$

it is sometimes called the *sophomore's dream* because, while the similar forms on each side of the equality 'look too good to be true,' it *is* a true statement.

If $c = 1$ and $a = 2$ then (6.1.2) becomes

$$\int_0^1 x^{x^2} dx = 1 - \frac{1}{3^2} + \frac{1}{5^3} - \frac{1}{7^4} + \frac{1}{9^5} - \cdots \qquad (6.1.5)$$

Summing the first six terms on the right gives $0.896488\ldots$ and MATLAB agrees as $integral(@(x)x.^(x.^2),0,1) = 0.896488\ldots$.

If $c = 1$ and $a = \frac{1}{2}$ then (6.1.2) becomes

$$\int_0^1 x^{\sqrt{x}} dx = 1 - \frac{1}{\left(\frac{3}{2}\right)^2} + \frac{1}{\left(\frac{4}{2}\right)^3} - \frac{1}{\left(\frac{5}{2}\right)^4} + \frac{1}{\left(\frac{6}{2}\right)^5} - \cdots$$

or,

$$\int_0^1 x^{\sqrt{x}} dx = 1 - \left(\frac{2}{3}\right)^2 + \left(\frac{2}{4}\right)^3 - \left(\frac{2}{5}\right)^4 + \left(\frac{2}{6}\right)^5 - \cdots \qquad (6.1.6)$$

Summing the first ten terms on the right gives $0.658582\ldots$ and MATLAB agrees as $integral(@(x)x.^sqrt(x),0,1) = 0.658582\ldots$.

6.2   Ahmed's Integral

In this section we'll do Ahmed's integral, named after the Indian mathematical physicist Zafar Ahmed who proposed it in 2002. Interesting in its own right, we'll also use it in the next section to do the derivation of Coxeter's integral that I mentioned in the original Preface. Ahmed's integral is,

$$\int_0^1 \frac{\tan^{-1}\left(\sqrt{2+x^2}\right)}{(1+x^2)\sqrt{2+x^2}}\,dx, \tag{6.2.1}$$

and it can be done using Feynman's favorite trick of differentiating under the integral sign. That is, we'll start with a 'u-parameterized' version of (6.2.1),

$$I(u) = \int_0^1 \frac{\tan^{-1}\left(u\sqrt{2+x^2}\right)}{(1+x^2)\sqrt{2+x^2}}\,dx, \tag{6.2.2}$$

and then differentiate it with respect to u. I(1) is Ahmed's integral.

Notice that if $u \to \infty$ then the argument of the inverse tangent also $\to \infty$ for all $x > 0$ and so, since $\tan^{-1}(\infty) = \frac{\pi}{2}$, we have

$$I(\infty) = \frac{\pi}{2}\int_0^1 \frac{dx}{(1+x^2)\sqrt{2+x^2}}. \tag{6.2.3}$$

The integral in (6.2.3) is easy to do once you recall the standard differentiation formula

$$\frac{d}{dx}\tan^{-1}\{f(x)\} = \frac{1}{1+f^2(x)}\left(\frac{df}{dx}\right).$$

If we use this formula to calculate

$$\frac{d}{dx}\tan^{-1}\left\{\frac{x}{\sqrt{2+x^2}}\right\},$$

then *you* should confirm that

$$\frac{d}{dx}\tan^{-1}\left\{\frac{x}{\sqrt{2+x^2}}\right\} = \frac{1}{(1+x^2)\sqrt{2+x^2}}$$

which is the integrand of (6.2.3). That is,

$$I(\infty) = \frac{\pi}{2} \int_0^1 \frac{d}{dx} \tan^{-1}\left\{ \frac{x}{\sqrt{2+x^2}} \right\} dx = \frac{\pi}{2} \left[\tan^{-1}\left\{ \frac{x}{\sqrt{2+x^2}} \right\} \right] \Big|_0^1$$

$$= \frac{\pi}{2} \left[\tan^{-1}\left\{ \frac{1}{\sqrt{3}} \right\} - \tan^{-1}\{0\} \right]$$

$$= \left(\frac{\pi}{2} \right)\left(\frac{\pi}{6} \right)$$

or,

$$I(\infty) = \frac{\pi^2}{12}.$$

Now, differentiate (6.2.2) with respect to u, again using

$$\frac{d}{dx} \tan^{-1}\{f(u)\} = \frac{1}{1 + f^2(u)} \left(\frac{df}{du} \right),$$

with $f(u) = u\sqrt{2 + x^2}$. Then, with just a bit of algebra,

$$\frac{dI}{du} = \int_0^1 \frac{dx}{(1+x^2)(1+2u^2+u^2x^2)}.$$

With a partial fraction expansion this becomes

$$\frac{dI}{du} = \int_0^1 \frac{1}{(1+u^2)} \left[\frac{1}{1+x^2} - \frac{u^2}{1+2u^2+u^2x^2} \right] dx$$

or,

$$\frac{dI}{du} = \frac{1}{(1+u^2)} \left[\int_0^1 \frac{dx}{1+x^2} - \int_0^1 \frac{dx}{\frac{1+2u^2}{u^2} + x^2} \right].$$

These last two integrals are each of the form

$$\int \frac{dx}{a^2 + x^2} = \frac{1}{a} \tan^{-1}\left(\frac{x}{a} \right)$$

and so, doing the integrals, we have

$$\frac{dI}{du} = \frac{1}{(1+u^2)} \left[\tan^{-1}(x) - \frac{u}{\sqrt{1+2u^2}} \tan^{-1}\left(\frac{xu}{\sqrt{1+2u^2}} \right) \right] \Big|_0^1$$

or,

$$\frac{dI}{du} = \frac{1}{(1 + u^2)} \left[\frac{\pi}{4} - \frac{u}{\sqrt{1 + 2u^2}} \tan^{-1}\left(\frac{u}{\sqrt{1 + 2u^2}}\right) \right]. \qquad (6.2.4)$$

Next, integrate both sides of (6.2.4) from 1 to ∞ with respect to u. On the left we get

$$\int_1^\infty \frac{dI}{du} du = \int_1^\infty dI = I(\infty) - I(1),$$

and on the right we get

$$\frac{\pi}{4} \int_1^\infty \frac{du}{1 + u^2} - \int_1^\infty \frac{u}{(1 + u^2)\sqrt{1 + 2u^2}} \tan^{-1}\left(\frac{u}{\sqrt{1 + 2u^2}}\right) du.$$

The first integral is easy:

$$\frac{\pi}{4} \int_1^\infty \frac{du}{1 + u^2} = \frac{\pi}{4} \left[\tan^{-1}(\infty) - \tan^{-1}(1) \right] = \frac{\pi}{4} \left[\frac{\pi}{2} - \frac{\pi}{4} \right] = \frac{\pi^2}{16}.$$

Thus,

$$I(\infty) - I(1) = \frac{\pi^2}{16} - \int_1^\infty \frac{u}{(1 + u^2)\sqrt{1 + 2u^2}} \tan^{-1}\left(\frac{u}{\sqrt{1 + 2u^2}}\right) du. \qquad (6.2.5)$$

That final integral looks pretty awful—but looks are deceiving. The integral yields with not even a whimper if we make the change of variable $t = \frac{1}{u}$ (and so $du = -\frac{1}{t^2} dt$) as follows:

$$\int_1^\infty \frac{u}{(1 + u^2)\sqrt{1 + 2u^2}} \tan^{-1}\left(\frac{u}{\sqrt{1 + 2u^2}}\right) du$$

$$= \int_1^0 \frac{\frac{1}{t}}{(1 + \frac{1}{t^2})\sqrt{1 + \frac{2}{t^2}}} \tan^{-1}\left(\frac{\frac{1}{t}}{\sqrt{1 + \frac{2}{t^2}}}\right) \left(-\frac{1}{t^2} dt\right)$$

$$= \int_0^1 \frac{\frac{1}{t}}{(t^2 + 1)\frac{\sqrt{t^2 + 2}}{t}} \tan^{-1}\left(\frac{\frac{1}{t}}{\frac{\sqrt{t^2 + 2}}{t}}\right) dt = \int_0^1 \frac{1}{(t^2 + 1)\sqrt{t^2 + 2}} \tan^{-1}\left(\frac{1}{\sqrt{t^2 + 2}}\right) dt.$$

Now, recall the identity

$$\tan^{-1}(s) + \tan^{-1}\left(\frac{1}{s}\right) = \frac{\pi}{2}$$

which becomes instantly obvious if you draw a right triangle with perpendicular sides of lengths 1 and s and remember that the two acute angles add to $\frac{\pi}{2}$. This says

$$\tan^{-1}\left(\frac{1}{\sqrt{t^2+2}}\right) = \frac{\pi}{2} - \tan^{-1}\left(\sqrt{t^2+2}\right)$$

and so we can write

$$\int_0^1 \frac{1}{(t^2+1)\sqrt{t^2+2}} \tan^{-1}\left(\frac{1}{\sqrt{t^2+2}}\right) dt = \frac{\pi}{2} \int_0^1 \frac{dt}{(t^2+1)\sqrt{t^2+2}}$$
$$- \int_0^1 \frac{\tan^{-1}\left(\sqrt{t^2+2}\right)}{(t^2+1)\sqrt{t^2+2}} dt.$$

That is, (6.2.5) becomes

$$I(\infty) - I(1) = \frac{\pi^2}{16} - \frac{\pi}{2} \int_0^1 \frac{dt}{(t^2+1)\sqrt{t^2+2}} + \int_0^1 \frac{\tan^{-1}\left(\sqrt{t^2+2}\right)}{(t^2+1)\sqrt{t^2+2}} dt,$$

and you should now see that *two* wonderful things have happened. First, if you look back at (6.2.3) you'll see that the first integral on the right is $I(\infty)$. Second, the rightmost integral, from (6.2.1), is just $I(1)$, that is, *Ahmed's integral*! So,

$$I(\infty) - I(1) = \frac{\pi^2}{16} - I(\infty) + I(1)$$

and so

$$2I(\infty) - \frac{\pi^2}{16} = 2I(1)$$

or, at last,

$$I(1) = I(\infty) - \frac{\pi^2}{32} = \frac{\pi^2}{12} - \frac{\pi^2}{32}$$

and we have our answer:

$$\int_0^1 \frac{\tan^{-1}\left(\sqrt{2+x^2}\right)}{(1+x^2)\sqrt{2+x^2}} dx = \frac{5\pi^2}{96} \qquad (6.2.6)$$

This is equal to 0.51404189 ..., and MATLAB agrees, as *integral(@(x)atan(sqrt (2 + x.^2))./((1 + x.^2).*sqrt(2 + x.^2)),0,1) = 0.51404189....*

6.3   Coxeter's Integral

In this section we'll do the integral that the young H. S. M. Coxeter pleaded for help with in the original Preface, a plea which the great Hardy answered. I don't know the details of what Hardy sent to Coxeter, and so here I'll show you an analysis that makes use of (6.2.6). The integral we are going to evaluate is

$$I = \int_0^{\pi/2} \cos^{-1}\left\{ \frac{\cos(x)}{1 + 2\cos(x)} \right\} dx, \qquad (6.3.1)$$

and it will be a pretty long haul; prepare yourself for the longest derivation in this book. I've tried to make every step crystal clear but still, in the immortal words of Bette Davis in her 1950 film *All About Eve*, "Fasten your seatbelts, it's going to be a bumpy night."

To start our analysis, we'll cast (6.3.1) into a different form. The double-angle formula from trigonometry for the cosine says that, for any θ,

$$\cos(2\theta) = 2\cos^2(\theta) - 1. \qquad (6.3.2)$$

If we write $u = \cos(\theta)$—and so $\theta = \cos^{-1}(u)$—then (6.3.2) says that

$$\cos(2\theta) = 2u^2 - 1,$$

from which it immediately follows that

$$\cos^{-1}(2u^2 - 1) = \cos^{-1}\{\cos(2\theta)\} = 2\theta = 2\cos^{-1}(u).$$

So, since u is simply an arbitrary variable (as is θ) we can write

$$\cos^{-1}(2\theta^2 - 1) = 2\cos^{-1}(\theta). \qquad (6.3.3)$$

Next, writing $\alpha = 2\theta^2 - 1$ (which means $\theta = \sqrt{\frac{1+\alpha}{2}}$), we have from (6.3.3) that

$$\cos^{-1}(\alpha) = 2\cos^{-1}\left(\sqrt{\frac{1+\alpha}{2}} \right).$$

Looking back at (6.3.1), let's write

$$\alpha = \frac{\cos(x)}{1 + 2\cos(x)}$$

and so we have

$$\cos^{-1}\left(\frac{\cos(x)}{1+2\cos(x)}\right) = 2\cos^{-1}\left(\sqrt{\frac{1+\frac{\cos(x)}{1+2\cos(x)}}{2}}\right)$$

$$= 2\cos^{-1}\left(\sqrt{\frac{1+3\cos(x)}{2+4\cos(x)}}\right). \qquad (6.3.4)$$

Now, if you apply the Pythagorean theorem to a right triangle with an acute angle whose cosine is $\sqrt{\frac{1+3\cos(x)}{2+4\cos(x)}}$, you'll see that the tangent of that same angle is $\sqrt{\frac{1+\cos(x)}{1+3\cos(x)}}$. That is,

$$\cos^{-1}\left(\frac{\cos(x)}{1+2\cos(x)}\right) = 2\tan^{-1}\left\{\sqrt{\frac{1+\cos(x)}{1+3\cos(x)}}\right\}$$

and so Coxeter's integral I in (6.3.1) becomes

$$I = 2\int_0^{\pi/2} \tan^{-1}\left\{\sqrt{\frac{1+\cos(x)}{1+3\cos(x)}}\right\} dx. \qquad (6.3.5)$$

Make the change of variable $x = 2y$ (and so $dx = 2dy$) which converts (6.3.5) to

$$I = 4\int_0^{\pi/4} \tan^{-1}\left\{\sqrt{\frac{1+\cos(2y)}{1+3\cos(2y)}}\right\} dy.$$

Using (6.3.2) again, and then applying a bit of algebraic simplification, we have

$$\sqrt{\frac{1+\cos(2y)}{1+3\cos(2y)}} = \frac{\cos(y)}{\sqrt{2-3\sin^2(y)}}$$

and so

$$I = 4\int_0^{\pi/4} \tan^{-1}\left\{\frac{\cos(y)}{\sqrt{2-3\sin^2(y)}}\right\} dy. \qquad (6.3.6)$$

Now, put (6.3.6) aside for the moment and notice the following (which may appear to be out of left-field, but be patient and you'll see its relevance soon):

$$\int_0^1 \frac{1}{1 + \left[\frac{\cos^2(y)}{2\,-3\sin^2(y)}\right]t^2}\,dt$$

is of the form

$$\int_0^1 \frac{1}{1 + b^2 t^2}\,dt = \frac{1}{b^2}\int_0^1 \frac{1}{\frac{1}{b^2} + t^2}\,dt = \frac{1}{b^2}\{b\,\tan^{-1}(bt)\}\bigg|_0^1 = \frac{1}{b}\tan^{-1}(b),\,b$$

$$= \frac{\cos(y)}{\sqrt{2 - 3\sin^2(y)}}.$$

Thus,

$$\int_0^1 \frac{1}{1 + \left[\frac{\cos^2(y)}{2\,-3\sin^2(y)}\right]t^2}\,dt = \frac{\sqrt{2 - 3\sin^2(y)}}{\cos(y)}\tan^{-1}\left\{\frac{\cos(y)}{\sqrt{2 - 3\sin^2(y)}}\right\}.$$

That is, the integrand of (6.3.6) is given by

$$\tan^{-1}\left\{\frac{\cos(y)}{\sqrt{2 - 3\sin^2(y)}}\right\} = \frac{\cos(y)}{\sqrt{2 - 3\sin^2(y)}}\int_0^1 \frac{1}{1 + \left[\frac{\cos^2(y)}{2\,-3\sin^2(y)}\right]t^2}\,dt$$

and so (6.3.6) is, itself, the *double* integral

$$I = 4\int_0^{\pi/4} \frac{\cos(y)}{\sqrt{2 - 3\sin^2(y)}}\left\{\int_0^1 \frac{1}{1 + \left[\frac{\cos^2(y)}{2\,-3\sin^2(y)}\right]t^2}\,dt\right\}dy. \qquad (6.3.7)$$

Wow! This may look like we've made things (a lot) worse. Well, hang in there because they're going to appear to get *even worse* before they get better—but they *will* get (a lot) better, although not for a while.

Continuing, we have

$$I = \int_0^{\pi/4}\int_0^1 \frac{4\cos(y)\{2 - 3\sin^2(y)\}}{\sqrt{2 - 3\sin^2(y)}\{2 - 3\sin^2(y) + t^2\cos^2(y)\}}\,dt\,dy$$

$$= \int_0^{\pi/4}\int_0^1 \frac{4\cos(y)\sqrt{2 - 3\sin^2(y)}}{2 - 3\sin^2(y) + t^2 - t^2\sin^2(y)}\,dt\,dy$$

or,

$$I = \int_0^{\pi/4} \int_0^1 \frac{4\cos(y)\sqrt{2 - 3\sin^2(y)}}{(t^2 + 2) - (t^2 + 3)\sin^2(y)}\, dt\, dy. \tag{6.3.8}$$

Next, make the change of variable $\sin(y) = \sqrt{\frac{2}{3}}\sin(w)$ in (6.3.8), and so $dy = \sqrt{\frac{2}{3}}\frac{\cos(w)}{\cos(y)}\, dw$. We have $w = 0$ when $y = 0$, and when $y = \frac{\pi}{4}$ we have $\sin\left(\frac{\pi}{4}\right) = \frac{1}{\sqrt{2}}$ and so $\sin(w) = \left(\sqrt{\frac{3}{2}}\right)\left(\frac{1}{\sqrt{2}}\right) = \frac{\sqrt{3}}{2}$ which says $w = \frac{\pi}{3}$. So

$$I = \int_0^{\pi/3} \int_0^1 \frac{4\cos(y)\sqrt{2 - 3\frac{2}{3}\sin^2(w)}}{(t^2 + 2) - (t^2 + 3)\frac{2}{3}\sin^2(w)}\, dt\, \sqrt{\frac{2}{3}}\frac{\cos(w)}{\cos(y)}\, dw$$

$$= \int_0^{\pi/3} \int_0^1 \frac{4\sqrt{2 - 2[1 - \cos^2(w)]}}{(t^2 + 2) - (t^2 + 3)\frac{2}{3}[1 - \cos^2(w)]}\, dt\, \sqrt{\frac{2}{3}}\cos(w)\, dw$$

$$= \int_0^{\pi/3} \int_0^1 \frac{4\sqrt{2}\cos(w)\sqrt{2}\cos(w)}{(t^2 + 2) - (t^2 + 3)\frac{2}{3}[1 - \cos^2(w)]}\, dt\, \frac{1}{\sqrt{3}}\, dw$$

and, after some straightforward algebra which I'm going to let you fill-in, we arrive at

$$I = \int_0^{\pi/3} \int_0^1 \frac{8\sqrt{3}\cos^2(w)}{t^2 + (2t^2 + 6)\cos^2(w)}\, dt\, dw. \tag{6.3.9}$$

Our next step is another change of variable, to $s = \tan(w)$. Thus, as $\tan(w) = \frac{\sin(w)}{\cos(w)}$, we have

$$\frac{ds}{dw} = \frac{\cos^2(w) + \sin^2(w)}{\cos^2(w)} = \frac{1}{\cos^2(w)},$$

and so $dw = \cos^2(w)\, ds$. Since

$$1 + s^2 = 1 + \tan^2(w) = 1 + \frac{\sin^2(w)}{\cos^2(w)} = \frac{1}{\cos^2(w)}$$

we have

$$\frac{1}{1 + s^2} = \cos^2(w)$$

and so

$$dw = \frac{ds}{1+s^2}.$$

Therefore, since $s = 0$ when $w = 0$, and $s = \sqrt{3}$ when $w = \frac{\pi}{3}$, we have

$$I = \int_0^{\sqrt{3}} \int_0^1 \frac{8\sqrt{3}\frac{1}{1+s^2}}{t^2 + (2t^2+6)\frac{1}{1+s^2}} \, dt \, \frac{ds}{1+s^2}$$

$$= \int_0^{\sqrt{3}} \int_0^1 \frac{8\sqrt{3}}{t^2(1+s^2)^2 + (2t^2+6)(1+s^2)} \, dt \, ds$$

or, after again some more straightforward algebra which you can do, we arrive at

$$I = \int_0^{\sqrt{3}} \int_0^1 \frac{8\sqrt{3}}{(1+s^2)(t^2s^2 + 3\,t^2 + 6)} \, dt \, ds. \qquad (6.3.10)$$

Krusty the Clown on the Simpson's TV cartoon-comedy show is fond of yelling, when frustrated, "Will it ever end?," and the answer here is, 'Not yet.' So, bravely plowing-on, let's now make a partial fraction expansion of the integrand in (6.3.10). That is, if we write

$$\frac{1}{(1+s^2)(t^2s^2 + 3\,t^2 + 6)} = \frac{A}{1+s^2} + \frac{B}{t^2s^2 + 3\,t^2 + 6}$$

it is then easy to confirm that

$$A = \frac{1}{2t^2+6}, B = -\frac{t^2}{2t^2+6}$$

and so

$$I = \int_0^{\sqrt{3}} \int_0^1 8\sqrt{3} \left[\frac{\frac{1}{2t^2+6}}{1+s^2} - \frac{\frac{t^2}{2t^2+6}}{t^2s^2 + 3\,t^2 + 6} \right] dt \, ds$$

which with just a little algebra (and a reversal of the order of integration) can be written as

$$I = \int_0^1 \frac{4\sqrt{3}}{t^2+3} \left\{ \int_0^{\sqrt{3}} \frac{ds}{1+s^2} - \int_0^{\sqrt{3}} \frac{ds}{s^2 + 3 + \frac{6}{t^2}} \right\} dt. \qquad (6.3.11)$$

The first inner integral on the right is easy:

$$\left\{ \tan^{-1}(s) \right\}\Big|_0^{\sqrt{3}} = \tan^{-1}\left(\sqrt{3}\right) = \frac{\pi}{3}.$$

The second inner integral is almost as easy, as it is equal to

$$\int_0^{\sqrt{3}} \frac{ds}{s^2 + \left[\sqrt{3 + \frac{6}{t^2}}\right]^2} = \frac{1}{\sqrt{3 + \frac{6}{t^2}}}$$

$$\times \left\{ \tan^{-1}\left(\frac{s}{\sqrt{3 + \frac{6}{t^2}}}\right) \right\}\Big|_0^{\sqrt{3}} = \tan^{-1}\left(\frac{st}{\sqrt{3}\sqrt{t^2 + 2}}\right)\Big|_0^{\sqrt{3}}$$

$$= \frac{t}{\sqrt{3}\sqrt{t^2 + 2}} \tan^{-1}\left(\frac{t}{\sqrt{t^2 + 2}}\right).$$

Thus,

$$I = \int_0^1 \frac{4\sqrt{3}}{t^2 + 3} \left\{ \frac{\pi}{3} - \frac{t}{\sqrt{3}\sqrt{t^2 + 2}} \tan^{-1}\left(\frac{t}{\sqrt{t^2 + 2}}\right) \right\} dt$$

$$= \frac{4\sqrt{3}\pi}{3} \int_0^1 \frac{dt}{t^2 + (\sqrt{3})^2} - 4\int_0^1 \left\{ \frac{t}{(t^2 + 3)\sqrt{t^2 + 2}} \tan^{-1}\left(\frac{t}{\sqrt{t^2 + 2}}\right) \right\} dt$$

$$= \frac{4\sqrt{3}\pi}{3} \left\{ \frac{1}{\sqrt{3}} \tan^{-1}\left(\frac{t}{\sqrt{3}}\right) \right\}\Big|_0^1 - 4\int_0^1 \left\{ \frac{t}{(t^2 + 3)\sqrt{t^2 + 2}} \tan^{-1}\left(\frac{t}{\sqrt{t^2 + 2}}\right) \right\} dt$$

or, as

$$\frac{4\sqrt{3}\pi}{3} \left\{ \frac{1}{\sqrt{3}} \tan^{-1}\left(\frac{t}{\sqrt{3}}\right) \right\}\Big|_0^1 = \frac{4\pi}{3} \tan^{-1}\left(\frac{1}{\sqrt{3}}\right) = \frac{4\pi}{3}\left(\frac{\pi}{6}\right) = \frac{2\pi^2}{9},$$

we have

$$I = \frac{2\pi^2}{9} - 4\int_0^1 \left\{ \frac{t}{(t^2 + 3)\sqrt{t^2 + 2}} \tan^{-1}\left(\frac{t}{\sqrt{t^2 + 2}}\right) \right\} dt. \qquad (6.3.12)$$

We are now in the homestretch, as we can do the integral in (6.3.12) by parts. To see this, let

$$u = \tan^{-1}\left(\frac{t}{\sqrt{t^2 + 2}}\right)$$

and

$$dv = \frac{t}{(t^2 + 3)\sqrt{t^2 + 2}} dt.$$

Then, remembering how to differentiate the inverse tangent from the opening discussion of this section, we have

$$\frac{du}{dt} = \frac{1}{(t^2 + 1)\sqrt{t^2 + 2}}.$$

And you can verify that

$$v = \tan^{-1}\left(\sqrt{t^2 + 2}\right)$$

by simply differentiating this v and observing that we get the above dv back. So, plugging all this into the integration by parts formula, we have

$$I = \frac{2\pi^2}{9} - 4\left[\left\{\tan^{-1}\left(\frac{t}{\sqrt{t^2 + 2}}\right)\tan^{-1}\left(\sqrt{t^2 + 2}\right)\right\}\bigg|_0^1 - \int_0^1 \frac{\tan^{-1}\left(\sqrt{t^2 + 2}\right)}{(t^2 + 1)\sqrt{t^2 + 2}} dt\right]$$

$$= \frac{2\pi^2}{9} - 4\left[\tan^{-1}\left(\frac{1}{\sqrt{3}}\right)\tan^{-1}\left(\sqrt{3}\right) - \int_0^1 \frac{\tan^{-1}\left(\sqrt{t^2 + 2}\right)}{(t^2 + 1)\sqrt{t^2 + 2}} dt\right]$$

$$= \frac{2\pi^2}{9} - 4\left[\left(\frac{\pi}{6}\right)\left(\frac{\pi}{3}\right) - \int_0^1 \frac{\tan^{-1}\left(\sqrt{t^2 + 2}\right)}{(t^2 + 1)\sqrt{t^2 + 2}} dt\right]$$

$$= \frac{2\pi^2}{9} - \frac{2\pi^2}{9} + 4\int_0^1 \frac{\tan^{-1}\left(\sqrt{t^2 + 2}\right)}{(t^2 + 1)\sqrt{t^2 + 2}} dt$$

and so,

$$I = 4\int_0^1 \frac{\tan^{-1}\left(\sqrt{t^2 + 2}\right)}{(t^2 + 1)\sqrt{t^2 + 2}} dt.$$

Now, look back at (6.2.6), our result for Ahmed's integral. *It is precisely the above integral.* Coxeter's integral is four times Ahmed's integral and so, *at last (!),*

$$\int_0^{\pi/2} \cos^{-1}\left\{\frac{\cos(x)}{1 + 2\cos(x)}\right\} dx = \frac{5\pi^2}{24}. \tag{6.3.13}$$

Wow! What a derivation! But is it correct? MATLAB says it is, as our theoretical answer of 2.05616758... matches *integral(@(x)acos(cos(x)./(1 + 2*cos(x))),0,pi/2)* = 2.05616758....

6.4 The Hardy-Schuster Optical Integral

In 1925 the German-born English physicist Arthur Schuster (1851–1934) published a paper on the theory of light. In that paper he encountered the intriguing integral

$$J = \int_0^\infty \{C^2(x) + S^2(x)\}dx, \tag{6.4.1}$$

where $C(x)$ and $S(x)$ are themselves integrals (called *Fresnel integrals*, and we'll see them again in the next chapter for the special case of $x = 0$):

$$C(x) = \int_x^\infty \cos\left(t^2\right)dt, \; S(x) = \int_x^\infty \sin\left(t^2\right)dt.$$

In fact, since we'll eventually need to know one of these two specific values—S (0)—to find J, here is its value now (we'll *derive* it in the next chapter as result (7.2.2)):

$$S(0) = \int_0^\infty \sin\left(t^2\right)dt = \frac{1}{2}\sqrt{\frac{\pi}{2}}. \tag{6.4.2}$$

Schuster was unable to evaluate J, but did write that the *physics* of the problem he was studying would be satisfied *if* J had a certain value. Alas, he couldn't show that J had that value and that is where he left matters.

Schuster's paper soon came to the attention of the great Hardy, who then (are you surprised and if so, why?) quickly computed J (confirming Schuster's conjecture) using *two* different approaches. One of them used sophisticated Fourier transform theory,[2] but in this section I'll show you an alternative 'freshman calculus' derivation that freely uses the physical interpretation of the Riemann integral. It is based on an idea that Hardy himself sketched in his other approach.

From the definitions of $C(x)$ and $S(x)$ we can write

$$C^2(x) = \int_x^\infty \cos\left(t^2\right)dt \int_x^\infty \cos\left(u^2\right)du = \int_x^\infty \int_x^\infty \cos\left(t^2\right)\cos\left(u^2\right)dt\,du$$

and

$$S^2(x) = \int_x^\infty \sin\left(t^2\right)dt \int_x^\infty \sin\left(u^2\right)du = \int_x^\infty \int_x^\infty \sin\left(t^2\right)\sin\left(u^2\right)dt\,du.$$

Thus,

---

[2]For a detailed discussion of Schuster's integral and Hardy's transform solution, see my book *Dr. Euler's Fabulous Formula*, Princeton 2006, pp. 263–274.

$$C^2(x) + S^2(x) = \int_x^\infty \int_x^\infty \left\{ \cos\left(t^2\right) \cos\left(u^2\right) + \sin\left(t^2\right) \sin\left(u^2\right) \right\} dt\, du.$$

From trigonometry we have the identity

$$\cos(a - b) = \cos(a)\cos(b) + \sin(a)\sin(b)$$

and so it immediately follows that

$$C^2(x) + S^2(x) = \int_x^\infty \int_x^\infty \cos\left(t^2 - u^2\right) dt\, du.$$

From (6.4.1) we then have

$$J = \int_0^\infty \left\{ \int_x^\infty \int_x^\infty \cos\left(t^2 - u^2\right) dt\, du \right\} dx. \tag{6.4.3}$$

(At this point you should take a look back at Sect. 3.6, where I showed you how to use MATLAB's Symbolic Math Toolbox to numerically evaluate this triple integral.)

Let's now agree to write $f(t, u) = \cos\left(t^2 - u^2\right)$, and so (6.4.3) becomes

$$J = \int_0^\infty \left\{ \int_x^\infty \int_x^\infty f(t, u) dt\, du \right\} dx. \tag{6.4.4}$$

We can express in words what (6.4.4) says as follows: The outermost integral (x) says 'starting with $x = 0$, evaluate $\left\{ \int_0^\infty \int_0^\infty f(t, u) dt\, du \right\} \Delta x$. Then, increment x by Δx to $x = \Delta x$ and evaluate $\left\{ \int_x^\infty \int_x^\infty f(t, u) dt\, du \right\} x$. Then increment x by Δx to $x = 2\Delta x$ and evaluate $\left\{ \int_{2x}^\infty \int_{2x}^\infty f(t, u) dt\, du \right\} x$. And so on. Then add all of these evaluations.'

We can reformulate this *mathematical* interpretation of (6.4.4) as a *physical* one by assigning the three variables x, t, and u to the axes of a three-dimensional Cartesian coordinate system, as shown in Fig. 6.4.1, where the x and u axes are in the plane of the page and the t-axis is perpendicular to the page (the positive t-axis is *into* the page).

Each of the individual integrals in our sum is simply the integral of $f(t,u)$ over the volume of an infinite 'slab' of thickness Δx, where the 'bottom' slab's corner[3] starts at the origin ($t = 0$, $u = 0$). The corners of the subsequent slabs lying above the bottom slab are gradually slid *up* (along the x-axis) and *away from* the origin along the line $t = u$. So, if you imagine yourself in space, hovering over the t, u-plane and

---

[3]Since the volume we are integrating over is infinite in extent in both the t and u directions, the corner at the origin is the *only* corner of the bottom slab.

Fig. 6.4.1 A coordinate
system

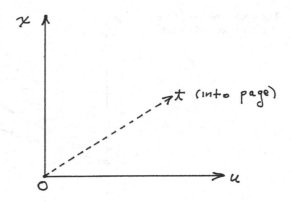

Fig. 6.4.2 A wedding cake
volume

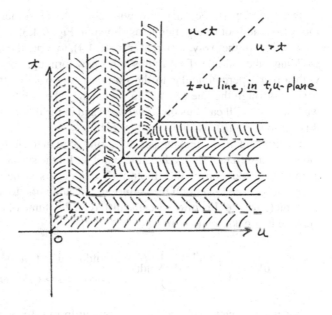

looking down along the x-axis, Fig. 6.4.2 is what you'd see. That is, the volume we
are integrating f(t, u) over looks like a *layered wedding cake*! The steps formed by
the layers have height and depth Δx. (For a carpenter building a staircase, these
would be the values of the *riser* and the *tread* of the steps, respectively.) That is,

$$J = \int_{\text{cake}} f(t, u)\, dV \qquad\qquad (6.4.5)$$

where dV is the differential volume of the wedding cake.

Fig. 6.4.3 The foot of the
wedding cake at the origin

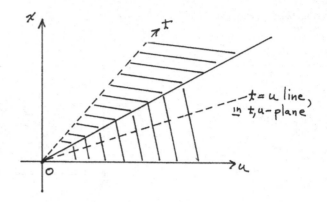

In a side view of the cake (now with $\Delta x \to 0$), the corner of the cake at the origin looks like the foot of a pyramid, as shown in Fig. 6.4.3.

There is another way, different from (6.4.4), to write the integral of $f(t, u)$ over the wedding cake volume. To start to set this new form up, notice that if we cut the cake with a plane perpendicular to the t,u-plane (the base of the cake) with the plane passing through the line $t = u$, then we cut the cake into two equal parts. In one half we have $u < t$ (I'll call this the *upper half*) and in the other half we have $u > t$ (I'll call this the *lower half*), as shown in Fig. 6.4.2.

We first pick a tiny rectangular 'footprint' dtdu in the t, u-plane, at (t, u). Then, we move upward (increasing x direction) until we hit the surface of the cake. For a given u and t (the location of the 'footprint') this occurs at height $x = \min(t, u)$. If the footprint is in the lower half then $\min(t, u) = t$, and if the footprint is in the upper half then $\min(t, u) = u$. In either case, the differential volume of this vertical plug through the wedding cake is given by

$$dV = \left\{ \int_0^{\min(t,u)} dx \right\} dtdu = \begin{cases} u \, dtdu, & u < t \; (\textit{upper half}) \\ t \, dtdu, & u > t \; (\textit{lower half}) \end{cases} \qquad (6.4.6)$$

For $u > t$ (lower half) we have u varying from 0 to ∞ and t varying from 0 to u. So, for the case of $u > t$ the integral in (6.4.5) is

$$J_{\text{lower}} = \int_0^\infty \left\{ \int_0^u t \, f(t, u) dt \right\} du.$$

For $u < t$ (upper half) we have t varying from 0 to ∞ and u varying from 0 to t. So, for the case of $u < t$ the integral in (6.4.5) is

$$J_{\text{upper}} = \int_0^\infty \left\{ \int_0^t u \, f(t, u) du \right\} dt.$$

Thus,

$$J = J_{\text{lower}} + J_{\text{upper}} = \int_0^\infty \left\{ \int_0^u t\, f(t, u)\, dt \right\} du + \int_0^\infty \left\{ \int_0^t u\, f(t, u)\, du \right\} dt$$

$$= \int_0^\infty \left\{ \int_0^u t\, \cos\left(t^2 - u^2\right) dt \right\} du + \int_0^\infty \left\{ \int_0^t u\, \cos\left(t^2 - u^2\right) du \right\} dt.$$

These last two integrals are obviously equal since, if we swap the two dummy variables t and u in either integral, we get the other integral. So,

$$J = 2 \int_0^\infty \left\{ \int_0^t u\, \cos\left(t^2 - u^2\right) du \right\} dt.$$

The inner integral can be integrated by inspection:

$$\int_0^t u \cos\left(t^2 - u^2\right) du = \left\{ -\frac{1}{2} \sin\left(t^2 - u^2\right) \right\} \Big|_0^t = \frac{1}{2} \sin\left(t^2\right),$$

and so

$$J = 2 \int_0^\infty \frac{1}{2} \sin\left(t^2\right) dt = \int_0^\infty \sin\left(t^2\right) dt$$

which is just S(0). That is,

$$\int_0^\infty \left\{ \int_x^\infty \int_x^\infty \cos\left(t^2 - u^2\right) dt\, du \right\} dx = \frac{1}{2} \sqrt{\frac{\pi}{2}}. \qquad (6.1.7)$$

This is 0.626657... and, as shown in Sect. 3.6, that's what the Symbolic Math Toolbox calculates as well.

6.5 The Watson/van Peype Triple Integrals

In 1939 the English mathematician George N. Watson (1886–1965) published an elegant paper[4] in which he showed how to evaluate the following three integrals:

---

[4]G. N. Watson, "Three Triple Integrals," *Quarterly Journal of Mathematics*, 1939, pp. 266–276.

$$I_1 = \frac{1}{\pi^3} \int_0^\pi \int_0^\pi \int_0^\pi \frac{du\,dv\,dw}{1 - \cos(u)\cos(v)\cos(w)},$$

$$I_2 = \frac{1}{\pi^3} \int_0^\pi \int_0^\pi \int_0^\pi \frac{du\,dv\,dw}{3 - \cos(v)\cos(w) - \cos(w)\cos(u) - \cos(u)\cos(v)},$$

$$I_3 = \frac{1}{\pi^3} \int_0^\pi \int_0^\pi \int_0^\pi \frac{du\,dv\,dw}{3 - \cos(u) - \cos(v) - \cos(w)}.$$

These three integrals were not pulled out of thin air, but rather all had appeared in a paper[5] published the previous year by W. F. van Peype, a student of the famous Dutch physicist H. A. Kramers (1894–1952). Van Peype was able to evaluate I_1, but not I_2 or I_3. Kramers was apparently sufficiently fascinated by these integrals that he sent them to the British physicist Ralph Fowler (1889–1944) who, apparently also stumped, passed them on into the hands of—who else?—G. H. Hardy, famous slayer of definite integrals.

At this point, matters become interesting beyond just mathematics. As Watson wrote in the opening of his paper, "The problem then became common knowledge first in Cambridge [Hardy's lair] and subsequently in Oxford, when it made the journey to Birmingham [Watson was on the faculty at the University of Birmingham] without difficulty." Perhaps I am reading too much into that, but I suspect this was one of those rare cases where Hardy *failed* to evaluate a definite integral, and Watson was only too happy to (not so subtly) allude to that as he started to present his solutions. Birmingham was small potatoes compared to Cambridge, and I think just a bit of friendly 'in your face' was involved here.

Watson's analyses of I_1, I_2, and I_3 are quite clever. Watson wrote that I_1 and I_2 "are easily expressible in terms of gamma functions whose arguments are simple fractions," but he was not able to do that for I_3. I_3, he suspected, *required* the use of elliptic integrals,[6] mathematical creatures I'll say a little bit more about in the next section. In this belief, Watson was wrong. In fact,

---

[5]W. F. van Peype, "Zur Theorie der Magnetischen Anisotropic Kubischer Kristalle Beim Absoluten Nullpunkt," *Physica*, June 1938, pp. 465–482. That is, van Peype was studying magnetic behavior in certain cubic crystalline lattice structures at very low temperatures (*low* means near absolute zero). The Watson/van Peype integrals turn-up not only in the physics of frozen magnetic crystals, but also in the pure mathematics of random walks. You can find a complete discussion of both the history and the mathematics of the integrals in I. J. Zucker, "70+ Years of the Watson Integrals," *Journal of Statistical Physics*, November 2011, pp. 591–612.

[6]Integrals with the integrands $\frac{1}{\sqrt{1-k^2\sin^2(\theta)}}$ and $\sqrt{1 - k^2\sin^2(\theta)}$ are *elliptic integrals* of the first and second kind, respectively (there is a third form, too). Such integrals occur in many important physical problems, such as the theory of the non-linear pendulum. As another example, the Italian mathematical physicist Galileo Galilei (1564–1642) studied the so-called "minimum descent time" problem, which involves an elliptic integral of the first kind, and its evaluation puzzled mathematicians for over a century. Eventually the French mathematician Adrien Marie Legendre (1752–1833) showed that the reason for the difficulty was that such integrals are *entirely new functions*, different from all other known functions. You can find more about Galileo's problem, and

$$I_1 = \frac{\Gamma^4\left(\frac{1}{4}\right)}{4\pi^3} = 1.393203929\ldots,$$

$$I_2 = \frac{3\Gamma^6\left(\frac{1}{3}\right)}{2^{14/3}\pi^4} = 0.448220394\ldots,$$

$$I_3 = \frac{\Gamma\left(\frac{1}{24}\right)\Gamma\left(\frac{5}{24}\right)\Gamma\left(\frac{7}{24}\right)\Gamma\left(\frac{11}{24}\right)}{16\sqrt{6}\pi^3} = 0.505462019\ldots.$$

To give you the 'flavor' of what Watson did, I'll take you through the derivation for I_1. Before starting, however, notice that all three integrals are volume integrals over a three-dimensional cube with edge-length π, normalized to the volume of that cube. I'll ignore the normalizing π^3 until we get to the end of the analysis.

We start with the change of variables

$$x = \tan\left(\frac{1}{2}u\right), \quad y = \tan\left(\frac{1}{2}v\right), \quad z = \tan\left(\frac{1}{2}w\right)$$

which convert the 0 to π integrations in u, v, and w in I_1 into 0 to ∞ integrations in x, y, and z, respectively. Now, since

$$u = 2\tan^{-1}(x)$$

then

$$\frac{du}{dx} = \frac{2}{1+x^2}$$

or,

$$du = \frac{2}{1+x^2}dx.$$

Similarly,

$$dv = \frac{2}{1+y^2}dy$$

and

---

the elliptic integral it encounters, in my book *When Least is Best*, Princeton 2007, pp. 200–210 and 347–351. A nice discussion of the non-linear pendulum, and the numerical evaluation of its elliptic integral, is in the paper by T. F. Zheng et al., "Teaching the Nonlinear Pendulum," *The Physics Teacher*, April 1994, pp. 248–251.

$$dw = \frac{2}{1+z^2}\,dz.$$

Also, from the half-angle formula for the tangent, we have

$$\tan\left(\frac{1}{2}u\right) = \sqrt{\frac{1-\cos(u)}{1+\cos(u)}} = x$$

and so, solving for cos(u),

$$\cos(u) = \frac{1-x^2}{1+x^2}.$$

Similarly,

$$\cos(v) = \frac{1-y^2}{1+y^2}$$

and

$$\cos(w) = \frac{1-z^2}{1+z^2}.$$

Putting these results for the differentials, and the cosines, into the I_1 integral (remember, we are temporarily ignoring the normalizing π^3), we have

$$\int_0^\pi\int_0^\pi\int_0^\pi \frac{du\,dv\,dw}{1-\cos(u)\cos(v)\cos(w)} = 8\int_0^\infty\int_0^\infty\int_0^\infty \frac{\left(\frac{dx}{1+x^2}\right)\left(\frac{dy}{1+y^2}\right)\left(\frac{dz}{1+z^2}\right)}{1-\left(\frac{1-x^2}{1+x^2}\right)\left(\frac{1-y^2}{1+y^2}\right)\left(\frac{1-z^2}{1+z^2}\right)} \cdots$$

$$= 8\int_0^\infty\int_0^\infty\int_0^\infty \frac{dx\,dy\,dz}{(1+x^2)(1+y^2)(1+z^2) - (1-x^2)(1-y^2)(1-z^2)}$$

which, after a bit of multiplying and combining of terms in the denominator, reduces to

$$4\int_0^\infty\int_0^\infty\int_0^\infty \frac{dx\,dy\,dz}{x^2+y^2+z^2+x^2y^2z^2}.$$

Notice, carefully, that this is a volume integral over the entire positive octant $(x \geq 0,\, y \geq 0,\, z \geq 0)$ in three-dimensional space.

Next, change variables again as follows:

$$x = r\sin(\theta)\cos(\phi), \quad y = r\sin(\theta)\sin(\phi), \quad z = r\cos(\theta)$$

which is, physically, simply a shift to spherical coordinates from the rectangular coordinates in our last integral.[7] To continue to be physically integrating over the entire positive octant in three-dimensional space, we see that our triple integral, with differentials dr, dθ, and dϕ, must be over the intervals 0 to ∞, 0 to $\frac{\pi}{2}$, and 0 to $\frac{\pi}{2}$, respectively. The differential volume element in rectangular coordinates (dx dy dz) becomes the differential volume element $r^2 \sin(\theta)d\phi\, d\theta\, dr$ in spherical coordinates. Thus, in this new coordinate system the I_1 integral becomes

$$4\int_0^{\pi/2}\int_0^{\pi/2}\int_0^{\pi/2}\frac{r^2\sin(\theta)d\phi\,d\theta\,dr}{r^2\sin^2(\theta)\cos^2(\phi)+r^2\sin^2(\theta)\sin^2(\phi)+r^2\cos^2(\theta)+r^6\sin^2(\theta)\cos^2(\phi)\sin^2(\theta)\sin^2(\phi)\cos^2(\theta)}$$

$$=4\int_0^{\pi/2}\int_0^{\pi/2}\int_0^{\infty}\frac{\sin(\theta)d\phi\,d\theta\,dr}{1+r^4\sin^4(\theta)\cos^2(\theta)\sin^2(\phi)\cos^2(\phi)}$$

or, with the order of integration explicitly displayed,

$$=4\int_0^{\pi/2}\left\{\int_0^{\pi/2}\left\{\int_0^{\infty}\frac{\sin(\theta)}{1+r^4\sin^4(\theta)\cos^2(\theta)\sin^2(\phi)\cos^2(\phi)}dr\right\}d\theta\right\}d\phi.$$

Now, define the variable

$$\psi = 2\phi.$$

Then, the double-angle formula for the sine says

$$\sin(\phi)\cos(\phi)=\frac{1}{2}\sin(2\phi)$$

and so

$$\sin^2(\phi)\cos^2(\phi)=\frac{1}{4}\sin^2(2\phi)=\frac{1}{4}\sin^2(\psi).$$

Since

$$d\phi=\frac{1}{2}d\psi$$

our integral then becomes

---

[7]In this notation, the angle ϕ is measured from the positive x-axis and θ is measured from the positive z-axis. Some authors reverse this convention, but of course if one maintains consistency from start to finish everything comes out the same. The symbols are, after all, just squiggles of ink.

$$2 \int_0^\pi \left\{ \int_0^{\pi/2} \left\{ \int_0^\infty \frac{\sin(\theta)}{1 + \frac{1}{4} r^4 \sin^4(\theta) \cos^2(\theta) \sin^2(\psi)} \, dr \right\} d\theta \right\} d\psi.$$

At this point things may superficially appear to be bordering on the desperate but, as the old saying goes, 'appearances can be deceptive'; we are actually almost done. First, the outer-most integration (with respect to ψ) is symmetrical in ψ around $\psi = \frac{\pi}{2}$. That is, as Watson says in his paper,

$$" \frac{1}{2} \int_0^\pi (\ldots) d = \int_0^{\pi/2} (\ldots) d\psi, "$$

which is, of course, the result of ψ appearing in the integrand only as $\sin^2(\psi)$. Thus, our integral becomes

$$4 \int_0^{\pi/2} \left\{ \int_0^{\pi/2} \left\{ \int_0^\infty \frac{\sin(\theta)}{1 + \frac{1}{4} r^4 \sin^4(\theta) \cos^2(\theta) \sin^2(\psi)} \, dr \right\} d\theta \right\} d\psi.$$

Second, Watson uses the fact that the inner-most (that is, the *first*) integration (with respect to r) is performed for θ and ψ held fixed. That is, in the change of variable

$$t = r \sin(\theta) \sqrt{\frac{1}{2} \cos(\theta) \sin(\psi)}$$

t is actually a function *only* of just r, and not of r, *and* θ, *and* ψ. So, making that change we have

$$dr = \frac{dt}{\sin(\theta) \sqrt{\frac{1}{2} \cos(\theta) \sin(\psi)}}$$

and

$$t^4 = \frac{1}{4} r^4 \sin^4(\theta) \cos^2(\theta) \sin^2(\psi),$$

which converts our integral to

$$4 \int_0^{\pi/2} \left\{ \int_0^{\pi/2} \left\{ \int_0^\infty \frac{\sin(\theta)}{\sin(\theta) \sqrt{\frac{1}{2} \cos(\theta) \sin(\psi)} (1 + t^4)} \, dt \right\} d\theta \right\} d\psi$$

$$= 4\sqrt{2} \int_0^{\pi/2} \left\{ \int_0^{\pi/2} \left\{ \int_0^\infty \frac{1}{\sqrt{\cos(\theta)}\sqrt{\sin(\psi)}(1+t^4)} \, dt \right\} d\theta \right\} d\psi,$$

a scary-looking object that—suddenly and with unspeakable joy to the analyst[8]—separates into the product of three *one*-dimension integrals, each of which we've already done! That is, we have

$$\int_0^\pi \int_0^\pi \int_0^\pi \frac{du \, dv \, dw}{1 - \cos(u)\cos(v)\cos(w)} = 4\sqrt{2} \int_0^\infty \frac{dt}{(1+t^4)} \int_0^{\pi/2} \frac{d\theta}{\sqrt{\cos(\theta)}} \int_0^{\pi/2} \frac{d\psi}{\sqrt{\sin(\psi)}}.$$

The first (t) integral is, from (2.3.4), equal to $\frac{\pi\sqrt{2}}{4}$, while the θ and ψ integrals are, from (4.2.14), each equal to $\frac{\Gamma^2\left(\frac{1}{4}\right)}{2\sqrt{2\pi}}$. Thus, remembering the π^3 normalizing factor, we have

$$I_1 = \frac{1}{\pi^3} 4\sqrt{2} \left(\frac{\pi\sqrt{2}}{4}\right) \left(\frac{\Gamma^2\left(\frac{1}{4}\right)}{2\sqrt{2\pi}}\right) \left(\frac{\Gamma^2\left(\frac{1}{4}\right)}{2\sqrt{2\pi}}\right)$$

and so

$$\frac{1}{\pi^3} \int_0^\pi \int_0^\pi \int_0^\pi \frac{du \, dv \, dw}{1 - \cos(u)\cos(v)\cos(w)} = \frac{\Gamma^4\left(\frac{1}{4}\right)}{4\pi^3}. \qquad (6.5.1)$$

This answer is given in van Peype's paper, *without* derivation.

We can use MATLAB's triple integration command *integral3* to check this. It does for triple integrals what *integral* does for one-dimensional integrals (and what *integral2* does for double integrals). For our problem, here, the syntax is:

integrnd = @(u,v,w) 1./(1-cos(u).*cos(v).*cos(w));
integral3(integrnd, 1e-5, pi, 1e-5, pi, 1e-5, pi)/(pi^3)

which produces the answer: 1.39312..., in pretty good agreement with the theoretical result I gave you earlier for I_1. (Notice that the integration limits start at just a little greater than zero, since when u = v = w = 0 the integrand blows-up.)

---

[8]I can only imagine what Watson's words to his cat must have been when he reached this point in his work. Perhaps, maybe, they were something like this: "By Jove, Lord Fluffy, I've done it! Cracked the damn thing wide-open, just like when that egg-head Humpty-Dumpty fell off his bloody wall!"

6.6   Elliptic Integrals in a Physical Problem

In this section I'll elaborate just a bit on the topic of elliptic integrals, which were mentioned in passing in the previous section (see note 6 again). Specifically,

$$F(k, \phi) = \int_0^\phi \frac{d\phi'}{\sqrt{1 - k^2 \sin^2(\phi')}} \tag{6.6.1}$$

and

$$E(k, \phi) = \int_0^\phi \sqrt{1 - k^2 \sin^2(\phi')} \, d\phi' \tag{6.6.2}$$

are the elliptic integrals of the first and second kind, respectively, where $0 \le k \le 1$ (the constant k is called the *modulus*). When $\phi < \frac{\pi}{2}$ the integrals are called *incomplete*, while when $\phi = \frac{\pi}{2}$ the integrals are called *complete*. Except for the two special cases of $k = 0$ and $k = 1$, $F(k, \phi)$ and $E(k, \phi)$ are *not* expressible in terms of any of the elementary functions we typically use (trigonometric, exponential, algebraic, and so on).

You might think it would require a complicated physical situation for either F or E to make an appearance, but that's not so. What I'll show you now is a *seemingly* simple physical problem that will nonetheless involve elliptic integrals (both E *and* F, in fact). Making the problem even more interesting is that it has appeared in textbooks for well over a century, but until 1989 was routinely analyzed incorrectly.[9]

Figure 6.6.1 shows a perfectly flexible, inextensible (that is, there is no stretching as in a bungee cord) rope with a constant mass density of μ per unit length. The rope has length L, with one end (the left end) permanently attached to a ceiling. The other end is temporarily held at the ceiling, too, until at time $t = 0$ that end is released. The released portion of the rope then, of course, begins to fall (the figure shows the situation after the falling end has descended a distance of x) until, at time $t = T$, the rope is hanging straight down. Our question is simple: what is T?

The one assumption we'll make in answering this question is that the falling rope conserves energy, which means there are no energy dissipation mechanisms in play (such as internal frictional heating losses in the rope caused by flexing at the bottom of the bend). At all times the sum of the rope's potential and kinetic energies will be a constant. When we get done with our analysis we'll find that the result for T has a *very* surprising aspect to it.

The center of mass of the left-hand-side of the rope is $\frac{1}{2}\left(\frac{L+x}{2}\right) = \frac{L+x}{4}$ below the ceiling, while the center of mass of the right-hand-side of the rope is $x + \frac{1}{2} \times$

---

[9]See M. G. Calkin and R. H. March, "The Dynamics of a Falling Chain. I," *American Journal of Physics*, February 1989, pp. 154–157.

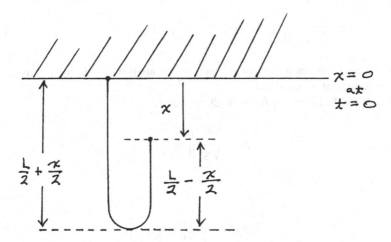

Fig. 6.6.1 The falling rope

$\left(\frac{L-x}{2}\right) = \frac{L+3x}{4}$ below the ceiling. So, taking the zero of potential energy (P.E.) at the ceiling, the P.E. of the rope when the released end of the rope has descended by x is given by (where g is the acceleration of gravity)

$$\text{P.E.} = -\left[\mu\left(\frac{L+x}{2}\right)\right]g\frac{L+x}{4} - \left[\mu\left(\frac{L-x}{2}\right)\right]g\frac{L+3x}{4}$$

which reduces (after just a little easy algebra) to

$$\text{P.E.} = -\frac{1}{4}\mu g\left[L^2 + 2xL - x^2\right]. \tag{6.6.3}$$

The kinetic energy (K.E.) of the rope is the K.E. of the descending right-hand-side of the rope, which is the only portion of the rope that is moving. Since the K.E. of a mass $m = \mu\left(\frac{L-x}{2}\right)$ moving at speed $v = \frac{dx}{dt}$ is $\frac{1}{2}mv^2$, we have the K.E. of the rope as

$$\text{K.E.} = \frac{1}{4}\mu(L-x)\left(\frac{dx}{dt}\right)^2. \tag{6.6.4}$$

From (6.6.3), the *initial* (when $x = 0$) P.E. is $-\frac{1}{4}\mu gL^2$. Also, since the rope starts its fall from rest, the *initial* K.E. is zero. Thus, the total initial energy of the rope is $-\frac{1}{4}\mu gL^2$ and by conservation of energy this is the total energy of the rope for all $t \geq 0$. So,

$$\frac{1}{4}\mu(L-x)\left(\frac{dx}{dt}\right)^2 - \frac{1}{4}\mu g\left[L^2 + 2xL - x^2\right] = -\frac{1}{4}\mu gL^2$$

which reduces to

$$\frac{1}{4}(L - x)\left(\frac{dx}{dt}\right)^2 = \frac{1}{4}g(2L - x)x. \tag{6.6.5}$$

Notice that μ has canceled away, and so our analysis holds for *any* rope with *arbitrary* (constant) mass density.

Solving (6.6.5) for the differential dt, we have

$$dt = \sqrt{\frac{L - x}{g(2L - x)x}}$$

or,

$$\sqrt{g}\, dt = \sqrt{\frac{L - x}{(2L - x)x}}. \tag{6.6.6}$$

If we integrate both sides of (6.6.6), where time runs from 0 to t and the descent distance of the falling end of the rope varies from 0 to x, we have (I've changed the dummy variables of integration from t to t' and from x to x' so that we can keep t and x as our final variables) then

$$\int_0^t \sqrt{g}\, dt' = \int_0^x \sqrt{\frac{L - x'}{(2L - x')x'}}\, dx' = t\, \sqrt{g}. \tag{6.6.7}$$

Now, x varies from 0 (at $t = 0$) to L (at $t = T$). So, if we define the variable ϕ as

$$\sin(\phi) = \sqrt{\frac{x}{L}}$$

then ϕ varies from 0 to $\frac{\pi}{2}$. Since $x = L\sin^2(\phi)$, let's now change variable in (6.6.7) to $x' = L\sin^2(\phi')$ and so

$$\frac{dx'}{d\phi'} = 2L\sin(\phi')\cos(\phi')$$

or,

$$dx' = 2L\sin(\phi')\cos(\phi')d\phi'.$$

Also, with just a bit of easy algebra you should be able to show that

$$\sqrt{\frac{L - x'}{(2L - x')x'}} = \frac{\cos(\phi')}{\sqrt{L}\,\sin(\phi')\sqrt{1 + \cos^2(\phi')}}.$$

Thus, (6.6.7) becomes

$$t \sqrt{g} = \int_0^{\phi} \frac{\cos(\phi')}{\sqrt{L} \sin(\phi') \sqrt{1 + \cos^2(\phi')}} 2L \sin(\phi') \cos(\phi') d\phi'$$

or, with a bit of cancelling and rearranging,

$$\frac{t}{\sqrt{\frac{2L}{g}}} = \sqrt{2} \int_0^{\phi} \frac{\cos^2(\phi')}{\sqrt{1 + \cos^2(\phi')}} d\phi'$$

and so, finally, since $t = T$ when $\phi = \frac{\pi}{2}$,

$$\frac{T}{\sqrt{\frac{2L}{g}}} = \sqrt{2} \int_0^{\frac{\pi}{2}} \frac{\cos^2(\phi')}{\sqrt{1 + \cos^2(\phi')}} d\phi'. \tag{6.6.8}$$

'Why the curious form of the left-hand-side of (6.6.8)?', you are no doubt wondering. To see the why of it, suppose you drop a point mass at time $t = 0$, starting from $x = 0$. How long does it take for that point mass to fall distance L? From freshman physics we know the mass will fall distance $\frac{1}{2}gt^2$ in time t and so $\frac{1}{2}gt^2 = L$ or, $t = \sqrt{\frac{2L}{g}}$, which is the curious denominator on the left in (6.6.8). Thus, if $T > \sqrt{\frac{2L}{g}}$, that is, if the rope fall takes longer than the free-fall time of the point mass, then the right-hand-side of (6.6.8) will be greater than 1, but if the rope falls *faster* than does the point mass then the right-hand-side of (6.6.8) will be *less* than 1. Finally, if the rope falls *with* the acceleration of gravity (as does the point mass) then the right-hand side of (6.6.8) will be exactly 1. So, which is it?

This is an easy question for MATLAB to answer for us, and in fact the rope falls *faster*(!) than does the point mass because *sqrt(2)*integral(@(x)(cos(x).^2)./sqrt ((1 + cos(x).^2)),0,pi/2) = 0.847213....* This is not an insignificant deviation from 1 (that is, it isn't round-off error), and the rope falls more than 15% faster than does the point mass. Are you surprised? If not, why not? After all, as I mentioned the point mass falls with the acceleration of gravity and so, for the rope to beat the point mass, it must fall 'faster than gravity!' How can that be?[10]

At this point an engineer or physicist would probably (after shaking their head in amused surprise) start searching for the physical reason behind this curious result.

---

[10]The reason is that the falling part of the rope does so under the influence of not just gravity alone, but also from the non-zero tension in it that joins with gravity in pulling the rope down. To pursue this point here would take us too far away from the theme of the book but, if interested, it's all worked out in the paper cited in note 9. The 'faster than gravity' prediction was experimentally confirmed, indirectly, in note 9 via tension measurements made during actual falls. In 1997 direct photographic evidence was published, and today you can find YouTube videos on the Web clearly showing 'faster than gravity' falls.

A mathematician,[11] however, would more likely first wonder just what nice mathematical expression is that curious number 0.847213... equal to? The answer is, as you might expect from the title of this section, *elliptic integrals*. Here's why.

Starting with just the integral on the right-hand-side of the equation just before (6.6.8)—we'll put the $\sqrt{2}$ factor in at the end—we have

$$\int_0^\phi \frac{\cos^2(\phi')}{\sqrt{1+\cos^2(\phi')}} d\phi' = \int_0^\phi \frac{1-\sin^2(\phi')}{\sqrt{2-\sin^2(\phi')}} d\phi' = \int_0^\phi \frac{1}{\sqrt{2-\sin^2(\phi')}} d\phi'$$

$$-\int_0^\phi \frac{\sin^2(\phi')}{\sqrt{2-\sin^2(\phi')}} d\phi' = \frac{1}{\sqrt{2}} \int_0^\phi \frac{1}{\sqrt{1-\frac{1}{2}\sin^2(\phi')}} d\phi' - \frac{1}{\sqrt{2}} \int_0^\phi \frac{\sin^2(\phi')}{\sqrt{1-\frac{1}{2}\sin^2(\phi')}} d\phi'$$

and so, recalling (6.6.1),

$$\int_0^\phi \frac{\cos^2(\phi')}{\sqrt{1+\cos^2(\phi')}} d\phi' = \frac{1}{\sqrt{2}} F\left(\frac{1}{\sqrt{2}},\phi\right) - \frac{1}{\sqrt{2}} \int_0^\phi \frac{\sin^2(\phi')}{\sqrt{1-\frac{1}{2}\sin^2(\phi')}} d\phi'. \quad (6.6.9)$$

Now, concentrate on the integral on the right-hand-side of (6.6.9).

$$\int_0^\phi \frac{\sin^2(\phi')}{\sqrt{1-\frac{1}{2}\sin^2(\phi')}} d\phi' - \int_0^\phi \frac{1-\frac{1}{2}\sin^2(\phi')}{\sqrt{1-\frac{1}{2}\sin^2(\phi')}} d\phi' + \int_0^\phi \frac{-1+\frac{3}{2}\sin^2(\phi')}{\sqrt{1-\frac{1}{2}\sin^2(\phi')}} d\phi'.$$

$$= \int_0^\phi \sqrt{1-\frac{1}{2}\sin^2(\phi')} d\phi' - \int_0^\phi \frac{d\phi'}{\sqrt{1-\frac{1}{2}\sin^2(\phi')}} + \frac{3}{2} \int_0^\phi \frac{\sin^2(\phi')}{\sqrt{1-\frac{1}{2}\sin^2(\phi')}} d\phi'.$$

Thus, using (6.6.2), and (6.6.1) again, we have

$$\int_0^\phi \frac{\sin^2(\phi')}{\sqrt{1-\frac{1}{2}\sin^2(\phi')}} d\phi' = E\left(\frac{1}{\sqrt{2}},\phi\right) - F\left(\frac{1}{\sqrt{2}},\phi\right) + \frac{3}{2} \int_0^\phi \frac{\sin^2(\phi)}{\sqrt{1-\frac{1}{2}\sin^2(\phi)}} d\phi$$

or,

---

[11]Mathematicians are just as interested in this problem, and in related problems, as are physicists. Indeed, the study of falling ropes and chains was initiated by the British *mathematician* Arthur Cayley (1821–1895): see his note "On a Class of Dynamical Problems," *Proceedings of the Royal Society of London* 1857, pp. 506–511, that opens with the words "There are a class of dynamical problems which, so far as I am aware, have not been considered in a general manner." That's certainly not the case today, with Cayley's problem in particular still causing debate over when energy is (and isn't) conserved: see Chun Wa Wong and Kosuke Yasui, "Falling Chains," *American Journal of Physics*, June 2006, pp. 490–496. Take a look, too, at Challenge Problem 6.4.

$$\frac{1}{2}\int_0^\phi \frac{\sin^2(\phi')}{\sqrt{1-\frac{1}{2}\sin^2(\phi')}}d\phi' = F\left(\frac{1}{\sqrt{2}},\phi\right) - E\left(\frac{1}{\sqrt{2}},\phi\right).$$

That is, we have the interesting identity

$$\int_0^\phi \frac{\sin^2(\phi')}{\sqrt{1-\frac{1}{2}\sin^2(\phi')}}d\phi' = 2F\left(\frac{1}{\sqrt{2}},\phi\right) - 2E\left(\frac{1}{\sqrt{2}},\phi\right), 0 \le \phi' \le \frac{\pi}{2}. \quad (6.6.10)$$

Using (6.6.10) in (6.6.9), we have

$$\int_0^\phi \frac{\cos^2(\phi')}{\sqrt{1+\cos^2(\phi')}}d\phi' = \frac{1}{\sqrt{2}}F\left(\frac{1}{\sqrt{2}},\phi\right) - \frac{1}{\sqrt{2}}\left[2F\left(\frac{1}{\sqrt{2}},\phi\right) - 2E\left(\frac{1}{\sqrt{2}},\phi\right)\right]$$

$$= \left(\frac{1}{\sqrt{2}} - \sqrt{2}\right)F\left(\frac{1}{\sqrt{2}},\phi\right) + \sqrt{2}E\left(\frac{1}{\sqrt{2}},\phi\right)$$

and so we have a second interesting identity

$$\int_0^\psi \frac{\cos^2(\phi')}{\sqrt{1+\cos^2(\phi')}}d\phi' = \sqrt{2}E\left(\frac{1}{\sqrt{2}},\phi\right) - \frac{1}{\sqrt{2}}F\left(\frac{1}{\sqrt{2}},\phi\right), 0 \le \phi$$

$$\le \frac{\pi}{2}. \quad (6.6.11)$$

Putting this result into (6.6.8), and including the $\sqrt{2}$ factor in front of the integral, we arrive at (for $\phi = \frac{\pi}{2}$),

$$\frac{T}{\sqrt{\frac{2L}{g}}} = 2E\left(\frac{1}{\sqrt{2}},\frac{\pi}{2}\right) - F\left(\frac{1}{\sqrt{2}},\frac{\pi}{2}\right). \quad (6.6.12)$$

Does (6.6.12) explain MATLAB's result of 0.847213...? Yes, because a quick look in math tables for the values of the complete elliptic integrals of the first and second kind tells us that

$$F\left(\frac{1}{\sqrt{2}},\frac{\pi}{2}\right) = 1.8540746\ldots$$

and

$$E\left(\frac{1}{\sqrt{2}},\frac{\pi}{2}\right) = 1.3506438\ldots.$$

The right-hand-side of (6.6.12) is thus numerically equal to

$$2(1.3506438\ldots) - (1.8540746\ldots) = 0.847213\ldots$$

and so (6.6.12) and MATLAB are in *excellent* agreement.

6.7   Ramanujan's Master Theorem

In this final section we'll develop a beautiful 'close cousin' to Riemann's integral formula of (5.3.4), a formula that also involves the gamma function. We start by defining the so-called *forward shift* operator[12] E as follows:

$$E\{\lambda(n)\} = \lambda(n+1). \tag{6.7.1}$$

Thus,

$$E\{\lambda(0)\} = \lambda(1),$$

$$E\{\lambda(1)\} = E\{E\{\lambda(0)\}\} = E^2\{\lambda(0)\} = \lambda(2),$$

$$E\{\lambda(2)\} = E\{E\{E\{\lambda(0)\}\}\} = E^3\{\lambda(0)\} = \lambda(3),$$

and so on. In general,

$$E^k\{\lambda(0)\} = \lambda(k). \tag{6.7.2}$$

Next, imagine that we have some function $\varphi(x)$ with the power series expansion

$$\varphi(x) = \sum_{n=0}^{\infty} \frac{(-1)^n}{n!}\lambda(n)x^n. \tag{6.7.3}$$

Then, continuing in a *formal* way (see note 12 again), we write

$$\int_0^{\infty} x^{s-1}\varphi(x)dx = \int_0^{\infty} x^{s-1}\sum_{n=0}^{\infty} \frac{(-1)^n}{n!}\lambda(n)x^n dx. \tag{6.7.4}$$

Using (6.7.2), we replace $\lambda(n)$ in (6.7.4) with $E^n\{\lambda(0)\}$, which we loosely write (by dropping the curly brackets) as $E^n\lambda(0)$. Continuing with our formal manipulations, then, we have

---

[12]*Operators* are highly useful concepts, and are actually quite common in 'higher math.' A study of differential equations, for example, quickly leads to the differential operator D, defined as $D\{\lambda\} = \frac{d\lambda}{dt}$. The fact that operators can often be manipulated just as if they are numbers makes their use intuitive, and that feature will let us arrive at useful results through the use of only so-called *formal* manipulations, rather than rigorous reasoning.

$$\int_0^\infty x^{s-1}\varphi(x)dx = \int_0^\infty x^{s-1}\sum_{n=0}^\infty \frac{(-1)^n}{n!}E^n\lambda(0)x^n dx$$

$$= \lambda(0)\int_0^\infty x^{s-1}\sum_{n=0}^\infty \frac{(-1)^n}{n!}(Ex)^n dx.$$

Well, what can I say: there is simply no denying that we've really played pretty fast-and-loose with the operator E in arriving at the last integral (in particular, notice how E^n goes from 'operating on' $\lambda(0)$ to 'operating on' x^n, while the $\lambda(0)$ slides outside to the front of the integral—this is symbol pushing at its most arrogant). Take a deep breath, however, as there is still *more* of such doings to come!

Recalling the power series expansion of e^{-t},

$$e^{-t} = 1 - t + \frac{t^2}{2!} - \frac{t^3}{3!} + \ldots = \sum_{n=0}^\infty \frac{(-1)^n}{n!}t^n,$$

and writing $t = Ex$, then we *formally* have

$$\int_0^\infty x^{s-1}\varphi(x)dx = \lambda(0)\int_0^\infty x^{s-1}e^{-Ex}dx. \tag{6.7.5}$$

Treating Ex as an ordinary variable, let's now change variable to $y = Ex$. Then,

$$dx = \frac{1}{E}dy$$

and (6.7.5) becomes

$$\int_0^\infty x^{s-1}\varphi(x)dx = \lambda(0)\int_0^\infty \left(\frac{y}{E}\right)^{s-1}e^{-y}\frac{1}{E}dy = \frac{\lambda(0)}{E^s}\int_0^\infty y^{s-1}e^{-y}dy. \tag{6.7.6}$$

From (4.1.1) we see that the right-most integral in (6.7.6) is $\Gamma(s)$ and so, continuing,

$$\int_0^\infty x^{s-1}\varphi(x)dx = \frac{\lambda(0)}{E^s}\Gamma(s) = E^{-s}\lambda(0)\Gamma(s).$$

Finally, recalling (6.7.2) with $k = -s$, we have

$$\int_0^\infty x^{s-1}\varphi(x)dx = \Gamma(s)\lambda(-s) \tag{6.7.7}$$

with λ denoting the coefficients in the power-series expansion of $\varphi(x)$, as given in (6.7.3), a result called *Ramanujan's Master Theorem* (or RMT), named after the

Indian mathematician Srinivasa Ramanujan (1887–1920), who discovered (6.7.7) sometime during the first decade of the twentieth century.[13]

This development of (6.7.7) almost surely strikes you—as it should—as being patently absurd, akin to arriving at the correct statement that $\frac{16}{64} = \frac{1}{4}$ by simply cancelling the two 6's! I've purposely taken you through what most mathematicians would call an outrageous 'derivation' for two reasons: (1) it closely resembles how Ramanujan himself did it (Euler would have loved it!), and (2) the result in (6.7.7) *can* be rigorously established using mathematics beyond AP-calculus (an approach I wish, however, to avoid in this book). You'll be relieved to learn that when Ramanujan made his famous journey to England in 1914 to come under the tutelage of his hero, the great G. H. Hardy (events dramatically described in Robert Kanigel's 1991 book *The Man Who Knew Infinity*), he learned both that his derivation was technically faulty as well as how to do it right.[14]

As an example of the RMT in action, suppose

$$\varphi(x) = (1 + x)^m.$$

By the binomial theorem,

$$(1 + x)^m = 1 + mx + \frac{m(m-1)}{2!}x^2 + \frac{m(m-1)(m-2)}{3!}x^3 + \dots.$$

If $m = -a$, where $a > 0$, then[15]

$$(1 + x)^{-a} = 1 - ax + \frac{-a(-a-1)}{2!}x^2 + \frac{-a(-a-1)(-a-2)}{3!}x^3 + \dots$$

$$= 1 + \frac{a}{1!}(-1)^1 x + \frac{(a)(a+1)}{2!}(-1)^2 x^2 + \frac{(a)(a+1)(a+2)}{3!}(-1)^3 x^3 + \dots.$$

We see that the general term is

---

[13]For a brief historical discussion of the RMT, and of how it was very nearly discovered decades before Ramanujan (in 1874) by the English mathematician J. W. L. Glaisher (1848–1928), see V. H. Moll et al., "Ramanujan's Master Theorem," *The Ramanujan Journal*, December 2012, pp. 103–120. For more on Ramanujan's discovery of the RMT, see Bruce C. Berndt, "The Quarterly Reports of S. Ramanujan," *The American Mathematical Monthly*, October 1983, pp. 505–516. In advanced math the integral in (6.7.7) is called the *Mellin transform* of $\varphi(x)$, after the Finnish mathematician Robert Hjalman Mellin (1854–1933). In computer science the Mellin transform is used in electronic image processing, an application that Mellin and Ramanujan couldn't have even imagined in their wildest dreams.

[14]See, for example, G. H. Hardy, *Ramanujan: Twelve Lectures on Subjects Suggested by His Life and Work*, Chelsea 1978, based on a series of lectures given in 1936 by Hardy at Harvard University.

[15]The binomial theorem, for m a positive integer, gives a finite number of terms and has been known since the French mathematician Blaise Pascal (1623–1662). Newton extended its use (without proof) to rational m (and even negative m), for which the expansion has an infinite number of terms. The proof of the theorem, for any m in general, wasn't done until the Norwegian mathematician Niels Henrik Abel (1802–1829) did it in 1826.

$$\frac{(a)(a+1)(a+2)\ldots(a+n-1)}{n!}(-1)^n x^n$$

or, writing the numerator product in reverse order,

$$\frac{(a+n-1)\ldots(a+2)(a+1)(a)}{n!}(-1)^n x^n$$

$$=\frac{(a+n-1)\ldots(a+2)(a+1)(a)(a-1)!}{n!(a-1)!}(-1)^n x^n$$

$$=\frac{(a+n-1)!}{(a-1)!}\frac{(-1)^n}{n!}x^n.$$

That is,

$$(1+x)^{-a}=\sum_{n=0}^{\infty}\frac{(a+n-1)!}{(a-1)!}\frac{(-1)^n}{n!}x^n$$

or, using the gamma function notation of (4.1.3),

$$(1+x)^{-a}=\sum_{n=0}^{\infty}\frac{\Gamma(a+n)}{\Gamma(a)}\frac{(-1)^n}{n!}x^n. \tag{6.7.8}$$

Comparing the right-hand-side of (6.7.8) with the right-hand-side of (6.7.3), we see that

$$\lambda(n)=\frac{\Gamma(a+n)}{\Gamma(a)}$$

and so

$$\lambda(-s)=\frac{\Gamma(a-s)}{\Gamma(a)}.$$

The RMT then gives us the very pretty

$$\int_0^{\infty}\frac{x^{s-1}}{(1+x)^a}dx=\frac{\Gamma(s)\Gamma(a-s)}{\Gamma(a)},\quad a>0. \tag{6.7.9}$$

In particular, if we let $a=1$ and $s-1=-p$ (and so $s=1-p$), then

$$\int_0^{\infty}\frac{x^{-p}}{1+x}dx=\int_0^{\infty}\frac{1}{(1+x)x^p}dx=\frac{\Gamma(1-p)\Gamma(p)}{\Gamma(1)}$$

or, remembering both the reflection formula, and that $\Gamma(1)=1$, we arrive at the following interesting integral (which is generally developed in textbooks as the result of a contour integration in the complex plane, and we will do it that way, too, in Chap. 8):

$$\int_0^\infty \frac{1}{(1+x)x^p}\,dx = \frac{\pi}{\sin(p\pi)}, \quad 0 < p < 1. \tag{6.7.10}$$

To end this section, I'll now show you a *really spectacular* use of the master theorem to do an integral that, frankly, I can't see how one would otherwise attack it. This will be a moderately long analysis, and so let me tell you straight-off what our result will be, a result so 'mysterious' that I think your interest will be sustained throughout the discussion simply because you'll want to see where it comes from!:

$$\int_0^\infty x^{s-1} \frac{\ln\{\Gamma(x+1)\} + \gamma x}{x^2}\,dx = \left\{\frac{\pi}{\sin(\pi s)}\right\}\frac{\zeta(2-s)}{2-s}, 0 < s < 1 \tag{6.7.11}$$

I use the word 'mysterious' to describe (6.7.11) because, on the left, we have the gamma *function* as well as the *number* gamma, and on the right we have the zeta function. To help convince you that (6.7.11) might actually mean something, we (as usual) turn to MATLAB. So, on the left we have (for $s = \frac{1}{2}$ and using 10^9 as a stand-in for the upper-limit of infinity).

   *integral(@(x)(gammaln(x + 1) + 0.57721566∗x)./x.^2.5,0,1e9) = 5.469...*while on the right we have (since $\sin\left(\frac{1}{2}\pi\right) = 1$).

   pi∗zeta(1.5)/1.5 = 5.47....

Pretty good agreement![16]

We start the derivation of (6.7.11) with the defining integral for the gamma function in (4.1.1), that I'll repeat here:

$$\Gamma(x) = \int_0^\infty e^{-t}t^{x-1}\,dt.$$

Now, recall that

$$e^{-t} = \lim_{n\to\infty}\left(1 - \frac{t}{n}\right)^n.$$

If we define

---

[16]The gamma function blows-up so fast as x increases that, when calculating $\ln\{\Gamma(x+1)\}$ as part of an integrand with x going from zero to infinity, one has to be careful when doing numerical work. Doing it in the direct, obvious way, that is by first calculating $\Gamma(x + 1)$ and then taking the logarithm, is certain to generate an error message informing you that an unbounded value has been encountered when evaluating the gamma function, causing the calculation to abort. For that reason, scientific programming languages like MATLAB include a special command that avoids the direct route: in MATLAB, instead of writing *log(gamma(x + 1))*, one should write *gammaln(x + 1)*, as I've done in the text.

$$\Gamma(x, n) = \int_0^n t^{x-1}\left(1 - \frac{t}{n}\right)^n dt$$

then

$$\Gamma(x) = \lim_{n\to\infty}\Gamma(x, n). \tag{6.7.12}$$

If we change variable to $y = \frac{t}{n}$ (and so $dt = n\,dy$), then

$$\Gamma(x, n) = \int_0^1 (ny)^{x-1}(1 - y)^n n\, dy$$

or,

$$\Gamma(x, n) = n^x \int_0^1 y^{x-1}(1 - y)^n\, dy. \tag{6.7.13}$$

Next, integrate by-parts. That is, let $u = (1 - y)^n$ and $dv = y^{x-1}dy$ in

$$\int_0^1 u\, dv = (uv)\Big|_0^1 - \int_0^1 v\, du.$$

Then $du = -n(1 - y)^{n-1}dy$ and $v = \frac{y^x}{x}$. Thus,

$$\Gamma(x, n) = n^x\left\{(1 - y)^n \frac{y^x}{x}\Big|_0^1 + \int_0^1 \frac{y^x}{x} n(1 - y)^{n-1}dy\right\} = \frac{1}{x}n^{x+1}\int_0^1 y^x (1 - y)^{n-1}dy.$$

Since (6.7.13) says

$$\Gamma(x + 1, n - 1) = (n - 1)^{x+1}\int_0^1 y^x(1 - y)^{n-1}\, dy$$

then

$$\int_0^1 y^x(1 - y)^{n-1}\, dy = \frac{\Gamma(x + 1,\ n - 1)}{(n - 1)^{x+1}}$$

and so

$$\Gamma(x, n) = \frac{1}{x}n^{x+1}\frac{\Gamma(x + 1,\ n - 1)}{(n - 1)^{x+1}}.$$

Thus, we have the recurrence

$$\Gamma(x, n) = \frac{1}{x} \left(\frac{n}{n-1} \right)^{x+1} \Gamma(x+1, n-1). \tag{6.7.14}$$

To actually *use* (6.7.14) we need something to *start* the recurrence. Putting $n = 1$ in (6.7.13), we get

$$\Gamma(x, 1) = \int_0^1 y^{x-1}(1-y)\, dy = \int_0^1 y^{x-1} dy - \int_0^1 y^x\, dy = \left(\frac{y}{x} \right)^x \Big|_0^1 - \left(\frac{y^{x+1}}{x+1} \right) \Big|_0^1$$

$$= \frac{1}{x} - \frac{1}{x+1}$$

or,

$$\Gamma(x, 1) = \frac{1}{x(x+1)}. \tag{6.7.15}$$

Next, set $n = 2$ in (6.7.14), to get

$$\Gamma(x, 2) = \frac{1}{x} \left(\frac{2}{1} \right)^{x+1} \Gamma(x+1, 1)$$

where, from (6.7.15),

$$\Gamma(x+1, 1) = \frac{1}{(x+1)(x+2)}$$

and so

$$\Gamma(x, 2) = \frac{1}{x} \left(\frac{2}{1} \right)^{x+1} \frac{1}{(x+1)(x+2)}$$

or

$$\Gamma(x, 2) = \frac{1}{x} 2^x \frac{(2)}{(x+1)(x+2)}. \tag{6.7.16}$$

Next, set $n = 3$ in (6.7.14) to get

$$\Gamma(x, 3) = \frac{1}{x} \left(\frac{3}{2} \right)^{x+1} \Gamma(x+1, 2)$$

where, from (6.7.16)

$$\Gamma(x+1,2) = \frac{1}{x+1} 2^{x+1} \frac{(2)}{(x+2)(x+3)}.$$

Thus,

$$\Gamma(x,3) = \frac{1}{x} \frac{3^{x+1}}{2^{x+1}} \frac{1}{(x+1)} 2^{x+1} \frac{(2)}{(x+2)(x+3)}$$

or,

$$\Gamma(x,3) = \frac{1}{x} 3^x \frac{(3)(2)}{(x+1)(x+2)(x+3)}. \tag{6.7.17}$$

I'll let you repeat this process a few more times (if you need convincing) but in general

$$\Gamma(x,n) = n^x \frac{n!}{x(x+1)(x+2)\dots(x+n)}.$$

That is, using (6.7.12),

$$\Gamma(x) = \lim_{n\to\infty} n^x \frac{n!}{x(x+1)(x+2)\dots(x+n)}, \tag{6.7.18}$$

an expression given by Euler in 1729. Continuing,

$$\Gamma(x) = \lim_{n\to\infty} n^x \frac{n!}{x(x+1)[2(\frac{x}{2}+1)][3(\frac{x}{3}+1)]\dots[n(\frac{x}{n}+1)]}$$

$$= \lim_{n\to\infty} n^x \frac{n!}{x(x+1)[(\frac{x}{2}+1)][(\frac{x}{3}+1)]\dots[(\frac{x}{n}+1)]n!}$$

or,

$$\Gamma(x) = \lim_{n\to\infty} \frac{n^x}{x\prod_{k=1}^{n}\left(1+\frac{x}{k}\right)}. \tag{6.7.19}$$

Now, notice that since

$$n+1 = \left(\frac{n+1}{n}\right)\left(\frac{n}{n-1}\right)\left(\frac{n-1}{n-1}\right)\dots\left(\frac{3}{2}\right)\left(\frac{2}{1}\right)$$

$$= \left(1+\frac{1}{n}\right)\left(1+\frac{1}{n-1}\right)\left(1+\frac{1}{n-2}\right)\dots\left(1+\frac{1}{2}\right)\left(1+\frac{1}{1}\right)$$

then

$$(n+1)^x = \left(1+\frac{1}{n}\right)^x \left(1+\frac{1}{n-1}\right)^x \left(1+\frac{1}{n-2}\right)^x \cdots \left(1+\frac{1}{2}\right)^x \left(1+\frac{1}{1}\right)^x$$

and so

$$(n+1)^x = \prod_{k=1}^{n} \left(1+\frac{1}{k}\right)^x. \tag{6.7.20}$$

Since

$$\lim_{n\to\infty} \frac{n^x}{(n+1)^x} = 1$$

we can, with vanishing error, replace n^x in (6.7.19) with $(n+1)^x$ from (6.7.20) to get, in the limit as $n \to \infty$,

$$\Gamma(x) = \frac{\prod_{k=1}^{\infty}\left(1+\frac{1}{k}\right)^x}{x\prod_{k=1}^{\infty}\left(1+\frac{x}{k}\right)}$$

and so, recalling (4.1.2),

$$x\Gamma(x) = \frac{\prod_{k=1}^{\infty}\left(1+\frac{1}{k}\right)^x}{\prod_{k=1}^{\infty}\left(1+\frac{x}{k}\right)} = \Gamma(x+1). \tag{6.7.21}$$

Taking the logarithm of (6.7.21), and then differentiating with respect to x,

$$\frac{d}{dx}[\ln\{\Gamma(x+1)\}] = \frac{d}{dx}\ln\left\{\frac{\prod_{k=1}^{\infty}\left(1+\frac{1}{k}\right)^x}{\prod_{k=1}^{\infty}\left(1+\frac{x}{k}\right)}\right\}$$

$$= \frac{d}{dx}\left[x\sum_{k=1}^{\infty}\ln\left(1+\frac{1}{k}\right) - \sum_{k=1}^{\infty}\ln\left(1+\frac{x}{k}\right)\right].$$

That is,

$$\frac{d}{dx}[\ln\{\Gamma(x+1)\}] = \sum_{k=1}^{\infty}\ln\left(1+\frac{1}{k}\right) - \sum_{k=1}^{\infty}\frac{\frac{1}{k}}{1+\frac{x}{k}}. \tag{6.7.22}$$

We can write (6.7.22) as

$$\frac{d}{dx}[\ln\{\Gamma(x+1)\}] = \lim_{n\to\infty}\sum_{k=1}^{n}\left\{\ln\left(\frac{k+1}{k}\right) - \frac{1}{x+k}\right)\right\}$$

or, since

$$\sum_{k=1}^{n} \ln\left(\frac{k+1}{k}\right) = \ln\left(\frac{2}{1}\right) + \ln\left(\frac{3}{2}\right) + \ln\left(\frac{4}{3}\right) + \cdots + \ln\left(\frac{n+1}{n}\right)$$

$$= \ln\left\{\left(\frac{2}{1}\right)\left(\frac{3}{2}\right)\left(\frac{4}{3}\right)\cdots\left(\frac{n+1}{n}\right)\right\} = \ln(n+1),$$

then

$$\frac{d}{dx}[\ln\{\Gamma(x+1)\}] = \lim_{n\to\infty}\left\{\ln(n+1) - \sum_{k=1}^{n}\frac{1}{x+k}\right\}$$

$$= \lim_{n\to\infty}\left\{\ln(n+1) - \sum_{k=1}^{n}\frac{1}{k} + \sum_{k=1}^{n}\left(\frac{1}{k} - \frac{1}{x+k}\right)\right\}$$

$$= \lim_{n\to\infty}\left\{\left(\ln(n+1) - \sum_{k=1}^{n}\frac{1}{k}\right) + x\sum_{k=1}^{n}\frac{1}{k(x+k)}\right\}.$$

Recalling the definition of gamma (Euler's constant) from Sect. 5.4, as $n \to \infty$ we arrive at

$$\frac{d}{dx}[\ln\{\Gamma(x+1)\}] = -\gamma + x\sum_{k-1}^{\infty}\frac{1}{k(x+k)}. \qquad (6.7.23)$$

Now, notice that

$$x\sum_{k=1}^{n}\frac{1}{k(x+k)} = x\sum_{k=1}^{\infty}\frac{1}{k(\frac{x}{k}+1)k} = x\sum_{k=1}^{\infty}\frac{1}{k^2(1+\frac{x}{k})}$$

$$= x\sum_{k=1}^{\infty}\frac{1}{k^2}\left[1 - \left(\frac{x}{k}\right) + \left(\frac{x}{k}\right)^2 - \left(\frac{x}{k}\right)^3 + \cdots\right], \quad -1 \le x \le 1$$

and so (6.7.23) becomes

$$\frac{d}{dx}[\ln\{\Gamma(x+1)\}] = -\gamma + \sum_{k=1}^{\infty}\frac{x}{k^2} - \sum_{k=1}^{\infty}\frac{x^2}{k^3} + \sum_{k=1}^{\infty}\frac{x^3}{k^4} - \cdots.$$

Thus, integrating indefinitely, with C an arbitrary constant,

$$\ln\{\Gamma(x+1)\} = -\gamma x + \frac{x^2}{2}\zeta(2) - \frac{x^3}{3}\zeta(3) + \frac{x^4}{4}\zeta(4) - \cdots + C.$$

For $x = 0$ we have $\ln\{\Gamma(1)\} = \ln(1) = 0 = 0 + C$ and so $C = 0$. Thus,

$$\ln\{\Gamma(x+1)\} + \gamma x = \sum_{k=2}^{\infty}\frac{\zeta(k)}{k}(-1)^k x^k, \quad -1 < x \le 1. \qquad (6.7.24)$$

The power series expansion in (6.7.24) is *almost* in the proper form for use in the RMT, *but not quite*, as you'll notice that the sum starts at $k = 2$ not $k = 0$, and that there is no k! in the denominator. So, we have just one last step to complete, as follows. Let

$$f(x) = \ln\left\{\Gamma(x+1)\right\} + \gamma x,$$

and then let's calculate the power series expansion for $\phi(x) = \frac{f(x)}{x^2}$. That is,

$$\phi(x) = \frac{f(x)}{x^2} = \frac{\sum_{k=2}^{\infty} \frac{\zeta(k)}{k}(-1)^k x^k}{x^2} = \sum_{k=2}^{\infty} \frac{(k-1)!\zeta(k)}{(k-1)!k}(-1)^k x^{k-2}$$

$$= \sum_{k=2}^{\infty} \frac{(k-1)!\zeta(k)}{k!}(-1)^k x^{k-2}.$$

Next, define the new summation index $j = k - 2$. Then,

$$\phi(x) = \frac{f(x)}{x^2} = \sum_{j=0}^{\infty} \frac{(j+1)!\zeta(j+2)}{(j+2)!}(-1)^{j+2} x^j$$

$$= \sum_{j=0}^{\infty} \frac{(j+1)!\zeta(j+2)}{j!(j+1)(j+2)}(-1)^j(-1)^2 x^j$$

or, as $(-1)^2 = 1$,

$$\phi(x) = \frac{f(x)}{x^2} = \sum_{j=0}^{\infty} \frac{\frac{(j+1)!}{(j+1)(j+2)}\zeta(j+2)}{j!}(-1)^j x^j.$$

Now we have a power series expansion that fits the form required by the RMT, with $\phi(x) = \frac{f(x)}{x^2}$, and

$$\lambda(j) = \frac{(j+1)!}{(j+1)(j+2)}\zeta(j+2).$$

Returning to the summation index symbol k (just to match the notation in the statement of the RMT), and simplifying,

$$\lambda(k) = \frac{k!(k+1)}{(k+1)(k+2)}\zeta(k+2) = \frac{k!}{(k+2)}\zeta(k+2)$$

and so the RMT says

$$\int_0^\infty x^{s-1} \frac{f(x)}{x^2}\,dx = \Gamma(s)\lambda(-s) = \Gamma(s)\lambda(-s)\frac{(-s)!}{(-s+2)}\zeta(-s+2) = \frac{s!}{s}(-s)!\frac{\zeta(2-s)}{2-s}$$

because, from (4.1.3), $s\Gamma(s) = s!$ Further, just before we derived (4.3.5) we found

$$(-z)! = \frac{\pi z}{z! \sin{(\pi z)}}$$

and so, writing s in place of z,

$$\int_0^\infty x^{s-1} \frac{\ln\{\Gamma(x+1)\} + \gamma x}{x^2} \, dx = \left\{\frac{s!}{s}\right\} \left\{\frac{\pi s}{s! \sin{(\pi s)}}\right\} \frac{\zeta(2-s)}{2-s}$$

$$= \left\{\frac{\pi}{\sin{(\pi s)}}\right\} \frac{\zeta(2-s)}{2-s}, 0 < s < 1$$

just as given in (6.7.11).

If you can derive (6.7.11) *without* using the RMT, please let me know *how*!

6.8 Challenge Problems

To end this chapter on tough integrals, it seems appropriate to challenge you with some more tough integrals.

(C6.1) Look back at Sect. 6.3, where we evaluated what I called Coxeter's integral. But that wasn't the *only* Coxeter integral. Coxeter was actually (as you'll recall from the original Preface) stumped by a number of integrals, all of which Hardy solved. So, here's another one of Coxeter's original integrals for you to try your hand at:

$$\int_0^{\pi/2} \cos^{-1}\left\{\frac{1}{1 + 2\cos{(x)}}\right\} dx.$$

When this is given to MATLAB we get *integral(@(x)acos(1./(1 + 2∗cos(x))),0, pi/2)* = 1.64493... which might suggest to you (after reading Sect. 5.3) that the exact value is Euler's $\zeta(2) = \frac{\pi^2}{6}$. Your challenge here is to prove that this is indeed so. Hint: this actually isn't all *that* tough—at least it shouldn't be after reading Sect. 6.3—and the integral will yield using the same approach we used in the text.

(C6.2) When we get to Chap. 8 on contour integration we'll do the integral $\int_0^\infty \frac{x^m}{x^n+1} dx$, where (to insure the integral exists) m and n are non-negative integers such that $n - m \geq 2$. For the special case of $m = 1$—see (8.7.8)—the result is $\int_0^\infty \frac{x}{x^n+1} dx = \frac{\pi/n}{\sin{\left(\frac{2\pi}{n}\right)}}, n \geq 3$. If we evaluate the right-hand-side for the first few values of permissible n, we get

n	Value of integral
3	1.2091995
4	0.7853981
5	0.6606532
6	0.6045997
7	0.5740354

Now, consider the integral $\int_0^\infty \frac{dx}{x^{n-1}+x^{n-2}+...+x+1}, n \geq 3$. If we use MATLAB's *integral* to numerically evaluate this integral for the same values of n we get

n	Value of integral
3	1.2091995
4	0.7853981
5	0.6606531
6	0.6045997
7	0.5740354

These numerical results strongly suggest that the two integrals might be equal. You *could* study this question directly, by 'simply' evaluating the second integral (but that might not be so 'simple'!). Another possibility is to first write the difference of the two integrands $f(x) = \frac{x}{x^n+1} - \frac{1}{x^{n-1}+x^{n-2}+...+x+1}$, and then (somehow) show that $\int_0^\infty f(x)dx = 0$. See if you can do this.

(C6.3) Here's a problem reminiscent of the 'sophomore's dream' of (6.1.4). There is a value of c with $0 < c < 1$ such that $\int_0^1 c^x \, dx = \sum_{k=1}^\infty c^k$. Calculate the value of c accurate to *at least* 13 decimal places and, for that value, what is the common value of the integral and the sum? Hint: to start, observe that since c is between 0 and 1 there must be some $\lambda > 0$ such that $c = e^{-\lambda}$. ($\lambda = 0$ gives $c = 1$, and $\lambda = \infty$ gives $c = 0$.) This notational trick makes both the integral and the sum easy to evaluate. Then, equate your expressions for the integral and the sum to get a transcendental equation for λ, an equation that can be solved numerically using any number of well-known algorithms (my suggestion: see any book on numerical analysis and look for a discussion of the *binary chop* algorithm) to give c to the required accuracy.

(C6.4) I mentioned 'Cayley's Problem' in the text, but didn't provide any details. It is the problem of computing how a uniform mass density (μ) linked-chain, initially heaped-up near the edge of a table, falls from the table as it slides without friction over the edge. If x is the length of the chain hanging over the edge at time t, then the problem is to find the differential equation of motion (involving x and t) and then to solve (that is, integrate) it for x as a function of t. In his 1943 book *Mechanics*, the German physicist Arnold Sommerfeld (1868–1951) stated without derivation that the equation of motion is (where g is the acceleration of gravity) $\frac{d}{dt}(x\dot{x}) = x\ddot{x} + \dot{x}^2 = gx$. In this dot notation (due to Newton) $\dot{x} = \frac{dx}{dt}$ and $\ddot{x} = \frac{d^2x}{dt^2}$. In Leibniz's more suggestive differential notation (you'll recall from Chap. 1 we get the modern integral sign ∫from him, too), Sommerfeld's equation of motion is $\frac{d}{dt} \times$ $\left(x\frac{dx}{dt}\right) = x\frac{d^2x}{dt^2} + \left(\frac{dx}{dt}\right)^2 = gx$, and of it he said "its integration is somewhat difficult." A very clever derivation of Sommerfeld's equation can be found in the paper by David Keiffer, "The Falling Chain and Energy Loss," *American Journal of Physics*, March 2001, pp. 385–386. Keiffer made no reference to Sommerfeld, but rather based his derivation on a direct analysis of how a chain slides off a table, link-by-link. Keiffer gave an interesting twist to problem by converting it to determining the speed of the falling chain as a function not of time, but rather as a function of the length of chain that has already slid off the table (that is, of x). Calling the speed v(x), his equation of motion is $\frac{d(v^2)}{dx} = -\frac{2}{x}v^2 + 2g$.

(a) Show that Keiffer's equation and Sommerfeld's equation are one-in-the-same;
(b) Noticing that Keiffer's equation is a first-order differential equation in v^2, use the same approach we used in Chap. 3 (Sect. 3.9) to integrate Keiffer's equation to find v^2. Hint: consider using x^2 as the integrating factor;
(c) Show that the chain falls with the constant acceleration $\frac{1}{3}g$ (this was Cayley's central result);
(d) Use your result from (b) to calculate T, the time for a chain of length L to completely slide off the table.

Now, in the interest of honesty, I have to tell you that while all your calculations in response to the above questions represent good, solid *math*, there has been much debate in recent years on whether or not it is good *physics*! The Cayley, Sommerfeld, Keiffer analyses involve a failure of energy conservation when a chain slides off a table. This is actually quite easy to show. The chain's initial P.E. and K.E. are both zero (the table top is our zero reference level for P.E., and the chain is initially at rest), and so the initial total energy is zero. When the chain has just finished sliding completely off the table, its speed is $\sqrt{\frac{2gL}{3}}$, a result you should have already come across in your earlier analyses. So, its K.E. is $\frac{1}{2} \times$ $(\mu L)\frac{2gL}{3} = \mu\frac{L^2g}{3}$. The chain's center of mass is at $\frac{L}{2}$ below the table top, and so its P.E. is $-\mu Lg\frac{L}{2} = -\mu\frac{L^2g}{2}$. Thus, at the completion of the fall its total energy is $\mu\frac{L^2g}{3} - \mu\frac{L^2g}{2} < 0$. Thus, energy was lost during the fall. (This is a puzzle in its own right, since the slide was said to be frictionless. So, *how* is energy dissipated? This was, in fact, the question that stimulated Keiffer to write his paper in the first place.) In recent years, other physicists have claimed that Cayley's falling chain *does* conserve energy. So, for your last question here,

(e) Assuming conservation of energy, show that the chain falls with the constant acceleration of $\frac{1}{2}g$ (not Cayley's $\frac{1}{3}g$).

(C6.5) This problem involves a very different appearance of integrals in yet another physics problem. (See James M. Supplee and Frank W. Schmidt, "Brachistochrone in a Central Force Field," *American Journal of Physics*, May 1991, pp. 402 and 467.) Imagine a tiny bead of mass m with a wire threaded through a hole in it, allowing the bead to slide *without friction* along the wire. The wire lies entirely in a horizontal plane, with one end (in terms of the polar coordinates r and θ) at $\left(R, \frac{\pi}{3}\right)$ and the other end on the horizontal axis at $(R, 0)$. The only force acting on the bead is the inverse-square gravitational force due to a point mass M located at the origin. That is, the bead slides on the wire because it experiences the attractive radial force $F = G\frac{Mm}{r^2}$, where G is the universal gravitational constant. The bead has initial speed $\sqrt{2\frac{GM}{R}}$. What shape function r(θ) should the wire have to minimize the travel time of the bead as it slides from $(R, 0)$ to $\left(R, \frac{\pi}{3}\right)$? To answer this question, fill-in the details of the following steps.

(a) Show that the initial potential energy (P.E.) of the bead is $-\frac{GMm}{R}$, and that the initial kinetic energy (K.E.) of the bead is $\frac{GMm}{R}$, and so the total initial energy is zero. Hint: the initial P.E. is the energy required to transport the mass m in from infinity along the positive horizontal axis to $r = R$, which is given by $\int_\infty^R F\, dr$. This energy is *negative* because gravity is *attractive*;

(b) If T is the total time for the bead to travel from one end to the other, then $T = \int dt = \int \frac{ds}{v}$ where v is the instantaneous speed of the bead and ds is the differential path length along the wire. In polar coordinates $(ds)^2 = (dr)^2 + (rd\theta)^2$ and thus $ds = d\theta\sqrt{\left(\frac{dr}{d\theta}\right)^2 + r^2}$. Invoking conservation of energy (K.E. + P.E. = 0, *always*, because there is no friction), show that $T = \frac{1}{\sqrt{2GM}}\int_0^{\pi/3}\sqrt{r\left(\frac{dr}{d\theta}\right)^2 + r^3}\, d\theta$;

(c) To minimize T is a problem beyond ordinary freshman differential calculus, where we try to find the value of a variable that gives the extrema of some function. Our problem here is to find a *function* that minimizes the integral for T. This is a problem in what is called the *calculus of variations*, and a fundamental result from that subject is the *Euler-Lagrange equation*: if we write the integrand of the T-integral as $K = \sqrt{rr'^2 + r^3}$, where $r' = \frac{dr}{d\theta}$, then $\frac{\partial K}{\partial r} - \frac{d}{d\theta} \times \left(\frac{\partial K}{\partial r'}\right) = 0$. This result is derived in any book on the calculus of variations (or see my book *When Least Is Best*, Princeton 2004, 2007, pp. 233–238). Euler knew it by 1736, but the derivation I give in WLIB is the now standard one developed by the French-Italian mathematical physicist Joseph Louise Lagrange (1736–1813), in a letter the then *teenage* (!) Lagrange wrote to Euler in 1755. Use the Euler-Lagrange equation to show that the required r(θ) satisfies the differential equation $5r'^2 + 3r^2 = 2rr''$, where $r'' = \frac{d^2r}{d\theta^2}$;

(d) Change variable to $u = \frac{r'}{r}$, and show that the differential equation in (c) becomes $u^2 + 1 = \frac{2}{3}u'$, where $u' = \frac{du}{d\theta}$. Hint: Start by writing the differential equation in (c) as $3r'^2 + 3r^2 = 2rr'' - 2r'^2$;

(e) The differential equation in (d) is $\frac{3}{2}d\theta = \frac{du}{1+u^2}$ which you should be able to easily integrate indefinitely to show that $u = \tan\left(\frac{3}{2}\theta + C_1\right)$, where C_1 is an arbitrary constant;

(f) The result in (e) says $\frac{1}{r}\frac{dr}{d\theta} = \tan\left(\frac{3}{2}\theta + C_1\right)$, which you should be able to integrate indefinitely to show that $r = C_2 \cos^{-2/3}\left(\frac{3}{2}\theta + C_1\right)$;

(g) Use the coordinates of the ends of the wire to evaluate the constants C_1 and C_2, thus arriving at $r(\theta) = \frac{R}{\sqrt[3]{2}\,\cos^{2/3}\left(\frac{3}{2}\theta - \frac{\pi}{4}\right)}$, a curve called a *brachistochrone*, from the Greek *brachistos* (shortest) and *chronos* (time). Note, carefully, that it is *not* the shortest *length* curve, which of course would be a straight line segment.

(C6.6) In this problem we return once more to the topic of how integrals arise in 'interesting' physics problems. Imagine a 1 kg mass on the real axis at $x = 0$ at time $t = 0$, moving with speed 1 m/s. That is, $x(0) = 0$ and $v(0) = 1$. The mass then experiences, for $t > 0$, a speed-dependent acceleration force $F = v^3$ newtons. Thus, $\frac{d^2x}{dt^2} = \frac{dv}{dt} = v^3$. Ignoring any objections from special relativity, show that the mass rather quickly accelerates to infinite speed, and determine where the mass is on the x-axis when that occurs. Now, a combination hint and thought-puzzler for you to mull over: Your calculations should show that the mass reaches infinite speed at a finite time (call it $t = T$) when still pretty close to $x = 0$. So, what is T, and what happens when $t > T$? You should find that then the mass location suddenly becomes *complex* (has a non-zero imaginary part), which means the mass is no longer on the real axis. So, just where *is* it? Does this bizarre result remind you of a famous car, powered electrically by Doctor Emmett Brown's amazing flux-capacitor?

(C6.7) Consider the following claim:

$$\int_0^{2\pi} \sqrt{a^2 \sin^2(t) + b^2 \cos^2(t)}\,dt \geq \sqrt{4\pi\left\{\pi ab + (a - b)^2\right\}},$$

with obvious equality when $a = b$. In my book *When Least Is Best* I give a *geometric* proof of this inequality, based on first dissecting and then rearranging an ellipse, and then applying the ancient isoperimetric theorem (a closed, non-self-intersecting curve of given length that encloses the maximum possible area is a circle) that wasn't proven until the nineteenth century. In that book I challenged readers to find a purely *analytical* proof of the inequality. That challenge remained unanswered until August 2008, when M. W. Perera (who lives near Colombo, Sri Lanka) sent me a beautifully simple and elegant derivation. Indeed, Perera's proof improves on the original inequality by showing how to make the integral's lower bound even larger. Can you repeat Perera's achievement? To convince you of the validity of the inequality (and to show you the dramatic performance of Perera's improvement, Fig. C6.7 shows plots of the relative error for the original inequality (solid curve), and for Perera's improvement (dashed curve), both curves as a function of $\frac{a}{b}$. That is, the vertical axis is $\left\{\int_0^{2\pi} - \text{lower bound}\right\}/\int_0^{2\pi} = 1 - \frac{\text{lower bound}}{\int_0^{2\pi}}$.

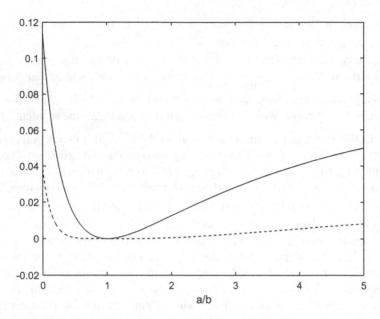

Fig. C6.7 Relative error of two lower bounds on an integral

Hint: You may find the following power series expansion of the complete elliptic integral of the second kind to be useful: if $0 \leq k \leq 1$, with $n = \frac{1-\sqrt{1-k^2}}{1+\sqrt{1-k^2}}$, then

$$\int_0^{\pi/2} \sqrt{1 - k^2 \sin^2(t)}\, dt = \frac{\pi}{2(1+n)} \left[1 + \frac{n^2}{4} + \frac{n^4}{64} + \frac{n^6}{256} + \cdots \right].$$

(C6.8) In 1833 the English mathematician Robert Murphy (1806–1843) solved the integral equation $\int_0^1 t^x f(t)\, dt = F(x)$ for the function $f(t)$ for each of several different assumed expressions for $F(x)$. If $F(x) = \frac{1}{(x+m)^n}$, for example (where $m > 0$ and n is a positive integer), Murphy found that $f(t) = \frac{1}{(n-1)!} t^{m-1} \ln^{n-1}\left(\frac{1}{t}\right)$. (For how he did this, see his paper "On the Inverse Method of Definite Integrals, with Physical Applications," in volume 4 of the *Transactions of the Cambridge Philosophical Society*, pp. 353–408.) Show that Murphy's solution reproduces the result of C4.2.

(C6.9) Continuing with the previous problem, if $F(x) = \ln\left(1 + \frac{a}{x+m}\right)$ where $m > 0$, $a > 0$, then Murphy showed that $f(t) = t^{m-1} \frac{t^a - 1}{\ln(t)}$. Show that Murphy's solution reproduces (3.4.4).

Chapter 7
Using $\sqrt{-1}$ to Evaluate Integrals

7.1 Euler's Formula

The use of $i = \sqrt{-1}$ to compute integrals is nicely illustrated with a quick example. Let's use i to do

$$\int_1^\infty \frac{dx}{x(x^2+1)}.$$

Making a partial fraction expansion of the integrand, we can write

$$\int_1^\infty \frac{dx}{x(x^2+1)} - \int_1^\infty \left\{ \frac{1}{x} - \frac{1}{2(x-i)} - \frac{1}{2(x+i)} \right\} dx$$

$$= \left\{ \ln(x) - \frac{1}{2}\ln(x-i) - \frac{1}{2}\ln(x+i) \right\} \Big|_1^\infty = \left\{ \ln(x) + \ln\left(\frac{1}{\sqrt{x-i}}\right) + \ln\left(\frac{1}{\sqrt{x+i}}\right) \right\} \Big|_1^\infty$$

$$= \ln\left\{ \frac{x}{\sqrt{(x-i)(x+i)}} \right\} \Big|_1^\infty = \ln\left\{ \frac{x}{\sqrt{x^2+1}} \right\} \Big|_1^\infty = \ln(1) - \ln\left(\frac{1}{\sqrt{2}}\right) = \ln\left(\sqrt{2}\right) = \frac{1}{2}\ln(2)$$

$$= 0.34657\ldots.$$

Checking with MATLAB, *integral(@(x)1./(x.*(x.^2 + 1)),1,inf) = 0.34657....*
This is nice, yes, but the utility of complex-valued quantities in doing definite integrals will *really* become clear when we get to the next chapter, on contour integration. The value of the complex can be immediately appreciated at a 'lower' level, however, without having to wait for contour integrals, and all we'll really need to get started is Euler's famous identity:

© Springer Nature Switzerland AG 2020
P. J. Nahin, *Inside Interesting Integrals*, Undergraduate Lecture Notes in Physics,
https://doi.org/10.1007/978-3-030-43788-6_7

$$e^{ibx} = \cos(bx) + i\,\sin(bx), \tag{7.1.1}$$

where b is any real quantity. I'll take (7.1.1) as known to you, but if you want to explore it further, both mathematically and historically, you can find more in two of my earlier books.[1]

A very straightforward and yet still quite interesting demonstration of Euler's identity can be found in the problem of calculating

$$\int_0^\infty \sin(bx)e^{-xy}\,dx.$$

This is usually done by-parts in freshman calculus, but using $\sqrt{-1}$ is easier. Since

$$e^{-ibx} = \cos(bx) - i\,\sin(bx)$$

it then follows that

$$\sin(bx) = \frac{e^{ibx} - e^{-ibx}}{2i}.$$

Thus

$$e^{-xy}\sin(bx) = \frac{e^{-x(y-ib)} - e^{-x(y+ib)}}{2i}$$

and so

$$\int_0^\infty e^{-xy}\sin(bx)dx = \frac{1}{2i}\int_0^\infty \left\{ e^{-x(y-ib)} - e^{-x(y+ib)} \right\}dx = \frac{1}{2i}\left\{ \frac{e^{-x(y-ib)}}{-(y-ib)} - \frac{e^{-x(y+ib)}}{-(y+ib)} \right\}\Big|_0^\infty$$

$$= \frac{1}{2i}\left\{ \frac{1}{y-ib} - \frac{1}{y+ib} \right\} = \frac{1}{2i}\left\{ \frac{y+ib - y + ib}{y^2 + b^2} \right\} = \frac{2ib}{2i(y^2 + b^2)}$$

and so

$$\int_0^\infty \sin(bx)e^{-xy}\,dx = \frac{b}{y^2 + b^2}. \tag{7.1.2}$$

---

[1] *An Imaginary Tale: the story of $\sqrt{-1}$*, and *Dr. Euler's Fabulous Formula: cures many mathematical ills*, both published by Princeton University Press (both in multiple editions).

7.2   The Fresnel Integrals

Well, that last calculation was certainly interesting but, to *really* demonstrate to you the power of Euler's identity, I'll now show you how to use it to derive two famous definite integrals named after the French scientist Augustin Jean Fresnel (1788–1827), who encountered them in an 1818 study of the illumination intensity of diffraction patterns. There is just a bit of irony in the fact that, despite the name, it was actually *Euler* who first found the values (in 1781)—before Fresnel was even born!—of the 'Fresnel' integrals. Euler, too, encountered the same integrals in a physical situation, but one that is entirely different from Fresnel's optical concerns.

For Euler, it was a study of the forces acting on an elastic spring coiled in the form of a spiral, with one end fixed and the other end free to move under the influence of an attached mass.[2] In any case, the integrals we are going to evaluate are

$$\int_0^\infty \cos\left(x^2\right)dx \text{ and } \int_0^\infty \sin\left(x^2\right)dx.$$

(You'll recall we used the second integral in Sect. 6.4, in the discussion there of the Hardy-Schuster optical integral). In his original analysis, Euler used his gamma function (from Chap. 4), but here I'll show you a different approach (in Challenge Problem 7.12 you'll find more on Euler's original approach).

We start with

$$G(x) = \left\{\int_0^x e^{it^2}dt\right\}^2 + i\int_0^1 \frac{e^{ix^2\left(t^2+1\right)}}{t^2+1}\,dt.$$

(I'll explain where this rather curious G(x) comes from in just a moment). Notice, in passing, that

$$G(0) = 0 + i\int_0^1 \frac{dt}{t^2+1} = i\,\tan^{-1}(1) = i\frac{\pi}{4},$$

which will be important for us to know a few steps from now. Differentiating G(x) with respect to x,

---

[2]A nice discussion of the history of the 'Fresnel' integrals is in R. C. Archibald, "Euler Integrals and Euler's Spiral—Sometimes Called Fresnel Integrals and the Clothoïde or Cornu's Spiral," *The American Mathematical Monthly*, June 1918, pp. 276–282.

$$\frac{dG}{dx} = 2\left\{ \int_0^x e^{it^2} dt \right\} e^{ix^2} + i \int_0^1 \frac{i2x(t^2+1)e^{ix^2(t^2+1)}}{t^2+1} dt$$

$$= 2e^{ix^2} \int_0^x e^{it^2} dt - 2x \int_0^1 e^{ix^2t^2} e^{ix^2} dt = 2e^{ix^2} \int_0^x e^{it^2} dt - 2xe^{ix^2} \int_0^1 e^{ix^2t^2} dt.$$

In the last integral change variable to u = tx (and so du = x dt or, dt = du/x). Then,

$$\frac{dG}{dx} = 2e^{ix^2} \int_0^x e^{it^2} dt - 2xe^{ix^2} \int_0^x e^{iu^2} \frac{du}{x} = 2e^{ix^2}\left[\int_0^x e^{it^2} dt - \int_0^x e^{iu^2} du \right] = 0.$$

That is, G(x) has zero rate of change with respect to x, for *all* x. G(x) is therefore a *constant* and, since $G(0) = i\frac{\pi}{4}$, *that's* the constant. This result explains the origin of the remarkably 'strange' G(x); it was specially created to have the property of a zero-everywhere derivative!

Now, as $x \to \infty$, we have

$$\lim_{x\to\infty} \int_0^1 \frac{e^{ix^2(t^2+1)}}{t^2+1} dt = 0,$$

a claim that I'll justify at the end of this section. For now, just accept it. Then,

$$G(\infty) = \left\{ \int_0^\infty e^{it^2} dt \right\}^2 = \left\{ \int_0^\infty \cos\left(t^2\right) dt + i \int_0^\infty \sin\left(t^2\right) dt \right\}^2.$$

Let's write this as

$$G(\infty) = (A + iB)^2$$

where A and B are the Fresnel integrals:

$$A = \int_0^\infty \cos\left(t^2\right) dt \text{ and } B = \int_0^\infty \sin\left(t^2\right) dt.$$

Then, since $G(\infty) = i\frac{\pi}{4}$ (remember, G(x) doesn't change as x changes and so $G(\infty) = G(0)$) we have

$$(A + iB)^2 = i\frac{\pi}{4} = A^2 + i2AB - B^2$$

and so, equating real and imaginary parts on both sides of the last equality, we have $A^2 - B^2 = 0$ (which means A = B) and $2AB = \frac{\pi}{4}$. So, $2A^2 = \frac{\pi}{4}$ and, suddenly, *we are done*: $A = \sqrt{\frac{\pi}{8}} = \frac{1}{2}\sqrt{\frac{\pi}{2}} = 0.6266\ldots$ and (because $A = B$) we have the Fresnel integrals:

$$\int_0^\infty \cos(x^2)dx = \frac{1}{2}\sqrt{\frac{\pi}{2}} \tag{7.2.1}$$

and

$$\int_0^\infty \sin(x^2)dx = \frac{1}{2}\sqrt{\frac{\pi}{2}}. \tag{7.2.2}$$

Checking, *integral(@(x)sin(x.^2),0,200)* $= 0.62585 \ldots$ and *integral(@(x) cos(x.^2),0,200)* $= 0.62902\ldots$, both suggestively close to $\frac{1}{2}\sqrt{\frac{\pi}{2}}$. And since

$$\int_0^\infty e^{ix^2}dx = \int_0^\infty \cos(x^2)dx + i\int_0^\infty \sin(x^2)dx$$

we have the interesting integral

$$\int_0^\infty e^{ix^2}dx = \frac{1}{2}\sqrt{\frac{\pi}{2}}(1+i). \tag{7.2.3}$$

To finish this discussion, I really should justify my earlier claim that $\lim_{x\to\infty}\int_0^1 \frac{e^{ix^2(t^2+1)}}{t^2+1}dt = 0$.

To show this, use Euler's identity to write

$$\int_0^1 \frac{e^{ix^2(t^2+1)}}{t^2+1}dt = \int_0^1 \frac{\cos\left\{x^2(t^2+1)\right\}}{t^2+1}dt + i\int_0^1 \frac{\sin\left\{x^2(t^2+1)\right\}}{t^2+1}dt.$$

Since

$$\cos\left\{x^2(t^2+1)\right\} = \cos(x^2t^2+x^2) = \cos(x^2t^2)\cos(x^2) - \sin(x^2t^2)\sin(x^2)$$

and

$$\sin\left\{x^2(t^2+1)\right\} = \sin(x^2t^2+x^2) = \sin(x^2t^2)\cos(x^2) + \cos(x^2t^2)\sin(x^2),$$

and because as $x \to \infty$ the factors $\cos(x^2)$ and $\sin(x^2)$ remain confined to the interval $[-1,1]$, our claim will be established if we can show that

$$\lim_{x\to\infty}\int_0^1 \frac{\sin\left\{x^2t^2\right\}}{t^2+1}dt = \lim_{x\to\infty}\int_0^1 \frac{\cos\left\{x^2t^2\right\}}{t^2+1}dt = 0.$$

This assertion follows from the famous *Riemann-Lebesgue lemma*, which says that if f(t) is *absolutely integrable* over the finite interval a to b, then

$\lim_{m \to \infty} \int_a^b f(t) \cos(mt)\, dt = 0$. In our case, $f(t) = \frac{1}{t^2+1}$ which *is* obviously absolutely integrable (its absolute value remains finite) for $0 \le t \le 1$. As an engineer would put it, the integral (over a finite interval) of the product of such a function with a sinusoid goes to zero as the frequency of the sinusoid becomes arbitrarily large.[3] In our case, $x^2 t$ (which certainly blows-up for every t in the integration interval as $x \to \infty$) plays the role of m. A similar conclusion, not surprisingly, is that

$$\lim_{x \to \infty} \int_0^1 \frac{\sin\left\{x^2 t^2\right\}}{t^2+1}\, dt = 0$$

as well.

As a final comment on the Fresnel integrals, it is interesting to note that the parametric equations

$$x(t) = \int_0^t \cos\left(x^2\right) dx, \, y(t) = \int_0^t \sin\left(x^2\right) dx, \, -\infty < t < \infty$$

produce the beautiful curve shown in Fig. 7.2.1. Given the origin of the integrals in Euler's study of a spiral spring, it is most appropriate that the curve is, itself, a spiral![4]

7.3 $\zeta(3)$ and More Log-Sine Integrals

You'll recall, from Chap. 5 (Sect. 5.3), Euler's fascination with the zeta function $\zeta(s) = \sum_{k=1}^{\infty} \frac{1}{k^s}$. He found explicit formulas for $\zeta(s)$ for any positive *even* integer value of s but he couldn't do the same for the odd values of s, even though he devoted enormous time and energy to the search. In 1772 he came as close as he ever would when he stated

$$\int_0^{\pi/2} x \ln\left\{\sin(x)\right\} dx = \frac{7}{16}(3) - \frac{\pi^2}{8} \ln(2).$$

The key to understanding how such an incredible result could be discovered, as you might expect from the earlier sections of this chapter, is Euler's identity. Here's how it goes.

---

[3]You can find a proof of the lemma (it's not difficult) in Georgi P. Tolstov's book *Fourier Series* (translated from the Russian by Richard A. Silverman), Dover 1976, pp. 70–71.

[4]The plot (created by a MATLAB code) in Figure 7.2.1 is yet another example of the naming of mathematical things after the wrong person. The curve is often called *Cornu's spiral*, after the French physicist Marie-Alfred Cornu (1841–1902). While Cornu did indeed make such a plot (in 1874), Euler had described the spiral *decades earlier*. Well, of course, if *everything* Euler did first was named after him maybe that *would* be a bit confusing.

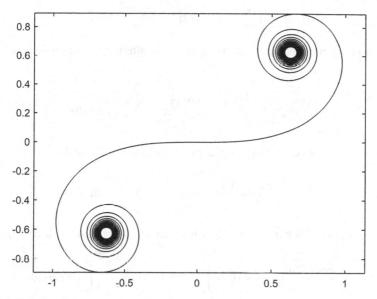

Fig. 7.2.1 Euler's spiral

Define the function S(y) as

$$S(y) = 1 + e^{iy} + e^{i2y} + e^{i3y} + \ldots + e^{imy}$$

where m is some finite integer. This looks like a geometric series and so, using the standard trick for summing such series, multiply through by the common factor e^{iy} that connects any two adjacent terms. Then,

$$e^{iy}S(y) = e^{iy} + e^{i2y} + e^{i3y} + \ldots + e^{imy} + e^{i(m+1)y}$$

and so

$$e^{iy}S(y) - S(y) = e^{i(m+1)y} - 1.$$

Solving for S(y),

$$S(y) = \frac{e^{i(m+1)y} - 1}{e^{iy} - 1} = \frac{e^{i(m+1)y} - 1}{e^{\frac{iy}{2}}\left(e^{\frac{iy}{2}} - e^{-\frac{iy}{2}}\right)} = \frac{e^{i\left(m+\frac{1}{2}\right)y} - e^{-\frac{iy}{2}}}{i2\sin\left(\frac{y}{2}\right)}$$

$$= \frac{\cos\left\{\left(m+\frac{1}{2}\right)y\right\} + i\sin\left\{\left(m+\frac{1}{2}\right)y\right\} - \cos\left(\frac{y}{2}\right) + i\sin\left(\frac{y}{2}\right)}{i2\sin\left(\frac{y}{2}\right)}$$

Now, looking back at the original definition of S(y), we see that it can also be written as

$$S(y) = 1 + \sum_{n=1}^{m} \cos(ny) + i\sum_{n=1}^{m} \sin(ny).$$

So, equating the imaginary parts of our two alternative expressions for S(y), we have

$$-\frac{\cos\left\{(m+\frac{1}{2})y\right\}}{2\sin\left(\frac{y}{2}\right)} + \frac{\cos\left(\frac{y}{2}\right)}{2\sin\left(\frac{y}{2}\right)} = \sum_{n=1}^{m} \sin(ny).$$

At this point it is convenient to change variable to y = 2t, and so

$$-\frac{\cos\left\{(2m+1)t\right\}}{\sin(t)} + \cot(t) = 2\sum_{n=1}^{m} \sin(2nt).$$

Then, integrate this expression, term-by-term, from t = x to t = $\frac{\pi}{2}$, getting

$$-\int_{x}^{\frac{\pi}{2}} \frac{\cos\left\{(2m+1)t\right\}}{\sin(t)}dt + \int_{x}^{\frac{\pi}{2}} \cot(t)dt = 2\sum_{n=1}^{m} \int_{x}^{\frac{\pi}{2}} \sin(2nt)dt.$$

The integral on the right of the equality sign is easy to do:

$$\int_{x}^{\frac{\pi}{2}} \sin(2nt)dt = \left\{-\frac{\cos(2nt)}{2n}\right\}\Big|_{x}^{\frac{\pi}{2}} = \frac{-\cos(n\pi) + \cos(2nx)}{2n} = \frac{\cos(2nx) - (-1)^n}{2n}.$$

The last integral on the left of the equality sign is just as easy:

$$\int_{x}^{\frac{\pi}{2}} \cot(t)dt = [\ln\{\sin(t)\}]\Big|_{x}^{\frac{\pi}{2}} = \ln\left\{\sin\left(\frac{\pi}{2}\right)\right\} - \ln\{\sin(x)\} = -\ln\{\sin(x)\}.$$

Thus,

$$-\int_{x}^{\frac{\pi}{2}} \frac{\cos\left\{(2m+1)t\right\}}{\sin(t)}dt - \ln\{\sin(x)\} = \sum_{n=1}^{m} \frac{\cos(2nx)}{n} - \sum_{n=1}^{m} \frac{(-1)^n}{n}.$$

You'll recall that at the start of Sect. 5.2 we had the power series

$$\ln(1+x) = x - \frac{x^2}{2} + \frac{x^3}{3} - \frac{x^4}{4} + \cdots$$

and so, with x = 1, this says

$$\ln(2) = \lim_{m\to\infty}\left\{-\sum_{n=1}^{m} \frac{(-1)^n}{n}\right\}.$$

So, if we let m → ∞ we have

$$- \lim_{m \to \infty} \int_x^{\frac{\pi}{2}} \frac{\cos\{(2m+1)t\}}{\sin(t)} dt - \ln\{\sin(x)\} = \sum_{n=1}^{\infty} \frac{\cos(2nx)}{n} + \ln(2).$$

Since the limit of the integral at the far left is zero,[5] we arrive at

$$\ln\{\sin(x)\} = -\sum_{n=1}^{\infty} \frac{\cos(2nx)}{n} - \ln(2).$$

The next step (one not particularly obvious!) is to first multiply through by x and then integrate from 0 to $\frac{\pi}{2}$. That is, to write

$$\int_0^{\frac{\pi}{2}} x \ln\{\sin(x)\} dx = -\sum_{n=1}^{\infty} \frac{1}{n} \int_0^{\frac{\pi}{2}} x \cos(2nx) dx - \ln(2) \int_0^{\frac{\pi}{2}} x\, dx$$

$$= -\sum_{n=1}^{\infty} \frac{1}{n} \int_0^{\frac{\pi}{2}} x \cos(2nx) dx - \frac{\pi^2}{8} \ln(2).$$

To do the integral on the right, use integration by parts, with $u = x$ and $dv = \cos(2nx)dx$. By this time in the book you should find this to be old hat, and so I'll let *you* fill-in the details to show that

$$\int_0^{\frac{\pi}{2}} x \cos(2nx) dx = \begin{cases} -\dfrac{1}{2n^2}, & \text{if n is odd} \\ 0, & \text{if n is even.} \end{cases}$$

Thus,

$$\int_0^{\frac{\pi}{2}} x \ln\{\sin(x)\} dx = \frac{1}{2}\left\{\sum_{n=1,n \text{ odd}}^{\infty} \frac{1}{n^3}\right\} - \frac{\pi^2}{8} \ln(2).$$

We are now almost done. All that's left to do is to note that

$$\sum_{n=1,n \text{ even}}^{\infty} \frac{1}{n^3} = \frac{1}{2^3} + \frac{1}{4^3} + \frac{1}{6^3} + \cdots = \frac{1}{(21)^3} + \frac{1}{(22)^3} + \frac{1}{(23)^3} + \cdots$$

$$= \frac{1}{8}\left(\frac{1}{1^3} + \frac{1}{2^3} + \frac{1}{3^3} + \cdots\right)$$

$$= \frac{1}{8} \sum_{n=1}^{\infty} \frac{1}{n^3} = \frac{1}{8}\zeta(3).$$

and so, since

---

[5]This assertion follows from the Riemann-Lebesgue lemma (see note 3). In our case here, $f(t) = \frac{1}{\sin(t)}$ which *is* absolutely integrable over $0 < x \le t \le \frac{\pi}{2}$ since over that interval $|f(t)| < \infty$.

$$\sum_{n=1,n\ odd}^{\infty} \frac{1}{n^3} + \sum_{n=1,n\ even}^{\infty} \frac{1}{n^3} = (3)$$

we have

$$\sum_{n=1,n\ odd}^{\infty} \frac{1}{n^3} = (3) - \sum_{n=1,n\ even}^{\infty} \frac{1}{n^3} = (3) - \frac{1}{8}(3) = \frac{7}{8}(3).$$

Thus, just as Euler declared,

$$\int_0^{\pi/2} x \ln\{\sin(x)\}dx = \frac{7}{16}(3) - \frac{\pi^2}{8} \ln(2). \tag{7.3.1}$$

The right-hand-side of (7.3.1) is $\left(\frac{7}{16}\right)(1.20205\ldots) - \frac{\pi^2}{8} \ln(2) = -0.32923\ldots$, while the integral on the left is equal to $integral(@(x)x.*log(sin(x)),0,pi/2) = -0.32923\ldots$.

As I mentioned in Chap. 2 (Sect. 2.4), years ago it was often claimed in textbooks that Euler's log-sine integral was best left to the massive power of contour integration. Yet, as you've seen in this book, classical techniques often do quite well. I'll now show you that even if we 'up-the-anty,' to log-sine integrals with a *squared* integrand, we can still do a lot. Specifically, let

$$I_1 = \int_0^{\pi/2} \ln^2\{a \sin(\theta)\}d\theta = \int_0^{\pi/2} \ln^2\{a \cos(\theta)\}d\theta$$

and

$$I_2 = \int_0^{\pi/2} \ln\{a \sin(\theta)\} \ln\{a \cos(\theta)\}d\theta$$

where a is a positive constant. These integrals were *not* evaluated by Euler, but rather are due to the now nearly forgotten English mathematician Joseph Wolstenholme (1829–1891). Here's how to do them.

$$\int_0^{\frac{\pi}{2}} [\ln\{a \sin(\theta)\} + \ln\{a \cos(\theta)\}]^2 d\theta = \int_0^{\frac{\pi}{2}} \ln^2\{a^2 \sin(\theta)\cos(\theta)\}d\theta$$

$$= \int_0^{\frac{\pi}{2}} \ln^2\{a \sin(\theta)\}d\theta + \int_0^{\frac{\pi}{2}} \ln^2\{a \cos(\theta)\}d\theta + 2\int_0^{\frac{\pi}{2}} \ln\{a \sin(\theta)\} \ln\{a \cos(\theta)\}d\theta$$

$$= 2I_1 + 2I_2.$$

But since

$$\int_0^{\pi/2} \ln^2\{a^2 \sin(\theta)\cos(\theta)\}\, d\theta = \int_0^{\pi/2} \ln^2\left\{a^2 \frac{\sin(2\theta)}{2}\right\} d\theta$$

then

$$2I_1 + 2I_2 = \int_0^{\pi/2} \ln^2\left\{a^2 \frac{\sin(2\theta)}{2}\right\} d\theta = \int_0^{\pi/2}\left[\ln\{a\sin(2\theta)\} - \ln\left\{\frac{2}{a}\right\}\right]^2 d\theta$$

$$= \int_0^{\pi/2} \ln^2\{a\sin(2\theta)\}\, d\theta - 2\ln\left\{\frac{2}{a}\right\}\int_0^{\pi/2} \ln\{a\sin(2\theta)\}d\theta + \ln^2\left\{\frac{2}{a}\right\}\int_0^{\pi/2} d\theta.$$

Write $\phi = 2\theta$. So, as $d\theta = \frac{1}{2}d\phi$, we have

$$\int_0^{\pi/2} \ln^2\{a\sin(2\theta)\}\, d\theta = \frac{1}{2}\int_0^\pi \ln^2\{a\sin(\phi)\}\, d\phi = \int_0^{\pi/2} \ln^2\{a\sin(\phi)\}\, d\phi = I_1$$

and, by (7.4.1),

$$\int_0^{\pi/2} \ln\{a\sin(2\theta)\}d\theta = \frac{1}{2}\int_0^\pi \ln\{a\sin(\phi)\}d\phi = \int_0^{\pi/2} \ln\{a\sin(\phi)\}d\phi = \frac{\pi}{2}\ln\left(\frac{a}{2}\right).$$

Thus,

$$2I_1 + 2I_2 = I_1 - 2\ln\left\{\frac{2}{a}\right\}\frac{\pi}{2}\ln\left(\frac{a}{2}\right) + \frac{\pi}{2}\ln^2\left\{\frac{2}{a}\right\}$$

or,

$$I_1 + 2I_2 = -2\ln\left\{\frac{2}{a}\right\}\frac{\pi}{2}\ln\left(\frac{a}{2}\right) + \frac{\pi}{2}\ln^2\left\{\frac{2}{a}\right\}.$$

Also,

$$\int_0^{\pi/2}[\ln\{a\sin(\theta)\} - \ln\{a\cos(\theta)\}]^2 d\theta = \int_0^{\pi/2} \ln^2\{a\sin(\theta)\}d\theta$$

$$+ \int_0^{\pi/2} \ln^2\{a\cos(\theta)\}d\theta - 2\int_0^{\frac{\pi}{2}} \ln\{a\sin(\theta)\}\ln\{a\cos(\theta)\}d\theta = 2I_1 - 2I_2$$

$$= \int_0^{\pi/2} \ln^2\left\{\frac{a\sin(\theta)}{a\cos(\theta)}\right\}d\theta = \int_0^{\pi/2} \ln^2\{\tan(\theta)\}d\theta = \frac{\pi^3}{8}$$

by (5.2.7). So, we have the simultaneous pair of equations

$$I_1 + 2I_2 = -2\ln\left\{\frac{2}{a}\right\}\frac{\pi}{2}\ln\left(\frac{a}{2}\right) + \frac{\pi}{2}\ln^2\left\{\frac{2}{a}\right\} = \frac{3\pi}{2}\ln^2\left\{\frac{2}{a}\right\}$$

$$I_1 - I_2 = \frac{\pi^3}{16}$$

which are easily solved to give

$$\int_0^{\pi/2}\ln^2\{a\ \sin(\theta)\}d\theta = \int_0^{\pi/2}\ln^2\{a\ \cos(\theta)\}d\theta = \frac{\pi^3}{24} + \frac{\pi}{2}\ln^2\left\{\frac{2}{a}\right\} \qquad (7.3.2)$$

and

$$\int_0^{\pi/2}\ln\{a\ \sin(\theta)\}\ln\{a\ \cos(\theta)\}d\theta = \frac{\pi}{2}\ln^2\left\{\frac{2}{a}\right\} - \frac{\pi^3}{48}. \qquad (7.3.3)$$

For example, if $a = 2$ (7.3.3) is equal to $-\frac{\pi^3}{48} = -0.645964\ldots$ and MATLAB agrees, as.

integral(@(x)log(2*sin(x)).*log(2*cos(x)),0,pi/2) $= -0.645964\ldots$, while for $a = 1$ (7.3.3) reduces to $\frac{\pi}{2}\ln^2\{2\} - \frac{\pi^3}{48} = 0.108729\ldots$ and *integral(@(x)log(sin (x)).*log(cos(x)),0,pi/2)* $= 0.108729\ldots$.

By now you're no doubt convinced that Euler loved doing log-sine integrals. How about the reverse, that is, sin-log integrals? Specifically, what if anything can we make of the integral (notice it's *indefinite*)

$$\int \sin\{\ln(x)\}dx = ?$$

This, actually, can *easily* be done with the aid of Euler's identity, as follows. Writing $y = \ln(x)$, and so $x = e^y$ and $dx = e^y dy$, we have

$$\int e^{i\ \ln(x)}dx = \int[\cos\{\ln(x)\} + i\ \sin\{\ln(x)\}]dx = \int e^{iy}e^y dy = \int e^{(1+i)y}dy$$

$$= \frac{e^{(1+i)y}}{1+i} + C$$

where C is the constant of indefinite integration. Continuing (with the C temporarily dropped),

$$\frac{e^{(1+i)y}}{1+i} = \frac{e^y[\cos(y) + i\ \sin(y)]}{1+i} = \frac{1}{2}e^y[\cos(y) + \sin(y)] + i\ \frac{1}{2}[\sin(y) - \cos(y)]$$

or, finally, equating real and imaginary parts (and putting C back),

$$\int \cos \{ \ln (x) \} dx = \frac{1}{2} x [\cos \{ \ln (x) \} + \sin \{ \ln (x) \}] + C$$

and, for the original question,

$$\int \sin \{ \ln (x) \} dx = \frac{1}{2} x [\sin \{ \ln (x) \} - \cos \{ \ln (x) \}] + C.$$

As you might suspect, Euler did such things, too. Consider, for example, a more difficult question he solved, that of evaluating

$$\int_0^1 \frac{\sin \{ \ln (x) \}}{\ln (x)} dx = ?$$

Euler started by using the power series expansion for the sine:

$$\sin (y) = y - \frac{y^3}{3!} + \frac{y^5}{5!} - \frac{y^7}{7!} + \cdots$$

which is easily established using Euler's identity (7.1.1) and the power series expansion for the exponential.[6] Thus,

$$\frac{\sin \{ \ln (x) \}}{\ln (x)} = \frac{\ln (x) - \frac{\{ \ln (x) \}^3}{3!} + \frac{\{ \ln (x) \}^5}{5!} - \frac{\{ \ln (x) \}^7}{7!} + \cdots}{\ln (x)}$$

$$= 1 - \frac{\ln^2(x)}{3!} + \frac{\ln^4(x)}{5!} - \frac{\ln^6(x)}{7!} + \cdots.$$

So,

$$\int_0^1 \frac{\sin \{ \ln (x) \}}{\ln (x)} dx = \int_0^1 dx - \frac{1}{3!} \int_0^1 \ln^2(x) dx + \frac{1}{5!} \int_0^1 \ln^4(x) dx$$

$$- \frac{1}{7!} \int_0^1 \ln^6(x) dx + \cdots. \tag{7.3.4}$$

The first integral on the right in (7.3.4) is pretty straightforward—it is, of course, equal to 1—and, in fact, all the other integrals on the right are easy, too, as they are special cases of a more general integral discussed earlier in C4.2:

---

[6]That is, since $e^{iy} = \cos (y) + i \sin (y) = \sum_{k=0}^{\infty} \frac{(iy)^k}{k!} = \frac{1}{0!} + \frac{iy}{1!} - \frac{y^2}{2!} - \frac{iy^3}{3!} + \cdots$ then setting the real and imaginary parts from Euler's identity equal, respectively, to the real and imaginary parts of the exponential expansion, we immediately get the power series expansions for cos(y) *and* sin(y).

$$\int_0^1 x^m \ln^n(x)dx = (-1)^n \frac{n!}{(m+1)^{n+1}}.$$

All the integrals on the right in (7.3.4) are for $m = 0$ and n even, which says

$$\int_0^1 \ln^n(x)dx = n!, n \text{ even}.$$

Thus, (7.3.4) becomes

$$\int_0^1 \frac{\sin\{\ln(x)\}}{\ln(x)}dx = 1 - \frac{2!}{3!} + \frac{4!}{5!} - \frac{6!}{7!} + \cdots = 1 - \frac{1}{3} + \frac{1}{5} - \frac{1}{7} + \cdots$$

which is Leibniz's famous series for $\frac{\pi}{4}$ (see note 5 in the original Preface and note 8 in Chap. 1). Therefore,

$$\int_0^1 \frac{\sin\{\ln(x)\}}{\ln(x)}dx = \frac{\pi}{4}. \tag{7.3.5}$$

MATLAB agrees, as *integral(@(x)sin(log(x))./log(x),0,1) = 0.785398....*

7.4 $\zeta(2)$, at Last (Three Times!)

A most impressive demonstration of the use of $\sqrt{-1}$ and Euler's identity is the derivation I have long promised you—the value of $\zeta(2)$, first calculated by Euler. We've already used this result ($\frac{\pi^2}{6}$) numerous times, but now I'll derive it in a way you almost surely have not seen before. Indeed, before we are through with this section I'll have shown you three derivations of $\zeta(2)$ that are most likely all new for you. In the spirit of this chapter's title, I'll start with a derivation in the complex plane.

Recall again the power series expansion from Chap. 5 (Sect. 5.2) for $\ln(1+z)$:

$$\ln(1+z) = z - \frac{z^2}{2} + \frac{z^3}{3} - \frac{z^4}{4} + \cdots$$

where now I'm taking z as a complex-valued quantity, and not simply a real quantity as I did in Chap. 5 where I wrote x instead of z.[7] If we write $z = e^{i\theta}$ where a and θ are real, then

---

[7]How do we know we can do this? This is a non-trivial question, and a mathematician would rightfully want to vigorously pursue it. But remember our philosophical approach—we'll just make the assumption that all is okay, see where it takes us, and then *check* the answers we eventually calculate with *integral*.

$$\ln\left(1 + ae^{i\theta}\right) = ae^{i\theta} - \frac{a^2 e^{i2\theta}}{2} + \frac{a^3 e^{i3\theta}}{3} - \frac{a^4 e^{i4\theta}}{4} + \cdots$$

or, expanding each term with Euler's identity and collecting real and imaginary parts together,

$$\ln\left(1 + ae^{i\theta}\right) = a\cos\left(\theta\right) - \frac{1}{2}a^2 \cos\left(2\theta\right) + \frac{1}{3}a^3 \cos\left(3\theta\right) - \frac{1}{4}a^4 \cos\left(4\theta\right) + \cdots$$

$$+ i\left\{ a\sin\left(\theta\right) - \frac{1}{2}a^2 \sin\left(2\theta\right) + \frac{1}{3}a^3 \sin\left(3\theta\right) - \cdots \right\}.$$

Now, $1 + ae^{i\theta} = 1 + a\cos\left(\theta\right) + i\, a\sin\left(\theta\right)$ is a complex quantity with magnitude and angle in the complex plane (with respect to the positive real axis)[8] and so we can write it as

$$1 + ae^{i\theta} = \sqrt{\left\{1 + a\cos\left(\theta\right)\right\}^2 + a^2 \sin^2\left(\theta\right)}\ e^{i\phi}$$

$$= \sqrt{1 + 2a\cos\left(\theta\right) + a^2 \cos^2\left(\theta\right) + a^2 \sin^2\left(\theta\right)}\ e^{i\phi}$$

$$= \sqrt{1 + 2a\cos\left(\theta\right) + a^2}\ e^{i\phi}.$$

Thus,

$$\ln\left(1 + ac^{i\theta}\right) = \ln\left\{\sqrt{1 + 2a\cos\left(0\right) + a^2}\right\} + i\phi$$

or, equating real parts,

$$\frac{1}{2}\ln\left\{1 + 2a\cos\left(0\right) + a^2\right\} = a\cos\left(\theta\right) - \frac{1}{2}a^2 \cos\left(2\theta\right) + \frac{1}{3}a^3 \cos\left(3\theta\right)$$

$$- \frac{1}{4}a^1 \cos\left(4\theta\right) + \cdots$$

or,

$$\ln\left\{1 + 2a\cos\left(\theta\right) + a^2\right\} = 2\left[a\cos\left(\theta\right) - \frac{1}{2}a^2 \cos\left(2\theta\right) + \frac{1}{3}a^3 \cos\left(3\theta\right) - \frac{1}{4}a^4 \cos\left(4\theta\right) + \cdots\right].$$

So, if we write $a = -x$ we have

---

[8]The angle is given by $\phi = \tan^{-1}\left\{\frac{a\,\sin\left(\phi\right)}{1 + a\,\cos\left(\phi\right)}\right\}$, but we'll never actually need to know this.

$$\ln\{1 - 2x\cos(\theta) + x^2\} = -2\left[x\cos(\theta) + \frac{1}{2}x^2\cos(2\theta) + \frac{1}{3}x^3\cos(3\theta) + \ldots\right]$$

and so

$$\ln^2\{1 - 2x\cos(\theta) + x^2\} = 4\left[x\cos(\theta) + \frac{1}{2}x^2\cos(2\theta) + \frac{1}{3}x^3\cos(3\theta) + \ldots\right]^2$$

$$= 4\left[x^2\cos^2(\theta) + \frac{x^4}{2^2}\cos^2(2\theta) + \frac{x^6}{3^2}\cos^2(3\theta) + \ldots\right] plus$$

all the cross-product terms of the form $\cos(m\theta)\cos(n\theta)$, $m \neq n$.

Integrals of these cross-product terms are easy to do, since

$$\int_0^\pi \cos(m\theta)\cos(n\theta)d\theta = \frac{1}{2}\int_0^\pi [\cos\{(m-n)\theta\} + \cos\{(m+n)\theta\}]d\theta.$$

These integrals are easy to do because the integral of a cosine with a non-zero argument (remember, $m \neq n$) gives a sine and so, between the given limits, every one of the cross-product integrals is zero. So,

$$\int_0^\pi \ln^2\{1 - 2x\ \cos(\theta) + x^2\}d\theta$$

$$= 4\left[x^2\int_0^\pi \cos^2(\theta)d\theta + \frac{x^4}{2^2}\int_0^\pi \cos^2(2\theta)d\theta + \frac{x^6}{3^2}\int_0^\pi \cos^2(3\theta)d\theta + \cdots\right].$$

From integral tables we have

$$\int_0^\pi \cos^2(k\theta)d\theta = \left[\frac{\theta}{2} + \frac{\sin(2k\theta)}{4k}\right]\Big|_0^\pi = \frac{\pi}{2}$$

and so

$$\int_0^\pi \ln^2\{1 - 2x\cos(\theta) + x^2\}\ d\theta = 4\left(\frac{\pi}{2}\right)\left[x^2 + \frac{x^4}{2^2} + \frac{x^6}{3^2} + \ldots\right]$$

or,

$$\frac{1}{2\pi}\int_0^\pi \ln^2\{1 - 2x\cos(\theta) + x^2\}\ d\theta = \frac{x^2}{1^2} + \frac{x^4}{2^2} + \frac{x^6}{3^2} + \ldots.$$

Since $\cos(\theta)$ varies from $+1$ to -1 over the interval of integration, we can change the sign in the integrand without changing the value of the integral to give

$$\frac{1}{2\pi}\int_0^\pi \ln^2\{1 + 2x\cos(\theta) + x^2\}\,d\theta = \frac{x^2}{1^2} + \frac{x^4}{2^2} + \frac{x^6}{3^2} + \ldots\,.$$

Thus, setting x = 1, we have

$$\frac{1}{2\pi}\int_0^\pi \ln^2\{2 + 2\cos(\theta)\}\,d\theta = \frac{1}{1^2} + \frac{1}{2^2} + \frac{1}{3^2} + \ldots = \zeta(2)$$

$$= \frac{1}{2\pi}\int_0^\pi \ln^2\{2\,[1 + \cos(\theta)]\}d\theta.$$

We are getting close!

If we can do the integral on the right, we'll have Euler's famous result. So, continuing, the double-angle formula from trigonometry says $\cos(2\alpha) = 2\cos^2(\alpha) - 1$ and so

$$\zeta(2) = \frac{1}{2\pi}\int_0^\pi \ln^2\left\{2\left[2\cos^2\left(\frac{\theta}{2}\right)\right]\right\}d\theta = \frac{1}{2\pi}\int_0^\pi \ln^2\left\{\left[2\cos\left(\frac{\theta}{2}\right)\right]^2\right\}d\theta$$

$$= \frac{1}{2\pi}\int_0^\pi 4\,\ln^2\left\{2\cos\left(\frac{\theta}{2}\right)\right\}d\theta = \frac{2}{\pi}\int_0^\pi \ln^2\left\{2\cos\left(\frac{\theta}{2}\right)\right\}d\theta.$$

Now, let $\alpha = \frac{\theta}{2}$ and so $d\theta = 2\,d\alpha$. Then,

$$\zeta(2) = \frac{2}{\pi}\int_0^{\pi/2} \ln^2\{2\cos(\alpha)\}2\,d\alpha = \frac{4}{\pi}\int_0^{\pi/2} \ln^2\{2\cos(\alpha)\}\,d\alpha.$$

From (7.3.2), with a = 2, we see that

$$\int_0^{\pi/2} \ln^2\{2\cos(\alpha)\}\,d\alpha = \frac{\pi^3}{24}$$

and so

$$\zeta(2) = \left(\frac{4}{\pi}\right)\left(\frac{\pi^3}{24}\right) = \frac{\pi^2}{6},$$

just as Euler showed (in a completely different way).

Well, you say, that was 'okay' but, boy, there is no question about it, it's a pretty demanding derivation for a high school AP-calculus student, with all of its use of complex quantities. Can we find a less 'complex' (pun intended) derivation? Euler's own original derivation of ζ(2) was in fact such a derivation but, while almost a supernaturally brilliant *tour de force*, it was also an analysis that didn't hesitate to make bold, leaping assumptions about what Euler took to be 'mathematically legal' arguments. Indeed, his original derivation received sufficient criticism that Euler searched for a less heroic proof. It took him a while but, being Euler, in 1743 he succeeded in finding such a demonstration. It starts with the well-known power series expansion for $\sin^{-1}(t)$:

$$\sin^{-1}(t) = t + \sum_{n=1}^{\infty} \frac{(1)(3)(5)\dots(2n-1)}{(2)(4)\dots(2n)} \left(\frac{t^{2n+1}}{2n+1} \right). \tag{7.4.1}$$

Then, observing (see note 7 in Chap. 1) that[9]

$$\sin^{-1}(t) = \int_0^t \frac{du}{\sqrt{1-u^2}},$$

it follows[10] that

$$\frac{1}{2}\{\sin^{-1}(x)\}^2 = \int_0^x \frac{\sin^{-1}(t)}{\sqrt{1-t^2}} dt. \tag{7.4.2}$$

Thus, putting (7.4.1) into the integral of (7.4.2), Euler had

$$\frac{1}{2}\{\sin^{-1}(x)\}^2 = \int_0^x \frac{t}{\sqrt{1-t^2}} dt + \sum_{n=1}^{\infty} \frac{(1)(3)(5)\dots(2n-1)}{(2)(4)\dots(2n)} \left(\frac{1}{2n+1}\right) \int_0^x \frac{t^{2n+1}}{\sqrt{1-t^2}} dt.$$
$$\tag{7.4.3}$$

Next, define

$$I_n(x) = \int_0^x \frac{t^{n+2}}{\sqrt{1-t^2}} dt = \int_0^x \frac{t^{n+1}}{\sqrt{1-t^2}} t \, dt \tag{7.4.4}$$

and integrate-by-parts. That is, for the right-most integral in (7.4.4) we have

$$\int_0^x u \, dv = (uv)\big|_0^x - \int_0^x v \, du$$

where

----

[9] It is from this integral (one that, as stated in note 7 of Chap. 1, I'm assuming you know) that you can derive (7.4.1), by using the binomial theorem to expand $(1-u^2)^{1/2}$ and then integrating term-by-term. I'll let you verify that.

[10] The reason (7.4.2) is correct is two-fold: (1) both sides are equal (to zero) at $x = 0$, and (2) the derivatives with respect to x of each side are equal (use Feynman's favorite trick to show this).

$$u = t^{n+1}, dv = \frac{t}{\sqrt{1-t^2}} dt.$$

Thus, $du = (n+1)t^n$ and $v = -\sqrt{1-t^2}$ and so

$$\int_0^x \frac{t^{n+2}}{\sqrt{1-t^2}} dt = -x^{n+1}\sqrt{1-x^2} + (n+1)\int_0^x t^n\sqrt{1-t^2}dt. \qquad (7.4.5)$$

Euler's next clever step was to multiply the integrand of the right-most integral in (7.4.5) by $1 = \left(\sqrt{1-t^2}\right)/\left(\sqrt{1-t^2}\right)$ to get

$$\int_0^x \frac{t^{n+2}}{\sqrt{1-t^2}} dt = -x^{n+1}\sqrt{1-x^2} + (n+1)\int_0^x \frac{t^n(1-t^2)}{\sqrt{1-t^2}} dt$$

$$= -x^{n+1}\sqrt{1-x^2} + (n+1)\left[\int_0^x \frac{t^n}{\sqrt{1-t^2}} dt - \int_0^x \frac{t^{n+2}}{\sqrt{1-t^2}} dt\right].$$

So, remembering (7.4.4), we have

$$I_n(x) = -x^{n+1}\sqrt{1-x^2} + (n+1)I_{n-2}(x) - (n+1)I_n(x)$$

which immediately gives us

$$I_n(x) = \frac{n+1}{n+2} I_{n-2}(x) - \frac{x^{n+1}}{n+2}\sqrt{1-x^2} = \int_0^x \frac{t^{n+2}}{\sqrt{1-t^2}} dt.$$

Setting $x = 1$, this reduces to

$$\int_0^1 \frac{t^{n+2}}{\sqrt{1-t^2}} dt = \frac{n+1}{n+2} \int_0^1 \frac{t^n}{\sqrt{1-t^2}} dt$$

and so, if we replace every n with $2n - 1$ (which preserves equality), we have

$$\int_0^1 \frac{t^{2n+1}}{\sqrt{1-t^2}} dt = \frac{2n}{2n+1} \int_0^1 \frac{t^{2n-1}}{\sqrt{1-t^2}} dt, \ n \geq 1. \qquad (7.4.6)$$

The recursion in (7.4.6), combined with the fact (easily established with the change of variable $u = 1 - t^2$) that

$$\int_0^1 \frac{t}{\sqrt{1-t^2}} dt = 1$$

allowed Euler to write

$$\int_0^1 \frac{t^{2n+1}}{\sqrt{1-t^2}} dt = \frac{(2n)(2n-2)\ldots(2)}{(2n+1)(2n-1)\ldots(3)}, \quad n \geq 1. \tag{7.4.7}$$

Using (7.4.7) for the value of the right-most integral in (7.4.3), we see all the factors in the numerators and denominators of the products completely cancel each other *except* for the $2n+1$ factors. That is, (7.4.3) becomes, for x = 1,

$$\frac{1}{2}\{\sin^{-1}(1)\}^2 = \int_0^1 \frac{t}{\sqrt{1-t^2}} dt + \sum_{n=1}^{\infty} \frac{1}{(2n+1)^2}.$$

But, since $\sin^{-1}(1) = \frac{\pi}{2}$, we have

$$\frac{\pi^2}{8} = 1 + \sum_{n=1}^{\infty} \frac{1}{(2n+1)^2}.$$

That is, the sum of the reciprocals of the odd integers squared is $\frac{\pi^2}{8}$, which you'll recall from the discussion just before the derivation of (5.2.1)— and also from the discussion of Hardy's integral in Sect. 3.5— is equivalent to $\zeta(2) = \frac{\pi^2}{6}$.

The Master's Hand is clearly displayed in this Euler gem!

So seductive is the fascination of finding an expression for $\zeta(2)$ that, even more than a century after Euler's death, the search went on for other derivations. I'll end this section with a third, particularly beautiful analysis dating from 1908. That year Paul Stäckel (1862–1919), a German university mathematician, published a brief note in which he observed what we've written as (5.3.1) (with a = 0):

$$\int_0^1 \int_0^1 \frac{dxdy}{1-xy} = \sum_{n=1}^{\infty} \frac{1}{n^2}.$$

Since Euler had already shown that the sum is $\zeta(2) = \frac{\pi^2}{6}$, it must be true that a *direct* evaluation of the double integral would provide a new, independent derivation of $\zeta(2)$. Could, asked Stäckel, someone do that *direct* evaluation? Stäckel died young but he lived long enough to see his challenge answered.

That success came in 1913 when a German *high school* math teacher, Franz Goldscheider (1852–1926), published a direct evaluation of the double integral through the use of an enormously ingenious sequence of changes of variables. It was a *tour de force* derivation that Euler, a master himself of devilishly clever symbolic manipulations that appear to come from seemingly out of nowhere, would have loved.

Starting with Stäckel's integral (called P), Goldscheider introduced a second, similar double integral that he called Q:

$$P = \int_0^1 \int_0^1 \frac{dxdy}{1-xy} \tag{7.4.8}$$

and

$$Q = \int_0^1 \int_0^1 \frac{dxdy}{1 + xy}.$$ (7.4.9)

He then formed $P - Q$ to get

$$P - Q = \int_0^1 \int_0^1 \left\{ \frac{1}{1 - xy} - \frac{1}{1 + xy} \right\} dxdy = \int_0^1 \int_0^1 \frac{2xy}{1 - x^2 y^2} dxdy.$$ (7.4.10)

Changing variables to $x^2 = u$ and $y^2 = v$ (and so $dx = \frac{du}{2x}$ and $dy = \frac{dv}{2y}$), he arrived at

$$dxdy = \frac{dudv}{4xy}$$

and so, since u and v also each vary from 0 to 1, (7.4.10) becomes

$$P - Q = \frac{1}{2} \int_0^1 \int_0^1 \frac{dudv}{1 - uv} = \frac{1}{2} P$$

from which it quickly follows that

$$P = 2Q.$$ (7.4.11)

Next, Goldscheider changed variable in P to $u = -y$ (and so $dy = -du$, with u varying from 0 to -1 as y varies from 0 to 1) to get

$$P = \int_0^1 \int_0^{-1} \frac{dx(-du)}{1 + xu} = \int_0^1 \int_{-1}^0 \frac{dxdu}{1 + xu} = \int_0^1 \int_{-1}^0 \frac{dxdy}{1 + xy}.$$

Thus, forming $P + Q$, he had

$$P + Q = \int_0^1 \int_{-1}^0 \frac{dxdy}{1 + xy} + \int_0^1 \int_0^1 \frac{dxdy}{1 + xy} = \int_0^1 \int_{-1}^1 \frac{dxdy}{1 + xy}$$

or,

$$P + Q = \int_{-1}^1 dy \left\{ \int_0^1 \frac{dx}{1 + xy} \right\}.$$ (7.4.12)

He then made yet another change of variables, to $u = y + \frac{1}{2} x(y^2 - 1)$, and so as y varies from -1 to 1 so does u. Since

$$\frac{du}{dy} = 1 + xy \left(\text{and so } dy = \frac{du}{1 + xy} \right)$$

he therefore saw (7.4.12) become

$$P + Q = \int_{-1}^{1} du \left\{ \int_{0}^{1} \frac{dx}{(1 + xy)^2} \right\} = \int_{-1}^{1} du \left\{ \int_{0}^{1} \frac{dx}{1 + 2xy + x^2y^2} \right\}. \qquad (7.4.13)$$

As unpromising as (7.4.13) might appear to be at first glance (those y's in the denominator of the integrand look troublesome), it actually isn't a disaster. That's because

$$1 + 2ux + x^2 = 1 + 2\left[y + \frac{1}{2}x(y^2 - 1)\right]x + x^2 = 1 + 2xy + x^2y^2$$

and so (7.4.13) becomes

$$P + Q = \int_{-1}^{1} du \left\{ \int_{0}^{1} \frac{dx}{1 + 2ux + x^2} \right\}. \qquad (7.4.14)$$

Well, you might say in response, (7.4.14) still doesn't really look all that terrific, either.

Have faith! Goldscheider saved the day by pulling a final change of variable out of his hat, with u = cos (φ) (and so du = −sin (φ)dφ). As u varies from −1 to 1 we see φ varying from π to 0, and so (7.4.14) becomes

$$P + Q = \int_{\pi}^{0} d\varphi \left\{ \int_{0}^{1} \frac{-\sin(\varphi)dx}{1 + 2x\cos(\varphi) + x^2} \right\} = \int_{0}^{\pi} d\varphi \left\{ \int_{0}^{1} \frac{\sin(\varphi)dx}{1 + 2x\cos(\varphi) + x^2} \right\}. \qquad (7.4.15)$$

We can avoid having (7.4.15) make us think we've driven over the edge of a cliff by recalling the classic differentiation formula

$$\frac{d}{d\theta} \tan^{-1}(s) = \left(\frac{1}{1 + s^2} \right) \frac{ds}{d\theta}.$$

If we apply this formula to the case of

$$s = \frac{x + \cos(\varphi)}{\sin(\varphi)}, \theta = x,$$

we have

$$\frac{d}{dx}\left[\tan^{-1}\left\{\frac{x+\cos(\varphi)}{\sin(\varphi)}\right\}\right] = \left(\frac{1}{1+\left\{\frac{x+\cos(\varphi)}{\sin(\varphi)}\right\}^2}\right)\frac{1}{\sin(\varphi)} = \frac{\sin(\varphi)}{1+2x\cos(\varphi)+x^2}.$$

Putting this into (7.4.15), we have

$$P+Q = \int_0^\pi d\varphi\left(\int_0^1 \frac{d}{dx}\left[\tan^{-1}\left\{\frac{x+\cos(\varphi)}{\sin(\varphi)}\right\}\right]dx\right)$$

$$= \int_0^\pi d\varphi\left(\int_0^1 d\left[\tan^{-1}\left\{\frac{x+\cos(\varphi)}{\sin(\varphi)}\right\}\right]\right) = \int_0^\pi d\varphi\left(\tan^{-1}\left\{\frac{x+\cos(\varphi)}{\sin(\varphi)}\right\}\Big|_{x=0}^{x=1}\right)$$

or,

$$P+Q = \int_0^\pi\left(\tan^{-1}\left\{\frac{1+\cos(\varphi)}{\sin(\varphi)}\right\} - \tan^{-1}\left\{\frac{\cos(\varphi)}{\sin(\varphi)}\right\}\right)d\varphi. \tag{7.4.16}$$

The integral in (7.4.16) may look like a tiger, but it's actually a pussycat. That's because, first of all,

$$\tan^{-1}\left\{\frac{\cos(\varphi)}{\sin(\varphi)}\right\} = \tan^{-1}\left\{\frac{1}{\tan(\varphi)}\right\} = \frac{\pi}{2} - \varphi,$$

a result that follows by simply drawing an arbitrary right triangle and observing that the tangents of the triangle's two acute angles are reciprocals of each other. If φ is one of those angles, then $\frac{\pi}{2} - \varphi$ is the other one. And second, after recalling a couple of identities from trigonometry, we have from a similar line of reasoning that

$$\tan^{-1}\left\{\frac{1+\cos(\varphi)}{\sin(\varphi)}\right\} = \tan^{-1}\left\{\frac{2\cos^2\left(\frac{\varphi}{2}\right)}{2\sin\left(\frac{\varphi}{2}\right)\cos\left(\frac{\varphi}{2}\right)}\right\} = \tan^{-1}\left\{\frac{1}{\tan\left(\frac{\varphi}{2}\right)}\right\}$$

$$= \frac{\pi}{2} - \frac{\varphi}{2}.$$

Thus, (7.4.16) becomes

$$P+Q = \int_0^\pi\left[\left(\frac{\pi}{2} - \frac{\varphi}{2}\right) - \left(\frac{\pi}{2} - \varphi\right)\right]d\varphi = \frac{1}{2}\int_0^\pi \varphi\,d\varphi = \frac{\pi^2}{4}.$$

Since P = 2Q from (7.4.11), it then immediately follows that $Q = \frac{\pi^2}{12}$ and so, just like that,

$$P = \int_0^1 \int_0^1 \frac{dxdy}{1-xy} = \frac{\pi^2}{6} = \zeta(2)$$

and Stäckel had the derivation he had requested 5 years earlier.[11]

7.5   The Probability Integral *Again*

Earlier in the book I've shown you how some of our tricks can be combined to really bring terrific force to attacking particularly difficult integrals. Here I'll do it again, but now we additionally have Euler's identity to join in the mix. First, let me remind you of (3.7.4) where, if we set a = 1 and b = 0 we have

$$\int_0^\infty e^{-x^2} dx = \frac{1}{2}\sqrt{\pi}. \tag{7.5.1}$$

Now, make the change of variable $x = u\sqrt{z}$, where z is a positive quantity. Then $dx = \sqrt{z}\,du$ and so

$$\int_0^\infty e^{-x^2} dx = \int_0^\infty e^{-u^2 z} \sqrt{z}\, du = \frac{1}{2}\sqrt{\pi}$$

or

$$\frac{1}{\sqrt{z}} = \frac{2}{\sqrt{\pi}} \int_0^\infty e^{-u^2 z} du.$$

The next step is the central trick: multiply both sides by e^{iz}, integrate with respect to z from a to b (where b > a > 0), and then reverse the order of integration in the resulting double integral on the right. When we do that we arrive at

$$\int_a^b \frac{e^{iz}}{\sqrt{z}} dz = \int_a^b e^{iz} \left\{ \frac{2}{\sqrt{\pi}} \int_0^\infty e^{-u^2 z} du \right\} dz = \frac{2}{\sqrt{\pi}} \int_0^\infty \left\{ \int_a^b e^{z(i-u^2)} dz \right\} du.$$

The inner integral is easy to do:

---

[11]It is easy to show that $\zeta(3) = \sum_{n=1}^\infty \frac{1}{n^3} = \int_0^1 \int_0^1 \int_0^1 \frac{dxdydz}{1-xyz}$, but a direct evaluation of this triple integral over the unit cube continues to elude all who have tried. MATLAB numerically calculates *integral3* $(@(x,y,z)1./(1-x.*y.*z),0,1,0,1,0,1) = 1.2020....$

$$\int_a^b e^{z(i-u^2)}\,dz = \left\{\frac{e^{z(i-u^2)}}{i-u^2}\right\}\bigg|_a^b = \frac{e^{b(i-u^2)} - e^{a(i-u^2)}}{i-u^2}.$$

Next, imagine that we let $b \to \infty$ and $a \to 0$. Then the first exponential on the right $\to 0$ and the second exponential on the right $\to 1$. That is,[12]

$$\int_0^\infty e^{z(i-u^2)}\,dz = \frac{-1}{i-u^2}$$

and so (remembering that $i^2 = -1$) we have

$$\int_0^\infty \frac{e^{iz}}{\sqrt{z}}\,dz = -\frac{2}{\sqrt{\pi}}\int_0^\infty \frac{1}{i-u^2}\,du = -\frac{2}{\sqrt{\pi}}\int_0^\infty \frac{-i-u^2}{(i-u^2)(-i-u^2)}\,du$$

$$= \frac{2}{\sqrt{\pi}}\int_0^\infty \frac{i+u^2}{1+u^4}\,du.$$

Or, if we use Euler's identity on the z-integral and equate real and imaginary parts, we arrive at the following pair of equations:

$$\int_0^\infty \frac{\cos(z)}{\sqrt{z}}\,dz = \frac{2}{\sqrt{\pi}}\int_0^\infty \frac{u^2}{1+u^4}\,du$$

and

$$\int_0^\infty \frac{\sin(z)}{\sqrt{z}}\,dz = \frac{2}{\sqrt{\pi}}\int_0^\infty \frac{1}{1+u^4}\,du.$$

We showed earlier, in (7.3.4), that the two u-integrals are each equal to $\frac{\pi\sqrt{2}}{4}$, and so we immediately have the beautiful results (where I've changed back to x as the dummy variable of integration)

$$\int_0^\infty \frac{\cos(x)}{\sqrt{x}}\,dx = \int_0^\infty \frac{\sin(x)}{\sqrt{x}}\,dx = \sqrt{\frac{\pi}{2}}. \qquad (7.5.2)$$

That is, both integrals are numerically equal to 1.253314..., and we check that conclusion with *integral(@(x)cos(x)./sqrt(x),0,1000)* = 1.279... and *integral(@(x) sin(x)./sqrt(x),0,1000)* = 1.235.... These numerical estimates by *integral* aren't as good as have been many of the previous checks we've done, and that's because the

---

[12]Mathematicians will want to check that the limiting operations $b \to \infty$, $a \to 0$, and that the reversal of the order of integration in the double integral, are valid, but again remember our guiding philosophy in this book: just *do it*, and check with *integral* at the end.

integrands are really not that small even at x = 1000 ($\sqrt{x}$ is not a fast-growing denominator, and the numerators don't decrease but rather simply oscillate endlessly between ±1). I'll say more on this issue in Chap. 8.

Now, to end this section, let me show you how to use i and Euler's identity to derive an impressive generalization of (7.5.2). Starting with the gamma function integral definition from (4.1.1),

$$\Gamma(n) = \int_0^\infty e^{-x} x^{n-1} dx, \quad n > 0, \tag{7.5.3}$$

change variable to u = x/(p + iq) where p and q are real, positive constants. Thus, (7.5.3) *formally* becomes

$$\Gamma(n) = \int_0^\infty e^{-(p+iq)u} \{(p + iq)u\}^{n-1} (p + iq) du \, ,$$

or

$$\Gamma(n) = \int_0^\infty (p + iq)^n u^{n-1} e^{-pu} e^{-iqu} du, \quad n > 0. \tag{7.5.4}$$

To be honest, we can't deny that there is a fair amount of 'by guess and by gosh' going on with this change of variable. While the original $\Gamma(n)$ integration path is confined to the real axis, the integration path of the transformed integral is now off the real axis and along a line in the complex plane at angle $\alpha = -\tan^{-1}\left(\frac{q}{p}\right)$ to the real axis. Is this an 'okay' alteration? As usual, for us, we'll ignore this very valid question and simply 'check' where we end-up when we are done, using MATLAB. When we get to Cauchy's theory of contour integration in the complex plane, in the next chapter, we'll pay a *lot* more attention to the question of where our integration paths are located. It was, in fact, concern over this sort of devil-may-care attitude with complex variable changes in the evaluation of definite integrals (by Euler, in particular), that motivated Cauchy's work. Anyway, changing the dummy variable of integration in (7.5.4) back to x, we have (we hope!)

$$\int_0^\infty x^{n-1} e^{-px} e^{-iqx} dx = \frac{\Gamma(n)}{(p + iq)^n}, \quad n > 0. \tag{7.5.5}$$

Next, write p + iq in polar form, that is, as a line segment of length r making angle α to the real axis, and so p + iq = r cos (α) + i r sin (α) or, by Euler's identity,

$$p + iq = re^{i\alpha}, \text{ where } r = \sqrt{p^2 + q^2} \text{ and } \alpha = \tan^{-1}\left(\frac{q}{p}\right).$$

Thus, (7.5.5) becomes

$$\int_0^\infty x^{n-1}e^{-px}e^{-iqx}dx = \frac{\Gamma(n)}{r^n}e^{-in\alpha}, \quad n > 0$$

and, using (yet again!) Euler's identity to equate the real and imaginary parts, respectively, of each side of this equation, we have

$$\int_0^\infty x^{n-1}e^{-px}\cos(qx)dx = \frac{\Gamma(n)}{r^n}\cos(n\alpha)$$

an

$$\int_0^\infty x^{n-1}e^{-px}\sin(qx)dx = \frac{\Gamma(n)}{r^n}\sin(n\alpha)$$

or, finally, we arrive at

$$\int_0^\infty x^{n-1}e^{-px}\cos(qx)dx = \frac{\Gamma(n)}{(p^2+q^2)^{n/2}}\cos\left\{n\,\tan^{-1}\left(\frac{q}{p}\right)\right\} \qquad (7.5.6)$$

and

$$\int_0^\infty x^{n-1}e^{-px}\sin(qx)dx = \frac{\Gamma(n)}{(p^2+q^2)^{n/2}}\sin\left\{n\,\tan^{-1}\left(\frac{q}{p}\right)\right\} \qquad (7.5.7)$$

In particular, if we set $n = \frac{1}{2}$, $p = 0$, and $q = 1$, we then see that (7.5.6) reduces to

$$\int_0^\infty \frac{\cos(x)}{\sqrt{x}}dx = \Gamma\left(\frac{1}{2}\right)\cos\left\{\frac{1}{2}\tan^{-1}(\infty)\right\} = \Gamma\left(\frac{1}{2}\right)\cos\left\{\frac{\pi}{4}\right\}$$

or, since (4.2.11) tells us that $\Gamma\left(\frac{1}{2}\right) = \sqrt{\pi}$ (and, of course, since $\cos\left\{\frac{\pi}{4}\right\} = \frac{1}{\sqrt{2}}$) we have

$$\int_0^\infty \frac{\cos(x)}{\sqrt{x}}dx = \sqrt{\frac{\pi}{2}}$$

which is just what we arrived at in (7.5.2). In the same way, (7.5.7) also correctly reduces to the other special case given in (7.5.2):

$$\int_0^\infty \frac{\sin(x)}{\sqrt{x}}dx = \sqrt{\frac{\pi}{2}}.$$

7.6 Beyond Dirichlet's Integral

Looking back at Dirichlet's integral that we derived in (3.2.1),

$$\int_0^\infty \frac{\sin{(ax)}}{x}\,dx = \frac{\pi}{2}, a > 0,$$

it might occur to you to ask what

$$\int_0^\infty \left\{\frac{\sin{(x)}}{x}\right\}^2 dx$$

is equal to? This is easy to answer if you recall (3.4.1):

$$\int_0^\infty \frac{\cos{(ax)} - \cos{(bx)}}{x^2}\,dx = \frac{\pi}{2}(b - a).$$

With $a = 0$ and $b = 2$ we have

$$\int_0^\infty \frac{1 - \cos{(2x)}}{x^2}\,dx = \pi,$$

and since $1 - \cos{(2x)} = 2\sin^2(x)$, we immediately have our answer:

$$\int_0^\infty \left\{\frac{\sin{(x)}}{x}\right\}^2 dx = \frac{\pi}{2}. \tag{7.6.1}$$

This is 1.57079 ..., and MATLAB agrees: *integral(@(x)(sin(x)./x).^2,0, inf)* = 1.57079. ...

Okay, that wasn't very difficult, and so the next obvious question is to ask what

$$\int_0^\infty \left\{\frac{\sin{(x)}}{x}\right\}^3 dx$$

is equal to? This is just a *bit* more difficult to answer, but certainly not impossibly so. Integrating by parts does the job. Let $u = \sin^3(x)$ and $dv = \frac{dx}{x^3}$ and so $\frac{du}{dx} = 3\sin^2(x)\cos{(x)}$ and $v = -\frac{1}{2x^2}$. Thus,

$$\int_0^\infty \left\{\frac{\sin{(x)}}{x}\right\}^3 dx = \left\{-\frac{\sin^3(x)}{2x^2}\right\}\Bigg|_0^\infty + \frac{3}{2}\int_0^\infty \frac{\sin^2(x)\cos{(x)}}{x^2}\,dx$$

$$= \frac{3}{2}\int_0^\infty \frac{\sin^2(x)\cos{(x)}}{x^2}\,dx.$$

Then, integrate by parts again, with $u = \sin^2(x) \cos(x)$ and $dv = \frac{dx}{x^2}$. Then $v = -\frac{1}{x}$ and $\frac{du}{dx} = 2\sin(x)\cos^2(x) - \sin^3(x) = 2\sin(x)[1 - \sin^2(x)] - \sin^3(x) = 2\sin(x) - 3\sin^3(x)$.

Thus,

$$\int_0^\infty \left\{ \frac{\sin(x)}{x} \right\}^3 dx = \frac{3}{2}\left[\left\{ -\frac{\sin^2(x)\cos(x)}{x} \right\}\Big|_0^\infty + \int_0^\infty \frac{2\sin(x) - 3\sin^3(x)}{x} dx \right]$$

$$= 3\int_0^\infty \frac{\sin(x)}{x} dx - \frac{9}{2}\int_0^\infty \frac{\sin^3(x)}{x} dx = 3\left(\frac{\pi}{2}\right) - \frac{9}{2}\int_0^\infty \frac{\sin^3(x)}{x} dx.$$

Since

$$\int_0^\infty \frac{\sin^3(x)}{x} dx = \int_0^\infty \frac{\frac{3}{4}\sin(x) - \frac{1}{4}\sin(3x)}{x} dx = \frac{3}{4}\int_0^\infty \frac{\sin(x)}{x} dx - \frac{1}{4}\int_0^\infty \frac{\sin(3x)}{x} dx$$

$$= \frac{3}{4}\left(\frac{\pi}{2}\right) - \frac{1}{4}\left(\frac{\pi}{2}\right) = \frac{\pi}{4},$$

then

$$\int_0^\infty \left\{ \frac{\sin(x)}{x} \right\}^3 dx = 3\left(\frac{\pi}{2}\right) - \frac{9}{2}\left(\frac{\pi}{4}\right) = \frac{12\pi}{8} - \frac{9\pi}{8}$$

or,

$$\int_0^\infty \left\{ \frac{\sin(x)}{x} \right\}^3 dx = \frac{3\pi}{8}. \tag{7.6.2}$$

This is equal to $1.178097\ldots$, and a MATLAB check agrees: *integral(@(x)(sin (x)./x).^3,0,inf) = 1.178097....*

To keep going in this way—what is $\int_0^\infty \left\{ \frac{\sin(x)}{x} \right\}^4 dx$, for example?—will soon prove to be onerous. Try it! With a more systematic approach, however, we can derive additional intriguing results with (hardly) any pain. We start with Euler's identity, and write

$$z = e^{ix} = \cos(x) + i\sin(x), i = \sqrt{-1}.$$

Then, for any integer $m \geq 0$,

$$z^m = e^{imx} = \cos(mx) + i\sin(mx)$$

and

$$\frac{1}{z^m} = z^{-m} = e^{-imx} = \cos(mx) - i\sin(mx)$$

and so

$$z^m - \frac{1}{z^m} = i\,2\sin(mx).$$

In particular, for m = 1

$$z - \frac{1}{z} = i\,2\sin(x).$$

So,

$$\{i\,2\sin(x)\}^{2n-1} = \left\{z - \frac{1}{z}\right\}^{2n-1}$$

or, expanding with the binomial theorem (where you'll recall the notation $\begin{pmatrix} a \\ b \end{pmatrix} = \frac{a!}{(a-b)!b!}$),

$$\{i\,2\sin(x)\}^{2n-1} = \sum_{r=0}^{2n-1} \begin{pmatrix} 2n-1 \\ r \end{pmatrix} z^{2n-1-r}\left(-\frac{1}{z}\right)^r$$

$$= \sum_{r=0}^{2n-1} (-1)^r \begin{pmatrix} 2n-1 \\ r \end{pmatrix} z^{(2n-1)-2r}.$$

When the summation index r runs from r = 0 to r = 2n − 1, it runs through 2n values. Half of those values are r = 0 to r = n − 1 (for which the exponent on z is greater than zero), and the other half (r = n to r = 2n − 1) give the same exponents but with negative signs. So, we can write

$$\{i\,2\sin(x)\}^{2n-1} = \sum_{r=0}^{n-1} (-1)^r \begin{pmatrix} 2n-1 \\ r \end{pmatrix} \left[z^{(2n-1)-2r} - \frac{1}{z^{(2n-1)-2r}}\right].$$

But

$$z^{(2n-1)-2r} - \frac{1}{z^{(2n-1)-2r}} = i\,2\sin\{(2n-2r-1)x\}$$

and so

$$\{i\,2\sin(x)\}^{2n-1} = \sum_{r=0}^{n-1} (-1)^r \begin{pmatrix} 2n-1 \\ r \end{pmatrix} 2i\,\sin\{(2n-2r-1)x\}.$$

$$= i^{2n-1}2^{2n-1}\sin^{2n-1}(x).$$

Now $i^{2n-1} = i(-1)^{n-1}$ and so[13]

$$i(-1)^{n-1}2^{2n-1} \sin^{2n-1}(x) = \sum_{r=0}^{n-1}(-1)^r \binom{2n-1}{r} 2i \sin\{(2n-2r-1)x\}$$

or,

$$\sin^{2n-1}(x) = \frac{(-1)^{n-1}}{2^{2n-1}} \sum_{r=0}^{n-1}(-1)^r \binom{2n-1}{r} 2 \sin\{(2n-2r-1)x\}. \quad (7.6.3)$$

Dividing both sides of (7.6.3) by x, and integrating from 0 to ∞, we have

$$\int_0^\infty \frac{\sin^{2n-1}(x)}{x} dx = \frac{(-1)^{n-1}}{2^{2n-1}} \sum_{r=0}^{n-1}(-1)^r \binom{2n-1}{r} 2 \int_0^\infty \frac{\sin\{(2n-2r-1)x\}}{x} dx.$$

In the integral on the right change variable to $y = (2n-2r-1)x$ and so

$$\int_0^\infty \frac{\sin\{(2n-2r-1)x\}}{x} dx = \int_0^\infty \frac{\sin(y)}{\frac{y}{2n-2r-1}} \left(\frac{dy}{2n-2r-1} \right) = \int_0^\infty \frac{\sin(y)}{y} dy = \frac{\pi}{2}.$$

Thus,

$$\int_0^\infty \frac{\sin^{2n-1}(x)}{x} dx = \frac{(-1)^{n-1}}{2^{2n-1}} \pi \sum_{r=0}^{n-1}(-1)^r \binom{2n-1}{r}.$$

We can simplify this by recalling the following combinatorial identity which you can confirm by expanding both sides:

$$\binom{s}{k} = \binom{s-1}{k-1} + \binom{s-1}{k}.$$

Here's how this result helps us. Consider the alternating sum

$$\sum_{k=0}^m (-1)^k \binom{s}{k} = \binom{s}{0} - \binom{s}{1} + \binom{s}{2} - \binom{s}{3} + \cdots + (-1)^m \binom{s}{m}.$$

The first term on the right is 1, and then expanding each of the remaining terms with our combinatorial identity, we have

---

[13]To see this, write $i^{2n-1} = ii^{2n-2} = i\frac{i^{2n}}{i^2} = i\frac{(i^2)^n}{(-1)} = i\frac{(-1)^n}{(-1)} = i(-1)^{n-1}$.

$$\sum_{k=0}^{m}(-1)^k\binom{s}{k} = 1 - \left[\binom{s-1}{0} + \binom{s-1}{1}\right] + \left[\binom{s-1}{1} + \binom{s-1}{2}\right]$$

$$- \left[\binom{s-1}{2} + \binom{s-1}{3}\right] + \cdots$$

$$+ (-1)^m\left[\binom{s-1}{m-1} + \binom{s-1}{m}\right].$$

Since $\binom{s-1}{0} = 1$, and since the last term in any square bracket cancels the first term in the next square bracket, we see that only the last term in the final square bracket survives. That is,

$$\sum_{k=0}^{m}(-1)^k\binom{s}{k} = (-1)^m\binom{s-1}{m}.$$

Thus, with $m = n-1$, $k = r$, and $s = 2n-1$, we have

$$\int_0^\infty \frac{\sin^{2n-1}(x)}{x}\,dx = \frac{(-1)^{n-1}}{2^{2n-1}}\pi(-1)^{n-1}\binom{2n-2}{n-1}$$

or, as $(-1)^{n-1}(-1)^{n-1} = (-1)^{2n-2} = 1$, we have the pretty result

$$\int_0^\infty \frac{\sin^{2n-1}(x)}{x}\,dx = \frac{\pi}{2^{2n-1}}\binom{2n-2}{n-1}. \qquad (7.6.4)$$

For $n = 5$, for example, this says

$$\int_0^\infty \frac{\sin^9(x)}{x}\,dx = \frac{35\pi}{256} = 0.4295\ldots$$

and a MATLAB check agrees, as *integral($@(x)(sin(x).^9)./x,0,1e4) = 0.4295.*...*
Now, looking back at (7.6.3) and multiplying through by $\cos(x)$,

$$\sin^{2n-1}(x)\cos(x) = \frac{(-1)^{n-1}}{2^{2n-1}}\sum_{r=0}^{n-1}(-1)^r\binom{2n-1}{r}2\sin\{(2n-2r-1)x\}\cos(x)$$

and remembering that $\sin(\alpha)\cos(\beta) = \frac{1}{2}[\sin(\alpha+\beta) + \sin(\alpha-\beta)]$, we have

$$\sin^{2n-1}(x)\cos(x) = \frac{(-1)^{n-1}}{2^{2n-1}} \sum_{r=0}^{n-1} (-1)^r \binom{2n-1}{r}$$
$$\times [\sin\{2(n-r)x\} + \sin\{2(n-r-1)x\}].$$

Thus,

$$\int_0^\infty \frac{\sin^{2n-1}(x)\cos(x)}{x}dx = \frac{(-1)^{n-1}}{2^{2n-1}} \sum_{r=0}^{n-1} (-1)^r \binom{2n-1}{r}$$
$$\times \left[\int_0^\infty \frac{\sin\{2(n-r)x\}}{x}dx + \int_0^\infty \frac{\sin\{2(n-r-1)x\}}{x}dx\right].$$

For every value of r from 0 to n − 1 the first integral on the right is $\frac{\pi}{2}$, from Dirichlet's integral. For every value of r from 0 to n1 *except* for r = n − 1 (where the argument of the sine function is zero) the second integral on the right is also $\frac{\pi}{2}$. For r = n − 1 the second integral is zero. So, if we 'pretend' the r = n − 1 case for the second integral also gives $\frac{\pi}{2}$ we can write the following (where the last term corrects for the 'pretend'!):

$$\int_0^\infty \frac{\sin^{2n-1}(x)\cos(x)}{x}dx = \left[\frac{(-1)^{n-1}}{2^{2n-1}} \pi \sum_{r=0}^{n-1} (-1)^r \binom{2n-1}{r}\right] - \frac{\pi}{2^{2n}}\binom{2n-1}{n-1}.$$

Now, just as before,

$$\sum_{r=0}^{n-1} (-1)^r \binom{2n-1}{r} = (-1)^{n-1}\binom{2n-2}{n-1}$$

and so

$$\int_0^\infty \frac{\sin^{2n-1}(x)\cos(x)}{x}dx = \frac{(-1)^{n-1}}{2^{2n-1}}\pi(-1)^{n-1}\binom{2n-2}{n-1} - \frac{\pi}{2^{2n}}\binom{2n-1}{n-1}$$
$$= \frac{\pi}{2^{2n-1}}\binom{2n-2}{n-1} - \frac{\pi}{2^{2n}}\binom{2n-1}{n-1} = \frac{\pi}{2^{2n}}\left[2\binom{2n-2}{n-1} - \binom{2n-1}{n-1}\right].$$

Or, since

$$2\binom{2n-2}{n-1} - \binom{2n-1}{n-1} = \frac{1}{n}\binom{2n-2}{n-1}$$

as you can easily verify by expanding the binomial coefficients, we have

$$\int_0^\infty \frac{\sin^{2n-1}(x) \cos(x)}{x} dx = \frac{\pi}{2^{2n}n} \binom{2n-2}{n-1}. \tag{7.6.5}$$

For n = 2, for example,

$$\int_0^\infty \frac{\sin^3(x) \cos(x)}{x} dx = \frac{\pi}{2^4 2} \binom{2}{1} = \frac{\pi}{16} = 0.19638\ldots$$

and a MATLAB check says *integral(@(x)(sin(x).^3).*cos(x)./x,0,1e3)* = 0.19637....

To finish this section, notice that

$$\int_0^\infty \frac{\sin^{2n}(x)}{x^2} dx$$

can be integrated by parts as follows. Let u = $\sin^{2n}(x)$ and dv = $\frac{dx}{x^2}$. Then v = $-\frac{1}{x}$ and $\frac{du}{dx}$ = $2n\sin^{2n-1}(x) \cos(x)$ and we have

$$\int_0^\infty \frac{\sin^{2n}(x)}{x^2} dx = \left\{ -\frac{\sin^{2n}(x)}{x} \right\}\Big|_0^\infty + \int_0^\infty \frac{2n\sin^{2n-1}(x) \cos(x)}{x} dx$$

$$= 2n\int_0^\infty \frac{\sin^{2n-1}(x) \cos(x)}{x} dx$$

and so, looking at (7.6.5), we see that

$$\int_0^\infty \frac{\sin^{2n}(x)}{x^2} dx = 2n\frac{\pi}{2^{2n}n} \binom{2n-2}{n-1} = \frac{\pi}{2^{2n-1}} \binom{2n-2}{n-1}.$$

Thus, looking back at (7.6.4) we have the interesting result that

$$\int_0^\infty \frac{\sin^{2n}(x)}{x^2} dx = \int_0^\infty \frac{\sin^{2n-1}(x)}{x} dx = \frac{\pi}{2^{2n-1}} \binom{2n-2}{n-1}. \tag{7.6.6}$$

For example, if n = 19 (7.6.6) says

$$\int_0^\infty \frac{\sin^{38}(x)}{x^2} dx = \int_0^\infty \frac{\sin^{37}(x)}{x} dx = \frac{\pi}{2^{37}} \binom{36}{18} = 0.20744\ldots$$

and MATLAB agrees, as *integral(@(x)(sin(x).^38)./(x.^2),0,1000)* = 0.20731... and *integral(@(x)(sin(x).^37)./x,0,1000)* = 0.20723....

7.7 Dirichlet Meets the Gamma Function

In this section we'll continue with the sequence of calculations we started in the previous section, that of evaluating

$$\int_0^\infty \left\{ \frac{\sin(x)}{x} \right\}^n dx,$$

integrals we've already done for the $n = 1$, 2, and 3 cases. To do the $n \geq 4$ cases becomes challenging— *unless we see a new trick.* That's what I'll show you now.

We start (you'll see *why*, soon) with the integral

$$\int_0^\infty u^{q-1} e^{-xu} du.$$

If we change variable to $y = xu$, then differentiating with respect to y (treating x as a positive 'constant') gives

$$du = \frac{1}{x} dy.$$

Thus,

$$\int_0^\infty u^{q-1} e^{-xu} du = \int_0^\infty \left(\frac{y}{x}\right)^{q-1} e^{-y} \frac{1}{x} dy = \int_0^\infty \frac{y^{q-1}}{x^q} e^{-y} dy = \frac{1}{x^q} \int_0^\infty y^{q-1} e^{-y} dy.$$

This last integral should look familiar—it is the gamma function, defined in (4.1.1), equal to $\Gamma(q) = (q-1)!$ Thus,

$$\frac{1}{x^q} = \frac{1}{(q-1)!} \int_0^\infty u^{q-1} e^{-xu} du$$

and so

$$\int_0^\infty \frac{\sin^P(x)}{x^q} dx = \frac{1}{(q-1)!} \int_0^\infty \sin^P(x) \left\{ \int_0^\infty u^{q-1} e^{-xu} du \right\} dx$$

or, reversing the order of integration (this should remind you of the sort of thing we did back in Sect. 4.3),

$$\int_0^\infty \frac{\sin^P(x)}{x^q} dx = \frac{1}{(q-1)!} \int_0^\infty u^{q-1} \left\{ \int_0^\infty e^{-xu} \sin^P(x) dx \right\} du.$$

The inner, x-integral is easily done with integration by parts, *twice*. I'll let *you* fill-in the details, with the result being

$$\int_0^\infty e^{-xu} \sin^P(x)dx = \frac{p(p-1)}{p^2 + u^2} \int_0^\infty e^{-xu} \sin^{P-2}(x)dx.$$

Suppose p is even (≥ 2). Then we can repeat the integration by parts, over and over, each time reducing the power of the sine function in the integrand by 2, until we reduce that power down to zero, giving a final integral of

$$\int_0^\infty e^{-xu}dx = \left\{\frac{e^{-xu}}{-u}\right\}\Big|_0^\infty = \frac{1}{u}.$$

and so

$$\int_0^\infty e^{-xu} \sin^P(x)dx = \frac{p!}{[p^2 + u^2]\left[(p-2)^2 + u^2\right] \ldots [2^2 + u^2]}\left(\frac{1}{u}\right), \text{p even.}$$

Similarly for p odd, except that we stop integrating when we get the power of the sine function down to the first power:

$$\int_0^\infty e^{-xu} \sin^P(x)dx = \frac{p!}{[p^2 + u^2]\left[(p-2)^2 + u^2\right] \ldots [1^2 + u^2]}, \text{p odd.}$$

So, we have our central results:

$$\int_0^\infty \frac{\sin^P(x)}{x^q} dx = \frac{p!}{(q-1)!} \int_0^\infty \frac{u^{q-2}}{[u^2 + 2^2]\left[u^2 + 4^2\right] \ldots [u^2 + p^2]} du, \text{p even} \quad (7.7.1)$$

and

$$\int_0^\infty \frac{\sin^P(x)}{x^q} dx = \frac{p!}{(q-1)!} \int_0^\infty \frac{u^{q-1}}{[u^2 + 1^2]\left[u^2 + 3^2\right] \ldots [u^2 + p^2]} du, \text{p odd.} \quad (7.7.2)$$

For the question of what is

$$\int_0^\infty \left\{\frac{\sin(x)}{x}\right\}^4 dx = ?,$$

the question we asked (but didn't answer) just after (7.6.2), we have $p = q = 4$ (and so p even) and (7.7.1) tells us that

$$\int_0^\infty \left\{\frac{\sin(x)}{x}\right\}^4 dx = \frac{4!}{3!} \int_0^\infty \frac{u^2}{\left[u^2 + 2^2\right]\left[u^2 + 4^2\right]} \, du = 4 \int_0^\infty \frac{u^2}{\left[u^2 + 4\right]\left[u^2 + 16\right]} \, du.$$

Making a partial fraction expansion, we have

$$\int_0^\infty \left\{\frac{\sin(x)}{x}\right\}^4 dx = 4\left[\int_0^\infty \frac{-\frac{1}{3}}{u^2 + 4} \, du + \int_0^\infty \frac{\frac{4}{3}}{u^2 + 16} \, du\right]$$

$$= -\frac{4}{3}\left\{\frac{1}{2}\tan^{-1}\left(\frac{u}{2}\right)\right\}\Big|_0^\infty + \frac{16}{3}\left\{\frac{1}{4}\tan^{-1}\left(\frac{u}{4}\right)\right\}\Big|_0^\infty = -\frac{2}{3}\tan^{-1}(\infty) + \frac{4}{3}\tan^{-1}(\infty)$$

$$= \frac{2}{3}\tan^{-1}(\infty) = \left(\frac{2}{3}\right)\left(\frac{\pi}{2}\right)$$

or, at last,

$$\int_0^\infty \left\{\frac{\sin(x)}{x}\right\}^4 dx = \frac{\pi}{3}. \tag{7.7.3}$$

This is 1.047197... and MATLAB agrees, as *integral(@(x)(sin(x)./x).*^4,0,1e4) = 1.047197...*

You might think from all of these calculations that our integrals will always turn-out to be some rational number times π. That, however, is not true. For example, suppose p = 3 and q = 2. Then we get a result of an entirely different nature. Using (7.7.2) because p is odd,

$$\int_0^\infty \frac{\sin^3(x)}{x^2} \, dx = \frac{3!}{1!} \int_0^\infty \frac{u}{\left[u^2 + 1^2\right]\left[u^2 + 3^2\right]} \, du$$

$$= 6\left[\int_0^\infty \frac{\frac{1}{8}u}{u^2 + 1^2} \, du - \int_0^\infty \frac{\frac{1}{8}u}{u^2 + 3^2} \, du\right]$$

$$= \frac{6}{8}\left[\frac{1}{2}\ln\left(u^2 + 1^2\right) - \frac{1}{2}\ln\left(u^2 + 3^2\right)\right]\Big|_0^\infty = \frac{3}{8}\ln\left(\frac{u^2 + 1^2}{u^2 + 3^2}\right)\Big|_0^\infty = -\frac{3}{8}\ln\left(\frac{1}{3^2}\right) = \frac{3}{8}\ln\left(3^2\right)$$

or, finally,

$$\int_0^\infty \frac{\sin^3(x)}{x^2} \, dx = \frac{3\ln(3)}{4}. \tag{7.7.4}$$

This is equal to 0.8239592..., and a MATLAB check agrees, as *integral(@(x)(sin(x).^3)./(x.^2),0,1e4) = 0.82396....*

7.8   Fourier Transforms and Energy Integrals

In this section we'll come full circle back to the opening section and its use of Euler's identity. Here I'll show you a new trick that illustrates how some *physical* considerations well-known to electrical engineers and physicists will allow us to derive some very interesting integrals.

In the study of electronic information processing circuitry, the transmission of pulse-like signals in time is at the heart of the operation of such circuits. So, let's start with the simplest such time signal, a single pulse that is finite in both amplitude and duration. For example, let

$$f(t) = \begin{cases} 1, & a < t < b \\ 0, & \text{otherwise} \end{cases} \qquad (7.8.1)$$

where a and b are both positive constants. (For electronics engineers, the particular time $t = 0$ is simply short-hand for some especially interesting event, like 'when we turned the power on to the circuits' or 'when we started to pay attention to the output signal.') The signal f(t) might, for example, be a voltage pulse of unit amplitude and duration $b - a$ (where, of course, $b > a$). If this voltage pulse is the voltage drop across a resistor, for example, then the *instantaneous power* of f(t) is proportional to $f^2(t)$, which is a direct consequence of Ohm's law for resistors, a law familiar to all high school physics students.[14] Since *energy* is the time integral of power, then the energy of this f(t), written as W_f, is (since $f^2(t) = 1$)

$$W_f = \int_{-\infty}^{\infty} f^2(t)dt = \int_a^b dt = b - a. \qquad (7.8.2)$$

All of these comments are admittedly 'engineery' in origin but, in fact, given (7.8.1) even a pure mathematician would, if asked for the energy of f(t) in (7.8.1), also immediately write (7.8.2). The physical terminology of *power* and *energy* has been adopted by mathematicians.

Now, further pondering on the issue of the energy of a time signal leads to the concept of the so-called *energy spectrum* of that signal. A time signal can be thought of as the totality of many (perhaps infinitely many) sinusoidal components of different amplitudes and frequencies (usually written as $\omega = 2\pi\upsilon$, where υ is in cycles per second or hertz, and ω is in radians per second). The energy spectrum of f(t) is a description of how the total energy W_f is distributed across the frequency components of f(t). To get an idea of where we are going with this, suppose we had the energy spectrum of f(t), which I'll write as $S_f(\omega)$, in-hand. If we integrate $S_f(\omega)$ over all ω we should arrive the total energy of f(t), that is, W_f. That means, using (7.8.2),

---

[14]I mention this only for completeness. If Ohm's law is of no interest to you, that's okay.

$$W_f = \int_{-\infty}^{\infty} f^2(t)dt = \int_{-\infty}^{\infty} S_f(\omega)d\omega = b - a \qquad (7.8.3)$$

and it's this equality of two integrals that can give us some quite 'interesting integrals.' So, that's our immediate problem: how do we calculate the energy spectrum for a given time signal? The answer is the Fourier transform.

One of the beautiful results from what is called *Fourier theory* (after the French mathematician Joseph Fourier (1768–1830)) is the so-called *Fourier transform*. If we call $G(\omega)$ the Fourier transform of an arbitrary time signal g(t), then

$$G(\omega) = \int_{-\infty}^{\infty} g(t)e^{-i\omega t}dt. \qquad (7.8.4)$$

We can recover g(t) from $G(\omega)$ by doing another integral, called the *inverse transform*:

$$g(t) = \frac{1}{2\pi}\int_{-\infty}^{\infty} G(\omega)e^{i\omega t}d\omega. \qquad (7.8.5)$$

Together, g(t) from $G(\omega)$ form what is called a *Fourier transform pair*,[15] usually written as

$$g(t) \leftrightarrow G(\omega).$$

In general, $G(\omega)$ will be complex, with real and imaginary parts $R(\omega)$ and $X(\omega)$, respectively. That is, $G(\omega) = R(\omega) + iX(\omega)$. If g(t) is a real-valued function of time (as of course are all the signals in any electronic circuitry that can actually be constructed) then $G(\omega)$ will have some special properties. In particular, $R(\omega)$ will be even and $X(\omega)$ will be odd: $R(-\omega) = R(\omega)$ and $X(-\omega) = -X(\omega)$. If, in addition to being real, g(t) has certain symmetry properties, then $G(\omega)$ will have additional corresponding special properties. If, for example, g(t) is even (as is $\cos(\omega t)$) then $G(\omega)$ will be real, and if g(t) is odd (as is $\sin(\omega t)$) then $G(\omega)$ will be imaginary: $X(\omega) = 0$ and $R(\omega) = 0$, respectively. All of these statements are easily established by simply writing-out the Fourier transform integral using Euler's identity and examining the integrands of the $R(\omega)$ and $X(\omega)$ integrals (these claims are so easy to verify, in fact, that they aren't at the level of being Challenge Problems, but you should be sure *you* can establish them).

What makes (7.8.4) and (7.8.5) so incredibly useful to us in this book is what is called *Rayleigh's theorem*, after the English mathematical physicist we encountered back in Chap. 1, in the problem of 'Rayleigh's rotating ring.' It tells us how to

---

[15]Where these defining integrals in a Fourier pair come from is explained in any good book on Fourier series and/or transforms. Or, for an 'engineer's treatment' in the same spirit as this book, see *Dr. Euler* (note 1), pp. 200–204.

calculate the energy spectrum of g(t) (and so, of course, of f(t), which is simply a particular g(t)). Rayleigh's theorem, dating from 1889, is analogous to *Parseval's theorem* (named after the French mathematician Antoine Parseval des Chenes (1755–1836), from the theory of Fourier series expansions of periodic functions, a subject we are *not* going to discuss in this book).

To prove Rayleigh's theorem, we start by writing

$$W_g = \int_{-\infty}^{\infty} g^2(t)dt = \int_{-\infty}^{\infty} g(t)g(t)dt$$

and then replace one of the g(t) factors in the second integral with its inverse transform form from (7.8.5). So,

$$W_g = \int_{-\infty}^{\infty} g(t)g(t)dt = \int_{-\infty}^{\infty} g(t)\left\{\frac{1}{2\pi}\int_{-\infty}^{\infty} G(\omega)e^{i\omega t}d\omega\right\}dt.$$

Reversing the order of integration,

$$W_g = \int_{-\infty}^{\infty} \frac{1}{2\pi}G(\omega)\left\{\int_{-\infty}^{\infty} g(t)e^{i\omega t}dt\right\}d\omega.$$

Since g(t) is a real-valued function of time—what would a *complex*-valued voltage pulse look like on an oscilloscope screen!?—then the inner t-integral on the right is the *conjugate* of $G(\omega)$ because that integral looks just like (7.8.4) except it has $+i$ in the exponential instead of $-i$. That is,

$$\int_{-\infty}^{\infty} g(t)e^{i\omega t}dt = G^*(\omega)$$

and so

$$W_g = \int_{-\infty}^{\infty} \frac{1}{2\pi}G(\omega)G^*(\omega)d\omega = \int_{-\infty}^{\infty} \frac{|G(\omega)|^2}{2\pi}d\omega.$$

Or, if we now specifically let g(t) = f(t), then Rayleigh's energy theorem is

$$W_f = \int_{-\infty}^{\infty} \frac{|F(\omega)|^2}{2\pi}d\omega = \int_{-\infty}^{\infty} f^2(t)dt, \quad F(\omega) = R(\omega) + iX(\omega),$$

and so the energy spectrum of the f(t) in (7.8.1) is, by (7.8.3)

$$S_f(\omega) = \frac{|F(\omega)|^2}{2\pi}, \quad -\infty < \omega < \infty.$$

For all real f(t), $|F(\omega)|^2 = R^2(\omega) + X^2(\omega)$ will be an even function because both $R^2(\omega)$ and $X^2(\omega)$ are even.

The Fourier transform of f(t) in (7.8.1) is

$$F(\omega) = \int_{-\infty}^{\infty} f(t)e^{-i\omega t}dt = \int_a^b e^{-i\omega t}dt = \left\{\frac{e^{-i\omega t}}{-i\omega}\right\}\bigg|_a^b = \frac{e^{-i\omega b} - e^{-i\omega a}}{-i\omega} = \frac{e^{-i\omega a} - e^{-i\omega b}}{i\omega}$$

and so the energy spectrum of f(t) is

$$S_f(\omega) = \frac{\left|e^{-i\omega a} - e^{-i\omega b}\right|^2}{2\pi\omega^2}. \tag{7.8.6}$$

Inserting (7.8.6) into (7.8.3), we get

$$\frac{1}{2\pi}\int_{-\infty}^{\infty} \frac{\left|e^{-i\omega a} - e^{-i\omega b}\right|^2}{\omega^2} d\omega = b - a. \tag{7.8.7}$$

Now, *temporarily* forget all the physics I've mentioned, that is, put aside for now all that business about time functions and energy distributed over frequency, and just treat (7.8.7) as a pure mathematical statement. To be absolutely sure we are now thinking 'purely mathematical,' let's change the dummy variable of integration from ω to x, to write

$$\frac{1}{2\pi}\int_{-\infty}^{\infty} \frac{\left|e^{-ixa} - e^{-ixb}\right|^2}{x^2} dx = b - a, \quad b > a. \tag{7.8.8}$$

Concentrate for the moment on the numerator of the integrand in (7.8.8): using Euler's identity we have

$$\begin{aligned}
\left|e^{-ixa} - e^{-ixb}\right|^2 &= |\{\cos(ax) - i\sin(ax)\} - \{\cos(bx) - i\sin(bx)\}|^2 \\
&= |\{\cos(ax) - \cos(bx)\} - i\{\sin(ax) - \sin(bx)\}|^2 \\
&= \{\cos(ax) - \cos(bx)\}^2 + \{\sin(ax) - \sin(bx)\}^2
\end{aligned}$$

which, if you multiply-out and combine terms, becomes

$$= 2[1 - \{\cos(ax)\cos(bx) + \sin(ax)\sin(bx)\}].$$

Putting this into (7.8.8), we arrive at

$$\frac{1}{\pi}\int_{-\infty}^{\infty} \frac{1 - \{\cos(ax)\cos(bx) + \sin(ax)\sin(bx)\}}{x^2} dx = b - a, \quad b > a,$$

or, in very slightly rearranged form (you'll see why, soon),

$$\int_{-\infty}^{\infty} \frac{1 - \cos(ax)\cos(bx)}{x^2} dx - \int_{-\infty}^{\infty} \frac{\sin(ax)\sin(bx)}{x^2} dx = \pi(b-a), \quad b > a.$$

$$(7.8.9)$$

Next, look back at (3.4.1), which we can use to write the integral (as the special case of the parameters there found by setting a to zero and b to one)

$$\int_{0}^{\infty} \frac{1 - \cos(u)}{u^2} du = \frac{\pi}{2}$$

or, as the integrand is even,

$$\int_{-\infty}^{\infty} \frac{1 - \cos(u)}{u^2} du = \pi. \qquad (7.8.10)$$

We now change variable in (7.8.10) to $u = (a+b)x$, and so

$$\int_{-\infty}^{\infty} \frac{1 - \cos(ax+bx)}{(a+b)^2 x^2}(a+b)dx = \pi$$

or, using the trigonometric identity for $\cos(ax+bx)$,

$$\int_{-\infty}^{\infty} \frac{1 - [\cos(ax)\cos(bx) - \sin(ax)\sin(bx)]}{x^2} dx = (a+b)\pi$$

and so, in very slightly rearranged form,

$$\int_{-\infty}^{\infty} \frac{1 - \cos(ax)\cos(bx)}{x^2} dx + \int_{-\infty}^{\infty} \frac{\sin(ax)\sin(bx)}{x^2} dx = \pi(a+b). \quad (7.8.11)$$

Finally, subtract (7.8.9) from (7.8.11), to get

$$2\int_{-\infty}^{\infty} \frac{\sin(ax)\sin(bx)}{x^2} dx = \pi(a+b) - \pi(b-a) = 2\pi a, b > a$$

or, at last,

$$\int_{-\infty}^{\infty} \frac{\sin(ax)\sin(bx)}{x^2} dx = \pi a, b > a.$$

Of course, by the symmetry of the integrand we could equally-well write

$$\int_{-\infty}^{\infty} \frac{\sin (ax) \sin (bx)}{x^2} dx = \pi b, a > b.$$

Both of these statements can be written as one, as

$$\int_{-\infty}^{\infty} \frac{\sin (ax) \sin (bx)}{x^2} dx = \pi \min (a, b). \tag{7.8.12}$$

As a special case, if $a = b$ then (7.8.12) reduces to

$$\int_{-\infty}^{\infty} \frac{\sin^2 (ax)}{x^2} dx = \pi a$$

which is just (7.6.1) with $a = 1$ (see also Challenge Problem 7.5).

7.9 'Weird' Integrals from Radio Engineering

In this section I'll show you some quite interesting (almost bizarre) integrals that arise in the theory of radio (although I'll limit the discussion to pure mathematics, and not a single transistor, capacitor, or antenna will make an appearance). Let's start by recalling Dirichlet's integral from (3.2.1):

$$\int_0^\infty \frac{\sin (t\omega)}{\omega} d\omega = \begin{cases} \dfrac{\pi}{2}, & t > 0 \\ -\dfrac{\pi}{2}, & t < 0 \end{cases} \tag{7.9.1}$$

where I've replaced the parameters a and x in (3.2.1) with the parameters t and ω, respectively. Since the integrand in (7.9.1) is even, we can double the integral by integrating from $-\infty$ to ∞. Thus,

$$\int_{-\infty}^{\infty} \frac{\sin (t\omega)}{\omega} d\omega = \begin{cases} \pi, & t > 0 \\ -\pi, & t < 0 \end{cases}. \tag{7.9.2}$$

In radio engineering analyses it is found that a time signal that is -1 for negative time and $+1$ for positive time is highly useful (you'll see why, soon). It is given the special name *signum*—written sgn(t)—for its property of being the *sign* function (not to be confused with the sine function!). So, (7.9.2) can be written as

$$\int_{-\infty}^{\infty} \frac{\sin (t\omega)}{\omega} d\omega = \pi \operatorname{sgn} (t).$$

Notice that using Euler's identity we can write

$$\int_{-\infty}^{\infty} \frac{e^{i\omega t}}{\omega} d\omega = \int_{-\infty}^{\infty} \frac{\cos{(\omega t)}}{\omega} d\omega + i \int_{-\infty}^{\infty} \frac{\sin{(\omega t)}}{\omega} d\omega$$

and, as the first integral on the right vanishes since its integrand is odd, we have

$$\int_{-\infty}^{\infty} \frac{e^{i\omega t}}{\omega} d\omega = i\pi \; \text{sgn}\,(t). \qquad (7.9.3)$$

Another time signal that radio engineers find useful, one closely related to sgn(t), is the so-called unit step, equal to zero for negative time and to +1 for positive time. Written as u(t), we can connect u(t) to sgn(t) by writing

$$u(t) = \frac{1 + \text{sgn}\,(t)}{2}. \qquad (7.9.4)$$

The unit step is a constant for all t *except* at t = 0 where it *instantly* jumps from 0 to 1 as t passes from being negative to being positive. This jump occurs in zero time, and so the 'derivative' of u(t) is infinite at t = 0 and zero for all t ≠ 0. For a long time mathematicians did not consider the derivative of u(t) to be a respectable function, but nonetheless the English physicist Paul Dirac (1902–1984) showed that working with such a thing—called an *impulse function* and written as δ(t)— could indeed be quite useful.[16] Dirac formally wrote

$$\delta(t) = \frac{d}{dt} u(t)$$

and so, formally differentiating (7.9.4), we have

$$\delta(t) = \frac{d}{dt} \left(\frac{1 + \text{sgn}\,(t)}{2} \right) = \frac{1}{2} \frac{d}{dt} \{ \text{sgn}\,(t) \}.$$

Now, if (as usual in this book) we boldly assume we can differentiate under the integral sign in (7.9.3), then

---

[16]Although Dirac won the 1933 Nobel Prize in *physics*, he was the Lucasian Professor of *mathematics* at Cambridge University. His physical insight into such a bizarre thing as an infinite derivative was powered (by his own admission) with his undergraduate training in *electrical engineering*: he graduated with first-class honors in EE from the University of Bristol in 1921. Dirac was clearly 'a man for all seasons'! The mathematics of impulses has been placed on a firm theoretical foundation since Dirac's intuitive use of them in quantum mechanics. The central figure in that great achievement is generally considered to be the French mathematician Laurent Schwartz (1915–2002), with the publication of his two books *Theory of Distributions* (1950, 1951). For that work, Schwartz received the 1950 Fields Medal, often called the 'Nobel Prize of mathematics.'

$$\frac{d}{dt} \int_{-\infty}^{\infty} \frac{e^{i\omega t}}{\omega} \, d\omega = \frac{d}{dt} \{ i\pi \, \text{sgn}\,(t) \} = i\pi \frac{d}{dt} \{ \, \text{sgn}\,(t) \} = \int_{-\infty}^{\infty} \frac{i\omega e^{i\omega t}}{\omega} \, d\omega = i \int_{-\infty}^{\infty} e^{i\omega t} \, d\omega.$$

That is,

$$\int_{-\infty}^{\infty} e^{i\omega t} d\omega = \pi \frac{d}{dt} \{ \, \text{sgn}\,(t) \}. \tag{7.9.5}$$

From our differentiation of (7.9.4) we have

$$\frac{d}{dt} u(t) = \delta(t) = \frac{1}{2} \frac{d}{dt} \{ \, \text{sgn}\,(t) \}$$

and so

$$\frac{d}{dt} \{ \, \text{sgn}\,(t) \} = 2\delta(t).$$

Putting this into (7.9.5), we have

$$\int_{-\infty}^{\infty} e^{i\omega t} \, d\omega = 2\pi\delta(t). \tag{7.9.6}$$

The statement in (7.9.6) is an astonishing one because the integral just doesn't exist if we attempt to actually evaluate it, since $e^{i\omega t}$ doesn't even approach a limit as $|\omega| \to \infty$. The real and imaginary parts of $e^{i\omega t}$ both simply oscillate for all t other than zero as ω varies. The only way we can make any sense of (7.9.6) is, as Dirac did, by interpreting the integral on the left as a collection of printed squiggles that denote the same *concept* as do the printed squiggles on the right (for which we at least have a physical feel). Any time we encounter the integral squiggles we'll just replace them with the squiggles '$2\pi\delta(t)$.' As you'll soon see, impulses can occur with arguments more complicated than just 't,' and the general rule is that an impulse goes to infinity when its argument vanishes. So, for example, $\delta(t - t_0)$ is zero for all $t \neq t_0$ and infinity *at* $t = t_0$.

With (7.9.3) and (7.9.6) we can now find the Fourier transforms of sgn(t), $\delta(t)$, and u(t). For sgn(t), I claim its transform is $\frac{2}{i\omega}$. To see this, put $\frac{2}{i\omega}$ into the inverse transform integral of (7.8.5) to get

$$\frac{1}{2\pi} \int_{-\infty}^{\infty} \frac{2}{i\omega} e^{i\omega t} \, d\omega = \frac{1}{\pi i} \int_{-\infty}^{\infty} \frac{e^{i\omega t}}{\omega} \, d\omega$$

and then recall (7.9.3) which says the integral on the right is $i\pi$ sgn (t). That is,

$$\frac{1}{\pi i} \int_{-\infty}^{\infty} \frac{e^{i\omega t}}{\omega} \, d\omega = \frac{1}{\pi i} i\pi \, \text{sgn}\,(t) = \text{sgn}\,(t).$$

So, we have the transform pair

$$\text{sgn}(t) \leftrightarrow \frac{2}{i\omega}. \tag{7.9.7}$$

The energy spectrum of sgn(t) is

$$S_f(\omega) = \frac{\left|\frac{2}{i\omega}\right|^2}{2\pi} = \frac{2}{\pi\omega^2}, \quad -\infty < \omega < \infty$$

and so, if we integrate this spectrum over all ω, we get infinity. That is, sgn(t) is an infinite energy signal (something obvious from the get-go, of course, for a signal whose magnitude is 1 for all time), a clear clue that it is impossible to actually generate it!

Next, I claim the Fourier transform of $\delta(t)$ is 1, and again you can see this by putting 1 into the inverse transform integral of (7.8.5) to get

$$\frac{1}{2\pi} \int_{-\infty}^{\infty} 1 \, e^{i\omega t} \, d\omega = \frac{1}{2\pi} \int_{-\infty}^{\infty} e^{i\omega t} \, d\omega$$

and then recall (7.9.6) which says the integral on the right is $2\pi\delta(t)$. So, we have the transform pair

$$\delta(t) \leftrightarrow 1. \tag{7.9.8}$$

The energy spectrum of $\delta(t)$ is uniform over all ω or, as radio engineers sometimes put it, $\delta(t)$ has a *flat* spectrum.[17] From Rayleigh's theorem we see that $\delta(t)$, like sgn(t), is an infinite energy signal and so is impossible to actually generate. Unlike sgn(t), this infinity is *not* obvious from the time behavior of the impulse (which property dominates, the infinite value at one instant of time, or the fact that it is just one instant of time?). It's the energy spectrum that gives us the answer.

Now, what is the transform of the unit step, u(t)? From (7.9.4) we can write the transform of u(t) as the sum of the transforms of $\frac{1}{2}$ and $\frac{1}{2}$ sgn (t). That is, using (7.9.7), we have

$$u(t) \leftrightarrow \int_{-\infty}^{\infty} \frac{1}{2} e^{-i\omega t} \, dt + \frac{1}{i\omega}. \tag{7.9.9}$$

'All' we have to do is figure-out what the integral on the right in (7.9.9) is—and to do that, let me show you a neat little trick in notation. Look back at (7.9.6). Since it is an equality it *remains* an equality if we perform exactly the same operations on both

---

[17] In an analogy with white light, in which all optical frequencies (colors) are uniformly present, such an energy distribution is also often said to be a *white* spectrum. To continue with this terminology, signals with energy spectrums that are not flat (not white) are said to have a *pink* (or *colored*) spectrum. Who says radio engineers aren't romantic souls?!

sides. So, on the left replace every ω with t, and every t with ω, and on the right do the same. Then,

$$\int_{-\infty}^{\infty} e^{it\omega} \, dt = 2\pi\delta(\omega) \qquad (7.9.10)$$

where $\delta(\omega)$ is an impulse in the ω-domain. Just as $\delta(t)$ is zero for all $t \neq 0$ and infinite *at* $t = 0$, $\delta(\omega)$ is zero for all $\omega \neq 0$ and infinite *at* $\omega = 0$. That is, all the infinite energy of a signal that is a *constant* for all time should have no energy at any non-zero frequency, because otherwise the signal wouldn't be *constant* but rather would have a time-varying component.

Next, change variable in (7.9.10) to $u = -t$ $(dt = -du)$. Then,

$$\int_{-\infty}^{\infty} e^{it\omega} \, dt = \int_{\infty}^{-\infty} e^{i(-u)\omega} \, (-du) = \int_{-\infty}^{\infty} e^{-i\omega u} \, du.$$

That is,

$$2\pi\delta(\omega) = \int_{-\infty}^{\infty} e^{-i\omega u} \, du$$

or, if we change the dummy variable of integration from u back to t,

$$\int_{-\infty}^{\infty} e^{-i\omega t} \, dt = 2\pi\delta(\omega). \qquad (7.9.11)$$

Notice, from (7.9.10) and (7.9.11), that we've shown

$$\int_{-\infty}^{\infty} e^{it\omega} \, dt = 2\pi\delta(\omega) = \int_{-\infty}^{\infty} e^{-i\omega t} \, dt.$$

That is, $\delta(-\omega) = \delta(\omega)$ and so the impulse function, mathematically, is *even*. In any case, the integral on the right in (7.9.9), the Fourier transform of $\frac{1}{2}$, is $\pi\delta(\omega)$, and so we now have the pair

$$u(t) \leftrightarrow \pi\delta(\omega) + \frac{1}{i\omega} \qquad (7.9.12)$$

and so of course, like sgn(t) and $\delta(t)$, u(t) is an infinite energy signal.

We can expand on the physical meaning of $\delta(\omega)$ by computing the Fourier transform of the pure sinusoidal signal $\cos(\omega_0 t)$, $-\infty < t < \infty$. By definition, the transform is

$$\int_{-\infty}^{\infty} \cos(\omega_0 t)e^{-i\omega t}dt = \int_{-\infty}^{\infty} \frac{e^{i\omega_0 t} + e^{-i\omega_0 t}}{2} e^{-i\omega t}dt$$

$$= \frac{1}{2}\int_{-\infty}^{\infty} e^{-i(\omega-\omega_0)t} \, dt + \frac{1}{2}\int_{-\infty}^{\infty} e^{-i(\omega+\omega_0)t} \, dt.$$

Recalling (7.9.11) we see that the first integral on the right is $2\pi\delta(\omega - \omega_0)$ and the second integral on the right is $2\pi\delta(\omega + \omega_0)$. So, we have the pair

$$\cos(\omega_0 t) \leftrightarrow \pi\delta(\omega - \omega_0) + \pi\delta(\omega + \omega_0). \tag{7.9.13}$$

Since an impulse 'occurs' where its argument is zero, we see that all the (infinite) energy of the pure sinusoidal time signal $\cos(\omega_0 t)$ is equally concentrated at the *two* frequencies[18] $\omega = \pm \omega_0$.

There is one final property of the impulse function that is important to state. Returning to Dirac's formal definition of the impulse as derivative of the step, that is to

$$\delta(t) = \frac{d}{dt}\{u(t)\},$$

then if we formally integrate this we get

$$\int_{-\infty}^{t} \delta(y)dy = \int_{-\infty}^{t} \frac{d}{dy}\{u(y)\}dy = \int_{-\infty}^{t} d\{u(y)\} = u(y)\Big|_{-\infty}^{t} = u(t) - u(-\infty)$$

or, because $u(-\infty) = 0$,

$$\int_{-\infty}^{t} \delta(y)dy = u(t) = \begin{cases} 1, & t > 0 \\ 0, & t < 0 \end{cases}.$$

That is, even though the impulse has zero duration, it is nonetheless 'so infinite' that it bounds unit area! This is impossible to justify in the framework of nineteenth century mathematics, which is why mathematicians for so long dismissed the impulse as being utter nonsense (electrical engineers and physicists, however, didn't have much of a problem with it because the impulse solved many of their 'real-life' problems)—until the work of Laurent Schwartz.

---

[18]There are *two* frequencies in the transform because of the two exponentials in the transform integral, each of which represents a rotating vector in the complex plane. One rotates counterclockwise at frequency $+\omega_0$ (making an instantaneous angle with the real axis of $\omega_0 t$) and the other rotates clockwise at frequency $-\omega_0$ (making an instantaneous angle with the real axis of $-\omega_0 t$). The imaginary components of these two vectors always cancel, while the real components add along the real axis to produce the real-valued signal $\cos(\omega_0 t)$.

One intuitive way to 'understand' this result (a view common among physicists and radio engineers), is to think of the impulse as a very narrow pulse of height $\frac{1}{\varepsilon}$ for $-\frac{\varepsilon}{2} < t < \frac{\varepsilon}{2}$ (where $\varepsilon \approx 0$), and with zero height for all other t. (Notice that this makes the impulse *even*, as we concluded it must be after deriving (7.9.11)). For *all* ε this pulse always bounds unit area, even as we let $\varepsilon \to 0$. So, suppose $\phi(t)$ is any function that is continuous at $t = 0$. Then, during the interval $-\frac{\varepsilon}{2} < t < \frac{\varepsilon}{2}$ $\phi(t)$ can't change by much and so is essentially *constant* over that entire interval (an approximation that gets ever better as we let $\varepsilon \to 0$) with value $\phi(0)$, and so we can write

$$\int_{-\infty}^{\infty} \delta(t)\phi(t)dt = \int_{-\frac{\varepsilon}{2}}^{\frac{\varepsilon}{2}} \frac{1}{\varepsilon}\phi(0)dt = \phi(0)\int_{-\frac{\varepsilon}{2}}^{\frac{\varepsilon}{2}} \frac{1}{\varepsilon}dt = \phi(0)\frac{1}{\varepsilon}\varepsilon = \phi(0).$$

More generally, if $\phi(t)$ is continuous at $t = a$ then

$$\int_{-\infty}^{\infty} \delta(t - a)\phi(t)dt = \phi(a). \tag{7.9.14}$$

The integral of (7.9.14) is often called the *sampling* property of the impulse.

Now, to finish this section I'll take you through three simple theoretical results in Fourier transform theory. We'll find all three highly useful in the next section. To start, we can use the same notational trick we used to get (7.9.10) to derive what is called the *duality theorem*. Suppose we have the transform pair $g(t) \leftrightarrow G(\omega)$. Then, from the inverse transform integral (7.8.5), we have

$$g(t) - \frac{1}{2\pi}\int_{-\infty}^{\infty} G(\omega)e^{i\omega t}d\omega$$

or, replacing t with –t *on both sides* of the equality (which leaves the equality as an equality), we have

$$g(-t) = \frac{1}{2\pi}\int_{-\infty}^{\infty} G(\omega)e^{-i\omega t}d\omega.$$

Then, using our symbol-swapping trick again (so it's a *method*!)—replace t with ω, and ω with t—we get

$$g(-\omega) = \frac{1}{2\pi}\int_{-\infty}^{\infty} G(t)e^{-it\omega}dt$$

or,

$$2\pi g(-\omega) = \int_{-\infty}^{\infty} G(t)e^{-it\omega}dt.$$

That is,

$$\text{if } g(t) \leftrightarrow G(\omega)$$
$$\text{then } G(t) \leftrightarrow 2\pi g(-\omega)\ . \tag{7.9.15}$$

Our second result, called the *time/frequency scaling theorem*, starts with the given transform pair $f(t) \leftrightarrow F(\omega)$. We then ask: what is the transform of $f(at)$, where a is a positive constant? The answer is of course

$$\int_{-\infty}^{\infty} f(at)e^{-i\omega t}dt$$

which, if we change variable to $u = at$ ($dt = \frac{du}{a}$), becomes

$$\int_{-\infty}^{\infty} f(u)e^{-i\omega \frac{u}{a}}\frac{du}{a} = \frac{1}{a}\int_{-\infty}^{\infty} f(u)e^{-i\left(\frac{\omega}{a}\right)u}\, du = \frac{1}{a}F\left(\frac{\omega}{a}\right).$$

That is,

$$\text{if } f(t) \leftrightarrow F(\omega)$$
$$\text{then } f(at) \leftrightarrow \frac{1}{a}F\left(\frac{\omega}{a}\right)\ . \tag{7.9.16}$$

and finally, given the two time functions $g(t)$ and $m(t)$, with Fourier transforms $G(\omega)$ and $F(\omega)$, respectively, what is the Fourier transform of $m(t)g(t)$? By definition, the transform is

$$\int_{-\infty}^{\infty} m(t)g(t)e^{-i\omega t}\, dt = \int_{-\infty}^{\infty} m(t)\left\{\frac{1}{2\pi}\int_{-\infty}^{\infty} G(u)e^{iut}\, du\right\}e^{-i\omega t}\, dt$$

where $g(t)$ has been written in the form of an inverse Fourier transform integral (I've used u as the dummy variable of integration in the inner integral, rather than ω, to avoid confusion with the outer ω). Continuing, if we reverse the order of integration we have the transform of $m(t)g(t)$ as

$$\int_{-\infty}^{\infty} \frac{1}{2\pi}G(u)\left\{\int_{-\infty}^{\infty} m(t)e^{iut}\, e^{-i\omega t}\, dt\right\}\, du = \frac{1}{2\pi}\int_{-\infty}^{\infty} G(u)\left\{\int_{-\infty}^{\infty} m(t)\, e^{-i(\omega-u)t}\, dt\right\}\, du$$

or, as the inner integral is just $M(\omega - u)$, we have the Fourier transform pair

$$m(t)g(t) \leftrightarrow \frac{1}{2\pi}\int_{-\infty}^{\infty} G(u)M(\omega - u)\, du. \tag{7.9.17}$$

The integral on the right in (7.9.17) occurs so often in mathematical physics that it has its own name: the *convolution integral*, written in short hand as $G(\omega) * M(\omega)$.[19] Since it is arbitrary which time function we call m(t) and which we call g(t), then in fact convolution is commutative and so $m(t)g(t) \leftrightarrow \frac{1}{2\pi}G(\omega) * M(\omega) = \frac{1}{2\pi}M(\omega) * G(\omega)$.

We'll use (7.9.17) in the next section as a purely mathematical result but, to finish this section, you may find it interesting to know that it is the reason radio works. Here's why. Imagine Alice and Bob are each talking into a microphone at radio stations A and B, respectively. Since the sounds produced by both are generated via the same physical process (vibrating human vocal chords), the energies of the two voice signals will be concentrated at essentially the same frequencies, typically a few tens of hertz up to a few thousand hertz. That is, the frequency interval occupied by the electrical signals produced on the wires emerging from Alice's microphone is the same as the frequency interval occupied by the electrical signals produced on the wires emerging from Bob's microphone. This common interval of frequencies determines what is called the *baseband spectrum*, centered on $\omega = 0$.

To apply the baseband electrical signal from a microphone directly to an antenna will *not* result in the efficient radiation of energy into space, as Maxwell's equations for the electromagnetic field tell us that for the efficient coupling of the antenna to space to occur the physical size of the antenna must be comparable to the wavelength of the radiation (if you're not an electrical engineer or a physicist, just take my word for this). At the baseband frequency of 1 kHz, for example, a wavelength of electromagnetic radiation is *one million feet*, which is pretty long. To get a reasonably sized antenna, we need to reduce the wavelength, that is, to *increase* the frequency.

What is done in commercial broadcast AM (amplitude modulated) radio to accomplish that is to up-shift the baseband spectrum of the microphone signal by between about 500 kHz and 1500 kHz, to the so-called AM radio band. (Each radio station receives a license from the Federal Communications Commission—the FCC—that gives it permission to do the frequency up-shift by a value that no other station in the same geographical area may use). At 1000 kHz, for example, the wavelength is a thousand times shorter than it is at 1 kHz—that is, a wavelength is now 1000 feet. If a station's antenna is constructed to be a quarter-wavelength, for example, then it will have an antenna 250 feet high, which is just about what you'll see when you next drive by your local AM radio station's transmitter site.

So, suppose that at station A Alice's baseband signal is up-shifted by 900 kHz, while at station B Bob's baseband signal is up-shifted by 1100 kHz. A radio receiver then selects which signal to listen to by using a tunable filter circuit centered on either 900 kHz or 1100 kHz (in AM radio, the bandwidth of this filter is 10 kHz, and knowing how to design such a filter is part of the skill-set of radio engineers). Note that radio uses a frequency up-shift for *two* reasons: (1) to move baseband energy up to so-called 'radio

---

[19]Note, *carefully*: the $*$ symbol denotes *complex conjugation* when used as a superscript as was done in Sect. 7.8 when discussing the energy spectrum, and convolution when used in-line. Equation (7.9.16) is called the *frequency* convolution integral, to distinguish it from its twin, the *time* convolution integral, which says $m(t) * g(t) \leftrightarrow G(\omega)M(\omega)$. We won't use that pair in what follows, but you should now be able to derive it for yourself. Try it!

frequency' to achieve efficient radiation of energy and (2) to *separate* the baseband energies of multiple radio stations by using a *different* up-shift frequency at each station. At a receiver we need a final frequency down-shift to place the energy of the selected station signal back at the baseband frequencies to which our ears respond.

To accomplish these frequency shifts, both up and down, is as simple as doing a multiplication.[20] Here's how (7.9.17) works for the transmitter up-shift. Let $M(\omega)$ be the Fourier transform of either Alice's or Bob's baseband microphone signal. Then, remembering (7.9.13), the transform of $\cos(\omega_0 t)$, (7.9.17) tells us that the transform of $m(t) \cos(\omega_0 t)$ is

$$M(\omega) * [\pi\delta(\omega - \omega_0) + \pi\delta(\omega + \omega_0)] = \frac{1}{2\pi} \int_{-\infty}^{\infty} [\pi\delta(u - \omega_0) + \pi\delta(u + \omega_0)] M(\omega - u) \, du$$

$$= \frac{1}{2} \int_{-\infty}^{\infty} \delta(u - \omega_0) M(\omega - u) du + \frac{1}{2} \int_{-\infty}^{\infty} \delta(u + \omega_0) M(\omega - u) du$$

and so, remembering (7.9.14), the sampling property of the impulse, the transform of $m(t) \cos(\omega_0 t)$ is $\frac{1}{2} M(\omega - \omega_0) + \frac{1}{2} M(\omega + \omega_0)$. That is, while the energy spectrum of $m(t)$ is centered on $\omega = 0$, the energy spectrum of $m(t) \cos(\omega_0 t)$ is centered on $\omega = \pm\omega_0$. The energy spectrum of the original baseband signal $m(t)$ now rides piggyback on $\cos(\omega_0 t)$ (picturesquely called the *carrier wave* by radio engineers), and is efficiently radiated into space by a physically 'short' antenna. When used in radio, (7.9.17) is called the *heterodyne theorem*.[21]

7.10   Causality and Hilbert Transform Integrals

You'll recall that in Sect. 7.8 I made the following statements about the general transform pair $g(t) \leftrightarrow G(\omega)$: In general, $G(\omega)$ will be complex, with real and imaginary parts $R(\omega)$ and $X(\omega)$, respectively. That is, $G(\omega) = R(\omega) + i X(\omega)$. If g (t) is a real-valued function of time (as of course are all the signals in any electronic circuitry that can actually be constructed) then $G(\omega)$ will have some special proper-

---

[20]To be honest, multiplying at radio frequencies is *not* easy. To learn how radio engineers accomplish multiplication *without* actually multiplying, see *Dr. Euler*, pp. 295–297, 302–305, or my book *The Science of Radio*, Springer 2001, pp. 233–249.

[21]The *mathematics* of all this was of course known long before AM radio was invented, but the name of the theorem is due to the American electrical engineer Reginald Fessenden (1866–1932), who patented the multiplication idea in 1901 for use in a radio circuit. The word 'heterodyne' comes from the Greek *heteros* (for external) and *dynamic* (for force). Fessenden thought of the $\cos(\omega_0 t)$ signal as the 'external force' being generated by the radio receiver circuitry itself for the final frequency down-shift of the received signal to baseband (indeed, radio engineers call that part of an AM radio receiver the *local oscillator* circuit).

ties. In particular, $R(\omega)$ will be even and $X(\omega)$ will be odd: $R(-\omega) = R(\omega)$ and $X(-\omega) = -X(\omega)$. If, in addition to being real, $g(t)$ has certain symmetry properties, then $G(\omega)$ will have additional corresponding special properties. If, for example, $g(t)$ is even (as is $\cos(\omega t)$) then $G(\omega)$ will be real, and if $g(t)$ is odd (as is $\sin(\omega t)$) then $G(\omega)$ will be imaginary: $X(\omega) = 0$ and $R(\omega) = 0$, respectively.

If we now impose even further restrictions on $g(t)$ then, as you'd expect, there will be even further restrictions on $R(\omega)$ and $X(\omega)$. One restriction that is fundamental to the real world is *causality*. To understand what that means, suppose we have what electronics engineers call a 'black box,' with an input and an output. (The term 'black box' means we don't know the details of the circuitry inside the box and, indeed, we don't care). All we know is that if we apply an input signal starting at time $t = 0$ then, whatever the output signal is, *it had better be zero for $t < 0$*. That is, there should be no output before we apply the input. Otherwise we have what is called an *anticipatory* output, which is a polite name for a time machine! So, there's our question: If $g(t)$ is the output signal, a signal real-valued and zero for all $t < 0$, what more can we say about its Fourier transform $G(\omega) = R(\omega) + i\,X(\omega)$?

To start our analysis, let's write $g(t)$ as the sum of even and odd functions of time, that is, as

$$g(t) = g_e(t) + g_o(t) \tag{7.10.1}$$

where, by definition,

$$g_e(-t) = g_e(t), g_o(-t) = -g_o(t).$$

That we can actually write $g(t)$ in this way is most directly shown by simply demonstrating what $g_e(t)$ and $g_o(t)$ are (we did this back in Chap. 1, but here it is again). From (7.10.1) we can write

$$g(-t) = g_e(-t) + g_o(-t) = g_e(t) - g_o(t) \tag{7.10.2}$$

and so, adding (7.10.1)–(7.10.2) we get

$$g_e(t) = \frac{1}{2}[g(t) + g(-t)], \tag{7.10.3}$$

and subtracting (7.10.2) from (7.10.1) we get

$$g_o(t) = \frac{1}{2}[g(t) - g(-t)]. \tag{7.10.4}$$

So, (7.10.1) is always possible to write.

Since $g(t)$ is to be causal (and so by definition $g(t) = 0$ for $t < 0$), we have from (7.10.3) and (7.10.4) that

$$g_e(t) = \frac{1}{2}g(t)$$
$$\text{if } t > 0$$
$$g_o(t) = \frac{1}{2}g(t)$$

and

$$g_e(t) = \frac{1}{2}g(-t)$$
$$\text{if } t < 0.$$
$$g_o(t) = -\frac{1}{2}g(-t)$$

That is,

$$g_e(t) = g_o(t), \quad t > 0$$
$$g_e(t) = -g_o(t), \quad t < 0$$

and so

$$g_e(t) = g_o(t)\, \text{sgn}\,(t). \tag{7.10.5}$$

In a similar way, we can also write

$$g_o(t) = g_e(t)\, \text{sgn}\,(t). \tag{7.10.6}$$

Now, because of (7.10.1) we can write

$$G(\omega) = G_e(\omega) + G_o(\omega)$$

and since $g_e(t)$ is even we know that $G_e(\omega)$ is purely real, while since $g_o(t)$ is odd we know that $G_o(\omega)$ is purely imaginary. Thus,

$$G_e(\omega) = R(\omega) \tag{7.10.7}$$

and

$$G_o(\omega) = i\,X(\omega). \tag{7.10.8}$$

Now, recall the transform pair from (7.9.7):

$$\text{sgn}\,(t) \leftrightarrow \frac{2}{i\omega}.$$

From (7.10.5), and the frequency convolution theorem of (7.9.17), we have

$$G_e(\omega) = \frac{1}{2\pi} G_o(\omega) * \frac{2}{i\omega}$$

and so, using (7.10.7) and (7.10.8),

$$R(\omega) = \frac{1}{2\pi} i\, X(\omega) * \frac{2}{i\omega} = \frac{1}{\pi}\, X(\omega) * \frac{1}{\omega}. \qquad (7.10.9)$$

Also, from (7.10.6) and the frequency convolution theorem we have

$$G_o(\omega) = \frac{1}{2\pi} G_e(\omega) * \frac{2}{i\omega}$$

and so, using (7.10.7) and (7.10.8),

$$i\, X(\omega) = \frac{1}{2\pi} R(\omega) * \frac{2}{i\omega}$$

or,

$$X(\omega) = -\frac{1}{\pi} R(\omega) * \frac{1}{\omega}. \qquad (7.10.10)$$

Writing (7.10.9) and (7.10.10) as integrals, we arrive at

$$
\begin{aligned}
R(\omega) &= \frac{1}{\pi} \int_{-\infty}^{\infty} \frac{X(u)}{\omega - u}\, du \\
X(\omega) &= -\frac{1}{\pi} \int_{-\infty}^{\infty} \frac{R(u)}{\omega - u}\, du
\end{aligned} \qquad (7.10.11)
$$

These two equations show that $R(\omega)$ and $X(\omega)$ each determine the other for a causal signal. The integrals that connect $R(\omega)$ and $X(\omega)$ are called *Hilbert transforms*,[22] a name introduced by our old friend throughout this book, G. H. Hardy. Hardy published the transform for the first time *in English* in 1909 but, when he later learned that the German mathematician David Hilbert (1862–1943) had known of these formulas since 1904, Hardy began to call them Hilbert transforms. But Hilbert had not been the first, either, as they had appeared decades earlier in the 1873 doctoral dissertation of the Russian mathematician Yulian-Karl Vasilievich Sokhotsky (1842–1927).

---

[22]They are also sometimes called the *Kramers-Kronig relations*, after the Dutch physicist Hendrik Kramers (we encountered him back in Sect. 6.5, when discussing the Watson/van Peype triple integrals), and the American physicist Ralph Kronig (1904–1995), who encountered (7.10.11) when studying the spectra of x-rays scattered by the atomic lattice structures of crystals. See Challenge Problem 7.9 for an alternative way to write (7.10.11).

Notice that the Hilbert transform does *not* change domain, as does the Fourier transform. That is, in (7.10.11) the Hilbert transform is in the same domain (ω) on both sides of the equations. One can also take the Hilbert transform of a time function $x(t)$, getting a new time time function[23] written as $\overline{x(t)}$:

$$\overline{x(t)} = \frac{1}{\pi} \int_{-\infty}^{\infty} \frac{x(u)}{t - u} \, du. \tag{7.10.12}$$

For example, if $x(t)$ is any constant time function (call it k) then its Hilbert transform is zero. To show this, I'll use our old 'sneak' trick to handle the integrand singularity at $u = t$. That is,

$$\overline{x(t)} = \frac{1}{\pi} \int_{-\infty}^{\infty} \frac{k}{t - u} \, du = \frac{k}{\pi} \lim_{\varepsilon \to 0, T \to \infty} \left[\int_{-T}^{t-\varepsilon} \frac{du}{t - u} + \int_{t+\varepsilon}^{T} \frac{du}{t - u} \right].$$

Changing variable in both integrals on the right to $s = t - u$ ($ds = -du$), then

$$\overline{x(t)} = \frac{k}{\pi} \lim_{\varepsilon \to 0, T \to \infty} \left[\int_{t+T}^{\varepsilon} \left(-\frac{ds}{s} \right) + \int_{-\varepsilon}^{t-T} \left(-\frac{ds}{s} \right) \right]$$

$$= \frac{k}{\pi} \lim_{\varepsilon \to 0, T \to \infty} \left[\int_{\varepsilon}^{t+T} \frac{ds}{s} + \int_{t-T}^{-\varepsilon} \frac{ds}{s} \right]$$

$$= \frac{k}{\pi} \lim_{\varepsilon \to 0, T \to \infty} \left[\ln(s) \big|_{\varepsilon}^{t+T} + \ln(s) \big|_{t-T}^{-\varepsilon} \right]$$

$$= \frac{k}{\pi} \lim_{\varepsilon \to 0, T \to \infty} \left[\ln(t + T) - \ln(\varepsilon) + \ln(-\varepsilon) - \ln(t - T) \right]$$

$$= \frac{k}{\pi} \lim_{\varepsilon \to 0, T \to \infty} \left[\ln \left(\frac{t + T}{\varepsilon} \right) + \ln \left(\frac{-\varepsilon}{t - T} \right) \right]$$

$$= \frac{k}{\pi} \lim_{\varepsilon \to 0, T \to \infty} \left[\ln \left\{ \left(\frac{t + T}{\varepsilon} \right) \left(\frac{\varepsilon}{T - t} \right) \right\} \right]$$

and so, noticing that the ε's cancel,

$$\overline{x(t)} = \frac{k}{\pi} \lim_{T \to \infty} \left[\ln \left\{ \left(\frac{T + t}{T - t} \right) \right\} \right] = \frac{k}{\pi} \ln(1) = 0.$$

This gives us the interesting

---

[23]Combining a time signal $x(t)$ with its Hilbert transform to form the complex signal $z(t) = x(t) + i\,\overline{x(t)}$, you get what the Hungarian-born electrical engineer Dennis Gabor (1900–1979)—he won the 1971 Nobel Prize in physics—called the *analytic signal*, of great interest to engineers who study single-sideband (SSB) radio. To see how the analytic signal occurs in SSB radio theory, see *Dr. Euler*, pp. 309–323.

$$\int_{-\infty}^{\infty} \frac{du}{t-u} = 0. \tag{7.10.13}$$

To finish this section, let me now take you through the analysis of a particular causal time signal, which will end with the discovery of yet another interesting integral. To start, recall the Fourier transform of the f(t) in (7.8.1):

$$F(\omega) = \frac{e^{-i\omega a} - e^{-i\omega b}}{i\omega}$$

for

$$f(t) = \begin{cases} 1, & a < t < b \\ 0, & \text{otherwise} \end{cases} .$$

Suppose we set $a = -\frac{1}{2}$ and $b = \frac{1}{2}$. We then have a signal that is important enough in radio engineering to have its own name (it is called the *gate* function), and its own symbol, $\pi(t)$. That is,

$$\pi(t) = \begin{cases} 1, & -\frac{1}{2} < t < \frac{1}{2} \\ 0, & \text{otherwise} \end{cases} . \tag{7.10.14}$$

The Fourier transform of $\pi(t)$ is

$$\Pi(\omega) = \frac{e^{i\omega\frac{1}{2}} - e^{-i\omega\frac{1}{2}}}{i\omega} = \frac{i2\sin\left(\frac{\omega}{2}\right)}{i\omega} = \frac{\sin\left(\frac{\omega}{2}\right)}{\left(\frac{\omega}{2}\right)}, \quad -\infty < \omega < \infty. \tag{7.10.15}$$

From the duality theorem of (7.9.15) we have, from (7.10.14) and (7.10.15), the pair[24]

$$\Pi(t) \leftrightarrow 2\pi\ \pi(-\omega)$$

and so

---

[24]Note, carefully, the dual use of the symbol "π"—once for the number, and again for the name of the gate function. There will never be any confusion, however, because the gate function will always appear with an argument while π *alone* is the number.

$$\frac{\sin\left(\frac{t}{2}\right)}{\left(\frac{t}{2}\right)} \leftrightarrow 2\pi\,\pi(-\omega) = 2\pi\,\pi(\omega) \qquad (7.10.16)$$

because $\pi(\omega)$ is an even function. Next, applying the time/frequency scaling theorem of (7.9.16) with $a = 2$, (7.10.16) says

$$\frac{\sin(t)}{t} \leftrightarrow \pi\,\pi\left(\frac{\omega}{2}\right).$$

Since the gate function is 1 over the interval for which its argument is in the interval $-\frac{1}{2}$ to $\frac{1}{2}$ (this interval is $-\frac{1}{2} < \frac{\omega}{2} < \frac{1}{2}$, or $-1 < \omega < 1$), we have

$$\frac{\sin(t)}{t} \leftrightarrow \begin{cases} \pi, & -1 < \omega < 1 \\ 0, & \text{otherwise} \end{cases}. \qquad (7.10.17)$$

Now, the time signal in (7.10.17) is not a causal signal since it exists for all t, but we can use it to make a causal signal (zero for $t < 0$) by multiplying it by the step function. That is,

$$g(t) = \frac{\sin(t)}{t} u(t)$$

is a causal time function. As we showed at the beginning of this section, we can always write any causal time function as the sum of an even function and an odd function, where the even function is $\frac{1}{2}g(t)$, which for our problem here is $\frac{\sin(t)}{2t}$. You'll also recall that we showed the real part of $G(\omega)$ is due entirely to this even time function. So,

$$\frac{\sin(t)}{2t} \leftrightarrow R(\omega).$$

Looking back at (7.10.17), we see that

$$\frac{\sin(t)}{2t} = \frac{1}{2}\left\{\frac{\sin(t)}{t}\right\} \leftrightarrow R(\omega) = \begin{cases} \dfrac{\pi}{2}, & -1 < \omega < 1 \\ 0, & \text{otherwise} \end{cases}. \qquad (7.10.18)$$

So, from (7.10.11) we can find $X(\omega)$ the imaginary part of $G(\omega)$, as

$$X(\omega) = -\frac{1}{\pi}\int_{-\infty}^{\infty}\frac{R(u)}{\omega - u}\,du = -\frac{1}{2}\int_{-1}^{1}\frac{du}{\omega - u}.$$

Doing this integral isn't quite as straightforward as it might initially appear. That's because there is a singularity at $u = \omega$. If $|\omega| > 1$ this singularity is not in the interval of integration and we can proceed in the obvious way. Change variable to $s = \omega - u$ and so $ds = -du$. Then,

$$X(\omega) = -\frac{1}{2}\int_{\omega+1}^{\omega-1}\left(\frac{-ds}{s}\right) = \frac{1}{2}\int_{\omega+1}^{\omega-1}\left(\frac{ds}{s}\right) = \frac{1}{2}\ln(s)\Big|_{\omega+1}^{\omega-1} = \frac{1}{2}\ln\left(\frac{\omega-1}{\omega+1}\right), |\omega| > 1.$$

For $|\omega| > 1$ the argument of the log function is positive and all is okay. For the case of $|\omega| < 1$, however, the argument is negative and the log function gives an imaginary result, which is *not* okay. The reason for this difficulty is that the singularity is in the interval of integration when $|\omega| < 1$.

The fix is to use our sneak trick. In the limit as $\varepsilon \to 0$ we write

$$X(\omega) = -\frac{1}{2}\left[\int_{-1}^{\omega-\varepsilon}\frac{du}{\omega-u} + \int_{\omega+\varepsilon}^{1}\frac{du}{\omega-u}\right] = -\frac{1}{2}\left[\int_{\omega+1}^{\varepsilon}\left(\frac{-ds}{s}\right) + \int_{-\varepsilon}^{\omega-1}\left(\frac{-ds}{s}\right)\right]$$

$$= \frac{1}{2}\left[\ln(s)\Big|_{\omega+1}^{\varepsilon} + \ln(s)\Big|_{-\varepsilon}^{\omega-1}\right] = \frac{1}{2}\left[\ln\left(\frac{\varepsilon}{\omega+1}\right) + \ln\left(\frac{\omega-1}{-\varepsilon}\right)\right]$$

$$= \frac{1}{2}\left[\ln\left(\frac{\varepsilon}{\omega+1}\right) + \ln\left(\frac{1-\omega}{\varepsilon}\right)\right]$$

or, as the ε's cancel even before we let $\varepsilon \to 0$,

$$X(\omega) = \frac{1}{2}\ln\left(\frac{1-\omega}{1+\omega}\right), |\omega| < 1.$$

We can handle both cases, $|\omega| > 1$ and $|\omega| < 1$, simultaneously, by writing

$$X(\omega) = \frac{1}{2}\ln\left(\left|\frac{1-\omega}{1+\omega}\right|\right), \quad -\infty < \omega < \infty. \tag{7.10.19}$$

Now, from Rayleigh's energy theorem and (7.10.7) we can write

$$\int_{-\infty}^{\infty} g_e^2(t)dt = \frac{1}{2\pi}\int_{-\infty}^{\infty} R^2(\omega)d\omega,$$

and from Rayleigh's energy theorem and (7.10.8) we can write

$$\int_{-\infty}^{\infty} g_o^2(t)dt = \frac{1}{2\pi}\int_{-\infty}^{\infty} X^2(\omega)d\omega.$$

From (7.10.5) or (7.10.6) we see that the time integrals are clearly equal, and so then must be the frequency integrals. That is, for a causal signal,

$$\int_{-\infty}^{\infty} R^2(\omega)d\omega = \int_{-\infty}^{\infty} X^2(\omega)d\omega.$$

Since (7.10.18) says

$$\int_{-\infty}^{\infty} R^2(\omega)d\omega = \int_{-1}^{1} \frac{\pi^2}{4}d\omega = \frac{\pi^2}{2},$$

then using (7.10.19) in the $X(\omega)$ integral we have

$$\int_{-\infty}^{\infty} \frac{1}{4}\ln^2\left(\left|\frac{1-\omega}{1+\omega}\right|\right)d\omega = \frac{\pi^2}{2}$$

or, since $\int_0^\infty = \frac{1}{2}\int_{-\infty}^{\infty}$, we have the pretty result

$$\int_0^\infty \ln^2\left(\left|\frac{1-x}{1+x}\right|\right)dx = \pi^2. \tag{7.10.20}$$

This is 9.8696..., and MATLAB agrees as *integral(@(x)log(abs((x-1)./(x + 1))).* *^2,0,inf) = 9.8692....*

7.11   Laplace Transform Integrals

A favorite mathematical tool of electrical engineers (and physicists and mathematicians, too) is the *Laplace transform*,[25] closely related to the Fourier transform and defined as follows: if f(t) is a time function defined for $t \geq 0$ (for $t < 0$, f(t) is usually taken as equal to zero, but in general it is simply undefined), then its Laplace transform is

$$\mathcal{L}\{f(t)\} = F(s) = \int_0^\infty f(t)e^{-st}dt. \tag{7.11.1}$$

In sophisticated treatments of the Laplace transform, authors are careful to observe that, in general, s is a complex-valued variable (with a *positive* real part to insure that the integral converges), but one can often (as we will) formally treat it as a real parameter without running into trouble.

Math tables are available that contain the Laplace transforms of literally *hundreds* of time functions, and new transforms are continually being discovered. What I'll

---

[25] Despite the name, the genesis of the transform can be traced back to (are you surprised and, if so, why?) Euler. See M. A. B. Deakin, "Euler's Version of the Laplace Transform," *The American Mathematical Monthly*, April 1980, pp. 264–269.

show you next is how to find the transforms of some particularly interesting time functions, of increasing complexity. These calculations will present us with some interesting integrals and, in addition, the transform will give us an alternative way to derive some of the integrals we've already done. I'll start with the Laplace transforms of the two most common and useful time functions, the unit step u(t), and the decaying exponential e^{-kt} with k a positive constant:

$$\mathcal{L}\{u(t)\} = \int_0^\infty e^{-st}dt = \left(\frac{e^{-st}}{-s}\right)\Big|_0^\infty = \frac{1}{s} \tag{7.11.2}$$

where $\lim_{t\to\infty} e^{-st} = 0$ because s has a positive real part, and

$$\mathcal{L}\{e^{-kt}\} = \int_0^\infty e^{-kt}e^{-st}dt = \int_0^\infty e^{-(s+k)t}dt = \left(\frac{e^{-(s+k)t}}{-(s+k)}\right)\Big|_0^\infty = \frac{1}{s+k}. \tag{7.11.3}$$

A time function that electrical engineers simply couldn't exist without is $\cos(\omega t)$, and it has the transform

$$\mathcal{L}\{\cos(\omega t)\} = \int_0^\infty \cos(\omega t)e^{-st}dt - \int_0^\infty \frac{e^{i\omega t} + e^{-i\omega t}}{2}e^{-st}dt$$

$$= \frac{1}{2}\int_0^\infty e^{-(s-i\omega)t}dt + \frac{1}{2}\int_0^\infty e^{-(s+i\omega)t}dt = \frac{1}{2}\left\{\frac{e^{-(s-i\omega)t}}{-(s-i\omega)} + \frac{e^{-(s+i\omega)t}}{-(s+i\omega)}\right\}\Big|_0^\infty$$

$$= \frac{1}{2}\left\{\frac{1}{s-i\omega} + \frac{1}{s+i\omega}\right\} = \frac{1}{2}\left\{\frac{2s}{s^2+\omega^2}\right\}$$

and so

$$\mathcal{L}\{\cos(\omega t)\} = \frac{s}{s^2+\omega^2}. \tag{7.11.4}$$

In the same way, it is easy to show (you should do this) that

$$\mathcal{L}\{\sin(\omega t)\} = \frac{\omega}{s^2+\omega^2}. \tag{7.11.5}$$

A somewhat more complicated time function is $\frac{1}{\sqrt{t}}$, but with the aid of the gamma function we can find its transform as follows. We start by writing

$$\mathcal{L}\left\{\frac{1}{\sqrt{t}}\right\} = \int_0^\infty \frac{1}{\sqrt{t}}e^{-st}dt = \int_0^\infty t^{-\frac{1}{2}}e^{-st}dt.$$

From (4.1.1) we have

$$\Gamma(n) = \int_0^\infty e^{-x} x^{n-1} dx$$

and so, with $n = \frac{1}{2}$, we have

$$\int_0^\infty x^{-\frac{1}{2}} e^{-x} dx = \Gamma\left(\frac{1}{2}\right) = \sqrt{\pi}.$$

Now, change variable to $x = st$ (and so $dx = s\, dt$), to get

$$\int_0^\infty s^{-\frac{1}{2}} t^{-\frac{1}{2}} e^{-st} s\, dt = s^{\frac{1}{2}} \int_0^\infty t^{-\frac{1}{2}} e^{-st}\, dt = \sqrt{\pi}.$$

That is,

$$\int_0^\infty t^{-\frac{1}{2}} e^{-st}\, dt = \sqrt{\frac{\pi}{s}}$$

and so

$$\pounds\left\{\frac{1}{\sqrt{t}}\right\} = \sqrt{\frac{\pi}{s}}. \qquad (7.11.6)$$

A rather odd-looking time function that occasionally occurs in theoretical analyses is the *exponential integral*, defined as

$$E_i(t) = \int_t^\infty \frac{e^{-u}}{u} du. \qquad (7.11.7)$$

Before calculating its Laplace transform, it's convenient to get the t out of the lower integration limit by making the change of variable $p = \frac{u}{t}$ (and so $du = t\, dp$). Then,

$$E_i(t) = \int_1^\infty \frac{e^{-pt}}{pt} t\, dp = \int_1^\infty \frac{e^{-pt}}{p} dp.$$

Thus,

$$\pounds\{E_i(t)\} = \int_0^\infty \left\{\int_1^\infty \frac{e^{-pt}}{p} dp\right\} e^{-st} dt.$$

Then, using our old trick of reversing the order of integration in a double integral,

$$\pounds\{E_i(t)\} = \int_1^\infty \frac{1}{p} \left\{ \int_0^\infty e^{-pt} e^{-st} \, dt \right\} dp = \int_1^\infty \frac{1}{p} \left\{ \int_0^\infty e^{-(p+s)t} \, dt \right\} dp$$

$$= \int_1^\infty \frac{1}{p} \left\{ \frac{e^{-(p+s)t}}{-(p+s)} \right\} \Big|_0^\infty dp = \int_1^\infty \frac{dp}{p(p+s)} = \frac{1}{s} \int_1^\infty \left(\frac{1}{p} - \frac{1}{p+s} \right) dp$$

$$= \frac{1}{s} \{ \ln(p) - \ln(p+s) \} \Big|_1^\infty = \frac{\ln\left(\frac{p}{p+s}\right)}{s} \Big|_1^\infty = \frac{-\ln\left(\frac{1}{1+s}\right)}{s}$$

or, finally,

$$\pounds\{E_i(t)\} = \frac{\ln(1+s)}{s}. \tag{7.11.8}$$

As another example, a time function that varies from minus infinity to plus infinity as time varies from zero to infinity is $f(t) = \ln(t)$. This has a Laplace transform that looks a bit like that for $E_i(t)$, plus a little extra twist all its own. We start with

$$\pounds\{\ln(t)\} = \int_0^\infty \ln(t) e^{-st} dt \tag{7.11.9}$$

and then make the change of variable $p = st$ (and so $dt = \frac{dp}{s}$). Thus,

$$\pounds\{\ln(t)\} = \int_0^\infty \ln\left(\frac{p}{s}\right) e^{-p} \frac{dp}{s} = \frac{1}{s} \left[\int_0^\infty \ln(p) e^{-p} dp - \int_0^\infty \ln(s) e^{-p} dp \right]$$

$$= \frac{1}{s} \int_0^\infty \ln(p) e^{-p} dp - \frac{\ln(s)}{s} \{ -e^{-p} \} \Big|_0^\infty$$

or,

$$\pounds\{\ln(t)\} = \frac{1}{s} \int_0^\infty \ln(p) e^{-p} dp - \frac{\ln(s)}{s}. \tag{7.11.10}$$

Now, recall (5.4.3), where we showed

$$\int_0^\infty \ln(x) e^{-x} dx = -\gamma$$

where γ is Euler's constant. We see this is precisely the integral in (7.11.10), and so

$$\pounds\{\ln(t)\} = -\frac{\gamma}{s} - \frac{\ln(s)}{s}. \tag{7.11.11}$$

Notice from (7.11.11) and (7.11.2) that

$$\pounds\{\ln(t)+\gamma\} = \left[-\frac{\gamma}{s}-\frac{\ln(s)}{s}\right]+\frac{\gamma}{s} = -\frac{\ln(s)}{s},$$

and so we have the curious time function $\ln(t)+\gamma$ (a time-varying function with what an electrical engineer would call a 'dc-shift') with a Laplace transform of

$$\pounds\{\ln(t)+\gamma\} = -\frac{\ln(s)}{s}. \qquad (7.11.12)$$

One of the reasons the Laplace transform is useful is its ability to transform 'difficult' operations in the time-domain into ones that, in some sense, are 'easier' in the s-domain. There are numerous such transformations, and here are three (just as examples):

$$\pounds\left\{\int_0^t f(x)dx\right\} = \frac{F(s)}{s}, \qquad (7.11.13)$$

and

$$\pounds\{tf(t)\} = -\frac{dF(s)}{ds} \qquad (7.11.14)$$

and

$$\pounds\left\{\frac{f(t)}{t}\right\} = \int_s^\infty F(x)dx. \qquad (7.11.15)$$

These are not difficult relations to establish (see Challenge Problem 7.13) and here I'll limit myself to showing you a couple of applications of (7.11.15) to doing integrals. Writing (7.11.15) out in detail, it says

$$\int_0^\infty \frac{f(t)}{t}e^{-st}dt = \int_s^\infty F(x)dx$$

and so, setting $s = 0$, we get the interesting result

$$\int_0^\infty \frac{f(t)}{t}dt = \int_0^\infty F(x)dx \qquad (7.11.16)$$

when both integrals exist, where $F(s)$ is the Laplace transform of $f(t)$.
    For example, from (7.11.3) we have

$$\pounds\{e^{-at}-e^{-bt}\} = \frac{1}{s+a}-\frac{1}{s+b}$$

and so (7.11.16) says

$$\int_0^\infty \frac{e^{-at} - e^{-bt}}{t} dt = \int_0^\infty \left(\frac{1}{x+a} - \frac{1}{s+b} \right) dx = \{\ln(x+a) - \ln(x+b)\}\big|_0^\infty$$

$$= \ln\left(\frac{x+a}{x+b}\right)\big|_0^\infty$$

$$= -\ln\left(\frac{a}{b}\right)$$

and so, just like that, we have the integral

$$\int_0^\infty \frac{e^{-at} - e^{-bt}}{t} dt = \ln\left(\frac{b}{a}\right),$$

a result we derived earlier in (3.3.3) by other (more complicated!) means.

For another example of this sort of calculation, from (7.11.5) we know that

$$\mathcal{L}\{\sin(at)\} - \frac{a}{s^2 + a^2}$$

and so (7.11.6) says

$$\int_0^\infty \frac{\sin(at)}{t} dt = a \int_0^\infty \frac{dx}{x^2 + a^2} = a\left\{ \frac{1}{a} \tan^{-1}\left(\frac{x}{a}\right) \right\}\big|_0^\infty = \tan^{-1}\left(\frac{x}{a}\right)\big|_0^\infty.$$

If $a > 0$ then $\tan^{-1}\left(\frac{x}{a}\right)\big|_0^\infty = \frac{\pi}{2}$, while if $a < 0$ then $\tan^{-1}\left(\frac{x}{a}\right)\big|_0^\infty = -\frac{\pi}{2}$. Clearly, if $a = 0$ then the integral vanishes. Thus,

$$\int_0^\infty \frac{\sin(at)}{t} dt = \begin{cases} \frac{\pi}{2}, & \text{if } a > 0 \\ 0, & \text{if } a = 0 \\ -\frac{\pi}{2}, & \text{if } a < 0 \end{cases}$$

which is Dirichlet's integral in (3.2.1).

I'll end this section with the calculation of the transforms of two more important time functions. Our first calculation is an easy one that has a bit of an ironic twist to it, in that it shows that the *weirdest* time function we so far considered, the impulse $\delta(t)$, has the *simplest* transform. If the impulse occurs at time $t = t_0 = 0$, then f $(t) = \delta(t - t_0)$ and

$$\mathcal{L}\{\delta(t - t_0)\} = \int_0^\infty \delta(t - t_0) e^{-st} dt = e^{-st_0}$$

by the sampling property of the impulse from (7.9.14). So, if the impulse occurs at time $t = 0$, then $t_0 = 0$ and we have

$$£\{\delta(t)\} = 1. \tag{7.11.17}$$

For our last transform calculation, I'll start by defining the *error function*, a function that is well-known to mathematicians and electrical engineers[26]:

$$\text{erf}(t) = \frac{2}{\sqrt{\pi}} \int_0^t e^{-u^2} du, \tag{7.11.18}$$

where the factor in front of the integral sign is there to normalize the value of erf(∞) to 1 (obviously, erf(0) = 0).[27] What I'll do next is the reverse of what we've been doing so far, in that I'll *start* with a transform and calculate the time function that goes with that transform (and, as you won't be surprised to learn, that time function will involve the error function). Specifically, the question we'll answer next is: what time function goes with the Laplace transform $\frac{1}{s\sqrt{s+a}}$?

To answer that question, I'll use two general theorems. First,
if $£\{f(t)\} = F(s)$
then $£\{e^{-at}f(t)\} = F(s + a)$.
This is easily established, as follows:

$$£\{e^{-at}f(t)\} = \int_0^\infty e^{-at}f(t)e^{-st}dt = \int_0^\infty f(t)e^{-(s+a)t}dt = F(s + a).$$

So, for example, if g(t) has the transform

$$g(t) \leftrightarrow G(s) = \frac{1}{s\sqrt{s+a}}$$

then the time function $e^{at}g(t)$ is part of the transform pair

$$e^{at}g(t) \leftrightarrow \frac{1}{(s-a)\sqrt{s}}. \tag{7.11.19}$$

This example, as you won't be surprised to learn, wasn't picked at random. You'll notice that the transform in (7.11.19) has the form of a product of two individual factors, each of which is the transform of a known time function:

---

[26]This name was given to (7.11.18) in 1871 by the English mathematician J. W. L. Glaisher (see note 13 in Chap. 6) because of its appearance in the probabilistic theory of measurement errors. Later, electrical engineers found it invaluable in their probabilistic studies of electronic systems in the presence of noise.

[27]If this isn't immediately clear, look back at the discussion just before (3.1.8).

$$e^{at} \leftrightarrow \frac{1}{(s-a)}$$

and

$$\frac{1}{\sqrt{\pi t}} \leftrightarrow \frac{1}{\sqrt{s}}$$

which follow from (7.11.3) and (7.11.6), respectively. From this, using the second theorem I alluded to just a moment ago, we find the time function that pairs with the product of those two transforms and, since that time function is $e^{at}g(t)$, we can then calculate $g(t)$. So, what is that second theorem?

If you look back at our earlier discussion of the Fourier transform, you'll see it mentions what is called *frequency convolution*. That is, the Fourier transform of a product of two time functions is the convolution of the individual transforms, as written in (7.4.11). Then, in note 19, mention is made of a related operation called *time convolution*, in which the Fourier transform of time function formed by convolving two other *time* functions is the product of the individual transforms. Time and frequency convolution are mirror-images of each other.

The Laplace transform has these same mathematical properties; in particular, if you multiply two Laplace transforms together their product is the Laplace transform of the time function formed by convolving the individual time functions. That is,

if $\pounds\{f_1(t)\} = F_1(s)$ and $\pounds\{f_2(t)\} = F_2(s)$
then

$$\int_0^t f_1(t-u)f_2(u)du \leftrightarrow F_1(s)F_2(s). \qquad (7.11.20)$$

For example, if we set

$$f_1(t) = e^{at} \leftrightarrow F_1(s) = \frac{1}{s-a}$$

and

$$f_2(t) = \frac{1}{\sqrt{\pi t}} \leftrightarrow F_2(s) = \frac{1}{\sqrt{s}}$$

then (7.11.19) and (7.11.20) say

$$e^{at}g(t) = \int_0^t e^{a(t-u)}\frac{1}{\sqrt{\pi u}}du \leftrightarrow \frac{1}{(s-a)\sqrt{s}} \qquad (7.11.21)$$

and so our next task is the evaluation of the integral in (7.11.21).

Writing

$$\int_0^t e^{a(t-u)} \frac{1}{\sqrt{\pi u}} du = \frac{e^{at}}{\sqrt{\pi}} \int_0^t \frac{e^{-au}}{\sqrt{u}} du,$$

we then make the change of variable $u = x^2$ (and so $du = 2x\,dx$). Thus,

$$\int_0^t e^{a(t-u)} \frac{1}{\sqrt{\pi u}} du = \frac{e^{at}}{\sqrt{\pi}} \int_0^{\sqrt{t}} \frac{e^{-ax^2}}{x} 2x\,dx = \frac{2e^{at}}{\sqrt{\pi}} \int_0^{\sqrt{t}} e^{-ax^2} dx.$$

Now, let $p = x\sqrt{a}$ (and so $dx = \frac{dp}{\sqrt{a}}$). Then,

$$\int_0^t e^{a(t-u)} \frac{1}{\sqrt{\pi u}} du = \frac{2e^{at}}{\sqrt{\pi}} \int_0^{\sqrt{at}} e^{-p^2} \frac{dp}{\sqrt{a}} = \frac{2e^{at}}{\sqrt{\pi a}} \int_0^{\sqrt{at}} e^{-p^2} dp.$$

Looking back at (7.11.18), we see that the last integral at the far-right is $\frac{\sqrt{\pi}}{2} \operatorname{erf}\left(\sqrt{at}\right)$ and so

$$\int_0^t e^{a(t-u)} \frac{1}{\sqrt{\pi u}} du = \frac{e^{at}}{\sqrt{a}} \operatorname{erf}\left(\sqrt{at}\right).$$

But in (7.11.21) we said this is $e^{at}g(t)$ and so we have the Laplace transform pair

$$g(t) = \frac{1}{\sqrt{a}} \operatorname{erf}\left(\sqrt{at}\right) \leftrightarrow \frac{1}{s\sqrt{s+a}}. \tag{7.11.22}$$

Here's a pretty little application of (7.11.22), showing how it helps us evaluate the seemingly challenging integral

$$\int_0^\infty \frac{e^{-tx^2}}{1+x^2} dx = ?$$

Taking the Laplace transform of this integral (which is, of course, a function of t) we have

$$\pounds\left\{\int_0^\infty \frac{e^{-tx^2}}{1+x^2} dx\right\} = \int_0^\infty \left\{\int_0^\infty \frac{e^{-tx^2}}{1+x^2} dx\right\} e^{-st} dt = \int_0^\infty \frac{1}{1+x^2} \left\{\int_0^\infty e^{-tx^2} e^{-st} dt\right\} dx$$

$$= \int_0^\infty \frac{1}{1+x^2} \left\{\int_0^\infty e^{-t\left(s+x^2\right)} dt\right\} dx.$$

Now,

$$\int_0^\infty e^{-t(s+x^2)} dt = \frac{e^{-t(s+x^2)}}{-(s+x^2)} \Big|_0^\infty = \frac{1}{s+x^2}$$

and so

$$\pounds\left\{\int_0^\infty \frac{e^{-tx^2}}{1+x^2} dx\right\} = \int_0^\infty \frac{dx}{(1+x^2)(s+x^2)} dx = \frac{1}{s-1}\left\{\int_0^\infty \frac{dx}{(1+x^2)} - \int_0^\infty \frac{dx}{(s+x^2)}\right\}$$

$$= \frac{1}{s-1}\left\{\tan^{-1}(x) - \frac{1}{\sqrt{s}}\tan^{-1}\left(\frac{x}{\sqrt{s}}\right)\right\}\Big|_0^\infty = \frac{1}{s-1}\left\{\frac{\pi}{2} - \frac{1}{\sqrt{s}}\frac{\pi}{2}\right\}$$

or,

$$\pounds\left\{\int_0^\infty \frac{e^{-tx^2}}{1+x^2} dx\right\} = \frac{\frac{\pi}{2}}{s-1} - \frac{\frac{\pi}{2}}{(s-1)\sqrt{s}}.$$

The time function that pairs with the first term on the right is, from (7.11.3), $\frac{\pi}{2}e^t$. And the time function that pairs with the second term on the right is, from (7.11.19) and (7.11.22) with a = 1, $\frac{\pi}{2}e^t\mathrm{erf}(\sqrt{t})$. Thus,

$$\int_0^\infty \frac{e^{-tx^2}}{1+x^2} dx = \frac{\pi}{2}e^t\{1 - \mathrm{erf}(\sqrt{t})\}. \tag{7.11.23}$$

MATLAB includes a command named *erf* that numerically evaluates the error function with the same ease that MATLAB does (for example) the sine and square-root functions. For example, for t = 2 the right-hand-side of (7.11.23) is
pi/2*exp(2)*(1-erf(sqrt(2))) = 0.52810801209 ,
while the integral on the left-hand-side is.
integral(@(x)exp(−2*(x.^2))./(1 + x.^2),0,inf) = 0.52810801209. . . .
As (7.11.23) shows, now that we've added the error function to our list of 'standard' functions we have also expanded the list of integrals we can do. Here's another pretty example of that, as our last calculations of this chapter. To start, let me remind you that in (3.1.5) we derived

$$\int_0^\infty e^{-\frac{1}{2}x^2}\cos(tx)dx = \sqrt{\frac{\pi}{2}}e^{-\frac{1}{2}t^2}$$

which, with the trivial change in notation of writing α for t (a symbol I now want to avoid since we've been using it, here, to specifically denote time), becomes

$$\int_0^\infty e^{-\frac{1}{2}x^2}\cos(\alpha x)dx = \sqrt{\frac{\pi}{2}}e^{-\frac{1}{2}\alpha^2}. \tag{7.11.24}$$

The expression in (7.11.24) has just one free parameter in it (α), but with a little algebra we can derive a much more general result that has *two* independent parameters. Specifically, instead of $e^{-\frac{1}{2}x^2}$ in the integrand, we'll have the factor $e^{-\beta x^2}$, and so (7.11.24) will reduce to being the special case of a second parameter (β) equal to $\frac{1}{2}$. We'll then use the generalized version of (7.11.24) to derive a beautiful integral formula involving the error function. We start by making the change of variable $x = y\sqrt{2\beta}$ ($dx = \sqrt{2\beta}dy$) and so (7.11.24) becomes

$$\int_0^\infty e^{-\frac{2\beta y^2}{2}} \cos\left(\alpha\sqrt{2\beta}y\right)\sqrt{2\beta}dy = \sqrt{\frac{\pi}{2}}e^{-\frac{1}{2}\alpha^2}$$

or,

$$\int_0^\infty e^{-\beta y^2} \cos\left(\alpha\sqrt{2\beta}y\right)dy = \frac{1}{\sqrt{2\beta}}\sqrt{\frac{\pi}{2}}e^{-\frac{1}{2}\alpha^2}.$$

Writing x for y as the dummy variable of integration, this becomes

$$\int_0^\infty e^{-\beta x^2} \cos\left(\alpha\sqrt{2\beta}x\right)dx = \frac{1}{2}\sqrt{\frac{\pi}{\beta}}e^{-\frac{1}{2}\alpha^2}. \qquad (7.11.25)$$

If we write $a = \alpha\sqrt{2\beta}$ ($\alpha^2 = a^2/2\beta$) then (7.11.25) becomes

$$\int_0^\infty e^{-\beta x^2} \cos(ax)dx = \frac{1}{2}\sqrt{\frac{\pi}{\beta}}e^{-\frac{a^2}{4\beta}}$$

or, writing α in place of a, we have a generalization of (3.1.5) to

$$\int_0^\infty e^{-\beta x^2} \cos(\alpha x)dx = \frac{1}{2}\sqrt{\frac{\pi}{\beta}}e^{-\frac{\alpha^2}{4\beta}}. \qquad (7.11.26)$$

If we integrate (7.11.26) with respect to α, from 0 to 1, we can write

$$\int_0^1\left\{\int_0^\infty e^{-\beta x^2} \cos(\alpha x)dx\right\}d\alpha = \frac{1}{2}\sqrt{\frac{\pi}{\beta}}\int_0^1 e^{-\frac{\alpha^2}{4\beta}}d\alpha$$

or, reversing the order of integration in the double integral, that integral becomes

$$\int_0^\infty e^{-\beta x^2}\left\{\int_0^1 \cos(\alpha x)d\alpha\right\}dx = \int_0^\infty e^{-\beta x^2}\left\{\frac{\sin(\alpha x)}{x}\right\}\Big|_0^1 dx = \int_0^\infty e^{-\beta x^2}\frac{\sin(x)}{x}dx.$$

That is,

$$\int_0^\infty e^{-\beta x^2} \frac{\sin (x)}{x} dx = \frac{1}{2} \sqrt{\frac{\pi}{\beta}} \int_0^1 e^{-\frac{\alpha^2}{4\beta}} d\alpha. \tag{7.11.27}$$

In the integral on the right in (7.11.27), let $u = \frac{\alpha}{2\sqrt{\beta}}$ (and so $d\alpha = 2\sqrt{\beta} du$). Then,

$$\int_0^\infty e^{-\beta x^2} \frac{\sin (x)}{x} dx = \frac{1}{2} \sqrt{\frac{\pi}{\beta}} \int_0^{1/2\sqrt{\beta}} e^{-u^2} 2\sqrt{\beta} du = \sqrt{\pi} \int_0^{1/2\sqrt{\beta}} e^{-u^2} du$$

$$= \sqrt{\pi} \left\{ \frac{\sqrt{\pi}}{2} \mathrm{erf} \left(\frac{1}{2\sqrt{\beta}} \right) \right\}$$

or, finally, we arrive at the very pretty

$$\int_0^\infty e^{-\beta x^2} \frac{\sin (x)}{x} dx = \frac{\pi}{2} \mathrm{erf} \left(\frac{1}{2\sqrt{\beta}} \right). \tag{7.11.28}$$

Yes, you almost surely agree, (7.11.28) *is* pretty. But, you also correctly ask, is it 'right'? A quick numerical check will greatly increase our confidence in it.

For $\beta = 1$, for example, we have *(pi/2)*erf(1/2) = 0.8175992961. . .* on the right-hand-side of (7.11.28), while a numerical evaluation of the integral on the left-hand-side gives, in excellent agreement, *integral(@(x)exp(−(x.^2)).*sin(x)./x,0, inf) = 0.8175992961. . . .*

7.12   Challenge Problems

(C7.1) What are the values of

$$\int_0^\infty \left\{ \frac{\sin (x)}{x} \right\}^n dx$$

for n = 5, 6, and 7?

(C7.2) Use Euler's identity to show that if

$$F(x) = \int_x^\infty \int_x^\infty \sin (t^2 - u^2) dt\, du$$

then F(x) is *identically zero* (that is, F(x) = 0 for all x). Hint: Look back at Sect. 6.4, to the discussion there of the Hardy-Schuster optical integral, where we used the trig formula for cos(a − b) to establish that if C(x) and S(x) are the Fresnel cosine and sine integrals, respectively, then $C^2(x) + S^2(x) = \int_x^\infty \int_x^\infty \cos (t^2 - u^2) dt\, du$. Start now with $\int_x^\infty e^{it^2}\,dt = C(x) + iS(x)$ and see where that takes you. (At some point, think about taking a conjugate). This is (I think) not an obvious result, and just to help give you confidence that it's correct, the following table shows what MATLAB's Symbolic Math Toolbox numerically calculated for the value of F (x) for various arbitrarily selected values of x, using the code.

syms t u

int(int(sin(t^2-u^2),t,x,inf),u,x,inf)

The numerical value of x was substituted into the second line *before* each execution.

x	F(x)
−7	0
−2	0
0	0
1	0
5	0
29	0

(C7.3) In a recent physics paper[28] integrals of the form

$$\int_{-\infty}^\infty \frac{dx}{\sqrt{1 - ix^3}}$$

occur. The authors state, without proof, that this integral exists. Show this claim is, in fact, true. Hint: use the theorem from calculus that says

$$\left| \int_{-\infty}^\infty f(x)\,dx \right| \le \int_{-\infty}^\infty |f(x)|\,dx$$

which, using the area interpretation of the Riemann integral, should be obvious to you for the case where f(x) is real. Using contour integrals in the complex plane (see Chap. 8), this theorem can be extended to the case where f(x) is complex.

(C7.4) Recall the integral we worked-out in the first chapter (Sect. 1.8), $\int_1^\infty \frac{\{x\}-\frac{1}{2}}{x}\,dx = -1 + \ln\left(\sqrt{2\pi}\right)$, as well as the integrals of the second and third challenge problems of Chap. 5: $\int_1^\infty \frac{\{x\}}{x^2}\,dx = 1 - \gamma$ where $\{x\}$ is the fractional part of x and γ is Euler's constant, and $\zeta(3) = \sum_{k=1}^\infty \frac{1}{k^3} = \frac{3}{2} - 3\int_1^\infty \frac{\{x\}}{x^4}\,dx$. See if you can apply the same trick we used there to show that

---

[28] Carl Bender, et al., "Observation of *PT* phase transitions in a simple mechanical system," *American Journal of Physics*, March 2013, pp. 173–179.

$$\int_1^\infty \frac{\{x\}}{x^3}\,dx = 1 - \frac{\pi^2}{12},$$

now that we've formally calculated $\zeta(2) = \frac{\pi^2}{6}$ (in Sect. 7.4). MATLAB may help give you confidence this is correct, as $1 - \frac{\pi^2}{12} = 0.177532\ldots$ while.
  $integral(@(x)(x\text{-}floor(x))./x.\wedge 3,1,inf) = 0.177532\ldots$

(C7.5) Derive a more general form of (7.6.1) by differentiating

$$I(a) = \int_0^\infty \frac{\sin^2(ax)}{x^2}\,dx$$

with respect to a. That is, show that $I(a) = \frac{\pi}{2}|a|$, and so (7.6.1) is the special case of $a = 1$.

(C7.6) Look back to Sect. 4.3, to the results in (4.3.10) and (4.3.11), and show how the Fresnel integrals immediately follow from them.

(C7.7) Apply Rayleigh's theorem to the time signal $f(t) = \begin{cases} e^{-at}, & 0 \le t \le m \\ 0, & \text{otherwise} \end{cases}$.

where a and m are both positive constants. Hint: you should find that you have re-derived (3.1.7) in a way far different from the method used in Chap. 3.

(C7.8) Calculate the Fourier transforms of the following time signals:

(a) $\frac{1}{t^2+1}$;

(b) $\frac{1}{t^2+1}$;

(c) $\frac{1}{2}\delta(t) + i\frac{1}{2\pi t}$;

(d) $E_i(t) = \begin{cases} \int_t^\infty \frac{e^{-u}}{u}\,du, & t \ge 0 \\ 0, & t < 0 \end{cases}$.

Hint: For (a) you may find (3.1.7) helpful, for (b) don't forget Feynman's favorite trick (of differentiating integrals), and for (c) keep (7.9.3) and (7.9.14) in mind. (The Fourier transform of the complex time signal in (c) is at the heart of SSB radio). And finally, for (d), make the change of variable $x - \frac{u}{t}$, write the Fourier transform integral (which will of course be a double integral), and then reverse the order of integration.

(C7.9) Looking at the first Hilbert transform integral in (7.10.11), we see that we can write $\pi R(\omega) = \int_{-\infty}^\infty \frac{X(u)}{\omega-u}\,du = \int_{-\infty}^0 \frac{X(u)}{\omega-u}\,du + \int_0^\infty \frac{X(u)}{\omega-u}\,du$. If $x(t)$ is real then $R(-\omega) = R(\omega)$ and $X(-\omega) = -X(\omega)$, and so if we make the change of variable $s = -u$ in the first integral on the right we have $\pi R(\omega) = \int_0^\infty \frac{X(-s)}{\omega+s}(-ds) +$
$\int_0^\infty \frac{X(u)}{\omega-u}\,du = -\int_0^\infty \frac{X(s)}{\omega+s}\,ds + \int_0^\infty \frac{X(u)}{\omega-u}\,du \quad = \int_0^\infty X(u)\left[\frac{1}{\omega-u} - \frac{1}{\omega+u}\right]du = \int_0^\infty X(u)\frac{2u}{\omega^2-u^2}\,du.$

That is, an alternative form for the first Hilbert transform integral in (7.10.11) is
$R(\omega) = \frac{2}{\pi}\int_0^\infty \frac{uX(u)}{\omega^2-u^2}du$. Use this same approach to find an alternative form for the
second Hilbert transform integral in (7.10.11), one that gives $X(\omega)$ as an integral of
$R(\omega)$ for the case of $x(t)$ real.

(C7.10) Suppose $x(t)$, $y(t)$, and $h(t)$ are time signals such that the following three
conditions hold:

(a) $x(t)$ has finite energy;
(b) $y(t) = x(t) * h(t) = \int_{-\infty}^\infty x(u)h(t-u)du$;
(c) $\int_{-\infty}^\infty |h(t)|dt < \infty$.

Show that $y(t)$ has finite energy. Hint: Use Fourier transforms, Rayleigh's energy
theorem, and a look back at the hint for (C7.3) may also help.

(C7.11) Calculate the Hilbert transforms of $\cos(\omega_0 t)$ and $\sin(\omega_0 t)$. Hint: Use Euler's
identity in the defining integral (7.10.12) for the Hilbert transform of a time signal,
and you may also find that recalling (7.9.3) is helpful.

(C7.12) The integral formulas (7.5.6) and (7.5.7) were derived in 1781 by Euler,
who then used them to evaluate the Fresnel integrals that we did in (7.2.1) and
(7.2.2). What he did was to reduce (7.5.6) and (7.5.7) to the special cases of (7.5.2),
and then make the obvious change of variable. See if you can follow in Euler's
footsteps and derive the Fresnel integrals from (7.5.2).

(C7.13) Establish the relations in (7.11.13), (7.11.14), and (7.11.15). Hint: Do
(7.11.13) by-parts, do (7.11.14) by applying Feynman's favorite trick to the definition of the Laplace transform, and do (7.11.15) by *starting* with $\int_s^\infty F(x)dx$ and then
reverse the order of integration in the resulting double integral to arrive at
$\int_0^\infty \frac{f(t)}{t}e^{-st}dt$.

(C7.14) A study of the electrical behavior of a very long communication cable
quickly leads to a consideration of the curious time function $f(t) = \frac{e^{-k^2/4t}}{\sqrt{\pi t^3}}, t \geq 0$,
where k is an arbitrary positive constant.[29] Show that $F(s) = \frac{2}{k}e^{-k\sqrt{s}}$. Hint: You'll
find that remembering (3.7.1) will be helpful.

(C7.15) In (3.1.7) we derived, using a fairly lengthy procedure, the integral formula
$\int_0^\infty \frac{\cos(ax)}{x^2+b^2}dx = \frac{\pi}{2b}e^{-ab}$. If we make the trivial notational change of writing t for a
(to make it explicit that the integral is a function of time), then (3.1.7) says
$\int_0^\infty \frac{\cos(tx)}{x^2+b^2}dx = \frac{\pi}{2b}e^{-bt}$. Use the Laplace transform to derive this result, and notice

---

[29]See, for example, my *Transients for Electrical Engineers*, Springer 2018, pp. 141–145, where the
"very long communication cable" is the famous mid-nineteenth century Trans-Atlantic electric
telegraph cable. (The mathematical physics *and* the history of the cable are presented at length in
my *Hot Molecules, Cold Electrons*, Princeton 2020).

how much shorter is the transform approach (indeed, *shorter by a lot*). Hint: Don't forget (7.11.4).

(C7.16) Recall the autocorrelation function of f(t), written as $R_f(\tau) = \int_{-\infty}^{\infty} f(t)f(t-\tau)dt$ in (1.8.6). Suppose we write R_f as a function of t (not τ), by simply swapping t and τ. That is, consider $R_f(t) = \int_{-\infty}^{\infty} f(\tau)f(\tau-t)d\tau$. Show that, to within a factor of 2π, the revamped autocorrelation function and the energy spectral density form a Fourier transform pair. This result is called the *Wiener-Khinchin theorem*, after the American mathematician Norbert Wiener (who we encountered back in Chap. 1 when we discussed the Wiener stochastic process in connection with Brownian motion) who discovered it in 1930, and the Russian mathematician Aleksandr Khinchin (1894–1959) who independently discovered it in 1934. Despite being named after Wiener and Khinchin, however, both men had been anticipated by many years by a physicist. In 1914 (the year before he dazzled the world of theoretical physics with the general theory of relativity, a world still grabbling with his special theory of 1905), Einstein delivered a paper at a Swiss Physical Society meeting in Basel. That paper, which contains the essence of the Wiener-Khinchin theorem, was then overlooked by mathematicians for the next 65 years, until the 1979 Einstein centenary. Einstein's interest in this topic (to which he never returned), one that is obviously *very* different from gravitational physics, was apparently motivated by necessity. Unable to find steady employment in the years just before the start of the First World War, he earned extra money by performing calculations for a researcher in sunspot activity (a highly irregular, stochastic process). You can read more about this little-known episode in mathematical physics in a 1985 paper by A. M. Yaglom ("Einstein's 1914 paper on the Theory of Irregularly Fluctuating Series of Observations"), originally published in Russian and reprinted (in English) in the *IEEE ASSP Magazine*, October 1987, pp. 7–11. Such was Einstein's youthful genius that, even when working as what amounted to being a 'hired-hand,' he produced a world-class result.

(C7.17) Show that $\int_0^{\infty} \cos\left(x^2 - \frac{1}{x^2}\right)dx = \int_0^{\infty} \sin\left(x^2 - \frac{1}{x^2}\right)dx = \frac{1}{2e^2}\sqrt{\frac{\pi}{2}}$. Hint 1: Follow Euler's fearless dare-doing and simply set parameters in an integral formula derived for real-valued parameters equal to imaginary values. Specifically, set $a = i$ and $b = -i$ in (3.7.4). Hint 2: At some point in your analysis you'll encounter the issue of calculating the square-root of an imaginary quantity, with the question of which of the two possible roots do you use? This question can be answered if you know the sign of the integral (is it positive or is it negative?). Doing the 'obvious' MATLAB integration doesn't work, but writing $\int_0^{\infty} = \int_0^1 + \int_1^{\infty}$ and then changing variable in the middle integral to $y = \frac{1}{x}$ to convert it to $\int_1^{\infty}$, converts the problem to the computation of $\int_1^{\infty} \frac{(1+x^2)\cos\left(x^2 - \frac{1}{x^2}\right)}{x^2}dx$. (You should confirm this). The MATLAB code *integral(@(x)(1 + x.^2).*cos(x.^2–1./(x.^2))./x.^2,1100)*, using 100 as a stand-in for the upper limit of infinity, gives an approximate value of 0.083 for the integral (the 'exact' value is *0.5*sqrt(pi/2)/ (2.718281828^2) = 0.0848...*). This 'works' because when x has reached 100 the integrand is oscillating very rapidly (as it is for 'small' values of x), with consecutive half-cycles mutually cancelling and so contributing essentially zero area to the integral. That is, the value of the integral is determined by the oscillations for a *finite* interval after the start of the total, infinite integration interval.

(C7.18) One can't do the previous problem without immediately wondering what $\int_0^\infty \cos\left(x^2 + \frac{1}{x^2}\right)dx$ and $\int_0^\infty \sin\left(x^2 + \frac{1}{x^2}\right)dx$ are. You can answer that by setting $a = b = i$ in (3.7.4), but the same issue of which root of an imaginary quantity do you use will again come up. To help you decide, using MATLAB as in the previous problem shows that $\int_0^\infty \cos\left(x^2 + \frac{1}{x^2}\right)dx \approx -0.83$ and $\int_0^\infty \sin\left(x^2 + \frac{1}{x^2}\right)dx \approx 0.31$. (Notice that $\int_0^\infty \cos\left(x^2 + \frac{1}{x^2}\right)dx \neq \int_0^\infty \sin\left(x^2 + \frac{1}{x^2}\right)dx$). Use your exact expressions for these two integrals to compute *accurate* values for them: you should get $\int_0^\infty \cos\left(x^2 + \frac{1}{x^2}\right)dx = -0.83059901675\ldots$ and $\int_0^\infty \sin\left(x^2 + \frac{1}{x^2}\right)dx = 0.30903630331\ldots$.

(C7.19) Here's one final example of fearlessly plugging imaginary quantities into equations derived for real ones. For $-1 < s < 0$, show that $\int_0^\infty x^s \cos(xt)dx = -\frac{\Gamma(s+1)}{t^{s+1}}\sin\left(\frac{\pi s}{2}\right)$. We've actually already derived this result, in (4.3.9), with the following changes in notation: (1) write $p = -s$ (and so $0 < p < 1$, as stated in (4.3.9), and (2) write $b = t$. Then, recall the reflection formula in (4.2.16) for the gamma function, and the double-angle formulas from trigonometry. (I'll let *you* confirm all this: it's not very hard to do). To do the integral the fearless way, recall the result from C4.11, where you showed (you *did*, right?) $\int_0^\infty x^{s-1}e^{-px}dx = \frac{\Gamma(s)}{p^s}$. Then, use Euler's identity to write $\cos(xt)$ in terms of complex exponentials, and set $p = \pm it$.

Chapter 8
Contour Integration

8.1 Prelude

In this, the penultimate chapter of the book, I'll give you a really fast, stripped-down, 'crash-course' presentation of the very beginnings of complex function theory, and the application of that theory to one of the gems of mathematics: *contour integration* and its use in doing definite integrals. As an historian of mathematics recently wrote, "A curious feature of mathematical analysis in the years around 1800 was the use of complex variables to evaluate real definite integrals. The practice had begun with Euler [recall the derivation of (7.5.6) and (7.5.7)] ... In his *Mémoire* on this topic that he presented in 1814 Cauchy commented that many of the integrals had been evaluated for the first time 'by means of a kind of induction' based on 'the passage from the real to the imaginary' and that no less a figure than Laplace had remarked that the method 'however carefully employed, leaves something to be desired in the proofs of the results.' Cauchy accordingly set himself the task of finding a 'direct and rigorous analysis' of this dubious passage."[1]

As we start this chapter on what came about from Cauchy's labors, I'll assume only that you are familiar with complex *numbers* and their manipulation. I've really already done that, of course, in Chap. 7, and so I think I am on safe ground here with that assumption. The first several sections will lay the theoretical groundwork and then, quite suddenly, you'll see how they all come together to give us the beautiful and powerful technique of contour integration. None of these preliminary sections is very difficult, but each is absolutely essential for understanding. Don't skip them!

In keeping with the spirit of this book, the presentation leans heavily on intuitive plausibility arguments and, while I don't think I do anything wildly outrageous, there will admittedly be occasions where professional mathematicians might feel tiny stabs of pain. (Mathematicians are a pretty hardy bunch, though, and they will

---

[1]Jeremy Gray, *The Real and the Complex: a history of analysis in the nineteenth century*, Springer 2015, pp. 59–60.

© Springer Nature Switzerland AG 2020
P. J. Nahin, *Inside Interesting Integrals*, Undergraduate Lecture Notes in Physics,
https://doi.org/10.1007/978-3-030-43788-6_8

survive!) This may be the appropriate place to quote the mathematician John Stalker (of Trinity College, Dublin), who once wrote "In mathematics, as in life, virtue is not always rewarded, *nor vice always punished* [my emphasis]."[2] As always, I'll feel vindicated when, after doing a series of manipulations, MATLAB's numerical calculations agree with whatever theoretical result we've just derived.

8.2   Line Integrals

Imagine two points, A and B, in the two-dimensional x,y plane. Further, imagine that A and B are the two end-points of the curve C in the plane, as shown in Fig. 8.2.1. A is the *starting* end-point and B is the *terminating* end-point. Now, suppose that we divide C into n parts (or arcs), with the k-th arc having length Δs_k (where k runs from 1 to n). Each of these arcs has a projection on the x-axis, where we'll write Δx_k as the x-axis projection of Δs_k. In the same way, we'll write Δy_k as the y-axis projection of Δs_k. Again, see Fig. 8.2.1. Finally, we'll assume, as n $\rightarrow \infty$, that $\Delta s_k \rightarrow 0$, that $\Delta x_k \rightarrow 0$, and that $\Delta y_k \rightarrow 0$, for each and every k (that is, the points along C that divide C into n arcs are distributed, loosely speaking, 'uniformly' along C).

Continuing, suppose that we have some function h(x, y) that is defined at every point along C. If we form the two sums $\sum_{k=1}^{n} h(x_k, y_k)\Delta x_k$ and $\sum_{k=1}^{n} h(x_k, y_k)\Delta y_k$

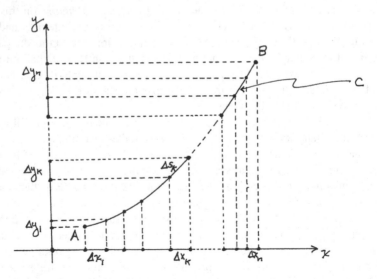

Fig. 8.2.1   A curve in the plane, and its projections on the x and y axes

[2]In his book *Complex Analysis: Fundamentals of the Classical Theory of Functions*, Birkhäuser 1998, p. 120.

where (x_k, y_k) is an arbitrary point in the arc Δs_k, then we'll write the limiting values of these sums as[3]

$$\lim_{n \to \infty} \sum_{k=1}^{n} h(x_k, y_k) \Delta x_k = \int_C h(x, y) dx = I_x \qquad (8.2.1)$$

and

$$\lim_{n \to \infty} \sum_{k=1}^{n} h(x_k, y_k) \Delta y_k = \int_C h(x, y) dy = I_y. \qquad (8.2.2)$$

The C's at the bottom of the integral signs in (8.2.1) and (8.2.2) are there to indicate that we are integrating *from A to B along* C. We'll call the two limits in (8.2.1) and (8.2.2) *line integrals* (sometimes the term *path integral* is used, commonly by physicists). If $A = B$ (that is, if C is a *closed loop*[4]) then the result is called a *contour integral*. When we encounter contour integrals it is understood that C never crosses itself (such a C is said to be *simple*). Further, it is understood that a contour integral is done in the counter-clockwise sense; to reverse the direction of integration will reverse the algebraic sign of the integral.

The value of a line integral depends, in general, on the coordinates of A and B, the function $h(x, y)$, *and on the specific path* C that connects A and B. For example, suppose that $A = (0, 0)$, $B = (1, 1)$, and that $h(x, y) = xy$. To start, let's suppose that $C = C_1$ is the broken path shown in Fig. 8.2.2. The first part is along the x-axis from $x = 0$ to $x = 1$, and then the second part is straight-up from $x = 1$ $(y = 0)$ to $x = 1$ $(y = 1)$. So, for this path we have $y = 0$ along the x-axis (and so $h(x, y) = 0$), and $x = 1$ on the vertical portion of C_1 (and so $h(x, y) = y$). Thus, our two line integrals *on this path* are

$$I_x = \int_{C_1} h(x, y) dx = \int_0^1 0 \, dx + \int_1^1 y \, dx = 0 + 0 = 0$$

and

---

[3]In keeping with the casual approach I'm taking in this book, I'll just *assume* that these two limits exist and then we'll see where that assumption takes us. Eventually we'll arrive at a new way to do definite integrals (contour integration) and *then* we'll check our assumption by seeing if our theoretical calculations agree with MATLAB's direct numerical evaluations.

[4]There are, of course, *two* distinct ways we can have $A = B$. The trivial way is if C simply has zero length, which immediately says $I_x = I_y = 0$. The non-trivial way is if C goes from A out into the plane, wanders around for a while, and then returns to A (which we re-label as B). It is this second way that gives us a closed loop.

Fig. 8.2.2 Two different
line integral paths

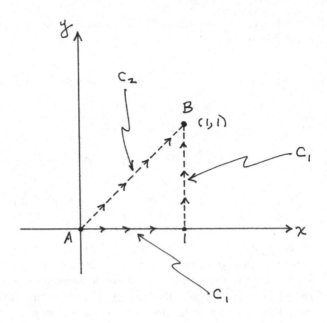

$$I_y = \int_{C_1} h(x, y) dy = \int_0^0 0 \, dy + \int_0^1 y \, dy = \left(\frac{1}{2}y^2\right)\Big|_0^{1^.} = \frac{1}{2}.$$

Along the path C_2, on the other hand, we have $y = x$ from A to B, and so $h(x, y) = x^2$ (or, equivalently, y^2). So, on *this* path the line integrals are

$$I_x = \int_{C_2} h(x, y) dx = \int_0^1 x^2 \, dx = \left(\frac{1}{3}x^3\right)\Big|_0^{1^.} = \frac{1}{3}$$

and

$$I_y = \int_{C_2} h(x, y) dy = \int_0^1 y^2 \, dy = \left(\frac{1}{3}y^3\right)\Big|_0^{1^.} = \frac{1}{3}.$$

Clearly, the values of the I_x, I_y line integrals are path-dependent and, for a given path, the I_x, I_y line integrals may or may not be equal. We can combine the I_x and I_y line integrals to write *the* line integral along C as $I_C = I_x + iI_y$, and so $I_{C_1} = i\frac{1}{2}$ while $I_{C_2} = \frac{1}{3} + i\frac{1}{3}$.

Looking back at the previous section, notice that in Fig. 8.2.2 we could write the unbroken line segment AB as $z = x + iy$ or, since $y = x$, $z = x + ix = x(1 + i)$ and so $dz = (1 + i)dx$. Then, as $h(x, y) = h(x, x) = x^2$, we have

$$I_{C_2} = \int_0^1 x^2(1+i)dx = (1+i)\left(\frac{1}{3}x^3\right)\Big|_0^1 = \frac{1}{3} + i\frac{1}{3}$$

which is just as we calculated before.

For now, we'll put aside these considerations and turn to expanding this book's discussion from functions of a real variable to functions of a complex variable. Soon, however, you'll see how this expanded view of functions will 'circle back'—how appropriate!—to closed contour line integrals, and what we've done in this section will prove to be *most* useful.

8.3 Functions of a Complex Variable

I will write the complex variable z as

$$z = x + iy \tag{8.3.1}$$

where x and y are each real with each varying over the doubly-infinite interval $-\infty$ to $+\infty$, and $i = \sqrt{-1}$. Geometrically, we'll interpret z as a point in an infinite, two-dimensional plane (called the *complex plane*) with x measured along a horizontal axis and y measured along a vertical axis. And we'll write a *complex function* of the complex variable z as

$$f(z) = f(x + iy) = u(x, y) + iv(x, y) \tag{8.3.2}$$

where u and v are each real-valued functions of the two real-valued variables x and y. For example, if

$$f(z) = z^2 = (x + iy)^2 = x^2 - y^2 + i2xy$$

then, in this case, $u = x^2 - y^2$ and $y = 2xy$. In x, y notation, we are said to be working in rectangular (or *Cartesian*) coordinates.

It is often convenient to work in polar coordinates, which means we write the complex variable z as

$$z = re^{i\theta} \tag{8.3.3}$$

where r and θ are each real: r is the radial distance from the origin of the coordinate system of the complex plane to the point z (and so $0 \le r < \infty$), and θ is the angle of the *radius vector* (of length r) measured counter-clockwise from the positive horizontal x-axis to the radius vector (and so we *generally* take $0 \le \theta < 2\pi$, although $-\pi \le \theta < \pi$ is also commonly assumed). You'll recall that we did this in deriving (7.5.6) and (7.5.7). Note, carefully, that θ is not *uniquely* determined, as we can add (or subtract) any multiple of 2π from θ and still be talking about the same physical point in the complex plane.

From Euler's identity we have from (8.3.3) that

$$z = r\{\cos(\theta) + i\,\sin(\theta)\}. \tag{8.3.4}$$

For example, if $f(z) = z^2$ then

$$f(z) = \left(re^{i\theta}\right)^2 = r^2\{\cos(\theta) + i\,\sin(\theta)\}^2$$

or, expanding both sides of the last equality,

$$r^2 e^{i2\theta} = r^2\{\cos(2\theta) + i\,\sin(2\theta)\} = r^2\{\cos^2(\theta) - \sin^2(\theta) + i2\cos(\theta)\sin(\theta)\}.$$

Since the real and imaginary parts of the expressions in the last equality must be separately equal, we conclude that $\cos(2\theta) = \cos^2(\theta) - \sin^2(\theta)$ as well as $\sin(2\theta) = 2\cos(\theta)\sin(\theta)$. These two formulas are, of course, the well-known *double-angle* formulas from trigonometry, and so already we have a nice illustration of the powerful ability of complex functions to do useful work for us.[5]

I'll end this section with two more spectacular demonstrations of that power. First, the calculation of

$$\int_0^{2\pi} e^{\cos(\theta)}\,d\theta,$$

an integral I am absolutely sure you have never seen done by the 'routine' integration techniques of freshman calculus. We'll do it here (using the polar form of z) with a contour integration in the complex plane. With $z = e^{i\theta}$, which puts z on the unit circle ($r = 1$) centered on the origin, we can write

$$\cos(\theta) = \frac{1}{2}\left(z + \frac{1}{z}\right)$$

because $\frac{1}{z} = e^{-i\theta}$ and Euler's identity says this is $\cos(\theta) - i\,\sin(\theta)$. Now, consider the complex function

$$f(z) = \frac{e^{\frac{1}{2}\left(z + \frac{1}{z}\right)}}{z}$$

which we'll integrate counter-clockwise *once* around the unit circle. That is, we'll compute

---

[5]If, instead, we had started with $f(z) = z^3 = (re^{i\theta})^3 = r^3 e^{i3\theta} = r^3\{\cos(3\theta) + i\,\sin(3\theta)\} = r^3\{\cos(\theta) + i\,\sin(\theta)\}^3$, then we could have just as easily have derived the *triple*-angle formulas that are not so easy to get by other means (just take a look at any high school trigonometry text).

$$\oint_C \frac{e^{\frac{1}{2}\left(z+\frac{1}{z}\right)}}{z}\, dz$$

where C is the circle $z = e^{i\theta}$. (The circle with the CCW arrowhead on the integral sign is there simply to emphasize that we are working with a *closed* line integral.)

The reason for the z in the denominator of the integrand is because $dz = ie^{i\theta}d\theta$ and we need an $e^{i\theta}$ in the denominator to cancel the $e^{i\theta}$ in dz. So,

$$\oint_C \frac{e^{\frac{1}{2}\left(z+\frac{1}{z}\right)}}{z}\, dz = \int_0^{2\pi} \frac{e^{\cos(\theta)}}{e^{i\theta}} ie^{i\theta}d\theta = i\int_0^{2\pi} e^{\cos(\theta)}d\theta. \qquad (8.3.5)$$

That is, the contour integral at the left in (8.3.5) is the integral we wish to calculate (multiplied by i). To directly calculate the contour integral, we start by expanding the exponential in the left-most integral in a power series. That is,

$$\oint_C \frac{e^{\frac{1}{2}\left(z+\frac{1}{z}\right)}}{z}\, dz = \oint_C \frac{1}{z} \sum_{n=0}^{\infty} \frac{\left\{\frac{1}{2}\left(z+\frac{1}{z}\right)\right\}^n}{n!}\, dz = \oint_C \frac{1}{z} \sum_{n=0}^{\infty} \frac{1}{2^n n!} \left(z+\frac{1}{z}\right)^n dz.$$

Using the binomial theorem to write

$$\left(z+\frac{1}{z}\right)^n = \sum_{k=0}^{n} \binom{n}{k} z^k \left(\frac{1}{z}\right)^{n-k} = \sum_{k=0}^{n} \binom{n}{k} z^k \frac{1}{z^{n-k}} = \sum_{k=0}^{n} \binom{n}{k} \frac{z^{2k}}{z^n},$$

we have

$$\oint_C \frac{e^{\frac{1}{2}\left(z+\frac{1}{z}\right)}}{z}\, dz = \oint_C \frac{1}{z} \sum_{n=0}^{\infty} \frac{1}{2^n n!} \left\{\sum_{k=0}^{n} \binom{n}{k} \frac{z^{2k}}{z^n}\right\} dz$$

$$= \oint_C \sum_{n=0}^{\infty} \frac{1}{2^n n!} \left\{\sum_{k=0}^{n} \binom{n}{k} z^{2k-n-1}\right\} dz$$

$$= \sum_{n=0}^{\infty} \frac{1}{2^n n!} \left\{\sum_{k=0}^{n} \binom{n}{k} \oint_C z^{2k-n-1} dz\right\}.$$

Now, concentrate on that last integral, where we'll replace z with $e^{i\theta}$ and dz with $ie^{i\theta}d\theta$:

$$\oint_C z^{2k-n-1}\, dz = \int_0^{2\pi} \left(e^{i\theta}\right)^{2k-n-1} ie^{i\theta}\, d\theta = i\int_0^{2\pi} e^{i(2k-n)\theta}d\theta = \begin{cases} 2\pi i \text{ if } 2k-n = 0 \\ \\ 0 \quad \text{if } 2k-n \neq 0 \end{cases}.$$

This is remarkable! Every one of these integrals on the right vanishes as n and k run through their values *except* for those cases where $k = \frac{n}{2}$. This has a profound implication, as then k can be an integer (which of course it is) only if n is *even*. For all odd values of n the integrals vanish, and in the cases of n even they vanish, too, if $k \neq \frac{n}{2}$. We can include all the integrals that don't vanish with the simple trick of writing n = 2 m, where m = 0, 1, 2, 3, ..., and so we have

$$\oint_C \frac{e^{\frac{1}{2}\left(z+\frac{1}{z}\right)}}{z} \, dz = \sum_{m=0}^{\infty} \frac{1}{2^{2m}(2m!)} \binom{2m}{m} 2\pi i = 2\pi i \sum_{m=0}^{\infty} \frac{1}{2^{2m}(2m!)} \frac{(2m!)}{m!m!}$$

$$= 2\pi i \sum_{m=0}^{\infty} \frac{1}{2^{2m}(m!)^2}.$$

From (8.3.5) we can now write

$$2\pi i \sum_{m=0}^{\infty} \frac{1}{2^{2m}(m!)^2} = i \int_0^{2\pi} e^{\cos(\theta)} \, d\theta$$

or, cancelling the *i*'s, we have our answer:

$$\int_0^{2\pi} e^{\cos(\theta)} \, d\theta = 2\pi \sum_{m=0}^{\infty} \frac{1}{2^{2m}(m!)^2}. \tag{8.3.6}$$

The terms in the series on the right decrease very rapidly and so the series quickly converges. Using just the first four terms the sum is $2\pi\left(1 + \frac{1}{4} + \frac{1}{64} + \frac{1}{2,304}\right) = 7.95488$ and MATLAB agrees, as *integral(@(x)exp(cos(x)),0,2*pi) = 7.95492....*

For the final demonstration in this section (this one from physics) of the amazing utility of complex functions, imagine a point mass m moving in a plane along the path given by (8.3.3),

$$z(t) = r(t)e^{i\theta(t)} \tag{8.3.7}$$

where now z, r and θ are specifically indicated to be functions of time (t). (The meaning of each of these variables is as given at the beginning of this section.) The motion of m is due entirely to a force acting along the line connecting the mass to the source of the force: the classic example of this situation is the Earth (the 'point' mass) moving under the influence of the gravitational field of the Sun (which we'll take as being at the origin of the x-y coordinate system). The *attractive* force on the Earth is, of course, always directed *radially inward* towards the Sun.

If we write the magnitude of the force on m as f(r, θ), Newton's famous second law of motion ('force is mass times acceleration') says

$$f(r, \theta)e^{i\theta} = m \frac{d^2z}{dt^2}. \tag{8.3.8}$$

From (8.3.7) we have

$$\frac{dz}{dt} = \frac{dr}{dt}e^{i\theta} + ir\frac{d\theta}{dt}e^{i\theta}$$

and so

$$\frac{d^2z}{dt^2} = \frac{d^2r}{dt^2}e^{i\theta} + \frac{dr}{dt}i\frac{d\theta}{dt}e^{i\theta} + i\left[\frac{dr}{dt}\frac{d\theta}{dt}e^{i\theta} + r\frac{d^2\theta}{dt^2}e^{i\theta} + r\frac{d\theta}{dt}i\frac{d\theta}{dt}e^{i\theta}\right]$$

or,

$$\frac{d^2z}{dt^2} = \left[\frac{d^2r}{dt^2} - r\left(\frac{d\theta}{dt}\right)^2\right]e^{i\theta} + i\left[2\frac{dr}{dt}\frac{d\theta}{dt} + r\frac{d^2\theta}{dt^2}\right]e^{i\theta}. \tag{8.3.9}$$

Using (8.3.9) in (8.3.8) and cancelling all the $e^{i\theta}$ (which are *never* zero), we arrive at

$$f(r,\theta) = m\left[\frac{d^2r}{dt^2} - r\left(\frac{d\theta}{dt}\right)^2\right] + i\,m\left[2\frac{dr}{dt}\frac{d\theta}{dt} + r\frac{d^2\theta}{dt^2}\right].$$

Equating real and imaginary parts of this last expression gives us the famous differential equations of motion in a radial force field:

$$f(r,\theta) = m\left[\frac{d^2r}{dt^2} - r\left(\frac{d\theta}{dt}\right)^2\right] \tag{8.3.10}$$

and

$$2\frac{dr}{dt}\frac{d\theta}{dt} + r\frac{d^2\theta}{dt^2} = 0. \tag{8.3.11}$$

Interestingly, the result in (8.3.11) was implicitly known long *before* Newton. Mathematics alone tells us that

$$\frac{d}{dt}\left\{r^2\frac{d\theta}{dt}\right\} = 2r\left(\frac{dr}{dt}\right)\frac{d\theta}{dt} + r^2\frac{d^2\theta}{dt^2} = r\left[2\left(\frac{dr}{dt}\right)\frac{d\theta}{dt} + r\frac{d^2\theta}{dt^2}\right]$$

and, since the expression in the square brackets is zero by (8.3.11), we have

$$\frac{d}{dt}\left\{r^2\frac{d\theta}{dt}\right\} = 0.$$

Fig. 8.3.1 Interpreting $r^2 \frac{d\theta}{dt}$

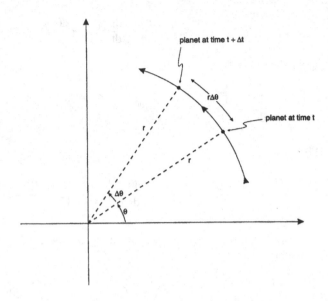

Thus, integration gives us

$$r^2 \frac{d\theta}{dt} = C \qquad (8.3.12)$$

where C is a constant. This result has a historically important physical interpretation in the theory of planetary motion.

Look at Fig. 8.3.1, which shows a planet's location at times t and $t + \Delta t$, with the Sun at the origin of our coordinate system. We assume Δt is so small that the angular change $\Delta\theta$ in the radius vector's angle is also very small, and that the length of the radius vector remains essentially unchanged. Then, the area between the two dashed lines is essentially that of an isosceles triangle with area ΔA given by

$$\Delta A = \frac{1}{2} \text{ base times height} \approx \frac{1}{2}(r\Delta\theta)r = \frac{1}{2}r^2\Delta\theta.$$

Dividing through by Δt gives

$$\frac{\Delta A}{\Delta t} \approx \frac{1}{2}r^2\frac{\Delta\theta}{\Delta t},$$

an expression that becomes exact as $\Delta t \to 0$. That is, replacing the delta quantities with differential ones, we have

$$\frac{dA}{dt} = \frac{1}{2}r^2\frac{d\theta}{dt}$$

or, from (8.3.12),

$$r^2 \frac{d\theta}{dt} = 2\frac{dA}{dt} = C.$$

This last expression is the mathematical form of the statement (given in 1609) by the German astronomer Johann Kepler (1671–1630) of his famous *area law*: the line joining the Sun to a planet sweeps over equal areas in equal time intervals. Kepler deduced this (the second of three general laws he discovered) not by physics or complex function theory, but rather from years of tedious observational data *made with the naked eye*.

8.4 The Cauchy-Riemann Equations and Analytic Functions

Complex function theory really starts with the study of what it means to talk of the *derivative* of f(z). In real function theory, the derivative of g(x) at $x = x_0$ is defined as

$$\frac{dg}{dx}\Big|_{x=x_0} = \lim_{x\to 0} \frac{g(x_0 + x) - g(x_0)}{x} - g'(x_0).$$

We do almost the same thing with a complex function. Indeed, the formal definition for the derivative of a complex f(z) at $z = z_0$ is

$$\frac{df}{dz}\Big|_{z=z_0} = \lim_{z\to 0} \frac{f(z_0 + z) - f(z_0)}{z} = f'(z_0).$$

The vanishing of $z = x + iy$ is, however, not quite as straightforward as it is in the case of a real variable. In that simpler case, where we let $x \to 0$ to calculate $g'(x_0)$, x only has to vanish along the *one*-dimensional real axis. That is, x can shrink to zero in just two ways: either from the left of x_0 or from the right of x_0. But in the complex case we must take into account that, since z_0 is a point in the complex, *two*-dimensional *plane*, then z can shrink to zero in an *infinity* of different ways (from the left of z_0, from the right of z_0, from below z_0, from above z_0 or, indeed, from *any* direction of the compass). So, just how *does* $z \to 0$?

Mathematicians consider the most condition-free definition possible for the derivative to be the best definition, and so their answer to our question is: we want $f'(z_0)$ to be the same *independent* of how $z \to 0$. To have this be the case, as you might suspect, comes with a price. If $f = u + iv$ then the price for a derivative at $z = z_0$ that doesn't depend on the precise nature of how $z \to 0$ is that u and v cannot be just *any* functions of x and y, but rather must satisfy certain conditions. If these conditions are satisfied at $z = z_0$ and at all points in a *region* (*domain* or *neighborhood* are terms that are also used) surrounding z_0, then we say that f(z) is an *analytic function* in that region (not to be confused with the analytic signal from radio theory that we encountered in the previous chapter).

The conditions for f(z) to be analytic are called the *Cauchy-Riemann (C-R)* equations,[6] which are actually pretty easy to state: at $z = z_0$ it must be true that

$$\frac{\partial u}{\partial x} = \frac{\partial v}{\partial y} \qquad\qquad (8.4.1)$$

and

$$\frac{\partial u}{\partial y} = -\frac{\partial v}{\partial x}. \qquad\qquad (8.4.2)$$

For example, suppose that $f(z) = z$. That is, $f(x, y) = x + iy = u(x, y) + i\, v(x, y)$ which means that $u(x, y) = x$ and that $v(x, y) = y$. Then,

$$\frac{\partial u}{\partial x} = 1, \; \frac{\partial u}{\partial y} = 0, \; \frac{\partial v}{\partial x} = 0 \; \frac{\partial v}{\partial y} = 1$$

and we see that the C-R equations are satisfied. Indeed, since the C-R equations are independent of z (of z_0) then $f(z) = z$ is analytic over the entire *finite* complex plane.[7] As a counter-example, of a f(z) that is *nowhere* analytic, consider $f(z) = \bar{z} = x - iy$, where $\bar{z}$ is the *conjugate* of z. Then,

$$\frac{\partial u}{\partial x} = 1, \; \frac{\partial u}{\partial y} = 0, \; \frac{\partial v}{\partial x} = 0, \; \frac{\partial v}{\partial y} = -1$$

and so (8.4.1) is *never* satisfied.

Under not particularly harsh requirements the C-R equations are *necessary and sufficient* conditions for f(z) to be analytic, and I'll refer you to any good text devoted to complex function theory for a proof of this.[8] To show that the C-R equations are *necessary* is not at all difficult, however. Since $z = x + iy$ then to have $z \to 0$ requires that both $x \to 0$ and $y \to 0$. That is, to speak of the derivative of f(z) at $z = z_0$ means to calculate

$$f'(z_0) = \lim_{\Delta x \to 0, \Delta y \to 0} \frac{f(x_0 + \Delta x, y_0 + \Delta y) - f(x_0, y_0)}{\Delta x + i\Delta y}.$$

---

[6] The C-R equations had, in fact, been known *before* either Cauchy or Riemann had been born, as the result of studies in hydrodynamics (see Gray, note 1, p. 60).

[7] The word *finite* is important: $f(z) = z$ blows-up as $|z| \to \infty$ and so f(z) is *not* said to be analytic at infinity. In fact, there is a theorem in complex function theory that says the only functions that are analytic over the entire complex plane, even at infinity, are constants. In those cases all four partial derivatives in the C-R equations are identically zero.

[8] See, for example, Joseph Bak and Donald J. Newman, *Complex Analysis* (third edition), Springer 2010, pp. 35–40. While the C-R equations *alone* are not sufficient for analyticity, if the partial derivatives in them are *continuous* then we do have sufficiency.

Now, out of the infinity of ways that both Δx and Δy can vanish, let's consider just two. First, assume that $\Delta y = 0$ and so $\Delta z = \Delta x$. That is, z approaches z_0 parallel to the x-axis. Second, assume that $\Delta x = 0$ and so $\Delta z = i\Delta y$. That is, z approaches z_0 parallel to the y-axis. If $f'(z_0)$ is to be unique, independent of the details of how $\Delta z \rightarrow 0$, then these two particular cases must give the same result. In the first case we have

$$f'(z_0) = \lim_{\Delta z = \Delta x \rightarrow 0} \frac{f(x_0 + \Delta x, y_0) - f(x_0, y_0)}{\Delta x}$$

$$= \lim_{\Delta x \rightarrow 0} \frac{\{u(x_0 + \Delta x, y_0) + i\, v(x_0 + \Delta x, y_0)\} - \{u(x_0, y_0) + i\, v(x_0, y_0)\}}{\Delta x}$$

$$= \lim_{\Delta x \rightarrow 0} \frac{\{u(x_0 + \Delta x, y_0) - u(x_0, y_0)\} + i\,\{v(x_0 + \Delta x, y_0) - v(x_0, y_0)\}}{\Delta x}$$

$$= \frac{\partial u}{\partial x} + i \frac{\partial v}{\partial x}.$$

And in the second case we have

$$f'(z_0) = \lim_{\Delta z = i\Delta y \rightarrow 0} \frac{f(x_0, y_0 + \Delta y) - f(x_0, y_0)}{i\Delta y}$$

$$= \lim_{\Delta y \rightarrow 0} \frac{\{u(x_0, y_0 + \Delta y) + i\, v(x_0, y_0 + \Delta y)\} - \{u(x_0, y_0) + i\, v(x_0, y_0)\}}{i\Delta y}$$

$$= \lim_{\Delta y \rightarrow 0} \frac{\{u(x_0, y_0 + \Delta y) - u(x_0, y_0)\} + i\,\{v(x_0, y_0 + \Delta y) - v(x_0, y_0)\}}{i\Delta y}$$

$$= \frac{1}{i} \frac{\partial u}{\partial y} + \frac{\partial v}{\partial y} = \frac{\partial v}{\partial y} - i \frac{\partial u}{\partial y}.$$

Equating the real and the imaginary parts of these two expressions for $f'(z_0)$ gives the C-R equations in (8.4.1) and (8.4.2).

Analytic functions are clearly a rather special subset of all possible complex functions, but certain broad classes are included. They are:

1. Every polynomial of z is analytic;
2. Every sum and product of two analytic functions is analytic;
3. Every quotient of two analytic functions is analytic *except* at those values where the denominator function is zero;
4. An analytic function of an analytic function is analytic.

So, from (1) $f(z) = z^2$ and $f(z) = e^z$ are both analytic, because in the first case $f(z)$ is a polynomial and in the second case because the exponential can be expanded in a power series. From (2) $f(z) = z^2 e^z$ is analytic, and from (3) $f(z) = e^z/(z^2 + 1)$ is analytic *except* at $z = \pm i$ which are called the *singularities* of f(z) because, at those values of z, f(z) blows-up.[9] And finally, from (4) $f(z) = e^{e^z}$ is analytic.

8.5   Green's Integral Theorem

In this section we'll continue our earlier discussion of line integrals to derive what is called *Green's theorem*.[10] We begin by imagining a closed path (*contour*) C that encloses a region R of the complex plane, as shown in Fig. 8.5.1. We further imagine that there are two real functions of the real variables x and y, P(x, y) and Q(x, y), defined at every point along C and in the region R (the *interior* of C). Then, Green's theorem says that

$$\oint_C \{P(x, y)\, dx + Q(x, y)\, dy\} = \iint_R \left\{ \frac{\partial Q}{\partial x} - \frac{\partial P}{\partial y} \right\} dx\, dy. \qquad (8.5.1)$$

The circle on the line integral in the left-hand side of (8.5.1) is there to emphasize that C is a *closed*, *non-self-intersecting* path (a simple curve traversed in the CCW

Fig. 8.5.1 A CCW contour
C and its interior R

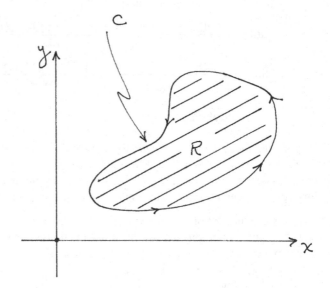

[9]If the function f(z) is analytic everywhere in some region except for a finite number of singularities, mathematicians say f(z) is *meromorphic* in that region and I tell you this simply so you won't be paralyzed by fear if you should ever come across that term.

[10]For the interesting history of this theorem, named after the English mathematician George Green (1793–1841), see my *An Imaginary Tale*, Princeton 2010, pp. 204–205.

sense, as mentioned in Sect. 8.2). R is called a *simply connected region*, which means every closed curve in R encloses only points in R. If a region is not simply connected then it is said to be *multiply-connected*: an example is a simply connected region that has a hole cut in it. The points 'in the hole' are considered to be in the *exterior* of C.

Green's theorem relates a contour integral along C to an *area* integral over the interior of C. For the contour of Fig. 8.5.1 it's pretty obvious where the interior of C is, but in just a bit we'll encounter contours whose interiors won't be so obvious. Here's an easy, low-level way to always locate the interior of a C: as you walk along C in the CCW sense, imagine you drag both hands along the ground. Your *left* hand will be in the *interior*, while your right hand will be in the *exterior* of C. (The idea that a simple, closed curve divides the plane into two regions—its interior and its exterior—seems pretty obvious. But not *so* obvious that mathematicians have nonetheless felt the need to term it the *Jordan Curve Theorem*, after the French mathematician Camille Jordan (1838–1922), who stated it in 1887.)

To prove Green's theorem isn't difficult, or at least it isn't if we make some highly simplifying assumptions. These assumptions are actually not required, but to remove them complicates the proof. To start, our first assumption is that R is a rectangular patch oriented parallel to the x and y axes, as shown in Fig. 8.5.2. (I've drawn the patch totally in the first quadrant, but that's just the way I drew it—in all that follows that is irrelevant.) The boundary edge of R is $C = C_1 + C_2 + C_3 + C_4$, which simply means that C is made of four sides. When we are done with this special R, I'll make some admittedly hand-waving (but plausible, too, I hope) arguments to try to convince you that far more complicated shapes for R are okay, too.

Starting with the $\iint_R -\frac{\partial P}{\partial y} dx\, dy$ term on the right-hand side of Green's theorem, we have

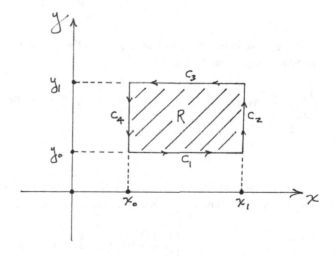

Fig. 8.5.2 When R is a rectangle

$$\iint_R -\frac{\partial P}{\partial y}\,dx\,dy = -\int_{x_0}^{x_1}\left\{\int_{y_0}^{y_1}\frac{\partial P}{\partial y}\,dy\right\}dx$$

$$= -\int_{x_0}^{x_1}\{P(x,y_1)-P(x,y_0)\}\,dx = \int_{x_0}^{x_1}P(x,y_0)dx + \int_{x_1}^{x_0}P(x,y_1)\,dx$$

$$= \int_{C_1}P(x,y)\,dx + \int_{C_3}P(x,y)\,dx.$$

Notice, carefully, that in the last two integrals I have dropped the subscripts on y_0 and y_1, subscripts that *were* included in the earlier integrals. I can do that because the subscripts were originally there to distinguish between integrating along the lower edge (y_0) or along the upper edge (y_1) of R, and that job is now done in the last two integrals by writing C_1 (the lower edge) and C_3 (the upper edge) beneath the appropriate integral sign. Notice, too, that writing $\int_{y_0}^{y_1}\frac{\partial P}{\partial y}\,dy = P(x,y_1)-P(x,y_0)$ makes the assumption that there is no discontinuity in $\frac{\partial P}{\partial y}$, that is, the partial derivative is *continuous*.

Similar integrals with respect to x can be written for the other two edges (C_2 and C_4) as well and, since those are *vertical* edges, we know that everywhere along them $dx = 0$. That is,

$$\int_{C_2}P(x,y)\,dx = \int_{x_1}^{x_1}P(x,y)dx = 0$$

and

$$\int_{C_4}P(x,y)\,dx = \int_{x_0}^{x_0}P(x,y)dx = 0.$$

Since those integrals vanish we can formally add them to our C_1 and C_3 integrals without changing anything. So,

$$\iint_R -\frac{\partial P}{\partial y}\,dx\,dy = \int_{C_1}P(x,y)\,dx + \int_{C_2}P(x,y)\,dx + \int_{C_3}P(x,y)\,dx + \int_{C_4}P(x,y)\,dx$$

$$= \oint_C P(x,y)\,dx.$$

If you repeat all the above for the $\iint_R \frac{\partial Q}{\partial x}\,dx\,dy$ term in Green's theorem, and observe $dy = 0$ along the horizontal edges C_1 and C_3, you should easily see that

$$\iint_R \frac{\partial Q}{\partial x}\,dx\,dy = \oint_C Q(x,y)\,dy,$$

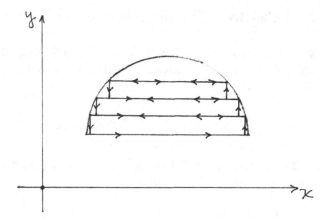

Fig. 8.5.3 Making a half-disk out of many thin rectangles

and that completes the proof of Green's theorem *for our nicely oriented rectangle* in Fig. 8.5.2. In fact, however, this proof extends rather easily to other much more complicated shapes for R.

In Fig. 8.5.3, for example, you see how a semicircular disk can be constructed from many very thin rectangles—the thinner they each are the more of them there are, yes, but that's okay; make them all each as thin as the finest onion-skin paper, if you like—the thinner they are the better they approximate the half-disk. If the boundary edge of the half-disk is denoted by C, and if the *complete* (all four edges) boundaries of the individual rectangles are denoted by $C_1, C_2, C_3, \ldots,$ then

$$\oint_C \{P\,dx + Q\,dy\} = \int_{C_1} \{P\,dx + Q\,dy\} + \int_{C_2} \{P\,dx + Q\,dy\} + \int_{C_3} \{P\,dx + Q\,dy\} + \ldots$$

because those edges of the individual rectangular boundaries that are parallel to the x-axis are traversed *twice*, once in each sense (CW and CCW), and so their contributions to the various line integrals on the right-hand-side of the above equation cancel. The only exception to this cancellation is the very bottom horizontal edge of the half-disk.

In addition, the integrations along the individual *vertical* edges of the thin rectangles avoid cancellation and, if the rectangles are *very* thin then the union of the vertical edges *is* the circular portion of the half-disk boundary. So, after integrating around all the rectangles, we are left with nothing more than integrating along the bottom of the half-disk and the circular portion. You can see that, using this same basic idea, we can build very complicated shapes out of appropriately arranged rectangles and, since Green's theorem works for each rectangle, then it works for all of them together and so Green's theorem works for their composite (and perhaps quite complicated) region R.

8.6   Cauchy's First Integral Theorem[11]

Well, it's taken a bit to get to this point, but it will soon be clear it was worth the effort. Our basic result is easy to state: if f(z) is analytic everywhere on and inside C then

$$\oint_C f(z)dz = 0. \tag{8.6.1}$$

To show this, recall (8.3.1) and (8.3.2). That is, with $f(z) = u(x, y) + i\, v(x, y)$ and writing $dz = dx + i\, dy$, we have

$$\oint_C f(z)dz = \oint_C (u + i\, v)(dx + i\, dy) = \oint_C (u\, dx + iu\, dy + iv\, dx - v\, dy)$$

or,

$$\oint_C f(z)dz = \oint_C (u\, dx - v\, dy) + i\oint_C (v\, dx + u\, dy). \tag{8.6.2}$$

Now, because of Green's theorem, the two contour integrals on the right are *each* equal to zero. To see this, consider the first integral on the right-hand side of (8.6.2), and look back at (8.5.1). You see that we have $P(x, y) = u(x, y)$ and $Q(x, y) = -v(x, y)$, and so the partial derivatives on the right-hand side of (8.5.1) are

$$\frac{\partial Q}{\partial x} = -\frac{\partial v}{\partial x}, \frac{\partial P}{\partial y} = \frac{\partial u}{\partial y}.$$

The C-R equation of (8.4.2), which holds here because we are assuming f(z) is analytic, says the integrand of the double integral in Green's theorem is

$$\frac{\partial Q}{\partial x} - \frac{\partial P}{\partial y} = -\frac{\partial v}{\partial x} - \frac{\partial u}{\partial y} = -\frac{\partial v}{\partial x} - \left(-\frac{\partial v}{\partial x}\right) = 0.$$

For the second integral on the right-hand side of (8.6.2) we have $P(x, y) = v(x, y)$ and $Q(x, y) = u(x, y)$. So now

---

[11]By convention, the theorem in this section is named after Cauchy who published it in 1814, but in a letter dated December 1811, written by the great Gauss to his fellow German mathematician Friedrich Wilhelm Bessel (1784–1846), he states (without proof) the theorem we will prove here. In mathematics, alas for Gauss (as if he really needed more to add to his enormous resumé), credit goes to the first to *publish*.

$$\frac{\partial Q}{\partial x} = \frac{\partial u}{\partial x}, \frac{\partial P}{\partial y} = \frac{\partial v}{\partial y}$$

and the C-R equation of (8.4.1) says the integrand of the double integral in Green's theorem is

$$\frac{\partial Q}{\partial x} - \frac{\partial P}{\partial y} = \frac{\partial u}{\partial x} - \frac{\partial v}{\partial y} = \frac{\partial v}{\partial y} - \frac{\partial v}{\partial y} = 0$$

because, again, f(z) is analytic. So, (8.6.1) is proven.

There is no denying that (8.6.1) looks pretty benign. But it has tremendous power. For example, consider the case of

$$f(z) = \frac{e^{iz}}{z},$$

which is analytic everywhere *except* at $z = 0$ because there f(z) blows-up. So, if we integrate f(z) around *any* C that avoids putting $z = 0$ in its interior, we know from (8.6.1) that we'll get zero for the integral. With that in mind, consider the contour C shown in Fig. 8.6.1, where $\varepsilon > 0$ and T is finite, and the two arcs are circular. In the notation of Fig. 8.6.1, we have

$$\oint_C f(z)dz = \int_{C_1} f(z)dz + \int_{C_2} f(z)dz + \int_{C_3} f(z)dz + \int_{C_4} f(z)dz = 0. \qquad (8.6.3)$$

For each of the four segments of C, we can write:
on C_1: $z = x$, $dz = dx$;
on C_2: $z = Te^{i\theta}$, $dz = iTe^{i\theta}d\theta$, $0 < \theta < \frac{\pi}{2}$;
on C_3: $z = i\,y$, $dz = i\,dy$;
on C_4: $z = \varepsilon e^{i\theta}$, $dz = i\varepsilon e^{i\theta}d\theta$, $\frac{\pi}{2} > \theta > 0$;

Fig. 8.6.1 contour that avoids a singularity at the origin

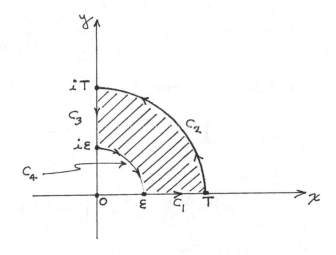

Thus, (8.6.3a) becomes

$$\int_\varepsilon^T \frac{e^{ix}}{x}dx + \int_0^{\frac{\pi}{2}} \frac{e^{iTe^{i\theta}}}{Te^{i\theta}}iTe^{i\theta}d\theta + \int_T^\varepsilon \frac{e^{i(iy)}}{iy}idy + \int_{\frac{\pi}{2}}^0 \frac{e^{i\varepsilon e^{i\theta}}}{\varepsilon e^{i\theta}}i\varepsilon e^{i\theta}d\theta = 0.$$

Then, doing all the obvious cancellations and reversing the direction of integration on the third and fourth integrals (and, of course, their algebraic signs, too), we arrive at

$$\int_\varepsilon^T \frac{e^{ix}}{x}dx + i\left\{\int_0^{\frac{\pi}{2}}\left(e^{iTe^{i\theta}} - e^{i\varepsilon e^{i\theta}}\right)d\theta\right\} - \int_\varepsilon^T \frac{e^{-y}}{y}dy = 0.$$

If, in the last integral, we change the dummy variable of integration from y to x, we then have

$$\int_\varepsilon^T \frac{e^{ix} - e^{-x}}{x}dx + i\left\{\int_0^{\frac{\pi}{2}}\left(e^{iTe^{i\theta}} - e^{i\varepsilon e^{i\theta}}\right)d\theta\right\} = 0.$$

Now, focus on the second integral and expand its integrand with Euler's identity:

$$e^{iTe^{i\theta}} - e^{i\varepsilon e^{i\theta}} = e^{iT\{\cos(\theta)+i\sin(\theta)\}} - e^{i\varepsilon\{\cos(\theta)+i\sin(\theta)\}}$$

$$= e^{-T\sin(\theta)}e^{iT\cos(\theta)} - e^{-\varepsilon\sin(\theta)}e^{i\varepsilon\cos(\theta)}.$$

If we let $T \to \infty$ and $\varepsilon \to 0$ then the first term on the right goes to zero because $\lim_{T\to\infty} e^{-T\sin(\theta)} = 0$ for all $0 < \theta < \frac{\pi}{2}$, while the second term goes to 1 because

$$\lim_{\varepsilon\to 0} e^{-\varepsilon\sin(\theta)} = \lim_{\varepsilon\to 0} e^{i\varepsilon\cos(\theta)} = e^0 = 1 \text{ for all } 0 < \theta < \frac{\pi}{2}.$$

Thus, (8.6.3) becomes, as $T \to \infty$ and $\varepsilon \to 0$,

$$\int_0^\infty \frac{e^{ix} - e^{-x}}{x}dx + i\left\{\int_0^{\frac{\pi}{2}}(-1)d\theta\right\} = 0$$

or, using Euler's identity again,

$$\int_0^\infty \frac{\cos(x) + i\sin(x) - e^{-x}}{x}dx - i\frac{\pi}{2} = 0$$

or,

$$\int_0^\infty \frac{\cos(x) - e^{-x}}{x}dx + i\int_0^\infty \frac{\sin(x)}{x}dx = i\frac{\pi}{2}.$$

Equating imaginary parts, we have

$$\int_0^\infty \frac{\sin(x)}{x}\,dx = \frac{\pi}{2}$$

which we've already derived in (3.2.1)—and it's certainly nice to see that our contour integration agrees—while equating real parts gives

$$\int_0^\infty \frac{\cos(x) - e^{-x}}{x}\,dx = 0. \tag{8.6.4}$$

MATLAB agrees, too, as *integral(@(x)(cos(x)-exp(−x))./x,0,1e4) = −3x10^{−5}* which, while not exactly zero, is pretty small. You'll recall this integral as a special case of (4.3.14), derived using 'normal' techniques.

You can see that the ability of contour integration in the *complex plane* to do improper *real* integrals, integrals like $\int_0^\infty$ and $\int_{-\infty}^\infty$, depends on the proper choice of the contour C. At the start, C encloses a finite region of the plane, with part of C lying on the real axis. Then, as we let C expand so that the real axis portion expands to $-\infty$ to ∞, or to 0 to ∞, the other portions of C result in integrations that are, in some sense, 'easy to do.'

The calculation of (8.6.4) was a pretty impressive example of this process, but here's another application of Cauchy's first integral theorem that, I think, tops it. Suppose a, b, and c are *any* real numbers ($a \neq 0$) such that $b^2 > 4ac$. What is the value of the integral

$$\int_{-\infty}^\infty \frac{dx}{ax^2 + bx + c} = ?$$

I think you'll be surprised by the answer. Here's how to calculate it, starting with the contour integral

$$\oint_C f(z)\,dz = \oint_C \frac{dz}{az^2 + bz + c}$$

where we notice that the integrand has two singularities that are both on the real axis, as shown in Fig. 8.6.2. That's because of the given $b^2 > 4ac$ condition, which says the denominator vanishes at the two *real* values (remember the quadratic formula!) of

$$x_2 = -\frac{b}{2a} - \frac{1}{2a}\sqrt{b^2 - 4ac} \text{ and } x_1 = -\frac{b}{2a} + \frac{1}{2a}\sqrt{b^2 - 4ac}.$$

The equality $b^2 = 4ac$ is the case where the two real roots have merged to form a *double* root which, as you'll soon see, *does* add a twist to our analysis.

In Fig. 8.6.2 I've shown the singularities as being on the negative real axis, but they could both be on the positive real axis—just where they are depends on the signs of a and b. All that actually matters, however, is that the singularities are *both*

Fig. 8.6.2 Avoiding
singularities with circular
indents

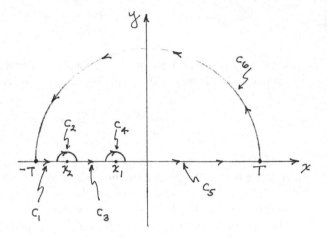

on the *real* axis. This means, when we select C, that we must arrange for its real axis
portion to *avoid* those singularities as you'll remember the big deal I made on that
very point in Chap. 1 with the discussion there of the 'sneaking up on a singularity'
trick. With contour integration we don't so much 'sneak up' on a singularity as
'swing around and avoid' it, which we do with the C_2 and C_4 portions of C shown in
Fig. 8.6.2 (and take a look back at Fig. 8.6.1, too, with its C_4 avoiding a singularity at
$z = 0$). Those circular swings (called *indents*) are such as to keep the singularities in
the *exterior* of C. Each indent has radius ε, which we'll eventually shrink to zero by
taking the limit $\varepsilon \to 0$.

So, here's what we have, with $C = C_1 + C_2 + C_3 + C_4 + C_5 + C_6$.
on C_1, C_3, C_5: $z = x$, $dz = dx$;
on C_2: $z = x_2 + \varepsilon e^{i\theta}$, $dz = i\varepsilon e^{i\theta} d\theta$, $\pi > \theta > 0$;
on C_4: $z = x_1 + \varepsilon e^{i\theta}$, $dz = i\varepsilon e^{i\theta} d\theta$, $\pi > \theta > 0$;
on C_6: $z = T e^{i\theta}$, $dz = i T e^{i\theta} d\theta$, $0 < \theta < \pi$.
Cauchy's first integral theorem says

$$\oint_C f(z)dz = \left\{ \int_{C_1} f(z)dz + \int_{C_3} f(z)dz + \int_{C_5} f(z)dz \right\}$$

$$+ \left\{ \int_{C_2} f(z)dz + \int_{C_4} f(z)dz + \int_{C_6} f(z)dz \right\} = 0.$$

When we eventually let $\varepsilon \to 0$ and $T \to \infty$, the first three line integrals will
combine to give us the real integral we are after. The value of that integral will
therefore be given by

$$- \left\{ \int_{C_2} f(z)dz + \int_{C_4} f(z)dz + \int_{C_6} f(z)dz \right\}.$$

So, let's now calculate each of these three line integrals.

For C_2,

$$\int_{C_2} f(z)dz = \int_\pi^0 \frac{i\varepsilon e^{i\theta}d\theta}{a(x_2 + \varepsilon e^{i\theta})^2 + b(x_2 + \varepsilon e^{i\theta}) + c}$$

$$= \int_\pi^0 \frac{i\varepsilon e^{i\theta}d\theta}{a(x_2^2 + 2x_2\varepsilon e^{i\theta} + \varepsilon^2 e^{i2\theta}) + b(x_2 + \varepsilon e^{i\theta}) + c}$$

$$= \int_\pi^0 \frac{i\varepsilon e^{i\theta}d\theta}{(ax_2^2 + bx_2 + c) + (2ax_2\varepsilon e^{i\theta} + a\varepsilon^2 e^{i2\theta} + b\varepsilon e^{i\theta})}.$$

Since $(ax_2^2 + bx_2 + c) = 0$ because x_2 is a zero of the denominator (*by definition*), and as $\varepsilon^2 \to 0$ faster than $\varepsilon \to 0$, then for very small ε we have

$$\lim{}_{\varepsilon \to 0} \int_{C_2} f(z)dz = \lim{}_{\varepsilon \to 0} \int_\pi^0 \frac{i\varepsilon e^{i\theta}d\theta}{(2ax_2 + b)\varepsilon e^{i\theta}} = -i\int_0^\pi \frac{d\theta}{(2ax_2 + b)}$$

$$= -\pi i \frac{1}{2ax_2 + b}.$$

In the same way,

$$\lim{}_{\varepsilon \to 0} \int_{C_4} f(z)dz = -\pi i \frac{1}{2ax_1 + b},$$

And finally,

$$\int_{C_6} f(z)dz = \int_0^\pi \frac{iTe^{i\theta}d\theta}{aT^2 e^{i2\theta} + bTe^{i\theta} + c}$$

and, since the integrand vanishes like $\frac{1}{T}$ as $T \to \infty$, then

$$\lim{}_{T \to \infty} \int_{C_6} f(z)dz = 0.$$

Thus,

$$\int_{-\infty}^\infty \frac{dx}{ax^2 + bx + c} = \pi i \left[\frac{1}{2ax_2 + b} + \frac{1}{2ax_1 + b} \right].$$

Since

$$2ax_2 + b = 2a\left[-\frac{b}{2a} - \frac{1}{2a}\sqrt{b^2 - 4ac} \right] + b = -\sqrt{b^2 - 4ac}$$

and

$$2ax_1 + b = 2a\left[-\frac{b}{2a} + \frac{1}{2a}\sqrt{b^2 - 4ac}\right] + b = \sqrt{b^2 - 4ac}$$

we see that the two singularities cancel each other and so we have the interesting result

$$\int_{-\infty}^{\infty} \frac{dx}{ax^2 + bx + c} = 0, \quad a \neq 0, b^2 > 4ac \qquad (8.6.5)$$

for *all possible values* of a, b, and c. This, I think, is not at all obvious! (In Challenge Problem 8.5 you are asked to do an integral that generalizes this result.)

An immediate question that surely comes to mind now is, what happens if $b^2 \leq 4ac$? Consider the two cases $b^2 = 4ac$ and $b^2 < 4ac$, separately. Suppose that $b^2 = 4ac$. Then, the denominator of (8.6.5) is $ax^2 + bx + \frac{b^2}{4a} = a\left(x^2 + \frac{b}{a}x + \frac{b^2}{4a^2}\right) = a\left(x + \frac{b}{2a}\right)^2$ and it is immediately obvious that $\int_{-\infty}^{\infty} \frac{dx}{a\left(x+\frac{b}{2a}\right)^2} \neq 0$ (indeed, the integral blows-up!)[12]

What if $b^2 < 4ac$? If that's the case the singularities of f(z) are no longer on the real axis but, instead, have non-zero imaginary parts. We'll come back to this question in the next section, where we'll find that the integral in (8.6.5) is, again, no longer zero under this new condition.

Contour indents around singularities are such a useful device that their application warrants another example. So, what I'll do next is use indents to derive a result that would be extremely difficult to get by other means: calculating the value of

$$\int_{-\infty}^{\infty} \frac{e^{ax}}{1 - e^x}\, dx, 0 < a < 1.$$

To evaluate this integral, we'll study the contour integral

$$\oint_C \frac{e^{az}}{1 - e^z}\, dz, \qquad (8.6.6)$$

using the curious contour C shown in Fig. 8.6.3.

The reasons for choosing this particular C (looking a bit like a block of cheese that mice have been nibbling on) probably require some explanation. The real-axis portions (C_1 and C_3) are perhaps obvious, as eventually we'll let $T \to \infty$, and these parts of C (where $z = x$) will give us the integral we are after. That is, the sum of the C_1 and C_3 integrals is

$$\int_{-T}^{-\varepsilon} \frac{e^{ax}}{1 - e^x}\, dx + \int_{\varepsilon}^{T} \frac{e^{ax}}{1 - e^x}\, dx.$$

---

[12]Can you show this? If not, go back and read Sect. 1.6 again.

Fig. 8.6.3 A curious contour

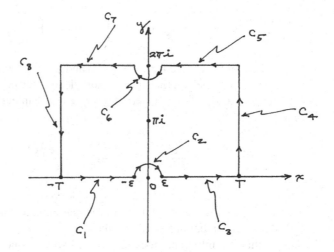

The semi-circular indent (C_2) with radius ε (we'll eventually let ε → 0) around the origin is also probably obvious because $z = 0$ is a singularity of the integrand, and so you can see I'm trying to set things up to use Cauchy's first integral theorem (which requires that C enclose *no* singularities). It's the other portions of C, the two vertical sides (C_4 and C_8), and the two sides parallel to the real axis (C_5 and C_7), that are probably the ones puzzling you right now.

Since I am trying to avoid enclosing *any* singularities, you can understand why I am not using our previous approach of including a semi-circular arc from T on the positive real axis back to –T on the negative real axis, an arc that then expands to infinity as T → ∞. That won't work here because the integrand has an *infinity* of singularities on the imaginary axis, spaced up and down at intervals of $2\pi i$ (because Euler's identity tells us that $1 - e^z = 0$ has the solutions $z = 2\pi i k$ for k any integer). A semi-circular arc would end-up enclosing an infinite number of singularities!

There is another issue, too. The $k = 0$ singularity is the one we've already avoided on the real-axis, but why (you might ask) are we intentionally running right towards the singularity for $k = 1$ (at $2\pi i$ on the imaginary axis)? Isn't the C in Fig. 8.6.3 just asking for trouble? Sure, we end-up avoiding that singularity with another semi-circular indent, but why not just run the top segment of C *below* the $k = 1$ singularity and so completely and automatically miss the singularity that way? Well, trust me— there *is* a reason, soon to be revealed.

Since we have arranged for there to be no singularities inside C we have, by Cauchy's first integral theorem,

$$\sum_{k=1}^{8} \int_{C_k} \frac{e^{az}}{1 - e^z} dz = 0$$

or, since on C_1 and C_3 we have $z = x$,

$$\int_{-T}^{-\varepsilon} \frac{e^{ax}}{1-e^x}\,dx + \int_{\varepsilon}^{T} \frac{e^{ax}}{1-e^x}\,dx = -\int_{C_2} - \int_{C_4} - \int_{C_5} - \int_{C_6} - \int_{C_7} - \int_{C_8}. \quad (8.6.7)$$

Soon, of course, we'll be letting $T \to \infty$ and $\varepsilon \to 0$ in these integrals. Let's now start looking at the ones on the right in more detail, starting with C_4.

On C_4 we have $z = T + iy$ where $0 \le y \le 2\pi$. The integrand of the C_4 integral is therefore

$$\frac{e^{az}}{1-e^z} = \frac{e^{a(T+iy)}}{1-e^{(T+iy)}} = \frac{e^{aT}e^{iay}}{1-e^Te^{iy}}.$$

As $T \to \infty$ we see that the magnitude of the numerator blows-up like e^{aT} ($|e^{iay}| = 1$), while the magnitude of the denominator blows-up like e^T. So, the magnitude of the integrand behaves like $e^{(a-1)T}$ as $T \to \infty$ which means, since $0 < a < 1$, that the integrand goes to zero and so we conclude that the C_4 integral vanishes as $T \to \infty$. In the same way, on C_8 we have $z = -T + iy$ with $2\pi > y > 0$. The integrand of the C_8 integral is

$$\frac{e^{az}}{1-e^z} = \frac{e^{a(-T+iy)}}{1-e^{(-T+iy)}} = \frac{e^{-aT}e^{iay}}{1-e^{-T}e^{iy}}$$

and so as $T \to \infty$ we see that the magnitude of the numerator goes to zero as e^{-aT} (because a is positive) while the magnitude of the denominator goes to 1. That is, the integrand behaves like e^{-aT} and so the C_8 integral also vanishes as $T \to \infty$.

Next, let's look at the C_5 and C_7 integrals. Be alert!—*this* is where you'll see why running C right towards the imaginary axis singularity at $2\pi i$ is a good idea, even though we are going to avoid it 'at the very last moment' (so to speak) with the C_6 semi-circular indent. On the C_5 and C_7 integrals we have $z = x + 2\pi i$ and so $dz = dx$ (just like on the C_1 and C_3 integrals). Writing-out the C_5 and C_7 integrals in detail, we have

$$\int_{T}^{\varepsilon} \frac{e^{a(x+2\pi i)}}{1-e^{x+2\pi i}}\,dx + \int_{-\varepsilon}^{-T} \frac{e^{a(x+2\pi i)}}{1-e^{x+2\pi i}}\,dx = \int_{T}^{\varepsilon} \frac{e^{ax}e^{2\pi ai}}{1-e^xe^{2\pi i}}\,dx + \int_{-\varepsilon}^{-T} \frac{e^{ax}e^{2\pi ai}}{1-e^xe^{2\pi i}}\,dx$$

or, because $e^{2\pi i} = 1$ (this is the crucial observation!) we have the sum of C_5 and C_7 integrals as

$$-e^{2\pi ai}\left[\int_{\varepsilon}^{T} \frac{e^{ax}}{1-e^x}\,dx + \int_{-T}^{-\varepsilon} \frac{e^{ax}}{1-e^x}\,dx\right].$$

Notice that, to within the constant factor $-e^{2\pi ai}$, this is the sum of the C_1 and C_3 integrals. We have this simplifying result *only* because we ran the top segment of C directly towards the $2\pi i$ singularity. All this means we can now write (8.6.7) as (because, don't forget, the C_4 and C_8 integrals vanish as $T \to \infty$):

$$\left(1 - e^{2\pi ai}\right)\left[\int_{\varepsilon}^{T} \frac{e^{ax}}{1 - e^x}\,dx + \int_{-T}^{-\varepsilon} \frac{e^{ax}}{1 - e^x}\,dx\right] = -\int_{C_2} - \int_{C_6}. \tag{8.6.8}$$

On C_2 we have $z = \varepsilon e^{i\theta}$ for $\pi \geq \theta \geq 0$ and so $dz = i\varepsilon e^{i\theta}d\theta$. Thus,

$$\int_{C_2} = \int_{\pi}^{0} \frac{e^{a\varepsilon e^{i\theta}}}{1 - e^{\varepsilon e^{i\theta}}} i\varepsilon e^{i\theta}d\theta.$$

Recalling the power series expansion of the exponential and keeping only the first-order terms in ε (because all the higher-order terms go to zero even faster than does ε), we have

$$1 - e^{\varepsilon e^{i\theta}} \approx 1 - \left[1 + \varepsilon e^{i\theta}\right] = -\varepsilon e^{i\theta}.$$

In the same way,

$$e^{a\varepsilon e^{i\theta}} \approx 1 + a\varepsilon e^{i\theta}$$

and so

$$\lim_{\varepsilon \to 0} \int_{C_2} = \lim_{\varepsilon \to 0} \int_{\pi}^{0} \frac{1 + a\varepsilon e^{i\theta}}{-\varepsilon e^{i\theta}} i\varepsilon e^{i\theta}d\theta = i\int_{0}^{\pi} d\theta = \pi i.$$

On C_6 we have $z = 2\pi i + \varepsilon e^{i\theta}$ for $0 \geq \theta \geq -\pi$ and so $dz = i\varepsilon e^{i\theta}d\theta$. Thus,

$$\int_{C_6} = \int_{0}^{-\pi} \frac{e^{a(2\pi i + \varepsilon e^{i\theta})}}{1 - e^{2\pi i + \varepsilon e^{i\theta}}} i\varepsilon e^{i\theta}d\theta = \int_{0}^{-\pi} \frac{e^{a2\pi i}e^{a\varepsilon e^{i\theta}}}{1 - e^{\varepsilon e^{i\theta}}} i\varepsilon e^{i\theta}d\theta$$

or, as we let $\varepsilon \to 0$,

$$\int_{C_6} = i e^{a2\pi i}\int_{0}^{-\pi} \frac{e^{a\varepsilon e^{i\theta}}}{-\varepsilon e^{i\theta}} \varepsilon e^{i\theta}d\theta = -i e^{a2\pi i}\int_{0}^{-\pi} d\theta = i\pi e^{a2\pi i}.$$

Plugging these two results for the C_2 and C_6 integrals into (8.6.8) and letting $T \to \infty$ and $\varepsilon \to 0$ we get

$$\int_{-\infty}^{\infty} \frac{e^{ax}}{1 - e^x}\,dx = -\frac{\pi i + i\pi e^{a2\pi i}}{1 - e^{2\pi ai}} = \pi i\frac{1 + e^{a2\pi i}}{e^{2\pi ai} - 1} = \pi i\frac{e^{a\pi i}\left[e^{-a\pi i} + e^{a\pi i}\right]}{e^{a\pi i}\left[e^{a\pi i} - e^{-a\pi i}\right]}$$

$$= \pi i\frac{2\cos(a\pi)}{2i\sin(a\pi)}$$

or

$$\int_{-\infty}^{\infty} \frac{e^{ax}}{1 - e^{x}} dx = \frac{\pi}{\tan (a\pi)}, \qquad 0 < a < 1. \tag{8.6.9}$$

Be sure to carefully note that the value of the integral in (8.6.9) comes *entirely* from the vanishingly small semi-circular paths around the two singularities. Singularities, and integration paths of 'zero' length around them, matter! If $a = \frac{1}{4}$ the integral is equal to π and MATLAB agrees because (using our old trick of 'sneaking up' on the singularity at x = 0), *integral(@(x)exp(x/4)./(1-exp(x)),-1e3, -.0001) + integral(@(x)exp(x/4)./(1-exp(x)),.0001,1e3)* = 3.14154. . . .

Before leaving this section I should tell you that not every use of Cauchy's first integral theorem is the calculation of the closed-form value of an integral. Another quite different and very nifty application is the transformation of an integral that is difficult to accurately calculate *numerically* into another equivalent integral that is much easier to calculate *numerically*. Two examples of this are

$$I(a) = \int_0^{\infty} \frac{\cos (x)}{x + a} dx = \lim_{T \to \infty} \int_0^T \frac{\cos (x)}{x + a} dx$$

and

$$J(a) = \int_0^{\infty} \frac{\sin (x)}{x + a} dx = \lim_{T \to \infty} \int_0^T \frac{\sin (x)}{x + a} dx,$$

where a is a positive constant. These two integrals have no closed-form values, and each has to be numerically evaluated for each new value of a.

To do that, *accurately*, using the usual numerical integration techniques is not easy, for the same reasons I gave in the last chapter when we derived (7.5.2). That is, the integrands of both I(a) and J(a) are really not that small even for 'large' T, as the denominators increase slowly and the numerators don't really decrease at all but simply *oscillate* endlessly between ±1. To numerically calculate I(a), for example, by writing *integral(@(x)cos(x)./(x + a),0,T)* with the numerical values of a and T inserted doesn't work well. For example, if a = 1 then for the four cases of T = 5, 10, 50, and 100 we get

T	I(1)
5	0.18366. . .
10	0.30130. . .
50	0.33786. . .
100	0.33828. . .

The calculated values of I(1) are not stable out to more than a couple of decimal places, even for T = 100. A similar table for J(1) is

T	J(1)
5	0.59977...
10	0.70087...
50	0.60264...
100	0.61296...

These values for J(1) are even more unstable than are those for I(1).

What I'll do now is show you how the first integral theorem can be used to get really excellent numerical accuracy, even with a 'small' value of T. What we'll do is consider the contour integral

$$\oint_C \frac{e^{iz}}{z+a} dz$$

where $C = C_1 + C_2 + C_3$ is the first quadrant circular contour shown in Fig. 8.6.4. The integrand has a lone singularity on the negative real axis at $z = -a < 0$, which lies *outside* of C. Thus, we immediately know from the first theorem that, for this C,

$$\oint_C \frac{e^{iz}}{z+a} dz = 0. \tag{8.6.10}$$

Now, for the three distinct sections of C, we have:
on C_1: $z = x$ and so $dz = dx$, $0 \le x \le T$;
on C_2: $z = Te^{i\theta}$, $dz = iTe^{i\theta}d\theta$, $0 < \theta < \frac{\pi}{2}$;
on C_3: $z = i\,y$, $dz = i\,dy$, $T \ge y \ge 0$.

So, starting at the origin and going around C in the counterclockwise sense, (8.6.10) becomes

Fig. 8.6.4 A contour that *excludes* a singularity

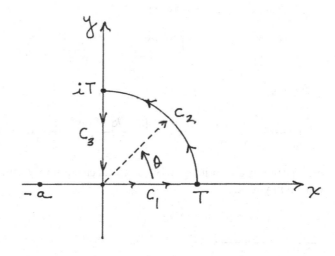

$$\int_0^T \frac{e^{ix}}{x+a}\,dx + \int_0^{\frac{\pi}{2}} \frac{e^{iTe^{i\theta}}}{Te^{i\theta}+a}\,iTe^{i\theta}d\theta + \int_T^0 \frac{e^{i(iy)}}{iy+a}\,i\,dy = 0$$

or,

$$\int_0^T \frac{e^{ix}}{x+a}\,dx - \int_0^T \frac{ie^{-y}}{iy+a}\,dy = -i\int_0^{\frac{\pi}{2}} e^{iTe^{i\theta}} e^{i\theta}\frac{T}{Te^{i\theta}+a}\,d\theta.$$

Our next step is to look at what happens when we let $T \to \infty$. Using Euler's identity, we have

$$e^{iTe^{i\theta}} = e^{iT\{\cos(\theta)+i\,\sin(\theta)\}} = e^{-T\sin(\theta)}e^{iT\cos(\theta)}$$

and so

$$e^{iTe^{i\theta}} e^{i\theta}\frac{T}{Te^{i\theta}+a} = e^{-T\sin(\theta)}e^{iT\cos(\theta)}e^{i\theta}\frac{T(Te^{-i\theta}+a)}{(Te^{i\theta}+a)(Te^{-i\theta}+a)}$$

$$= e^{-T\sin(\theta)}e^{iT\cos(\theta)}\frac{T^2+aTe^{i\theta}}{T^2+aT(e^{i\theta}+e^{-i\theta})+a^2}$$

and so

$$\lim_{T\to\infty}\left|e^{iTe^{i\theta}} e^{i\theta}\frac{T}{Te^{i\theta}+a}\right| = \lim_{T\to\infty}\left|e^{-T\sin(\theta)}\right|\left|e^{iT\cos(\theta)}\right|\left|\frac{T^2+aTe^{i\theta}}{T^2+aT(e^{i\theta}+e^{-i\theta})+a^2}\right|$$

$$= \lim_{T\to\infty}\left|e^{-T\sin(\theta)}\right| = 0.^{13}$$

Thus, as $T \to \infty$ we arrive at

$$\int_0^\infty \frac{e^{ix}}{x+a}\,dx - \int_0^\infty \frac{ie^{-y}}{iy+a}\,dy = 0.$$

That is,

$$\int_0^\infty \frac{\cos(x)+i\,\sin(x)}{x+a}\,dx = \int_0^\infty \frac{ie^{-y}(-iy+a)}{(iy+a)(-iy+a)}\,dy = \int_0^\infty \frac{ye^{-y}+i\,ae^{-y}}{y^2+a^2}\,dy$$

or, equating real and imaginary parts, and changing the dummy variable of integration from y to x,

---

[13]Because $\left|e^{iT\cos(\theta)}\right| = 1$ for *all* T, and $\lim_{T\to\infty}\left|\frac{T^2+aTe^{i\theta}}{T^2+aT(e^{i\theta}+e^{-i\theta})+a^2}\right| = 1$.

$$I(a) = \int_0^\infty \frac{\cos{(x)}}{x+a}\,dx = \int_0^\infty \frac{xe^{-x}}{x^2+a^2}\,dx$$

and

$$J(a) = \int_0^\infty \frac{\sin{(x)}}{x+a}\,dx = a\int_0^\infty \frac{e^{-x}}{x^2+a^2}\,dx.$$

The new integrals on the right for I(a) and J(a) have integrands that decrease rapidly as x increases from zero.

Calculating I(1) and J(1) again, using these alternative integrals, we have the following new tables:

T	I(1)
5	0.342260...
10	0.343373...
50	0.343378...
100	0.343378...

T	J(1)
5	0.621256...
10	0.621449...
50	0.621450...
100	0.621450...

You can see from these tables the *vastly* improved numerical performance of our calculations, and we can now say *with confidence* that

$$\int_0^\infty \frac{\cos{(x)}}{x+1}\,dx = 0.34337\ldots$$

$$\int_0^\infty \frac{\sin{(x)}}{x+1}\,dx = 0.62145\ldots.$$

(8.6.11)

8.7   Cauchy's Second Integral Theorem

When we try to apply Cauchy's first integral theorem we may find it is not possible to construct a useful contour C such that a portion of it lies along the real axis and yet does not have a singularity in its interior. The integral of (8.6.5) for the case of $b^2 < 4ac$ will prove to be an example of that situation, and I'll show you some other examples in this section, as well. The presence of singularities inside C means that Cauchy's first integral theorem no longer applies. 'Getting around' (pun intended!)

this complication leads us to Cauchy's second integral theorem: if f(z) is analytic everywhere on and inside C then, if z_0 is inside C,

$$\oint_C \frac{f(z)}{z - z_0}\, dz = 2\pi i\, f(z_0). \tag{8.7.1}$$

By successively differentiating with respect to z_0 under the integral sign, it can be shown that *all* the derivatives of an analytic f(z) exist (we'll use this observation in the next section):

$$f^{(n)}(z_0) = \frac{n!}{2\pi i} \oint_C \frac{f(z)}{(z - z_0)^{n+1}}\, dz,$$

where z_0 is any point inside C and $f^{(n)}$ denotes the n-*th* derivative of f.

While f(z) itself has no singularities (because it's analytic) inside C, the integrand of (8.7.1) *does* have a *first-order singularity*[14] at $z = z_0$. Now, before I prove (8.7.1) let me show you a pretty application of it, so you'll believe it will be well-worth your time and effort to understand the proof. What we'll do is evaluate the contour integral

$$\oint_C \frac{e^{iaz}}{b^2 + z^2}\, dz$$

where C is the contour shown in Fig. 8.7.1, and a and b are each a positive constant. When we are nearly done, we'll let $T \to \infty$ and you'll see we will have derived a famous result (one we've already done, in fact, in (3.1.7)), with the difference being that using Cauchy's second integral theorem will be the *easier* of the two derivations! Along the real axis part of C we have z = x, and along the semicircular arc we have $z = Te^{i\theta}$, where $\theta = 0$ at x = T and $\theta = \pi$ at x = −T. So,

$$\oint_C \frac{e^{iaz}}{b^2 + z^2}\, dz = \int_{-T}^{T} \frac{e^{iax}}{b^2 + x^2}\, dx + \int_0^{\pi} \frac{e^{ia(Te^{i\theta})}}{b^2 + z^2}\, i\, Te^{i\theta} d\theta.$$

The integrand of the contour integral can be written in a partial fraction expansion as

$$\frac{e^{iaz}}{b^2 + z^2} = \frac{e^{iaz}}{(z + ib)(z - ib)} = \frac{e^{iaz}}{i2b} \left[\frac{1}{z - ib} - \frac{1}{z + ib} \right],$$

and so we have

---

[14]The singularity in (8.7.1) is called first-order because it appears to the first power. By extension, $\frac{f(z)}{(z-z_0)^2}$ has a *second-order* singularity, and so on. I'll say much more about high-order singularities in the next section.

Fig. 8.7.1 A contour enclosing a single, first-order singularity

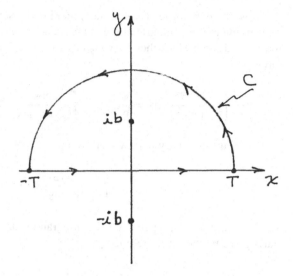

$$\frac{1}{i2b}\left[\oint_C \frac{e^{iaz}}{z-ib}dz - \oint_C \frac{e^{iaz}}{z+ib}dz\right] = \int_{-T}^{T}\frac{e^{iax}}{b^2+x^2}dx + \int_0^{\pi}\frac{e^{ia\left(Te^{i\theta}\right)}}{b^2+T^2e^{i2\theta}}iTe^{i\theta}d\theta.$$

Since the integrand of the second contour integral on the left-hand side is analytic everywhere inside of C—that integrand does have a singularity, yes, but it's at $z = -ib$ which is *outside* of C, as shown in Fig. 8.7.1—then we know from Cauchy's first integral theorem that the second contour integral on the left-hand side is zero. And once $T > b$ (remember, eventually we are going to let $T \to \infty$) then the singularity for the remaining contour integral on the left is *inside* C, at $z = ib$. Thus,

$$\frac{1}{i2b}\oint_C \frac{e^{iaz}}{z-ib}dz = \int_{-T}^{T}\frac{e^{iax}}{b^2+x^2}dx + \int_0^{\pi}\frac{e^{ia\left(Te^{i\theta}\right)}}{b^2+T^2e^{i2\theta}}iTe^{i\theta}d\theta.$$

The integrand of the contour integral on the left looks exactly like $f(z)/(z - z_0)$, with $f(z) = e^{iaz}$ and $z_0 = ib$. Cauchy's second integral theorem tells us that, if $T > b$, the contour integral is equal to $2\pi i\, f(z_0)$, and so the left-hand side of the last equation is equal to

$$\frac{1}{i2b}2\pi i\, e^{ia(ib)} = \frac{\pi}{b}e^{-ab}.$$

That is,

$$\int_{-T}^{T}\frac{e^{iax}}{b^2+x^2}dx + \int_0^{\pi}\frac{e^{ia\left(Te^{i\theta}\right)}}{b^2+T^2e^{i2\theta}}iTe^{i\theta}d\theta = \frac{\pi}{b}e^{-ab},\ T > b.$$

Now, if we at last let $T \to \infty$ then, making the same sort of argument that we did concerning the line integral along the circular arc in the previous section, we see that the second integral on the left vanishes like $\frac{1}{T}$. And so, using Euler's formula, we have

$$\int_{-\infty}^{\infty} \frac{e^{iax}}{b^2 + x^2} dx = \frac{\pi}{b} e^{-ab} = \int_{-\infty}^{\infty} \frac{\cos(ax)}{b^2 + x^2} dx + i \int_{-\infty}^{\infty} \frac{\sin(ax)}{b^2 + x^2} dx.$$

Equating imaginary parts we arrive at

$$\int_{-\infty}^{\infty} \frac{\sin(ax)}{b^2 + x^2} dx = 0$$

which is surely no surprise since the integrand is an odd function of x. Equating real parts gives us the far more interesting

$$\int_{-\infty}^{\infty} \frac{\cos(ax)}{b^2 + x^2} dx = \frac{\pi}{b} e^{-ab}.$$

This is a result we've already derived using 'routine' methods—see (3.1.7). We also did it, using the concept of the energy spectrum of a time signal, in Challenge Problem 7.7 (you *did* do that problem, right?). As I've said before, it's good to see contour integration in agreement with previous analysis.

Okay, here's how to see what's behind Cauchy's second integral theorem. The proof is beautifully elegant. In Fig. 8.7.2 I have drawn the contour C and, in its

Fig. 8.7.2 A simple curve C (enclosing point z_0) connected to an inner circle C^* via the cross-cut ab

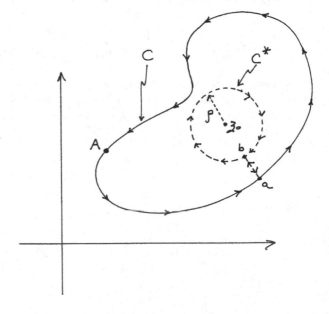

interior, marked the point z_0. In addition, centered on z_0 I've drawn a *circle* C^* with a radius ρ that is sufficiently small that C^* lies completely in the interior of C. Now, imagine that, starting on C at some arbitrary point (call it A), we begin to travel along C in the positive (CCW) sense until we reach point a, whereupon we then travel inward to point b on C^*. Once at point b we travel CW (that is, in the *negative* sense) along C^* until we return to point b. We then travel back out to C along the same path we traveled inward on until we return to point a. We then continue on along C in the CCW sense until we return to our starting point A.

Here's the first of two *crucially important* observations on what we've just done. *The complete path we've followed has always kept the annular region between C and C^* to our left.* That is, this path is the edge of a region which does *not* contain the point z_0. So, in that annular region from which $z = z_0$ has been excluded by construction, $f(z)/(z - z_0)$ is analytic everywhere. Thus, by Cauchy's first integral theorem, since $z = z_0$ is *outside* C we have

$$\oint_{C,ab,-C^*,ba} \frac{f(z)}{z - z_0} \, dz = 0. \tag{8.7.2}$$

The reason for writing $-C^*$ in the path description of the contour integral is that we went around C^* in the *negative* sense.

Here's the second of our two *crucially important* observations. The two trips along the ab-connection between C and C^* (mathematicians call this two-way connection a *cross-cut*) are in opposite directions and so *cancel each other*. That means we can write (8.7.2) as

$$\oint_{C,-C^*} \frac{f(z)}{z - z_0} \, dz = 0 = \oint_C \frac{f(z)}{z - z_0} \, dz - \oint_{C^*} \frac{f(z)}{z - z_0} \, dz. \tag{8.7.3}$$

The reason for the minus sign in front of the C^* contour integral at the far-right of (8.7.3) is, again, because we went around C^* in the negative sense. The two far-right integrals in (8.7.3) are in the *positive* sense, however, and so the minus sign has been moved from the $-C^*$ path descriptor at the bottom of the integral sign to the front of the integral, itself.

Now, while C is an arbitrary simple curve enclosing z_0, C^* is a *circle* with radius ρ centered on z_0. So, on C^* we can write $z = z_0 + \rho e^{i\theta}$ (which means $dz = i\rho e^{i\theta} d\theta$) and, therefore, as θ varies from 0 to 2π on our one complete trip around C^*, (8.7.3) becomes

$$\oint_C \frac{f(z)}{z - z_0} \, dz = \oint_{C^*} \frac{f(z)}{z - z_0} \, dz = \int_0^{2\pi} \frac{f(z_0 + \rho e^{i\theta})}{\rho e^{i\theta}} i\rho e^{i\theta} d\theta = i \int_0^{2\pi} f(z_0 + \rho e^{i\theta}) d\theta.$$

If the integral on the far left is to have a value then, whatever it is must be *independent* of ρ. After all, the integral at the far left has no ρ in it! So, the integral on the far right must be independent of ρ, too, even though it *does* have ρ in it. That means we must be able to use any value of ρ we wish. So, let's use a value for ρ that is convenient.

In particular, let's use a very small value, indeed one so small as to make the difference between $f(z)$ and $f(z_0)$, for all z on C^*, as small as we like. We can do this because $f(z)$ is assumed to be analytic, and so has a derivative everywhere inside C (including at $z = z_0$), and so is certainly continuous there. Thus, as $\rho \rightarrow 0$ we can argue $f(z) \rightarrow f(z_0)$ all along C^* and thus

$$\oint_C \frac{f(z)}{z - z_0} dz = i \int_0^{2\pi} f(z_0) d\theta.$$

Finally, pulling the *constant* $f(z_0)$ out of the integral, we have

$$\oint_C \frac{f(z)}{z - z_0} dz = i f(z_0) \int_0^{2\pi} d\theta = 2\pi i \, f(z_0),$$

which is (8.7.1) and our proof of Cauchy's second integral theorem is done.

We can now do the integral in (8.6.5) for the case of $b^2 < 4ac$. That is, we'll now study the contour integral

$$\oint_C \frac{dz}{az^2 + bz + c}, a \neq 0, b^2 < 4ac.$$

The integrand of this integral has two singularities, neither of which is on the real axis. Since $b^2 < 4ac$ these singularities are complex, and are given by

$$z_2 = -\frac{b}{2a} - i\frac{1}{2a} \sqrt{4ac - b^2}$$

and

$$z_1 = -\frac{b}{2a} + i\frac{1}{2a} \sqrt{4ac - b^2}.$$

In Fig. 8.7.3 I've shown these singular points having negative real parts, but they could be positive, depending on the signs of a and b. It really doesn't matter, however: all that matters is that with the contour C drawn in the figure only one of the singular points is inside C (arbitrarily selected to be z_1) while the other singularity (z_2) is in the exterior of C.

Now, write the integrand as a partial fraction expansion:

$$\frac{1}{az^2 + bz + c} = \frac{1}{a}\left[\frac{A}{z - z_1} + \frac{B}{z - z_2}\right] = \frac{1}{a}\left[\frac{-i\frac{a}{\sqrt{4ac - b^2}}}{z - z_1} + \frac{i\frac{a}{\sqrt{4ac - b^2}}}{z - z_2}\right].$$

Thus,

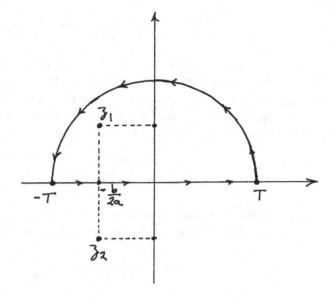

Fig. 8.7.3 A contour enclosing *one* of two singularities

$$\oint_C \frac{dz}{az^2 + bz + c} = \frac{1}{a} \oint_C \frac{-i\frac{a}{\sqrt{4ac-b^2}}}{z - z_1} dz + \frac{1}{a} \oint_C \frac{i\frac{a}{\sqrt{4ac\ b^2}}}{z - z_2} dz.$$

The second integral on the right is zero by Cauchy's first integral theorem (the singularity z_2 is not enclosed by C) and so

$$\oint_C \frac{dz}{az^2 + bz + c} = -i\frac{1}{\sqrt{4ac - b^2}} \oint_C \frac{dz}{z - z_1}.$$

From Cauchy's second integral theorem that we just proved (with $f(z) = 1$) we have

$$\oint_C \frac{dz}{z - z_1} = 2\pi i$$

and so

$$\oint_C \frac{dz}{az^2 + bz + c} = \frac{2\pi}{\sqrt{4ac - b^2}}.$$

But, the line integral around C is

$$\int_{-T}^{T} \frac{dx}{ax^2 + bx + c} + \int_0^\pi \frac{iTe^{i\theta}}{a\left(Te^{i\theta}\right)^2 + b\left(Te^{i\theta}\right) + c} d\theta$$

and the θ-integral clearly vanishes like $\frac{1}{T}$ as $T \to \infty$. Thus,

$$\int_{-\infty}^{\infty} \frac{dx}{ax^2 + bx + c} = \frac{2\pi}{\sqrt{4ac - b^2}}, a \neq 0, b^2 < 4ac. \tag{8.7.4}$$

For example, if $a = 5$, $b = 7$, and $c = 3$ (notice that $b^2 = 49 < 4ac = 4(5)(3) = 60$) then (8.7.4) says our integral is equal to $\frac{2\pi}{\sqrt{11}} = 1.8944\ldots$ and MATLAB agrees, as $integral(@(x)1./(5*x.^2 + 7*x + 3),-1e5,1e5) = 1.8944\ldots$.

For a dramatic illustration of the first and second theorems, I'll now use them to calculate an entire *class* of integrals:

$$\int_0^{\infty} \frac{x^m}{x^n + 1} dx,$$

where m and n are any non-negative integers such that (to insure the integral exists) $n - m \geq 2$. What we'll do is study the contour integral

$$\oint_C \frac{z^m}{z^n + 1} dz \tag{8.7.5}$$

with an appropriately chosen C. The integrand in (8.7.5) has n first-order singularities, at the n *n-th* roots of -1. These singular points are uniformly spaced around the unit circle in the complex plane. Since Euler's formula tells us that

$$-1 = e^{i(1+2k)\pi}.$$

for k *any* integer, then these singular points are located at

$$z_k = (-1)^{\frac{1}{n}} = e^{i\left(\frac{1+2k}{n}\right)\pi}, \quad k = 0, 1, 2, \ldots, n - 1.$$

For other values of k, of course, these same n points simply repeat. Now, let's concentrate our attention on just one of these singular points, the one for $k = 0$. We'll pick C to enclose *just that one* singularity, at $z = z_0 = e^{i\frac{\pi}{n}}$, as shown in Fig. 8.7.4. The central angle of the wedge is $\frac{2\pi}{n}$ and the singularity is at half that angle, $\frac{\pi}{n}$.

As we go around C to do the integral in (8.7.5), the descriptions of the contour's three portions are:

on C_1: $z = x$, $dz = dx$, $0 \leq x \leq T$;
on C_2: $z = Te^{i\theta}$, $dz = iTe^{i\theta}d\theta$, $0 \leq \theta \leq \frac{2\pi}{n}$;
on C_3: $z = re^{i\frac{2\pi}{n}}$, $dz = e^{i\frac{2\pi}{n}}dr$, $T \geq r \geq 0$;
So,

$$\oint_C \frac{z^m}{z^n + 1} dz = \int_0^T \frac{x^m}{x^n + 1} dx + \int_0^{\frac{2\pi}{n}} \frac{(Te^{i\theta})^m}{(Te^{i\theta})^n + 1} iTe^{i\theta}d\theta + \int_T^0 \frac{r^m e^{i\frac{m2\pi}{n}}}{r^n e^{i2\pi} + 1} e^{i\frac{2\pi}{n}} dr$$

$$= \int_0^T \frac{x^m}{x^n + 1} dx - \int_0^T \frac{r^m e^{i(m+1)\frac{2\pi}{n}}}{r^n + 1} dr + \int_0^{\frac{2\pi}{n}} \frac{T^{m+1} e^{im\theta}}{T^n e^{in\theta} + 1} ie^{i\theta}d\theta.$$

Fig. 8.7.4 A pie-shaped contour enclosing *one* of n singularities

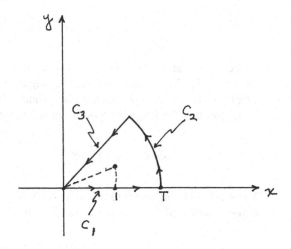

Now, clearly, as $T \to \infty$ the θ-integral goes to zero because $m + 1 < n$. Also,

$$\int_0^T \frac{r^m e^{i(m+1)\frac{2\pi}{n}}}{r^n + 1} \, dr = e^{i(m+1)\frac{2\pi}{n}} \int_0^T \frac{r^m}{r^n + 1} \, dr = e^{i(m+1)\frac{2\pi}{n}} \int_0^T \frac{x^m}{x^n + 1} \, dx.$$

So, as $T \to \infty$

$$\oint_C \frac{z^m}{z^n + 1} \, dz = \int_0^\infty \frac{x^m}{x^n + 1} \, dx \left[1 - e^{i(m+1)\frac{2\pi}{n}} \right].$$

Or, as

$$\left[1 - e^{i(m+1)\frac{2\pi}{n}} \right] = e^{i(m+1)\frac{\pi}{n}} \left[e^{-i(m+1)\frac{\pi}{n}} - e^{i(m+1)\frac{\pi}{n}} \right] = -2i \, \sin\left\{ (m+1)\frac{\pi}{n} \right\} e^{i(m+1)\frac{\pi}{n}},$$

we have

$$\oint_C \frac{z^m}{z^n + 1} \, dz = -2i \, \sin\left\{ (m+1)\frac{\pi}{n} \right\} e^{i(m+1)\frac{\pi}{n}} \int_0^\infty \frac{x^m}{x^n + 1} \, dx. \qquad (8.7.6)$$

Since

$$z^n + 1 = \prod_{k=0}^{n-1} (z - z_k)$$

we can write the integrand of the contour integral in (8.7.5) as a partial fraction expansion:

$$\frac{z^m}{z^n + 1} = \frac{N_0}{z - z_0} + \frac{N_1}{z - z_1} + \frac{N_2}{z - z_2} + \frac{N_3}{z - z_3} + \ldots + \frac{N_{n-1}}{z - z_{n-1}}$$

where the N's are constants. Integrating this expansion term-by-term, we get

$$\oint_C \frac{z^m}{z^n + 1} dz = N_0 \oint_C \frac{dz}{z - z_0}$$

since Cauchy's first integral theorem says all the other integrals are zero because, *by construction*, C does not enclose the singularities $z_1, z_2, \ldots, z_{n-1}$. The only singularity C encloses is z_0. Cauchy's second integral theorem in (8.7.1), with $f(z) = 1$, says that the integral on the right is $2\pi i$, and so (8.7.6) becomes

$$-2i \sin\left\{(m + 1)\frac{\pi}{n}\right\} e^{i(m+1)\frac{\pi}{n}} \int_0^\infty \frac{x^m}{x^n + 1} dx = 2\pi i N_0. \tag{8.7.7}$$

Our next (and final) step is to calculate N_0. To do that, multiply through the partial fraction expansion of the integrand in (8.7.5) by $z - z_0$ to get

$$\frac{(z - z_0)z^m}{z^n + 1} = N_0 + \frac{(z - z_0)N_1}{z - z_1} + \frac{(z - z_0)N_2}{z - z_2} + \frac{(z - z_0)N_3}{z - z_3} + \cdots$$

and then let $z \to z_0$. This causes all the terms on the right after the first to vanish, and so

$$N_0 = \lim_{z \to z_0} \frac{(z - z_0)z^m}{z^n + 1} = \lim_{z \to z_0} \frac{z^{m+1} - z_0 z^m}{z^n + 1} = \lim_{z \to z_0} \frac{z^{m+1} - z_0 z^m}{\left(e^{\frac{i\pi}{n}}\right)^n + 1} = \frac{0}{0}.$$

So, to resolve this indeterminacy, we'll use L'Hôpital's rule:

$$N_0 = \lim_{z \to z_0} \frac{(m + 1)z^m - m z_0 z^{m-1}}{n z^{n-1}} = \lim_{z \to z_0} \frac{m z^m + z^m - m z_0 z^{m-1}}{n z^{n-1}}$$

$$= \frac{z_0^{m-n+1}}{n}$$

or, with $z_0 = e^{\frac{i\pi}{n}}$,

$$N_0 = \frac{e^{\frac{i\pi}{n}(m-n+1)}}{n} = \frac{e^{\frac{im\pi}{n} - i\pi + \frac{i\pi}{n}}}{n} = \frac{e^{-i\pi} e^{i\left(\frac{m+1}{n}\right)\pi}}{n} = -\frac{e^{i\left(\frac{m+1}{n}\right)\pi}}{n}.$$

Inserting this result into (8.7.7),

$$\int_0^\infty \frac{x^m}{x^n + 1} dx = \frac{2\pi i\left\{-\frac{e^{i\left(\frac{m+1}{n}\right)\pi}}{n}\right\}}{-2i \sin\left\{(m + 1)\frac{\pi}{n}\right\} e^{i(m+1)\frac{\pi}{n}}},$$

and so we have the beautiful result

$$\int_0^\infty \frac{x^m}{x^n+1}dx = \frac{\frac{\pi}{n}}{\sin\left\{(m+1)\frac{\pi}{n}\right\}}, n-m \geq 2, m \geq 0. \tag{8.7.8}$$

For a specific example, you can confirm that $m = 0$ and $n = 4$ reproduces our result in (2.3.4). As a new result, if $m = 0$ and $n = 3$ then (8.7.8) says that.

$\int_0^\infty \frac{dx}{x^3+1} = \frac{\frac{\pi}{3}}{\sin\left\{\frac{\pi}{3}\right\}} = \frac{\frac{\pi}{3}}{\frac{\sqrt{3}}{2}} = \frac{2\pi}{3\sqrt{3}} = 1.209199\ldots$ and MATLAB agrees:

integral(@(x)1./(x.^3 + 1),0,inf) = 1.209199....

The result of (8.7.8) can be put into at least three alternative forms that commonly appear in the math literature. First, define $t = x^n$ and so $\frac{dt}{dx} = nx^{n-1}$ which means

$$dx = \frac{dt}{nx^{n-1}} = \frac{dt}{n(t)^{\frac{n-1}{n}}}.$$

Thus, (8.7.8) becomes

$$\int_0^\infty \frac{(t)^{\frac{m}{n}}}{t+1}\left(\frac{dt}{n(t)^{\frac{n-1}{n}}}\right) = \frac{\frac{\pi}{n}}{\sin\left\{(m+1)\frac{\pi}{n}\right\}} = \frac{1}{n}\int_0^\infty \frac{(t)^{\frac{m}{n}-\frac{n-1}{n}}}{t+1}dt$$

or,

$$\int_0^\infty \frac{(t)^{\frac{m-n+1}{n}}}{t+1}dt = \frac{\pi}{\sin\left\{(m+1)\frac{\pi}{n}\right\}} = \int_0^\infty \frac{(t)^{\frac{m+1}{n}-1}}{t+1}dt.$$

Now, define

$$a = \frac{m+1}{n},$$

which says[15]

$$\int_0^\infty \frac{x^{a-1}}{x+1}dx = \frac{\pi}{\sin(a\pi)}, \quad 0 < a < 1. \tag{8.7.9}$$

For example, if $a = \frac{1}{2}$ then

---

[15]The limits on a are because, first, since $n - m \geq 2$ it follows that $m + 1 \leq n - 1$ and so $a < 1$. Also, for $x \ll 1$ the integrand in (8.7.9) behaves as x^{a-1} which integrates to $\frac{x^a}{a}$ and this blows-up at the lower limit of integration if $a < 0$. So, $0 < a$.

$$\int_0^\infty \frac{dx}{\sqrt{x}\,(x+1)} = \frac{\pi}{\sin\left(\frac{\pi}{2}\right)} = \pi$$

and MATLAB agrees as *integral(@(x)1./(sqrt(x).*(x + 1)),0,inf) = 3.14159265....*
    Another way to reformulate (8.7.8) is to start with (8.7.9) and define t = ln (x), and so

$$\frac{dt}{dx} = \frac{1}{x} = \frac{1}{e^t}.$$

Thus, (8.7.9) becomes (because x = 0 means t = − ∞)

$$\int_{-\infty}^\infty \frac{e^{t(a-1)}}{1+e^t}e^t dt = \int_{-\infty}^\infty \frac{e^{at}}{1+e^t}\,dt.$$

That is,

$$\int_{-\infty}^\infty \frac{e^{ax}}{1+e^x}\,dx = \frac{\pi}{\sin{(a\pi)}}, \quad 0 < a < 1. \tag{8.7.10}$$

For example, if a = $\frac{1}{3}$ the integral equals $\frac{\pi}{\sin{(\pi/3)}} = \frac{\pi}{\sqrt{3}/2} = 2\frac{\pi}{\sqrt{3}} = 3.62759\ldots$ and MATLAB agrees as *integral(@(x)exp(x/3)./(1 + exp(x)),-1e3,1e3) = 3.62759....*
It's interesting to compare (8.7.10) with (8.6.9).
    And finally, in (8.7.9) make the change of variable

$$u = \frac{1}{x}$$

and so

$$dx = -\frac{du}{u^2}.$$

Then,

$$\int_0^\infty \frac{x^{a-1}}{x+1}\,dx = \int_\infty^0 \frac{\left(\frac{1}{u}\right)^{a-1}}{\left(\frac{1}{u}+1\right)}\left(-\frac{du}{u^2}\right) = \int_0^\infty \frac{du}{\left(\frac{1+u}{u}\right)(u^{a-1})u^2} = \int_0^\infty \frac{du}{\left(\frac{1+u}{u}\right)\left(\frac{u^a}{u}\right)u^2},$$

and so

$$\int_0^\infty \frac{dx}{(1+x)x^a} = \frac{\pi}{\sin{(a\pi)}}, \quad 0 < a < 1. \tag{8.7.11}$$

I'll end this section with two examples of the use of Cauchy's second integral theorem. In the first one multiple singularities of first-order appear when, for the arbitrary positive constant a, we'll calculate the value of

$$\int_0^{2\pi} \frac{d\theta}{a + \sin^2(\theta)}.$$

We can handle the multiple singularities by simply using the cross-cut idea from earlier in this section. That is, as we travel around a contour C, just move inward along a cross-cut to the first singularity and then travel around it on a tiny circle with radius ρ and then back out along the cross-cut to C. Then, after traveling a bit more on C do the same thing with a new cross-cut to the second singularity. And so on, for all the rest of the singularities. ('Tiny' means pick ρ small enough that none of the singularity circles intersect, and that all are always inside C.) For each singularity we'll pick-up a value of $2\pi i\, f(z_0)$, where the integrand of the contour integral we are studying is $\frac{f(z)}{(z-z_0)}$.

So, what contour integral *will* we be studying? On the unit circle C we have

$$z = e^{i\theta} = \cos(\theta) + i\,\sin(\theta),\ \frac{1}{z} = e^{-i\theta} = \cos(\theta) - i\,\sin(\theta) \qquad (8.7.12)$$

and so

$$dz = ie^{i\theta}\,d\theta = iz\,d\theta \left(\text{and so } d\theta = \frac{dz}{iz}\right)$$

as well as

$$\sin(\theta) = \frac{z - \frac{1}{z}}{2i} = \frac{z^2 - 1}{2zi}.$$

Thus,

$$\sin^2(\theta) = -\frac{(z^2 - 1)^2}{4z^2}$$

and so the contour integral we'll study is

$$\oint_C \frac{\frac{dz}{iz}}{a - \frac{(z^2-1)^2}{4z^2}} = 4i\oint_C \frac{z\,dz}{z^4 - z^2(2 + 4a) + 1}. \qquad (8.7.13)$$

The integrand clearly has four first-order singularities, all located on the real axis at:

$$z = \pm\sqrt{u}, u = (1 + 2a) + 2\sqrt{a(a+1)}$$

and

$$z = \pm\sqrt{u}, u = (1 + 2a) - 2\sqrt{a(a+1)}.$$

By inspection it is seen that for the first pair of singularities $|z| > 1$ and so both lie *outside* C, while for the second pair $|z| < 1$ and so both lie *inside* C. Specifically, let's write z_1 and z_2 as the *inside* singularities where

$$z_1 = \sqrt{(1 + 2a) - 2\sqrt{a(a+1)}}$$

and

$$z_2 = -\sqrt{(1 + 2a) - 2\sqrt{a(a+1)}} = -z_1,$$

while z_3 and z_4 are the *outside* singularities where

$$z_3 = \sqrt{(1 + 2a) + 2\sqrt{a(a+1)}}$$

and

$$z_4 = -\sqrt{(1 + 2a) + 2\sqrt{a(a+1)}} = -z_3.$$

The integrand of the contour integral on the right in (8.7.13) is

$$\frac{z}{(z - z_1)(z - z_2)(z - z_3)(z - z_4)} = \frac{z}{(z - z_1)(z + z_1)(z - z_3)(z + z_3)}$$

$$= \frac{z}{(z - z_1)(z + z_1)(z^2 - z_3^2)}$$

or, making a partial fraction expansion,

$$\frac{z}{(z - z_1)(z + z_1)(z^2 - z_3^2)} = \frac{N_1}{z - z_1} + \frac{N_2}{z + z_1} + \frac{N_3}{z - z_3} + \frac{N_4}{z + z_3}. \qquad (8.7.14)$$

Thus, the contour integral from (8.7.13) is

$$\oint_C \frac{z\,dz}{z^4 - z^2(2 + 4a) + 1} = N_1 \oint_C \frac{dz}{z - z_1} + N_2 \oint_C \frac{dz}{z + z_1} + N_3 \oint_C \frac{dz}{z - z_3}$$

$$+ N_4 \oint_C \frac{dz}{z + z_3}. \qquad (8.7.15)$$

Since the singularities $z = \pm z_3$ lie outside the unit circle C, Cauchy's first integral theorem tells us that the last two contour integrals on the right of (8.7.15) each vanish, and so the values of N_3 and N_4 are of no interest. That's not the case for the first two contour integrals, however, as the singularities $z = \pm z_1$ lie inside the unit circle C. So, we need to calculate N_1 and N_2, and they can be easily determined as follows. For N_1, multiply through (8.7.14) by $(z - z_1)$ and then let $z \to z_1$, to get

$$N_1 = \lim_{z \to z_1} \frac{z}{(z + z_1)(z^2 - z_3^2)} = \frac{z_1}{2z_1(z_2^2 - z_3^2)} = \frac{1}{2(z_1^2 - z_3^2)}$$

$$= \frac{1}{2\left[\{(1 + 2a) - 2\sqrt{a(a+1)}\} - \{(1 + 2a) + 2\sqrt{a(a+1)}\}\right]}$$

$$= -\frac{1}{8\sqrt{a(a+1)}}.$$

To find N_2, multiply through (8.7.14) by $(z + z_1)$ and then let $z \to -z_1$, to get

$$N_2 = \lim_{z \to -z_1} \frac{z}{(z - z_1)(z^2 - z_3^2)}$$

$$= \frac{-z_1}{-2z_1\left[\{(1 + 2a) - 2\sqrt{a(a+1)}\} - \{(1 + 2a) + 2\sqrt{a(a+1)}\}\right]}$$

$$= -\frac{1}{8\sqrt{a(a+1)}} = N_1.$$

Thus, Cauchy's second integral theorem tells us that

$$N_1 \oint_C \frac{dz}{z - z_1} + N_2 \oint_C \frac{dz}{z + z_1} = -\frac{1}{8\sqrt{a(a+1)}}(2\pi i + 2\pi i) = -\frac{\pi i}{2\sqrt{a(a+1)}}.$$

Looking back at (8.7.13) one last time, the integral we are after is $4i$ times this last result, and so

$$\int_0^{2\pi} \frac{d\theta}{a + \sin^2(\theta)} = \frac{2\pi}{\sqrt{a(a+1)}}, a > 0. \tag{8.7.16}$$

If $a = 3$ this is $1.81379936\ldots$, and MATLAB agrees because *integral(@(x)1./ (3 + (sin(x).^2)),0,2*pi)* $= 1.81379936\ldots$.

For a second example of Cauchy's second integral theorem, here's a little twist to the calculations. We'll evaluate

$$\int_0^\infty \frac{\cos(x)}{e^x + e^{-x}} dx$$

and so the obvious contour integral for us to study is

$$\oint_C \frac{\left(\frac{e^{iz}+e^{-iz}}{2}\right)}{e^z + e^{-z}}\,dz$$

as the integrand reduces to our desired integrand when z is on the real axis (where z = x).

One immediate question is, of course, what's C? And a second concern, which may (or may not) occur to you right away is that there is no apparent '(z − z_0)' in the denominator of the integrand (which the second theorem explicitly contains in its statement). The answers to these two concerns are, fortunately, not difficult to find. To start, let's determine where the singularities (if any) of the integrand are located. Any singularities are the solutions to $e^z + e^{-z} = 0$, and thus $e^{2z} = -1$. Since Euler's identity tells us that, for k any integer, $e^{i(\pi + 2\pi k)} = -1$, then there *are* singularities, infinite in number, all located on the imaginary axis at $z = i\pi\left(k + \frac{1}{2}\right)$, k = ... − 2, − 1, 0, 1, 2, ... This result helps us in deciding what contour C we should use.

If you recall the discussion of the contour of Fig. 8.6.3, we avoided using a semi-circular contour then because of the very same situation we have now—an infinite number of singularities on the imaginary axis that an expanding semi-circular contour would enclose. Instead, we earlier used a *rectangular* contour of fixed height that, as we expanded its *width* (along the real axis), always enclosed just *one* singularity. We'll do the same here, as well, with the contour of Fig. 8.7.5, with the single (k = 0) singularity $z_0 = i\frac{\pi}{2}$ enclosed by $C = C_1 + C_2 + C_3 + C_4$. This is actually an easier contour with which to work than was Fig. 8.6.3, as now we have no indents to consider.

To use Cauchy's second theorem, we'll write

$$\oint_C \frac{g(z)}{z - z_0}\,dz = 2\pi i g(z_0), z_0 = i\frac{\pi}{2}, \tag{8.7.17}$$

where

Fig. 8.7.5 A rectangular contour enclosing one singularity

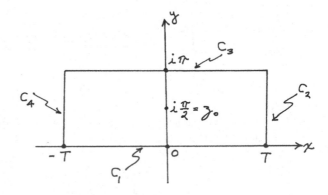

$$\frac{g(z)}{z - z_0} = \frac{\left(\frac{e^{iz}+e^{-iz}}{2}\right)}{e^z + e^{-z}}.$$

That is, $g(z_0)$ in (8.7.17) is given by

$$g(z_0) = \lim_{z\to z_0} g(z) = \lim(z - z_0)_{z\to z_0}\frac{e^{iz} + e^{-iz}}{2(e^z + e^{-z})} = \lim_{z\to i\frac{\pi}{2}}\frac{z - i\frac{\pi}{2}}{2\left(\frac{e^z+e^{-z}}{e^{iz}+e^{-iz}}\right)}$$

$$= \frac{0}{2\left(\frac{e^{\frac{i\pi}{2}}+e^{-\frac{i\pi}{2}}}{e^{-\frac{\pi}{2}}+e^{\frac{\pi}{2}}}\right)}$$

$$= \frac{0}{\cos\left(\frac{\pi}{2}\right)} = \frac{0}{0}.$$

To resolve this indeterminacy, we'll use L'Hôpital's rule. Thus,

$$g(z_0) = \lim_{z\to i\frac{\pi}{2}}\frac{1}{2\left[\frac{(e^{iz}+e^{-iz})(e^z-e^{-z})-(e^z+e^{-z})(ie^{iz}-i\,e^{-iz})}{(e^{iz}+e^{-iz})^2}\right]}$$

$$= \frac{\frac{1}{2}\left(e^{-\frac{\pi}{2}} + e^{\frac{\pi}{2}}\right)^2}{(e^{-\frac{\pi}{2}} + e^{\frac{\pi}{2}})(e^{\frac{i\pi}{2}} - e^{-\frac{i\pi}{2}}) - (e^{\frac{i\pi}{2}} + e^{-\frac{i\pi}{2}})i(e^{-\frac{\pi}{2}} - e^{\frac{\pi}{2}})}$$

$$= \frac{\frac{1}{2}\left(e^{-\frac{\pi}{2}} + e^{\frac{\pi}{2}}\right)^2}{(e^{-\frac{\pi}{2}} + e^{\frac{\pi}{2}})2i\sin\left(\frac{\pi}{2}\right) - 2\cos\left(\frac{\pi}{2}\right)i\,(e^{-\frac{\pi}{2}} - e^{\frac{\pi}{2}})}.$$

Since $\sin\left(\frac{\pi}{2}\right) = 1$ and $\cos\left(\frac{\pi}{2}\right) = 0$, then we have

$$g(z_0) = \frac{e^{-\frac{\pi}{2}} + e^{\frac{\pi}{2}}}{4i}$$

and (8.7.17) becomes

$$\oint_C \frac{e^{iz} + e^{-iz}}{2(e^z + e^{-z})}\,dz = 2\pi i g(z_0) = \frac{\pi}{2}\left(e^{-\frac{\pi}{2}} + e^{\frac{\pi}{2}}\right). \qquad (8.7.18)$$

So, writing out (8.7.18) explicitly, we have, starting with C_1 on the real axis (where $z = x$),

$$\int_{-T}^{T}\frac{\cos(x)}{e^x + e^{-x}}\,dx + \int_{C_2} + \int_{C_3} + \int_{C_4} = \frac{\pi}{2}\left(e^{-\frac{\pi}{2}} + e^{\frac{\pi}{2}}\right). \qquad (8.7.19)$$

Our next task is the evaluation of the C_2, C_3, and C_4 integrals. On C_2 we have $z = T + iy$ ($dz = idy$), on C_3 we have $z = x + i\pi$ ($dz = dx$), and on C_4 we have $z = -T + iy$ ($dz = idy$). In turn, then:

On C_2,

$$\int_{C_2} = \frac{1}{2} \int_0^\pi \frac{e^{i(T+iy)} + e^{-i(T+iy)}}{e^{(T+iy)} + e^{-(T+iy)}} idy = \frac{i}{2} \int_0^\pi \frac{e^{-y}e^{iT} + e^y e^{-iT}}{e^T e^{iy} + e^{-T}e^{-iy}} dy \to_{T\to\infty} = 0.$$

On C_4,

$$\int_{C_4} = \frac{1}{2} \int_\pi^0 \frac{e^{i(-T+iy)} + e^{-i(-T+iy)}}{e^{(-T+iy)} + e^{-(-T+iy)}} idy = \frac{i}{2} \int_\pi^0 \frac{e^{-y}e^{-iT} + e^y e^{iT}}{e^{-T} e^{iy} + e^T e^{-iy}} dy \to_{T\to\infty} = 0.$$

On C_3,

$$\int_{C_3} = \frac{1}{2} \int_T^{-T} \frac{e^{i(x+i\pi)} + e^{-i(x+i\pi)}}{e^{(x+i\pi)} + e^{-(x+i\pi)}} dx = -\frac{1}{2} \int_{-T}^T \frac{e^{-\pi}e^{ix} + e^\pi e^{-ix}}{e^x e^{i\pi} + e^{-x}e^{-i\pi}} dx$$

or, as $e^{i\pi} = e^{-i\pi} = -1$,

$$\int_{C_3} = \frac{1}{2} \int_{-T}^T \frac{e^{-\pi}e^{ix} + e^\pi e^{-ix}}{e^x + e^{-x}} dx \to_{T\to\infty} = \frac{1}{2} \int_{-\infty}^\infty \frac{e^{-\pi}e^{ix} + e^\pi e^{-ix}}{e^x + e^{-x}} dx.$$

Putting these results into (8.7.19), with $T \to \infty$, we have

$$\int_{-\infty}^\infty \frac{\cos(x)}{e^x + e^{-x}} dx + \frac{1}{2} \int_{-\infty}^\infty \frac{e^{-\pi}e^{ix} + e^\pi e^{-ix}}{e^x + e^{-x}} dx = \frac{\pi}{2}\left(e^{-\frac{\pi}{2}} + e^{\frac{\pi}{2}}\right).$$

Or, replacing $\cos(x)$ with complex exponentials,

$$\int_{-\infty}^\infty \frac{\frac{1}{2}e^{ix} + \frac{1}{2}e^{-ix} + \frac{1}{2}e^{-\pi}e^{ix} + \frac{1}{2}e^\pi e^{-ix}}{e^x + e^{-x}} dx = \frac{\pi}{2}\left(e^{-\frac{\pi}{2}} + e^{\frac{\pi}{2}}\right)$$

$$= \int_{-\infty}^\infty \frac{\frac{1}{2}e^{ix}(1 + e^{-\pi}) + \frac{1}{2}e^{-ix}(1 + e^\pi)}{e^x + e^{-x}} dx$$

$$= \int_{-\infty}^\infty \frac{\frac{1}{2}e^{ix}e^{-\frac{\pi}{2}}\left(e^{\frac{\pi}{2}} + e^{-\frac{\pi}{2}}\right) + \frac{1}{2}e^{-ix}e^{\frac{\pi}{2}}\left(e^{-\frac{\pi}{2}} + e^{\frac{\pi}{2}}\right)}{e^x + e^{-x}} dx$$

and so, cancelling the common $\left(e^{-\frac{\pi}{2}} + e^{\frac{\pi}{2}}\right)$ factors,

$$\frac{1}{2} \int_{-\infty}^\infty \frac{e^{ix}e^{-\frac{\pi}{2}} + e^{-ix}e^{\frac{\pi}{2}}}{e^x + e^{-x}} dx = \frac{\pi}{2}.$$

Finally, equating real parts on each side of the equality,[16]

$$\frac{1}{2}\int_{-\infty}^{\infty} \frac{\cos{(x)}e^{-\frac{x}{2}} + \cos{(x)}e^{\frac{x}{2}}}{e^{x} + e^{-x}} = \frac{\pi}{2}$$

or,

$$\int_{-\infty}^{\infty} \frac{\cos{(x)} + \cos{(x)}}{e^{x} + e^{-x}} dx = \frac{\pi}{(e^{\frac{\pi}{2}} + e^{-\frac{\pi}{2}})}$$

or, as $\int_{-\infty}^{\infty} = 2\int_{0}^{\infty}$ since the integrand is even, we at last have our answer:

$$\int_{0}^{\infty} \frac{\cos{(x)}}{e^{x} + e^{-x}} dx = \frac{\frac{\pi}{2}}{(e^{\frac{\pi}{2}} + e^{-\frac{\pi}{2}})}. \tag{8.7.20}$$

This equals $0.31301008281\ldots$ and MATLAB agrees, as $integral(@(x)cos(x)./$ $(exp(x) + exp(-x)),0,inf) = 0.31301008281\ldots$.

8.8 Singularities and the Residue Theorem

In this section we'll derive the wonderful *residue theorem*, which will reduce what appear to be astoundingly difficult definite integrals to ones being simply of 'routine' status. We start with a $f(z)$ that is analytic everywhere in some region **R** in the complex plane *except* at the point $z = z_0$ which is a singularity of order $m \geq 1$. That is,

$$f(z) = \frac{g(z)}{(z - z_0)^m} \tag{8.8.1}$$

where $g(z)$ is analytic throughout **R**. Because $g(z)$ is analytic we know it is 'well-behaved,' which is math-lingo for 'all the derivatives of $g(z)$ exist.' (Take a look back at (8.7.1) and the comment that follows it.) That means $g(z)$ has a *Taylor series expansion* (discussed in more detail in the next chapter, after (9.2.18)) about $z = z_0$ and so we can write

$$g(z) = c_0 + c_1(z - z_0) + c_2(z - z_0)^2 + c_3(z - z_0)^3 + \ldots. \tag{8.8.2}$$

Putting (8.8.2) into (8.8.1) gives

---

[16]If you equate imaginary parts you get $\int_{-\infty}^{\infty} \frac{\sin{(x)}}{e^{x} + e^{-x}} dx = 0$, which is trivially true since the integrand is odd.

$$f(z) = \frac{c_0}{(z - z_0)^m} + \frac{c_1}{(z - z_0)^{m-1}} + \frac{c_2}{(z - z_0)^{m-2}} + .. + c_m + c_{m+1}(z - z_0)$$

$$+ c_{m+1}(z - z_0)^2 + \ldots$$

or, as it is usually written,

$$f(z) = \sum_{n=0}^{\infty} a_n(z - z_0)^n + \sum_{n=1}^{\infty} \frac{b_n}{(z - z_0)^n} \tag{8.8.3}$$

where in the second sum all the $b_n = 0$ for $n > m$.

The unique existence of a Taylor series expansion for every distinct, infinitely differentiable function was long an accepted mathematical fact. That is, it was until 1823, when Cauchy gave an astounding counterexample. Consider what has to be the most trivial function that one can imagine: $f(x) = 0$ for all x. The Taylor series expansion for *any* differentiable function around $x = 0$ is $f(x) = \sum_{k=0}^{\infty} \frac{x^k}{k!} f^{(k)}(0)$, where $f^{(k)}(x)$ is the kth derivative of the function. For the trivial $f(x) = 0$ every one of those derivatives is obviously zero. Cauchy's counterexample was the demonstration of another function that is clearly *not* identically zero and yet has every one of its infinity of derivatives at $x = 0$ also equal to zero! Cauchy's function, $f(x) = e^{-\frac{1}{x^2}}$, is easy to differentiate, and you should be able to convince yourself that, for any k, $f^{(k)}(x)$ is a polynomial in $\frac{1}{x}$, multiplied by $e^{-\frac{1}{x^2}}$, and that $x^{-k} e^{-\frac{1}{x^2}} \to 0$ as $x \to 0$ for all values of k. None of this will cause any trouble for anything we do in this book, but you should know that things are not quite so benign as I may have led you to believe.

The series expansion in (8.8.3) of $f(z)$, an expansion about a singular point that involves both positive *and* negative powers of $(z - z_0)$, is called the *Laurent series* of $f(z)$, named after the French mathematician Pierre Alphonse Laurent (1813–1854) who developed it in 1843. (In books dealing with complex analysis in far more detail than I am doing here, it is shown that the Laurent series expansion is unique.) We can find formulas for the a_n and b_n coefficients in (8.8.3) as follows. Begin by observing that if k is any integer (negative, zero, or positive), then if C is a circle of radius ρ centered on z_0 (which means that on C we have $z = z_0 + \rho e^{i\theta}$), then

$$\oint_C (z - z_0)^k dz = \int_0^{2\pi} \rho^k e^{ik\theta} i\rho e^{i\theta} d\theta = i\rho^{k+1} \int_0^{2\pi} e^{i(k+1)\theta} d\theta$$

$$= i\rho^{k+1} \left\{ \frac{e^{i(k+1)\theta}}{i(k+1)} \right\} \Big|_0^{2\pi} = \frac{\rho^{k+1}}{k+1} \left\{ e^{i(k+1)\theta} \right\} \Big|_0^{2\pi}.$$

As long as $k \neq -1$ this last expression is 0. If, on the other hand, $k = -1$ our expression becomes the indeterminate $\frac{0}{0}$. To get around that, for $k = -1$ simply back-up a couple of steps and write

$$\oint_C (z - z_0)^{-1} dz = \int_0^{2\pi} \frac{1}{\rho e^{i\theta}} i\rho e^{i\theta} d\theta = i \int_0^{2\pi} d\theta = 2\pi i.$$

That is, for k any integer,

$$\oint_C (z - z_0)^k dz = \begin{cases} 0, & k \neq -1 \\ 2\pi i, & k = -1 \end{cases}. \tag{8.8.4}$$

So, to find a particular a-coefficient (say, a_j) in the Laurent series for f(z), simply divide through (8.8.3) by $(z - z_0)^{j+1}$ and integrate term-by-term. All of the integrals will vanish because of (8.8.4) *with a single exception*:

$$\oint_C \frac{f(z)}{(z - z_0)^{j+1}} dz = \oint_C \frac{a_j}{z - z_0} dz = 2\pi i a_j.$$

That is,

$$a_j = \frac{1}{2\pi i} \oint_C \frac{f(z)}{(z - z_0)^{j+1}} dz, \quad j = 0, 1, 2, 3, \dots. \tag{8.8.5}$$

And to find a particular b-coefficient (say, b_j), simply multiply by $(z - z_0)^{j-1}$ through (8.8.3) and integrate term-by-term. All of the integrals will vanish because of (8.8.4) *with a single exception*:

$$\oint_C f(z)(z - z_0)^{j-1} dz = \oint_C b_j (z - z_0)^{-1} dz = \oint_C \frac{b_j}{z - z_0} dz = 2\pi i b_j.$$

That is,

$$b_j = \frac{1}{2\pi i} \oint_C \frac{f(z)}{(z - z_0)^{-j+1}} dz, \quad j = 1, 2, 3, \dots. \tag{8.8.6}$$

One of the true miracles of contour integration is that, of the potentially infinite number of coefficients given by the formulas (8.8.5) and (8.8.6), only *one* will be of interest to us. That chosen one is b_1 and here's why. If we set $j = 1$ in (8.8.6) then[17]

$$\oint_C f(z) dz = 2\pi i b_1, \tag{8.8.7}$$

---

[17]The contour C in (8.8.7) has been a *circle* of radius ρ up to this point, but in fact by using the cross-cut idea of Figure 8.7.2 we can think of C as being *any* contour enclosing z_0 such that f(z) is everywhere analytic on and within C (except at z_0, of course).

which is almost precisely (to within a factor of $2\pi i$) the ultimate quest of our calculations, the determination of

$$\oint_C f(z)dz.$$

But of course we don't do the integral to find b_1 (if we could directly do the integral, who cares about b_1?!), but rather we *reverse the process* and calculate b_1 *by some means other than integration* and then *use* that result in (8.8.7) to find the integral. The value of b_1 is called the *residue* of $f(z)$ at the singularity $z = z_0$.

What does 'some means other than integration' mean? As it turns out, it is not at all difficult to get our hands on b_1. Let's suppose (as we did at the start of this section) that $f(z)$ has a singularity of order m. That is, writing-out (8.8.3) in just a bit more detail,

$$f(z) = \ldots + a_1(z - z_0) + a_0 + \frac{b_1}{z - z_0} + \frac{b_2}{(z - z_0)^2} + \ldots + \frac{b_m}{(z - z_0)^m}.$$

So, multiplying through by $(z - z_0)^m$ gives

$$(z - z_0)^m f(z) = \ldots + a_1(z - z_0)^{m+1} + a_0(z - z_0)^m + b_1(z - z_0)^{m-1}$$
$$+ b_2(z - z_0)^{m-2} + \ldots + b_m.$$

Next, differentiate with respect to z a total of $m - 1$ times. That has three effects: (1) all the a-coefficient terms will retain a factor of $(z - z_0)$ to at least the first power; (2) the b_1 term will be multiplied by $(m - 1)!$, but will have no factor involving $(z - z_0)$; and (3) all the other b-coefficient terms will be differentiated to zero. Thus, if we then let $z \to z_0$ the a-coefficient terms will vanish and we'll be left with nothing but $(m - 1)! \, b_1$. Therefore,

$$b_1 = \frac{1}{(m - 1)!} \lim_{z \to z_0} \frac{d^{m-1}}{dz^{m-1}} \{(z - z_0)^m f(z)\} \tag{8.8.8}$$

where z_0 is a m-order singularity of $f(z)$.

For a first-order singularity ($m = 1$) the formula in (8.8.8) reduces, with the interpretation of $\frac{d^{m-1}}{dz^{m-1}} = 1$ if $m = 1$, to

$$b_1 = \lim_{z \to z_0} (z - z_0)f(z).$$

Alternatively, write

$$f(z) = \frac{g(z)}{h(z)}$$

where, as before, $g(z)$ is analytic at the singularity $z = z_0$ (which is then, of course, a first-order *zero* of $h(z)$). That is,

$$h(z_0) = 0.$$

Then,

$$\lim_{z \to z_0}(z - z_0)f(z) = \lim_{z \to z_0}(z - z_0)\frac{g(z)}{h(z)} = \lim_{z \to z_0}\frac{g(z)}{\frac{h(z)-h(z_0)}{z-z_0}}$$

where the denominator (which you no doubt recognize as the definition of the derivative of $h(z)$) on the far-right follows because $h(z) = h(z) - h(z_0)$ because $h(z_0) = 0$. So,

$$\lim_{z \to z_0}(z - z_0)f(z) = \frac{g(z_0)}{\frac{d}{dz}h(z)\big|_{z=z_0}}.$$

That is, the residue for a *first-order* singularity at $z = z_0$ in the integrand $f(z) = \frac{g(z)}{h(z)}$ can be computed as

$$b_1 = \frac{g(z_0)}{h'(z_0)}. \tag{8.8.9}$$

I'll show you an example of the use of (8.8.9) in the next section of this chapter.

Sometimes you can use other 'tricks' to get the residue of a singularity. Here's one that directly uses the Laurent series without requiring *any* differentiation at all. Let's calculate

$$\int_0^{2\pi} \cos^k(\theta)d\theta$$

for k an *even* positive integer. (The integral is, of course, zero for k an odd integer because the cosine is symmetrical about the θ-axis over the interval 0 to 2π and so bounds zero area.) Using (8.7.12) again, on the unit circle C we have

$$\cos(\theta) = \frac{z^2 + 1}{2z}$$

and $dz = iz\,d\theta$. So, let's study the contour integral

$$\oint_C \left(\frac{z^2 + 1}{2z} \right)^k \frac{dz}{iz} = \frac{1}{i2^k} \oint_C \frac{(z^2 + 1)^k}{z^{k+1}} \, dz. \qquad (8.8.10)$$

Here we have a singularity at $z = 0$ of order $m = k + 1$. That can be a *lot* of differentiations, using (8.8.8), if k is a large number!

An easy way to get the residue of this high-order singularity is to use the binomial theorem to expand the integrand as

$$\frac{(z^2 + 1)^k}{z^{k+1}} = \frac{1}{z^{k+1}} \sum_{j=0}^{k} \binom{k}{j} (z^2)^j (1)^{k-j} = \sum_{j=0}^{k} \binom{k}{j} z^{2j-k-1}$$

which is a series expansion in negative and positive powers of z around the singular point at zero. It must be, that is, the Laurent expansion of the integrand (remember, such expansions are unique) from which we can literally read off the residue (the coefficient of the z^{-1} term). Setting $2j - k - 1 = -1$, we find that $j = \frac{k}{2}$ and so the residue is

$$\binom{k}{\frac{k}{2}} = \frac{k!}{(\frac{k}{2})! (\frac{k}{2})!} = \frac{k!}{[(\frac{k}{2})!]^2}.$$

Thus,

$$\oint_C \frac{(z^2 + 1)^k}{z^{k+1}} \, dz = 2\pi i \frac{k!}{[(\frac{k}{2})!]^2}$$

and so, from (8.8.10),

$$\int_0^{2\pi} \cos^k(\theta) d\theta = \frac{2\pi}{2^k} \frac{k!}{[(\frac{k}{2})!]^2}, \text{ k even.} \qquad (8.8.11)$$

For example, if $k = 18$ the integral is equal to $1.165346\ldots$ and MATLAB agrees because *integral(@(x)cos(x).^18,0,2*pi)* = 1.165346. . . .

Next, let's do an example using (8.8.8). In (3.4.8) we derived the result (where a and b are each a positive constant, with $a > b$)

$$\int_0^{\pi} \frac{1}{a + b\cos(\theta)} \, d\theta = \frac{\pi}{\sqrt{a^2 - b^2}},$$

which is equivalent to

$$\int_0^{2\pi} \frac{1}{a + b\cos(\theta)} \, d\theta = \frac{2\pi}{\sqrt{a^2 - b^2}}$$

because $\cos(\theta)$ from π to 2π simply runs through the same values it does from 0 to π. Now, suppose we set $a = 1$ and write $b = k < 1$. Then (3.4.8) says

$$\int_0^{2\pi} \frac{1}{1 + k\cos(\theta)} d\theta = \frac{2\pi}{\sqrt{1 - k^2}}, k < 1.$$

This result might prompt one to 'up the ante' and ask for the value of

$$\int_0^{2\pi} \frac{1}{\{1 + k\cos(\theta)\}^2} d\theta = ?$$

If we take C to be the unit circle centered on the origin, then on C we have as in the previous section that $z = e^{i\theta}$ and so, from Euler's identity, we can write

$$\cos(\theta) = \frac{1}{2}\left(z + \frac{1}{z}\right) = \frac{z^2 + 1}{2z}.$$

So, on C we have

$$\frac{1}{1 + k\cos(\theta)} = \frac{1}{1 + k\left(\frac{z^2+1}{2z}\right)} = \frac{2z}{2z + kz^2 + k}$$

and therefore

$$\frac{1}{\{1 + k\cos(0)\}^2} = \left(\frac{4}{k^2}\right)\frac{z^2}{\left(z^2 + \frac{2}{k}z + 1\right)^2}.$$

Also, as before,

$$dz = ie^{i\theta}d\theta = iz \, d\theta$$

and so

$$d\theta = \frac{dz}{iz}.$$

All this suggests that we consider the contour integral

$$\oint_C \left(\frac{4}{k^2}\right) \frac{z^2}{\left(z^2 + \frac{2}{k}z + 1\right)^2} \frac{dz}{iz} = \frac{1}{i}\left(\frac{4}{k^2}\right)\oint_C \frac{z}{\left(z^2 + \frac{2}{k}z + 1\right)^2} dz.$$

We see that the integrand has two singularities, and that each is *second*-order. That is, $m = 2$ and the singularities are at

$$z = \frac{1}{2}\left(-\frac{2}{k} \pm \sqrt{\frac{4}{k^2} - 4}\right) = \frac{-1 \pm \sqrt{1 - k^2}}{k}.$$

Since $k < 1$ the singularity at

$$z = z_{01} = \frac{-1 - \sqrt{1 - k^2}}{k}$$

is *outside* C, while the singularity at

$$z = z_{02} = \frac{-1 + \sqrt{1 - k^2}}{k}$$

is *inside* C. That is, z_{02} is the only singularity for which we need to compute the residue as given by (8.8.8).

So, with m = 2, that residue is

$$b_1 = \frac{1}{1!} \lim_{z \to z_{02}} \frac{d}{dz}\left\{(z - z_{02})^2 \frac{z}{\left(z^2 + \frac{2}{k}z + 1\right)^2}\right\}.$$

Since

$$(z - z_{02})^2 \frac{z}{\left(z^2 + \frac{2}{k}z + 1\right)^2} = (z - z_{02})^2 \frac{z}{(z - z_{01})^2(z - z_{02})^2} = \frac{z}{(z - z_{01})^2}$$

$$= \frac{z}{\left(z + \frac{1 + \sqrt{1 - k^2}}{k}\right)^2},$$

we have

$$\frac{d}{dz}\left\{(z - z_{02})^2 \frac{z}{\left(z^2 + \frac{2}{k}z + 1\right)^2}\right\} = \frac{d}{dz}\left\{\frac{z}{\left(z + \frac{1 + \sqrt{1 - k^2}}{k}\right)^2}\right\}$$

which, after just a bit of algebra that I'll let you confirm, reduces to

$$\frac{-z + \frac{1 + \sqrt{1 - k^2}}{k}}{\left(z + \frac{1 + \sqrt{1 - k^2}}{k}\right)^3}.$$

Then, finally, we let $z \to z_{02} = \frac{-1 + \sqrt{1 - k^2}}{k}$ and so

$$b_1 = \frac{\frac{1-\sqrt{1-k^2}}{k} + \frac{1+\sqrt{1-k^2}}{k}}{\left(\frac{-1+\sqrt{1-k^2}}{k} + \frac{1+\sqrt{1-k^2}}{k}\right)^3} = \frac{\frac{2}{k}}{\frac{8(1-k^2)^{3/2}}{k^3}} = \frac{k^2}{4(1-k^2)^{3/2}}.$$

Thus,

$$\oint_C \frac{z}{\left(z^2 + \frac{2}{k}z + 1\right)^2} dz = 2\pi i \frac{k^2}{4(1-k^2)^{3/2}} = \frac{\pi i k^2}{2(1-k^2)^{3/2}}$$

and so

$$\int_0^{2\pi} \frac{1}{\{1 + k\cos(\theta)\}^2} d\theta = \frac{1}{i}\left(\frac{4}{k^2}\right)\frac{\pi i k^2}{2(1-k^2)^{3/2}}.$$

That is,

$$\int_0^{2\pi} \frac{1}{\{1 + k\cos(\theta)\}^2} d\theta = \frac{2\pi}{(1-k^2)^{3/2}}. \tag{8.8.12}$$

For example, if $k = \frac{1}{2}$ then our result is 9.67359..., and MATLAB agrees because *integral(@(x)1./(1 + 0.5*cos(x)).^2,0,2*pi) = 9.67359....*

To finish this section, I'll now formally state what we've been doing all through it: if $f(z)$ is analytic on and inside contour C with the exception of N singularities, and if R_j is the residue of the j-*th* singularity, then

$$\oint_C f(z)\, dz = 2\pi i \sum_{j=1}^N R_j. \tag{8.8.13}$$

This is the famous *residue theorem*. For each singularity we'll pick-up a contribution to the integral of $2\pi i$ times the residue of that singularity, with the residue calculated according to (8.8.8), or (8.8.9) if m = 1, using the value of m that goes with each singularity. That's it! In the next (and final) section of this chapter I'll show you an example of (8.8.13) applied to an integral that has one additional complication we haven't yet encountered.

8.9 Integrals with Multi-Valued Integrands

All of the wonderful power of contour integration comes from the theorems that tell us what happens when we travel once around a *closed* path in the complex plane. The theorems apply *only* for paths that are *closed*. I emphasize this point—

particularly the word *closed*—because there is a subtle way in which closure can fail, so subtle in fact that it is all too easy to miss. Recognizing the problem, and then understanding the way to get around it, leads to the important concepts of *branch cuts* and *branch points* in the complex plane.

There are numerous examples that one could give of how false closed paths can occur, but the classic one involves integrands containing the logarithmic function. Writing the complex variable as we did in (8.3.3) as $z = re^{i\theta}$, we have log $(z) = \ln (z) = \ln (re^{i\theta}) = \ln (r) + i\theta, 0 \leq \theta < 2\pi$. Notice, *carefully*, the $\leq$ sign to the left of θ but that it is the strict $<$ sign on the right. As was pointed out in Sect. 8.3, θ is not uniquely determined, as we can add (or subtract) any multiple of 2π from θ and still seemingly be talking about the same physical point in the complex plane. That is, we should really write $\log(z) = \ln (r) + i(\theta \pm 2\pi n), 0 \leq \theta < 2\pi, n = 0, 1, 2,$ .... The logarithmic function is said to be *multi-valued* as we loop endlessly around the origin. The mathematical problem we run into with this more complete formulation of the logarithmic function is that it is *not continuous* on any path that crosses the positive real axis! Here's why.

Consider a point $z = z_0$ on the positive real axis. At that point, $r = x_0$ and $\theta = 0$. But the imaginary part of $\log(z)$ is not a continuous function of z at x_0 because its value, in all tiny neighborhoods 'just below' the positive real axis at x_0, is arbitrarily near 2π, not 0. The crucial implication of this failure of continuity is that the *derivative* of $\log(z)$ fails to exist as we cross the positive real axis, which means analyticity fails there, too. And that means all our wonderful integral theorems are out the window!

What is happening, geometrically, as we travel around what *seems* to be a closed circular path (starting at x_0 and then winding around the origin) is that we do *not* return to the starting point x_0. Rather, when we cross the positive real axis we enter a new *branch* of the log function. An everyday example of this occurs when you travel along a spiral path in a multi-level garage looking for a parking space and move from one level (branch) to the next level (another branch) of the garage.[18] Your spiral path 'looks closed' to an observer on the roof looking downward (just like you looking down on your math paper as you draw what *seems* to be a closed contour in a flat complex plane), but your parking garage trajectory is *not* closed. And neither is that apparently 'closed' contour. There is no problem for your car with this, of course, but it seems to be a fatal problem for our integral theorems.

Or, perhaps not. Remember the old saying: "If your head hurts because you're banging it on the wall, then stop banging your head on the wall!" We have the same situation here: "If crossing the positive real axis blows-up the integral theorems, well then, *don't cross the positive real axis*." What we need to do here, when constructing

---

[18]Each of these branches exists for each new interval of θ of width 2π, with each branch lying on what is called a *Riemann surface* The logarithmic function has an infinite number of branches, and so an infinite number of Riemann surfaces. The surface for $0 \leq \theta < 2\pi$ is what we observe as the usual complex plane (the entry level of our parking garage). The concept of the Riemann surface is a very deep one, and my comments here are meant only to give you an 'elementary geometric feel' for it.

a contour involving the logarithmic function, is to simply avoid crossing the positive real axis. What we'll do, instead, is label the positive real axis, from the origin out to plus-infinity, as a so-called *branch cut* (the end-points of the cut, $x = 0$ and $x = +\infty$, are called *branch points*), and then avoid crossing that line. Any contour that we draw satisfying this restriction is *absolutely guaranteed* to be closed (that is, to always remain on a single branch) and thus our integral theorems remain valid.

Another commonly encountered multi-valued function that presents the same problem is the fractional power $z^p = r^p e^{ip\theta}$, where $-1 < p < 1$ and, as before, we take $0 \leq \theta < 2\pi$. Suppose, for example, we have the function $\sqrt{z}$ and so $p = \frac{1}{2}$. Any point *on* the positive real axis has $\theta = 0$, but in a tiny neighborhood 'just below' the positive real axis the angle of z is arbitrarily near to 2π and so the angle of $\sqrt{z}$ is $\frac{2\pi}{2} = \pi$. That is, *on* the positive real axis the function value at a point is $\sqrt{r}$ while an arbitrarily tiny downward shift of the point into the fourth quadrant gives a function value of $\sqrt{r}e^{i\pi} = -\sqrt{r}$. The function value is not continuous across the positive real axis. The solution for handling z^p is, again, to define the positive real axis as a branch cut and to avoid using a contour C that crosses that cut.

The fact that I've taken $0 \leq \theta < 2\pi$ is the reason the branch cut is along the *positive* real axis. If, instead, I'd taken $-\pi < \theta \leq \pi$ we would have run into the failure of continuity problem as we crossed the *negative* real axis, and in that case we would simply make the negative real axis the branch cut and avoid using any C crossing it. In both cases $z = 0$ would be the branch point. Indeed, in the examples I've discussed here we could pick any direction we wish, starting at $z = 0$, draw a straight from there out to infinity, and call *that* our branch cut.

Let's see how this all works. For the final calculation of this chapter, using these ideas, I'll evaluate

$$\int_0^\infty \frac{\ln(x)}{(x+a)^2 + b^2}\, dx, \quad a \geq 0, b > 0, \tag{8.9.1}$$

where a and b are constants. We've already done two special cases of (8.9.1). In (1.5.1), for $a = 0$ and $b = 1$, we found that

$$\int_0^\infty \frac{\ln(x)}{x^2 + 1}\, dx = 0,$$

and in (2.1.3) we generalized this just a bit to the case of arbitrary b:

$$\int_0^\infty \frac{\ln(x)}{x^2 + b^2}\, dx = \frac{\pi}{2b} \ln(b), \quad b > 0.$$

In (8.9.1) we'll now allow a, too, to have any non-negative value. The contour C we'll use is shown in Fig. 8.9.1, which you'll notice avoids crossing the branch cut (the positive real axis), as well as circling around the branch point at the origin. This insures that C lies entirely on a single branch of the logarithmic function, and so C is truly closed and the conditions for Cauchy's integral theorems remain satisfied.

Fig. 8.9.1 A closed contour
that avoids crossing a
branch cut

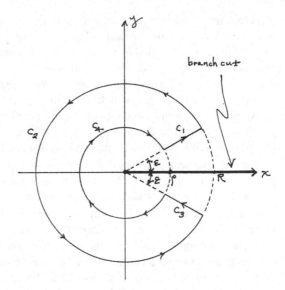

The contour C consists of four parts, where ρ and R are the radii of the small (C_4)
and large (C_2) circular portions, respectively, and ε is a small *positive* angle:

on C_1: $z = re^{i\varepsilon}$, $dz = e^{i\varepsilon}dr$, $\rho < r < R$;

on C_2: $z = Re^{i\theta}$, $dz = iRe^{i\theta}d\theta$, $\varepsilon < \theta < 2\pi - \varepsilon$;

on C_3: $z = re^{i(2\pi - \varepsilon)}$, $dz = e^{i(2\pi - \varepsilon)}dr$, $R > r > \rho$;

on C_4: $z = \rho e^{i\theta}$, $dz = i\rho e^{i\theta}d\theta$, $2\pi - \varepsilon > \theta > \varepsilon$.

We will, eventually, let $\rho \to 0$, $R \to \infty$, and $\varepsilon \to 0$.

Our integrand will be

$$f(z) = \frac{\{\ln(z)\}^2}{(z+a)^2 + b^2},\tag{8.9.2}$$

and you are almost surely wondering why the numerator is ln(z) *squared*? Why not
just ln(z)? The answer, which I think isn't obvious until you do the calculations, is
that if we use just ln(z) we won't get the value of (8.9.1), but rather that of a different
integral. If we use ln(z) *squared*, however, we *will* get the value of (8.9.1). I'll now
show you how it all goes with ln(z) *squared*, and you should verify my comments
about using just ln(z).

The integrand has three singularities: one at $z = 0$ (the branch point) where the
numerator blows-up, and two at $z = -a \pm ib$ where the denominator vanishes. Only
the last two are inside C as ρ and ε each go to zero, and as R goes to infinity, and each
is first-order. From the residue theorem, (8.8.13), we therefore have

$$\oint_C f(z)\,dz = 2\pi i \sum_{j=1}^{2} R_j$$

where R_1 is the residue of the first-order singularity at $z = -a + ib$ and R_2 is the residue of the first-order singularity at $z = -a - ib$. As we showed in (8.8.9), the residue of a first-order singularity at $z = z_0$ in the integrand function

$$f(z) = \frac{g(z)}{h(z)}$$

is given by

$$\frac{g(z_0)}{h'(z_0)}.$$

For our problem,

$$g(z) = \{ \ln (z) \}^2$$

and

$$h(z) = (z + a)^2 + b^2.$$

Since

$$h'(z_0) = \frac{d}{dz} h(z)|_{z=z_0} = 2(z + a)|_{z=z_0}$$

then, as

$$2(-a + ib + a) = i2b$$

and

$$2(-a - ib + a) = -i2b,$$

we have

$$R_1 = \frac{\{ \ln (-a + ib) \}^2}{i2b}$$

and

$$R_2 = \frac{\{ \ln (-a - ib) \}^2}{-i2b}.$$

Since a and b are both non-negative, the $-a + ib$ singularity is in the second quadrant, and the $-a - ib$ singularity is in the third quadrant. In polar form, then, the second quadrant singularity is at

$$-a + ib = \sqrt{a^2 + b^2} e^{i\{\pi - \tan^{-1}\left(\frac{b}{a}\right)\}}$$

and the third quadrant singularity is at

$$-a - ib = \sqrt{a^2 + b^2} e^{i\{\pi + \tan^{-1}\left(\frac{b}{a}\right)\}}.$$

Therefore,

$$R_1 = \frac{\left\{\ln\left(\sqrt{a^2 + b^2} e^{i\{\pi - \tan^{-1}\left(\frac{b}{a}\right)\}}\right)\right\}^2}{i2b} = \frac{\left[\ln\left(\sqrt{a^2 + b^2}\right) + i\{\pi - \tan^{-1}\left(\frac{b}{a}\right)\}\right]^2}{i2b}$$

and

$$R_2 = \frac{\left\{\ln\left(\sqrt{a^2 + b^2} e^{i\{\pi + \tan^{-1}\left(\frac{b}{a}\right)\}}\right)\right\}^2}{-i2b} = \frac{\left[\ln\left(\sqrt{a^2 + b^2}\right) + i\{\pi + \tan^{-1}\left(\frac{b}{a}\right)\}\right]^2}{-i2b}.$$

Thus, $2\pi i$ times the sum of the residues is

$$2\pi i(R_1 + R_2) = 2\pi i\left(\frac{\left[\ln\left(\sqrt{a^2+b^2}\right) + i\{\pi - \tan^{-1}\left(\frac{b}{a}\right)\}\right]^2}{i2b} - \frac{\left[\ln\left(\sqrt{a^2+b^2}\right) + i\{\pi + \tan^{-1}\left(\frac{b}{a}\right)\}\right]^2}{i2b}\right)$$

$$= \frac{2\pi i}{i2b}\left(\left[\ln\left(\sqrt{a^2+b^2}\right) + i\left\{\pi - \tan^{-1}\left(\frac{b}{a}\right)\right\}\right]^2 - \left[\ln\left(\sqrt{a^2+b^2}\right) + i\left\{\pi + \tan^{-1}\left(\frac{b}{a}\right)\right\}\right]^2\right)$$

$$= \frac{\pi}{b}\left(\begin{array}{l} 2\ln\left(\sqrt{a^2+b^2}\right)i\left\{\pi - \tan^{-1}\left(\frac{b}{a}\right)\right\} - \{\pi - \tan^{-1}\left(\frac{b}{a}\right)\}^2 \\ -2\ln\left(\sqrt{a^2+b^2}\right)i\left\{\pi + \tan^{-1}\left(\frac{b}{a}\right)\right\} + \{\pi + \tan^{-1}\left(\frac{b}{a}\right)\}^2 \end{array}\right)$$

$$= \frac{\pi}{b}\left(\begin{array}{l} -4i\ln\left(\sqrt{a^2+b^2}\right)\tan^{-1}\left(\frac{b}{a}\right) - \left\{\pi^2 - 2\pi\tan^{-1}\left(\frac{b}{a}\right) + \left[\tan^{-1}\left(\frac{b}{a}\right)\right]^2\right\} \\ + \left\{\pi^2 + 2\pi\tan^{-1}\left(\frac{b}{a}\right) + \left[\tan^{-1}\left(\frac{b}{a}\right)\right]^2\right\} \end{array}\right)$$

$$= \frac{\pi}{b}\left(4\pi\tan^{-1}\left(\frac{b}{a}\right) - 4i\ln\left(\sqrt{a^2+b^2}\right)\tan^{-1}\left(\frac{b}{a}\right)\right) = \frac{4\pi}{b}\tan^{-1}\left(\frac{b}{a}\right)\left[\pi - i\ln\left(\sqrt{a^2+b^2}\right)\right].$$

So, for the f(z) in (8.9.2) and the C in Fig. 8.9.1, we have

$$\oint_C f(z)\, dz = \frac{4\pi}{b} \tan^{-1}\left(\frac{b}{a}\right)\left[\pi - i \ln\left(\sqrt{a^2 + b^2}\right)\right]$$

$$= \int_{C_1} + \int_{C_2} + \int_{C_3} + \int_{C_4}. \tag{8.9.3}$$

Based on our earlier experiences, we expect our final result is going to come from the C_1 and C_3 integrals because, as we let ρ, ε, and R go to their limiting values (of 0, 0, and ∞, respectively), we expect the C_2 and C_4 integrals will each vanish. To see that this is, indeed, the case, let's do the C_2 and C_4 integrals first. For the C_2 integral we have

$$\int_{C_2} = \int_{\varepsilon}^{2\pi - \varepsilon} \left[\frac{\{\ln\left(Re^{i\theta}\right)\}^2}{\left(Re^{i\theta} + a\right)^2 + b^2}\right]\left[iRe^{i\theta}\right]d\theta.$$

Now, as $R \to \infty$ consider the expression in the left-most square-brackets in the integrand. The numerator blows-up like $\ln^2(R)$ for any given θ in the integration interval, while the denominator blows-up like R^2. That is, the left-most square-brackets behave like $\frac{\ln^2(R)}{R^2}$. The expression in the right-most square-brackets blows-up like R. Thus, the integrand behaves like

$$\frac{\ln^2(R)}{R^2}R = \frac{\ln^2(R)}{R}$$

and so the C_2 integral behaves like

$$2\pi\frac{\ln^2(R)}{R}$$

as $R \to \infty$. Now

$$\lim_{R\to\infty} 2\pi\frac{\ln^2(R)}{R} = \frac{\infty}{\infty}$$

which is, of course, indeterminate, and so let's use L'Hospital's rule:

$$\lim_{R\to\infty} 2\pi\frac{\ln^2(R)}{R} = \lim_{R\to\infty} 2\pi\frac{\frac{d}{dR}\ln^2(R)}{\frac{d}{dR}R} = 2\pi \lim_{R\to\infty}\frac{2\ln(R)\frac{1}{R}}{1}$$

$$= 4\pi \lim_{R\to\infty}\frac{\ln(R)}{R} = 0.$$

So, our expectation of the vanishing of the C_2 integral is justified.

Turning next to the C_4 integral, we have

$$\int_{C_4} = \int_{2\pi - \varepsilon}^{\varepsilon} \left[\frac{\{\ln(\rho e^{i\theta})\}^2}{(\rho e^{i\theta} + a)^2 + b^2} \right] \left[i\rho e^{i\theta} \right] d\theta.$$

As $\rho \to 0$ the expression in the left-most square-brackets in the integrand behaves like $\frac{\ln^2(\rho)}{a^2 + b^2}$ while the expression in the right-most square-brackets behaves like ρ. So, the C_4 integral behaves like

$$2\pi \frac{\ln^2(\rho)}{a^2 + b^2} \rho$$

as $\rho \to 0$. Now,

$$\lim_{\rho \to 0} 2\pi \frac{\ln^2(\rho)}{a^2 + b^2} \rho = \frac{2\pi}{a^2 + b^2} \lim_{\rho \to 0} \rho \ln^2(\rho).$$

Define $u = \frac{1}{\rho}$. Then, as $\rho \to 0$ we have $u \to \infty$ and so

$$\lim_{\rho \to 0} \rho \ln^2(\rho) = \lim_{u \to \infty} \frac{1}{u} \ln^2\left(\frac{1}{u}\right) = \lim_{u \to \infty} \frac{\ln^2(u)}{u},$$

which we've just shown (in the C_2 integral analysis) goes to zero. So, our expectation of the vanishing of the C_4 integral is also justified.

Turning our attention at last to the C_1 and C_3 integrals, we have

$$\int_{C_1} + \int_{C_3} = \int_{\rho}^{R} \frac{\{\ln(re^{i\varepsilon})\}^2}{(re^{i\varepsilon} + a)^2 + b^2} e^{i\varepsilon} dr + \int_{R}^{\rho} \frac{\{\ln(re^{i(2\pi - \varepsilon)})\}^2}{(re^{i(2\pi - \varepsilon)} + a)^2 + b^2} e^{i(2\pi - \varepsilon)} dr$$

$$= \int_{\rho}^{R} \frac{\{\ln(r) + i\varepsilon\}^2}{(re^{i\varepsilon} + a)^2 + b^2} e^{i\varepsilon} dr - \int_{\rho}^{R} \left\{ \frac{\ln(r) + i(2\pi - \varepsilon)\}^2}{(re^{i(2\pi - \varepsilon)} + a)^2 + b^2} \right\} e^{i(2\pi - \varepsilon)} dr$$

or, as $\rho \to 0$, $R \to \infty$, and $\varepsilon \to 0$,

$$\int_{C_1} + \int_{C_3} = \int_{0}^{\infty} \frac{\ln^2(r)}{(r + a)^2 + b^2} dr - \int_{0}^{\infty} \frac{\{\ln(r) + i2\pi\}^2}{(re^{i2\pi} + a)^2 + b^2} e^{i2\pi} dr$$

$$= \int_{0}^{\infty} \frac{\ln^2(r) - \{\ln(r) + i2\pi\}^2}{(r + a)^2 + b^2} dr = \int_{0}^{\infty} \frac{-i4\pi \ln(r) + 4\pi^2}{(r + a)^2 + b^2} dr.$$

(Notice, carefully, how the $\ln^2(r)$ terms cancel in these last calculations, leaving just $\ln(r)$ in the final expression.) Inserting these results into (8.9.3), we have

$$4\pi^2 \int_0^\infty \frac{dr}{(r+a)^2 + b^2} - i4\pi \int_0^\infty \frac{\ln(r)}{(r+a)^2 + b^2} dr$$

$$= \frac{4\pi^2}{b} \tan^{-1}\left(\frac{b}{a}\right) - i\frac{4\pi}{b} \tan^{-1}\left(\frac{b}{a}\right) \ln\left(\sqrt{a^2 + b^2}\right).$$

Equating real parts, we get

$$\int_0^\infty \frac{dx}{(x+a)^2 + b^2} = \frac{1}{b} \tan^{-1}\left(\frac{b}{a}\right)$$

which shouldn't really be a surprise.[19](This is the *lone* integral you'll get if you use just ln(z) instead of ln(z) *squared* in (8.9.2). Try it and see.) Equating imaginary parts is what gives us our prize for all this work:

$$\int_0^\infty \frac{\ln(x)}{(x+a)^2 + b^2} dx = \frac{1}{b} \tan^{-1}\left(\frac{b}{a}\right) \ln\left(\sqrt{a^2 + b^2}\right). \qquad (8.9.4)$$

This reduces to our earlier results for particular values of a and b. To see (8.9.4) in action, if both a and b equal 1 (for example) then

$$\int_0^\infty \frac{\ln(x)}{(x+1)^2 + 1} dx = \frac{\pi}{4} \ln\left(\sqrt{2}\right) = \frac{\pi}{8} \ln(2) = 0.272198\ldots$$

and MATLAB agrees, as *integral(@(x)log(x)./((x+1).^2+1),0,inf)* = 0.272198. . . .

8.10   A Final Calculation

In the new Preface for this edition I promised you a 'routine methods' derivation of (8.6.9), as an example of how an integral done by contour integration *can* (if one is simply clever enough) be done using less powerful means. This is a point I made in the original edition (see Challenge Problems 8.7 and 8.8), but there my point was that a freshman calculus solution might be the *easier* of the two approaches. Here I'll show you that, yes, we can indeed get (8.6.9) with just freshman calculus ideas, but it's a toss-up value judgement if that is, in some sense, the 'easier' approach.

Our analysis starts with Euler's gamma function integral in (4.1.1), that I'll rewrite here:

---

[19]You should be able to show that this result immediately follows from the indefinite integral $\int \frac{du}{u^2+b^2} = \frac{1}{b} \tan^{-1}\left(\frac{u}{b}\right)$ followed by the change of variable $u = x + a$.

$$\Gamma(a) = \int_0^\infty e^{-x} x^{a-1} dx. \tag{8.10.1}$$

Now, as a preliminary calculation, we are going to find $\Gamma'(a)$, the derivative of $\Gamma(a)$ with respect to a (using Feynman's favorite trick of differentiating an integral with respect to a parameter—in this case, a). To be sure this calculation is crystal clear, I'll write (8.10.1) as

$$\Gamma(a) = \int_0^\infty e^{-x} e^{\ln\left(x^{a-1}\right)} dx = \int_0^\infty e^{-x} e^{(a-1)\ln(x)} dx = \int_0^\infty e^{-x} e^{-\ln(x)} e^{a\,\ln(x)} dx$$

$$= \int_0^\infty \frac{e^{-x}}{e^{\ln(x)}} e^{a\,\ln(x)} dx = \int_0^\infty \frac{e^{-x}}{x} e^{a\,\ln(x)} dx.$$

Thus,

$$\Gamma'(a) = \int_0^\infty \frac{e^{-x}}{x} \ln(x) e^{a\,\ln(x)} dx = \int_0^\infty e^{-x} \frac{\ln(x)}{x} e^{\ln(x^a)} dx = \int_0^\infty e^{-x} \frac{\ln(x)}{x} x^a dx$$

$$= \int_0^\infty e^{-x} x^{a-1} \ln(x) dx$$

or, changing the dummy variable of integration from x to s (which of course changes nothing but it will help keep the notation absolutely clear and free of confusion as I reference other results from earlier in the book), we have

$$\Gamma'(a) = \int_0^\infty e^{-s} s^{a-1} \ln(s) ds. \tag{8.10.2}$$

Next, remembering the result we derived in (3.3.3),

$$\int_0^\infty \frac{e^{-pz} - e^{-qz}}{z} dz = \ln\left(\frac{q}{p}\right),$$

we set $p = 1$ and $q = s$ to get

$$\ln(s) = \int_0^\infty \frac{e^{-z} - e^{-sz}}{z} dz. \tag{8.10.3}$$

Putting (8.10.3) into (8.10.2), we have

$$\Gamma'(a) = \int_0^\infty e^{-s} s^{a-1} \left\{ \int_0^\infty \frac{e^{-z} - e^{-sz}}{z} dz \right\} ds$$

or, reversing the order of integration,

$$\Gamma'(a) = \int_0^\infty \left\{ e^{-z} \int_0^\infty e^{-s} s^{a-1} ds - \int_0^\infty e^{-s(1+z)} s^{a-1} ds \right\} \frac{dz}{z}$$

and so, remembering (8.10.1),

$$\Gamma'(a) = \int_0^\infty \left\{ e^{-z} \Gamma(a) - \int_0^\infty e^{-s(1+z)} s^{a-1} ds \right\} \frac{dz}{z}. \qquad (8.10.4)$$

Now, concentrate your attention on the inner, right-most integral, and change variable to $u = s(1 + z)$. Thus, $s = \frac{u}{1+z}$ and $ds = \frac{du}{1+z}$, and therefore

$$\int_0^\infty e^{-s(1+z)} s^{a-1} ds = \int_0^\infty e^{-u} \frac{u^{a-1}}{(1+z)^{a-1}} \frac{du}{1+z} = \frac{1}{(1+z)^a} \int_0^\infty e^{-u} u^{a-1} du = \frac{\Gamma(a)}{(1+z)^a}.$$

Using this in (8.10.4),

$$\Gamma'(a) = \int_0^\infty \left\{ e^{-z} \Gamma(a) - \frac{\Gamma(a)}{(1+z)^a} \right\} \frac{dz}{z}$$

or,

$$\Gamma'(a) = \Gamma(a) \left[\int_0^\infty \frac{e^{-z}}{z} dz - \int_0^\infty \frac{dz}{(1+z)^a z} \right]. \qquad (8.10.5)$$

You'll recall from Sect. 5.4 that we called $\Gamma'(a)/\Gamma(a)$ the *digamma function* (see note 6 in Chapter 5, and Challenge Problem 8.9), which I'll write here as

$$\psi(a) = \frac{\Gamma'(a)}{\Gamma(a)} = \int_0^\infty \frac{e^{-z}}{z} dz - \int_0^\infty \frac{dz}{(1+z)^a z}. \qquad (8.10.6)$$

If we now change variable to $1 + z = e^y$, we have $\frac{dz}{dy} = e^y$ or, $dz = e^y dy$. Also, $z = e^y - 1$ and so $y = 0$ when $z = 0$ and $y = \infty$ when $z = \infty$. Thus, the right-most integral in (8.10.6) is

$$\int_0^\infty \frac{dz}{(1+z)^a z} = \int_0^\infty \frac{e^y}{e^{ay}(e^y - 1)} dy = -\int_0^\infty \frac{e^{(1-a)y}}{1 - e^y} dy$$

and (8.10.6) becomes (using x as the dummy variable of integration))

$$\psi(a) = \int_0^\infty \frac{e^{-x}}{x} dx + \int_0^\infty \frac{e^{(1-a)x}}{1 - e^x} dx. \qquad (8.10.7)$$

Okay, at last, we are ready to tackle the integral we are after, which I'll write as

$$\int_{-\infty}^{\infty} \frac{e^{ax}}{1-e^x}\,dx = \int_{-\infty}^{0} \frac{e^{ax}}{1-e^x}\,dx + \int_{0}^{\infty} \frac{e^{ax}}{1-e^x}\,dx.$$

If we make the change of variable $y = -x$ in the first integral on the right (and so $dx = -dy$), we have

$$\int_{-\infty}^{\infty} \frac{e^{ax}}{1-e^x}\,dx = \int_{\infty}^{0} \frac{e^{-ay}}{1-e^{-y}}(-dy) + \int_{0}^{\infty} \frac{e^{ax}}{1-e^x}\,dx$$

$$= \int_{0}^{\infty} \frac{e^{-ay}e^{y}}{e^{y}-1}\,dy + \int_{0}^{\infty} \frac{e^{ax}}{1-e^x}\,dx = \int_{0}^{\infty} \frac{e^{(1-a)y}}{e^{y}-1}\,dy + \int_{0}^{\infty} \frac{e^{ax}}{1-e^x}\,dx$$

and so

$$\int_{-\infty}^{\infty} \frac{e^{ax}}{1-e^x}\,dx = \int_{0}^{\infty} \frac{e^{ax}}{1-e^x}\,dx - \int_{0}^{\infty} \frac{e^{(1-a)x}}{1-e^x}\,dx. \qquad (8.10.8)$$

Now, let's do something perhaps just a bit unexpected. Let's add and subtract $\int_{0}^{\infty} \frac{e^{-x}}{x}\,dx$ to the right-hand-side of (8.10.8), which of course changes nothing. This gives us

$$\int_{-\infty}^{\infty} \frac{e^{ax}}{1-e^x}\,dx = \left\{ \int_{0}^{\infty} \frac{e^{-x}}{x}\,dx + \int_{0}^{\infty} \frac{e^{ax}}{1-e^x}\,dx \right\} - \left\{ \int_{0}^{\infty} \frac{e^{-x}}{x}\,dx + \int_{0}^{\infty} \frac{e^{(1-a)x}}{1-e^x}\,dx \right\}. \qquad (8.10.9)$$

If you compare the two expressions in the two pairs of curly brackets in (8.10.9) with the expression for $\psi(a)$ in (8.10.7), you see that the first expression on the right in (8.10.9) is $\psi(1-a)$, while the second expression on the right in (8.10.9) is $\psi(a)$. That is,

$$\int_{-\infty}^{\infty} \frac{e^{ax}}{1-e^x}\,dx = \psi(1-a) - \psi(a). \qquad (8.10.10)$$

We can get our hands on $\psi(1-a) - \psi(a)$ from the reflection formula for the gamma function, derived in (4.2.16). That is, as we showed there,

$$\Gamma(a)\Gamma(1-a) = \frac{\pi}{\sin(\pi a)}$$

and so, taking logarithms on both sides,

$$\ln\{\Gamma(a)\} + \ln\{\Gamma(1-a)\} = \ln(\pi) - \ln\{\sin(\pi a)\}$$

and then differentiating with respect to a,

$$\frac{\Gamma'(a)}{\Gamma(a)} - \frac{\Gamma'(1-a)}{\Gamma 1 - a)} = \psi(a) - \psi(1-a) = -\frac{\pi\cos(\pi a)}{\sin(\pi a)}$$

or,

$$\psi(1-a) - \psi(a) = \frac{\pi}{\tan(\pi a)}. \tag{8.10.11}$$

Putting (8.10.11) into (8.10.10) gives us our result, the one we derived in (8.6.9) using contour integration:

$$\int_{-\infty}^{\infty} \frac{e^{ax}}{1 - e^x} dx = \frac{\pi}{\tan(\pi a)}.$$

After all of this, perhaps contour integration doesn't look quite so difficult anymore!

Historical Note: For many years the general feeling among mathematicians was that there were some integrals that were, ironically, simply 'too complex' to be evaluated by complex variables (contour integration). The probability integral $\int_{-\infty}^{\infty} e^{-x^2} dx$ was the usual case put forth in support of that view. Then, in 1947, the British mathematician James Cadwell (1915–1982) published a brief note (*The Mathematical Gazette*, October) where he showed how to do that integral by performing two successive contour integrations, one around a rectangle, followed by another around a pie-shaped sector of a circle. Shortly after that, the Russian-born English mathematician Leon Mirsky (1918–1983) showed how to do it in an even shorter note (*The Mathematical Gazette*, December 1949) using just a single contour (a parallelogram). In a footnote, however, Cadwell had noted that he had learned, since writing his original note, that the American-born British mathematician Louis Joel Mordell (1888–1972) had, decades earlier, already evaluated via contour methods the much more general $\int_{\infty}^{\infty} \frac{e^{at^2+bt}}{e^{ct}+d} dt$ which reduces to the probability integral for the special case of a = 1 and b = c = d = 0. You can find Mordell's quite difficult (in my opinion) analysis in the *Quarterly Journal of Pure and Applied Mathematics* (48) 1920, pp. 329–342. The prevailing view today is that any integral that *can* be done, can be cracked using either real *or* complex methods.

8.11 Challenge Problems

(C8.1) Suppose f(z) is analytic everywhere in some region **R** in the complex plane, with an m-th order zero at z = z_0. That is, f(z) = g(z)(z − z_0)^m, where g(z) is analytic everywhere in **R**. Let C be any simple, closed CCW contour in **R** that encircles z_0. Explain why

$$\frac{1}{2\pi i}\oint_C \frac{f'(z)}{f(z)}\,dz = m.$$

(**C8.2**) Back in Challenge Problem C3.9 I asked you to accept that $\int_0^\infty \frac{\sin(mx)}{x(x^2+a^2)}\,dx = \frac{\pi}{2}\left(\frac{1-e^{-am}}{a^2}\right)$ for $a > 0$, $m > 0$. Here you are to derive this result using contour integration. Hint: Notice that since the integrand is even, $\int_0^\infty = \frac{1}{2}\int_{-\infty}^\infty$. Use $f(z) = \frac{e^{imz}}{z(z^2+a^2)}$, notice where the singularities are (this should suggest to you the appropriate contour to integrate around) and then, at some point, think about taking an imaginary part.

(**C8.3**) Derive the following integration formulas:

(a) $\int_0^{2\pi} \frac{d\theta}{1-2a\,\cos(\theta)+a^2} = \frac{2\pi}{1-a^2}, 0 < a < 1$;

(b) $\int_{-\infty}^\infty \frac{\cos(x)}{(x+a)^2+b^2}\,dx = \frac{\pi}{b}e^{-b}\cos(a)$ and $\int_{-\infty}^\infty \frac{\sin(x)}{(x+a)^2+b^2}\,dx = -\frac{\pi}{b}e^{-b}\sin(a),\ a > 0, b > 0$;

(c) $\int_{-\infty}^\infty \frac{\cos(x)}{(x^2+a^2)\,(x^2+b^2)}\,dx = \frac{\pi}{a^2-b^2}\left(\frac{e^{-b}}{b} - \frac{e^{-a}}{a}\right), a > b > 0$;

(d) $\int_0^\infty \frac{\cos(ax)}{\left(x^2+b^2\right)^2}\,dx = \frac{\pi}{4\,b^3}(1+ab)e^{-ab}, a > 0, b > 0$.

In (a), use the approach of Sect. 8.3 to convert the integral into a contour integration around the unit circle. In (b), (c), and (d), use the contour in Fig. 8.7.1.

(**C8.4**) Using the contour in Fig. 8.9.1, show that $\int_0^\infty \frac{x^k}{(x^2+1)^2}\,dx = \frac{\pi(1-k)}{4\cos\left(\frac{k\pi}{2}\right)}, -1 < k < 3$. Before doing any calculations, explain the limits on k. Hint: Use $f(z) = \frac{z^k}{(z^2+1)^2}$, notice that the singularities at $z = \pm i$ are both second-order, and write $z^k = e^{\ln\left(z^k\right)} = e^{k\,\ln(z)}$.

(**C8.5**) Show that $\int_{-\infty}^\infty \frac{\cos(mx)}{ax^2+bx+c}\,dx = -2\pi\,\dfrac{\cos\left(\frac{mb}{2a}\right)\sin\left(\frac{m\sqrt{b^2-4ac}}{2a}\right)}{\sqrt{b^2-4ac}}$ when $b^2 > 4ac$. Notice that this result contains (8.6.5) as the special case of $m = 0$.

(**C8.6**) Show that $\int_0^\infty \frac{x^p}{(x+1)(x+2)}\,dx = (2^p - 1)\frac{\pi}{\sin(p\pi)}, -1 < p < 1$. For $p = \frac{1}{2}$ this is $(\sqrt{2} - 1)\pi = 1.30129\ldots$, and MATLAB agrees as $integral(@(x)sqrt(x)./((x+1).*(x+2)),0,inf) = 1.30129\ldots$ Use the contour in Fig. 8.9.1.

(**C8.7**) In his excellent 1935 book *An Introduction to the Theory of Functions of a Complex Variable*, Edward Copson (1901–1980), who was professor of mathematics at the University of St. Andrews in Scotland, wrote "A definite integral which can be evaluated by Cauchy's method of residues can always be evaluated by other means, though generally not so simply." Here's an example of what Copson meant, an integral attributed to the great Cauchy himself. It is easily done with contour integration, but would (I think) otherwise be pretty darn tough: show that $\int_0^\infty \frac{e^{\cos(x)}\,\sin\{\sin(x)\}}{x}\,dx = \frac{\pi}{2}(e-1)$. MATLAB agrees with Cauchy, as this is $2.69907\ldots$ and $integral(@(x)exp(cos(x)).*sin(sin(x))./x,0,1e6) = 2.69595\ldots$. Hint: Look back at how we derived (8.6.4)—in particular the contour in Fig. 8.6.1—and try to construct the proper f(z) to integrate on that contour. (In the new Preface for this edition you'll recall that there is a freshman calculus derivation of Cauchy's integral that is, in fact, much easier to do than is the contour integration.)

(C8.8) Here's an example of an integral that Copson himself assigned as an end-of-chapter problem to be done by contour integration and residues, but which is actually easier to do by freshman calculus: show that $\int_{-\infty}^\infty \frac{x^2}{(x^2+a^2)^3}\,dx = \frac{\pi}{8a^3}, a > 0$. The two singularities in the integrand are each third-order and, while not a really terribly difficult computation (you should do it), here's a simpler and more general approach. You are to fill-in the missing details.

(a) Start with $\int_{-\infty}^\infty \frac{x^2}{(x^2+a^2)(x^2+b^2)}\,dx$, with $a \neq b$, make a partial fraction expansion, and do the resulting two easy integrals; (b) let $b \to a$ and so arrive at the value for $\int_{-\infty}^\infty \frac{x^2}{(x^2+a^2)^2}\,dx$; (c) finally, use Feynman's favorite trick of differentiating an integral to get Copson's answer. Notice that you can now continue to differentiate endlessly to calculate $\int_{-\infty}^\infty \frac{x^2}{(x^2+a^2)^n}\,dx$ for any $n > 3$ you wish.

(C8.9) In Sect. 5.4 we argued that the derivative of the gamma function $\Gamma(x)$ at $x = 1$ is $\Gamma'(1) - -\gamma$, where γ is Euler's constant. Show that the integral form of the digamma function derived in (8.10.6) is consistent with that assertion. That is, show, since $\Gamma(1) = 0 ! = 1$, that $\psi(1) = \frac{\Gamma'(1)}{\Gamma(1)} = \Gamma'(1) = \int_0^\infty \frac{e^{-z}}{z}\,dz - \int_0^\infty \frac{dz}{(1+z)z} = -\gamma$. Hint: Do the first integral by-parts, and the second one with a partial fraction expansion, with both integrals written as $\lim_{\varepsilon \to 0}\int_\varepsilon^\infty$. This limiting operation is necessary because the two integrals, individually, diverge, but you'll find that their *difference* is finite. And don't forget (5.4.3).

The final four challenge problems of this chapter are not contour integrals, themselves, but rather are included here because they show how our earlier contour integrations can be further manipulated to give us even more quite interesting results using just standard freshman calculus techniques.

(C8.10) Suppose m and n are non-negative real numbers such that $n > m + 1$. Show that $\int_0^\infty \frac{x^m}{1-x^n}\,dx = \frac{\pi/n}{\tan\left(\frac{m+1}{n}\pi\right)}$. Hint: Start with (8.6.9), that is with $\int_{-\infty}^\infty \frac{e^{ay}}{1-e^y}\,dy = \frac{\pi}{\tan(a\pi)}$ with $0 < a < 1$, and make the change of variable $e^y = x^n$.

(C8.11) Show that $\int_0^\infty \frac{x^m}{x^n+b}\,dx = \frac{\pi}{nb^{(n-m-1)/n}\sin\left\{\frac{(m+1)}{n}\pi\right\}}$. Hint: Try the change of variable $x = ya^{1/n}$ in (8.7.8) (and then, a bit later, you'll find $b = 1/a$ helpful).

(C8.12) Show that $\int_0^\infty \frac{x^a}{(x+b)^2} dx = b^{a-1} \frac{\pi a}{\sin(\pi a)}$. Hint: Using the result from C8.11, with $m = a$ and $n = 1$, apply Feynman's trick of differentiating with respect to a parameter (in this case, b).

(C8.13) Show that $\int_0^\infty \frac{\sqrt{x}\, \ln(x)}{(x+b)^2} dx = \left[1 + \frac{1}{2} \ln(b)\right] \frac{\pi}{\sqrt{b}}$. Hint: Differentiate the result from C8.12 with respect to a, and then set $a = \frac{1}{2}$. This is a generalization of an integral that appeared in an 1896 textbook by the French mathematician Félix Tisserand (1845–1896), who did the b = 1 special case.

Chapter 9
Epilogue

9.1 Riemann, Prime Numbers, and the Zeta Function

Starting with the frontispiece photo, this entire book, devoted to doing Riemannian definite integrals, has been a continuing ode to the genius of Bernhard Riemann. He died far too young, of tuberculosis, at age 39. And yet, though he was just reaching the full power of his intellect when he left this world, you can appreciate from the numerous times his name has appeared here how massively productive he was. He left us with many brilliant results, but he also left the world of mathematics its greatest *unsolved* problem, too, a problem that has often been described as the Holy Grail of mathematics. It's a problem *so* difficult, and *so* mysterious, that many mathematicians have seriously entertained the possibility that it *can't* be solved. And it is replete with interesting integrals!

This is the famous (at least in the world of mathematics) *Riemann Hypothesis* (RH), a conjecture which has so far soundly defeated, since Riemann formulated it in 1859, all the efforts of the greatest mathematical minds in the world (including his) to either prove or disprove it. Forty years after its conjecture, and with no solution in sight, the great German mathematician David Hilbert (we discussed his transform back in Chap. 7) decided to add some incentive. In 1900, at the Second International Congress of Mathematicians in Paris, he gave a famous talk titled "Mathematical Problems." During that talk he discussed a number of problems that he felt represented potentially fruitful directions for future research. The problems included, for example, the transcendental nature (or not) of $2^{\sqrt{2}}$, Fermat's Last Theorem (FLT), and the RH, in *decreasing* order of difficulty (in Hilbert's estimation).

All of Hilbert's problems became famous overnight, and to solve one brought instant celebrity among fellow mathematicians. Hilbert's own estimate of the difficulty of his problems was slightly askew, however, as the $2^{\sqrt{2}}$ issue was settled by 1930 (it *is* transcendental!), and FLT was laid to rest by the mid-1990s. The RH, however, the presumed 'easiest' of the three, has proven itself to be the toughest. Hilbert eventually came to appreciate this. A well-known story in mathematical lore

© Springer Nature Switzerland AG 2020
P. J. Nahin, *Inside Interesting Integrals*, Undergraduate Lecture Notes in Physics,
https://doi.org/10.1007/978-3-030-43788-6_9

says he once remarked that, if he awoke after sleeping for five hundred years, the first question he would ask is 'Has the Riemann hypothesis been proven?' The answer is currently still *no* and so, a century after Hilbert's famous talk in Paris, the Clay Mathematics Institute in Cambridge, MA proposed in 2000 seven so-called "Millennium Prize Problems," with each to be worth a one million dollar award to its solver. The RH is one of those elite problems and, as I write in 2020, the one million dollars for its solution remains unclaimed.

The RH is important for more than just being famous for being unsolved; there are numerous theorems in mathematics, all of which mathematicians believe to be correct, that are based on the assumed truth of the RH. If the RH is someday shown to be false, the existing proofs of all those theorems collapse and they will have to be revisited and new proofs (hopefully) found. To deeply discuss the RH is far beyond the level of this book, but since it involves complex numbers and functions, bears Riemann's name, abounds in integrals, and is *unsolved*, it nonetheless seems a fitting topic with which to end this book.

Our story begins, as do so many of the fascinating tales in mathematics, with an amazing result from Euler. In 1748 he showed that if s is real and greater than 1, and if we write the zeta function (see Sect. 5.3 again)

$$\zeta(s) = \sum_{n=1}^{\infty} \frac{1}{n^s} = 1 + \frac{1}{2^s} + \frac{1}{3^s} + \cdots, \tag{9.1.1}$$

then

$$\zeta(s) = \frac{1}{\prod_{j=1}^{\infty} \left\{ 1 - \frac{1}{p_j^s} \right\}} \tag{9.1.2}$$

where p_j is the j-*th* prime ($p_1 = 2$, $p_2 = 3$, $p_3 = 5$, $p_4 = 7$, and so on). That is, Euler showed that there is an intimate, *surprising* connection between $\zeta(s)$, a continuous function of s, and the primes which as integers are the very signature of discontinuity.[1]

Riemann was led to the zeta function because of Euler's connection of it to the primes (he called it his "point of departure"), with the thought that studying $\zeta(s)$ would aid in his quest for a formula for $\pi(x)$, defined to be the number of primes not greater than x. $\pi(x)$ is a measure of how the primes are distributed among the integers. It should be obvious that $\pi(\frac{1}{2}) = 0$, that $\pi(2) = 1$, and that $\pi(6) = 3$, but perhaps it is not quite so obvious that $\pi(10^{18}) = 24{,}739{,}954{,}287{,}740{,}860$. When Riemann started his studies of the distribution of the primes, one known *approximation* to $\pi(x)$ is the so-called *logarithmic integral*, written as

---

[1]To derive (9.1.2) is not difficult, just 'devilishly' clever; you can find a proof in any good book on number theory. Or see my book, *An Imaginary Tale: the story of* $\sqrt{-1}$, Princeton 2016, pp. 150–152.

$$\text{li}(x) = \int_2^x \frac{du}{\ln(u)} \qquad (9.1.3)$$

which is actually a pretty good approximation.[2] For example,

$$\frac{\pi(1,000)}{\text{li}(1,000)} = \frac{168}{178} = 0.94\ldots,$$

$$\frac{\pi(100,000)}{\text{li}(100,000)} = \frac{9,592}{9,630} = 0.99\ldots,$$

$$\frac{\pi(100,000,000)}{\text{li}(100,000,000)} = \frac{5,761,455}{5,762,209} = 0.999\ldots,$$

$$\frac{\pi(1,000,000,000)}{\text{li}(1,000,000,000)} = \frac{50,847,478}{50,849,235} = 0.9999\ldots.$$

In an 1849 letter the great German mathematician C. F. Gauss, who signed-off on Riemann's 1851 doctoral dissertation with a glowing endorsement, claimed to have known of this behavior of li(x) since 1791 or 1792, when he was just fourteen. With what is known of Gauss' genius, there is no doubt that is true!

Numerical calculations like those above immediately suggest the conjecture

$$\lim_{x \to \infty} \frac{\pi(x)}{\text{li}(x)} = 1,$$

which is a statement of what mathematicians call the *prime number theorem*. Although highly suggestive, such numerical calculations of course *prove* nothing, and in fact it wasn't until 1896 that mathematical proofs of the prime number theorem were simultaneously and independently discovered by Charles-Joseph de la Vallée-Poussin (1866–1962) in Belgium and Jacques Hadamard (1865–1963) in France. Each man used very advanced techniques from complex function theory, applied to the zeta function. It was a similar quest (using the zeta function as well) that Riemann was on in 1859, years before either Vallée-Poussin or Hadamard had been born.

Riemann's fascination with the distribution of the primes is easy to understand. The primes have numerous properties which, while easy to state, are almost paradoxical. For example, it has been known since Euclid that the primes are infinite in number and yet, if one looks at a list of consecutive primes from 2 on up to very high values, it is evident that they become, *on average*, ever less frequent. I emphasize the 'on average' because, every now and then, one does encounter *consecutive* odd integers (every prime but 2 is, of course, odd) that are both primes. Such a pair forms

---

[2]A slight variation is the modern li(x) = $\int_2^x \frac{du}{\ln(u)}$ + 1.045... which, for x = 1,000 (for example), MATLAB computes *integral(@(x)1./log(x),2,1000)+1.045 = 177.6*.

a *twin prime*. It is not known if the twin primes are infinite in number. Mathematicians believe they are, but can't *prove* it.

If one forms the sum of the reciprocals of all the positive integers then we have the harmonic series which, of course, diverges. That is, $\zeta(1) = \infty$. If you then go through and eliminate all terms in that sum except for the reciprocals of the primes, the sum *still* diverges (proven by Euler in 1737), a result that almost always surprises when first seen demonstrated. (This also proves, in a way different from Euclid's proof, that the primes are infinite in number.) In 1919 the Norwegian mathematician Viggo Brun (1885–1975) showed that if one further eliminates from the sum all terms but the reciprocals of the twin primes, *then* the sum is finite. Indeed, the sum isn't very large at all. The value, called *Brun's constant*, is

$$\left(\frac{1}{3}+\frac{1}{5}\right) + \left(\frac{1}{5}+\frac{1}{7}\right) + \left(\frac{1}{11}+\frac{1}{13}\right) + \left(\frac{1}{17}+\frac{1}{19}\right) + \ldots \approx 1.90216\ldots.$$

The finite value of Brun's constant does *not* preclude the twin primes from being infinite in number, however, and so that question remains open.

Here's another example of the curious nature of the distribution of the primes: for any a > 1 there is always at least one prime between a and 2a. And yet, for any a > 1 there is also always a stretch of at least a − 1 consecutive integers that is *free* of primes! This is true, no matter how large a may be. The first statement is due to the French mathematician Joseph Bertrand (1822–1900), whose conjecture of it in 1845 was proven (it's *not* an 'easy' proof) by the Russian mathematician P. L. Chebyshev (1821–1894) in 1850. For the second statement, simply notice that every number in the consecutive sequence a ! + 2, a ! + 3, a ! + 4, . . ., a ! + a is divisible and so none are prime.

In 1837 a really remarkable result was established by Dirichlet (the same Dirichlet we first meet in Chap. 1): if a and b are relatively prime positive integers (that means their greatest common factor is 1) then the arithmetic progression a, a + b, a + 2b, a + 3b, ... contains an infinite number of primes. This easy-to-state theorem immediately gives us the not very obvious conclusion that there are infinitely many primes ending with 999 (as do 1999, 100,999, and 1,000,999). That's because *all* the numbers in the progression (not *all* of which are prime) formed from a = 999 and b = 1000 (which have only the factor 1 in common) end in 999.

With examples like these in mind, it should now be easy to understand what the mathematician Pál Erdös (1913–1996) meant when he famously (at least in the world of mathematics) declared "It will be another million years, at least, before we understand the primes." With properties like the ones above, how could *any* mathematician, including Riemann, *not* be fascinated by the primes?[3]

To start his work, Riemann immediately tackled a technical issue concerning the very definition of $\zeta(s)$ as given in (9.1.1), namely, the sum converges only if s > 1.

---

[3]The English mathematician A. E. Ingham (1900–1967) opens his book *The Distribution of the Primes* (Cambridge University Press 1932) with the comment "A problem *at the very threshold of mathematics* [my emphasis] is the question of the distribution of the primes among the integers."

More generally, if we extend the s in Euler's definition of the zeta function from being real to being complex (that is, $s = \sigma + i\,t$) then $\zeta(s)$ as given in (9.1.1) makes sense only if $\sigma > 1$. Riemann, however, wanted to be able to treat $\zeta(s)$ as defined *everywhere* in the complex plane or, as he put it, he wanted a formula for $\zeta(s)$ "which remains valid for all s." Such a formula would give the same values for $\zeta(s)$ as does (9.1.1) when $\sigma > 1$, but would also give sensible values for $\zeta(s)$ even when $\sigma < 1$. Riemann was fabulously successful in discovering how to do that.

He did it by discovering what is called the *functional equation* of the zeta function and, just to anticipate things a bit, here it is (we'll derive it in the next section):

$$\zeta(s) = 2(2\pi)^{s-1} \sin\left(\frac{\pi s}{2}\right)\Gamma(1-s)\zeta(1-s). \qquad (9.1.4)$$

Riemann's functional equation is considered to be one of the gems of mathematics. Here's how it works. What we have is

$$\zeta(s) = F(s)\zeta(1-s), \quad F(s) = 2(2\pi)^{s-1}\sin\left(\frac{\pi s}{2}\right)\Gamma(1-s).$$

F(s) is a well-defined function for all σ. So, if we have an s with $\sigma > 1$ we'll use (9.1.1) to compute $\zeta(s)$, but if $\sigma < 0$ we'll use (9.1.4) (along with (9.1.1) to compute $\zeta(1-s)$ because the real part of $1 - s$ is >1 if $\sigma < 0$).

There is, of course, the remaining question of computing $\zeta(s)$ for the case of $0 < \sigma < 1$, where s is in the so-called *critical strip* (a vertical band with width 1 extending from $-i\infty$ to $+i\infty$). The functional equation doesn't help us now, because if s is in the critical strip then so is $1 - s$. This is actually a problem we've already solved, however, as you can see by looking back at (5.3.7), where we showed

$$\sum_{k=1}^{\infty} \frac{(-1)^{k-1}}{k^s} = [1 - 2^{1-s}]\zeta(s).$$

So, for example, right in the middle of the critical strip, on the real axis, we have $s = \frac{1}{2}$ and so

$$\sum_{k=1}^{\infty} \frac{(-1)^{k-1}}{k^{1/2}} = \left[1 - 2^{1-(\frac{1}{2})}\right]\zeta\left(\frac{1}{2}\right) = \left[\frac{1}{\sqrt{1}} - \frac{1}{\sqrt{2}} + \frac{1}{\sqrt{3}} - \frac{1}{\sqrt{4}} - \cdots\right].$$

Thus,

$$\zeta\left(\frac{1}{2}\right) = \frac{1}{1 - \sqrt{2}}\left[1 - \frac{1}{\sqrt{2}} + \frac{1}{\sqrt{3}} - \frac{1}{\sqrt{4}} - \cdots\right].$$

If we keep the first one million terms—a well-known theorem in freshman calculus tells us that the partial sums of an alternating series (with monotonically

decreasing terms) converge, and the maximum error we make is less than the first
term we neglect and so our error *for the sum* should be less than 10^{-3} —we get

$$\zeta\left(\frac{1}{2}\right) \approx -1.459147\ldots.$$

The actual value is known to be

$$\zeta\left(\frac{1}{2}\right) = -1.460354\ldots.$$

This is obviously *not* a very efficient way (a *million* terms!?) to calculate $\zeta\left(\frac{1}{2}\right)$, but
the point here is that (5.3.7) is correct.

For Euler's case of s purely real, the plots in Fig. 9.1.1 show the general behavior
of $\zeta(s)$. For $s > 1$, $\zeta(s)$ smoothly decreases from $+\infty$ towards 1 as s increases from
1, while for $s < 0$ $\zeta(s)$ oscillates, eventually heading off to $-\infty$ as s approaches
1 from below. Fig. 9.1.1 indicates that $\zeta(0) = -0.5$, and in the next section I'll show
you how to *prove* that $\zeta(0) = -\frac{1}{2}$ using the functional equation. Notice, too, that
Fig. 9.1.1 hints at $\zeta(s) = 0$ for s a *negative*, even integer, another conclusion
supported by the functional equation.

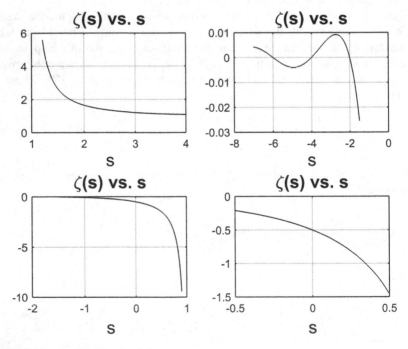

Fig. 9.1.1 The zeta function for real s

To make that last observation crystal-clear, let's write s $= -2n$, where n $= 0, 1, 2,$ 3, ... Then, (9.1.4) becomes

$$\zeta(-2n) = -2^{-2n}\pi^{-(2n+1)}\Gamma(1 + 2n)\sin(n\pi)\zeta(1 + 2n) = 0$$

because all of the factors on the right of the first equality are finite, for all n, including $\sin(n\pi)$ which is, of course, zero for all integer n. We must exclude the case of $n = 0$, however, because then $\zeta(1 + 2n) = \zeta(1) = \infty$, and this infinity is sufficient to overwhelm the zero of $\sin(0)$. We know this because, as stated above, and as will be shown in the next section, $\zeta(0) \neq 0$ but rather $\zeta(0) = -\frac{1}{2}$.

When a value of s gives $\zeta(s) = 0$ then we call that value of s a *zero of the zeta function*. Thus, all the even, negative integers are zeros of $\zeta(s)$, and because they are so easy to 'compute' they are called the *trivial zeros* of $\zeta(s)$. There are other zeros of $\zeta(s)$, however, which are not so easy to compute,[4] and where they are in the complex plane is what the RH is all about.

Here is what Riemann correctly believed about the non-trivial zeros (even if he couldn't prove all the following in 1859):

(1) they are infinite in number;
(2) all are complex (of the form $\sigma + i\,t, t \neq 0$);
(3) all are in the critical strip ($0 < \sigma < 1$);
(4) they occur in pairs, symmetrically displaced around the vertical $\sigma = \frac{1}{2}$ line (called the *critical line*), that is, if $\frac{1}{2} - \varepsilon + i\,t$ is a zero, then so is $\frac{1}{2} + \varepsilon + i\,t$ for some ε in the interval $0 \leq \varepsilon < \frac{1}{2}$;
(5) they are symmetrical about the real axis ($t = 0$), that is, if $\sigma + i\,t$ is a zero then so is $\sigma - i\,t$ (the zeros appear as conjugate pairs).

The RH is now easy to state: $\varepsilon = 0$. That is, *all* of the complex zeros are *on* the critical line and so have real part $\frac{1}{2}$. More precisely, Riemann conjectured "it is *very probable* [my emphasis] that all the [complex zeros are on the critical line]." Since 1859, all who have tried to prove the RH have failed, including Riemann, who wrote "Certainly one would wish [for a proof]; I have meanwhile temporarily put aside the search for [a proof] after some fleeting futile attempts, as it appears unnecessary for [finding a formula for $\pi(x)$]."

---

[4]The methods used to compute the non-trivial zeros are far from 'obvious' and beyond the level of this book. If you are interested in looking further into how such computations are done, I can recommend the following four books: (1) H. M. Edwards, *Riemann's Zeta Function*, Academic Press 1974; (2) E. C. Titchmarsh, *The Theory of the Riemann Zeta-function* (2nd edition, revised by D. R. Heath-Brown), Oxford Science Publications 1986; (3) Aleksandar Ivić, *The Riemann Zeta-Function*, John Wiley & Sons 1985; and (4) *The Riemann Hypothesis* (Peter Borwein *et al.*, editors), Springer 2008.

There *have* been some impressive partial results since 1859. In 1914 it was shown by the Dane Harald Bohr (1887–1951) and the German Edmund Landau (1877–1938) that all but an infinitesimal proportion of the complex zeros are arbitrarily close to the critical line (that is, they are in the 'arbitrarily thin' vertical strip $\frac{1}{2} - \varepsilon < \sigma < \frac{1}{2} + \varepsilon$ for any $\varepsilon > 0$). That same year Hardy proved that an infinity of complex zeros are *on* the critical line (this does *not* prove that *all* of the complex zeros are on the critical line). In 1942 it was shown by the Norwegian Atle Selberg (1917–2007) that an (unspecified) fraction of the zeros are on the critical line. In 1974 Selberg's fraction was shown to be greater than $\frac{1}{3}$ by the American Norman Levinson (1912–1975), and then in 1989 the American J. B. Conrey showed the fraction is greater than $\frac{2}{5}$.

There is also what appears, *at first glance*, to be quite substantial computational support for the truth of the RH. Ever since Riemann himself computed the locations of the first three complex zeros,[5] the last few decades have seen that accomplishment vastly surpassed. In 2004, the first 10^{13} (yes, ten *trillion*!) zeros were shown to all be on the critical line. Since even a single zero off the critical line is all that is needed to disprove the RH, this looks pretty impressive—but mathematicians are, frankly, *not* impressed. As Ivić wrote in his book (see note 4), "No doubt the numerical data will continue to accrue, but number theory is unfortunately one of the branches of mathematics where numerical evidence does not count for much."

Fig. 9.1.2 The first six non-trivial zeros of $\zeta\left(\frac{1}{2} + it\right)$

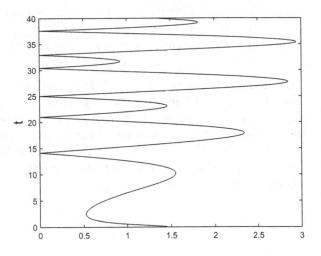

---

[5]Because of the symmetry properties of the complex zero locations, one only has to consider the case of t > 0. The value of t for a zero is called the *height* of the zero, and the zeros are ordered by increasing t (the first six zeros are shown in Fig. 9.1.2, where a zero occurs each place $\left|\zeta\left(\frac{1}{2} + it\right)\right|$ touches the vertical t-axis). The heights of the first six zeros are 14.134725, 21.022040, 25.010856, 30.424878, 32.935057, and 37.586176. In addition to the first 10^{13} zeros, billions more zeros at heights as large as 10^{24} and beyond have also all been confirmed to be on the critical line.

There are, in fact, lots of historical examples in mathematics where initial, massive computational 'evidence' has prompted conjectures which later proved to be false. A particularly famous example involves $\pi(x)$ and li(x). For all values of x for which $\pi(x)$ and li(x) are known, $\pi(x) <$ li(x). Further, the difference between the two increases as x increases and, for 'large' x the difference is significant; for $x = 10^{18}$, for example, $d(x) =$ li$(x) - \pi(x) \approx 22,000,000$. Based on this impressive numerical 'evidence' it was commonly believed for a long time that $d(x) > 0$ for all x. Gauss believed this (as did Riemann) all his life. *But it's not true.*

In 1912, Hardy's friend and collaborator J. E. Littlewood (1885–1977) proved that there is some x for which $d(x) < 0$. Two years later he extended his proof to show that as x continues to increase the sign of $d(x)$ flips back-and-forth endlessly. The value of x at which the first change in sign of $d(x)$ occurs is not known, only that it is *very* big. In 1933 the South African Stanley Skewes (1899–1988) derived a stupendously huge upper-bound on the value of that first x: $e^{e^{e^{79}}} \approx 10^{10^{10^{34}}}$. This has become famous (at least in the world of mathematics) as the *first Skewes number*. In his derivation Skewes assumed the truth of the RH, but in 1955 he dropped that assumption to calculate a new upper-bound for the first x at which $d(x)$ changes sign: this is the *second Skewes number* equal to $10^{10^{10^{1,000}}}$ and it is, of course, *much* larger than the first one. In 2000 the upper-bound was dropped to 'just' 1.39×10^{316}. All of these numbers are far beyond anything that can be numerically studied on a computer, and the number of complex zeros that have been found on the critical line is minuscule in comparison. It is entirely possible that the first complex zero off the critical line (thus disproving the RH) may not occur until a vastly greater height is reached than has been examined so far.

Some mathematicians have been markedly less than enthusiastic about the RH. Littlewood, in particular, was quite blunt, writing "I believe [the RH] to be false. There is no evidence for it . . . One should not believe things for which there is no evidence ... I have discussed the matter with several people who know the problem in relation to electronic calculation; they are all agreed that the chance of finding a zero off the line in a lifetime's calculation is millions to one against it. It looks then as if we may never know."[6] A slightly more muted (but perhaps not by much) position is that of the American mathematician H. M. Edwards (born 1936), who wrote in his classic book on the zeta function (see note 4) "Riemann based his hypothesis on no insights ... which are not available to us today ... and that, on the contrary, had he known some of the facts which have since been discovered, he might well have been led to reconsider ... unless some basic cause is operating which has eluded mathematicians for 110 years [161 years now, as I write in 2020], occasional [complex zeros] off the [critical] line are altogether possible ... Riemann's insight was stupendous, but it was not supernatural, and what seemed 'probable' to him in 1859 might seem less so today."

---

[6]From Littlewood's essay "The Riemann Hypothesis," in *The Scientist Speculates: An Anthology of Partly-Baked Ideas* (I. J. Good, editor), Basic Books 1962.

9.2   Deriving the Functional Equation for $\zeta(s)$

The derivation of the functional equation for $\zeta(s)$ that appears in Riemann's famous 1859 paper involves a contour integral in which a branch cut is involved. We've already been through an example of that, however, and so here I'll show you a different derivation (one also due to Riemann) that makes great use of results we've already derived in the book.

We start with the integral

$$\int_0^\infty x^{m-1}e^{-ax}dx, m \geq 1, a > 0,$$

and make the change of variable $u = ax$ (and so $dx = \frac{du}{a}$). Thus,

$$\int_0^\infty x^{m-1}e^{-ax}\,dx = \int_0^\infty \left(\frac{u}{a}\right)^{m-1}e^{-u}\,\frac{du}{a} = \frac{1}{a^m}\int_0^\infty u^{m-1}e^{-u}\,du.$$

The right-most integral is, from (4.1.1), $\Gamma(m)$, and so

$$\int_0^\infty x^{m-1}e^{-ax}dx = \frac{\Gamma(m)}{a^m}. \tag{9.2.1}$$

Now, if we let

$$m - 1 = \frac{1}{2}s - 1 \left(m = \frac{1}{2}s\right)$$

and

$$a = n^2\pi,$$

then (9.2.1) becomes

$$\int_0^\infty x^{\frac{1}{2}s-1}e^{-n^2\pi x}\,dx = \frac{\Gamma\left(\frac{1}{2}s\right)}{(n^2\pi)^{\frac{1}{2}s}} = \frac{\Gamma\left(\frac{1}{2}s\right)}{\pi^{\frac{1}{2}s}n^s}. \tag{9.2.2}$$

Then, summing (9.2.2) over all positive integer n, we have

$$\sum_{n=1}^\infty \int_0^\infty x^{\frac{1}{2}s-1}e^{-n^2\pi x}\,dx = \sum_{n=1}^\infty \frac{\Gamma\left(\frac{1}{2}s\right)}{\pi^{\frac{1}{2}s}n^s}$$

or, reversing the order of summation and integration on the left,

$$\int_0^\infty x^{\frac{1}{2}s-1} \sum_{n=1}^\infty e^{-n^2\pi x}\, dx = \pi^{-\frac{1}{2}s} \Gamma\left(\frac{1}{2}s\right) \sum_{n=1}^\infty \frac{1}{n^s} = \pi^{-\frac{1}{2}s} \Gamma\left(\frac{1}{2}s\right) \zeta(s). \quad (9.2.3)$$

At this point Riemann defined the function

$$\psi(x) = \sum_{n=1}^\infty e^{-n^2\pi x}, \quad (9.2.4)$$

and then used the identity[7]

$$\sum_{n=-\infty}^\infty e^{-n^2\pi x} = \frac{1}{\sqrt{x}} \sum_{n=-\infty}^\infty e^{-n^2\pi/x}. \quad (9.2.5)$$

The left-hand side of (9.2.5) is (because $n^2 > 0$ for n negative *or* positive)
$$\sum_{n=-\infty}^{-1} e^{-n^2\pi x} + 1 + \sum_{n=1}^\infty e^{-n^2\pi x} = \sum_{n=1}^\infty e^{-n^2\pi x} + 1 + \sum_{n=1}^\infty e^{-n^2\pi x} = 2\psi(x) + 1.$$
The right-hand side of (9.2.5) is (for the same reason)

$$\frac{1}{\sqrt{x}} \sum_{n=-\infty}^\infty e^{-n^2\pi/x} = \frac{1}{\sqrt{x}} \left\{ 2\psi\left(\frac{1}{x}\right) + 1 \right\}.$$

Thus,

$$2\psi(x) + 1 = \frac{1}{\sqrt{x}} \left\{ 2\psi\left(\frac{1}{x}\right) + 1 \right\}$$

or, solving for $\psi(x)$,

$$\psi(x) = \frac{1}{\sqrt{x}} \psi\left(\frac{1}{x}\right) + \frac{1}{2\sqrt{x}} - \frac{1}{2} - \sum_{n=1}^\infty e^{-n^2\pi x}. \quad (9.2.6)$$

Now, putting (9.2.4) into (9.2.3) gives us

$$\pi^{-\frac{1}{2}s} \Gamma\left(\frac{1}{2}s\right) \zeta(s) = \int_0^\infty x^{\frac{1}{2}s-1} \psi(x)\, dx$$

or, breaking the integral into two parts,

---

[7]In my book *Dr. Euler's Fabulous Formula*, Princeton 2017, pp. 246–253, you'll find a derivation of the identity $\sum_{k=-\infty}^\infty e^{-\alpha k^2} = \sqrt{\frac{\pi}{\alpha}} \sum_{n=-\infty}^\infty e^{-\pi^2 n^2/\alpha}$. If you write $\alpha = \pi x$ then (9.2.5) immediately results. The derivation in *Dr. Euler* combines Fourier theory with what mathematicians call *Poisson summation*, all of which might sound impressively exotic. In fact, it is all at the level of nothing more than the end of freshman calculus. If you can read this book then you can easily follow the derivation of (9.2.5) in *Dr. Euler*.

$$\pi^{-\frac{1}{2}s}\Gamma\left(\frac{1}{2}s\right)\zeta(s) = \int_0^1 x^{\frac{1}{2}s-1}\psi(x)dx + \int_1^\infty x^{\frac{1}{2}s-1}\psi(x)dx. \qquad (9.2.7)$$

Using (9.2.6) in the first integral on the right of (9.2.7), we have

$$\pi^{-\frac{1}{2}s}\Gamma\left(\frac{1}{2}s\right)\zeta(s) = \int_0^1 x^{\frac{1}{2}s-1}\left\{\frac{1}{\sqrt{x}}\psi\left(\frac{1}{x}\right)+\frac{1}{2\sqrt{x}}-\frac{1}{2}\right\}dx + \int_1^\infty x^{\frac{1}{2}s-1}\psi(x)dx$$

$$= \int_0^1 x^{\frac{1}{2}s-1}\left\{\frac{1}{2\sqrt{x}}-\frac{1}{2}\right\}dx + \int_0^1 x^{\frac{1}{2}s-\frac{3}{2}}\psi\left(\frac{1}{x}\right)dx + \int_1^\infty x^{\frac{1}{2}s-1}\psi(x)dx.$$

The first integral on the right is easy to do (for s > 1):

$$\int_0^1 x^{\frac{1}{2}s-1}\left\{\frac{1}{2\sqrt{x}}-\frac{1}{2}\right\}dx = \frac{1}{2}\int_0^1 x^{\frac{1}{2}s-\frac{3}{2}}dx - \frac{1}{2}\int_0^1 x^{\frac{1}{2}s-1}dx$$

$$= \frac{1}{2}\left(\frac{x^{\frac{1}{2}s-\frac{1}{2}}}{\frac{1}{2}s-\frac{1}{2}}\right)\Big|_0^1 - \frac{1}{2}\left(\frac{x^{\frac{1}{2}s}}{\frac{1}{2}s}\right)\Big|_0^1 = \frac{1}{2}\left(\frac{1}{\frac{1}{2}s-\frac{1}{2}}\right) - \frac{1}{2}\left(\frac{1}{\frac{1}{2}s}\right) = \frac{1}{s-1}-\frac{1}{s}$$

$$= \frac{1}{s(s-1)}.$$

Thus,

$$\pi^{-\frac{1}{2}s}\Gamma\left(\frac{1}{2}s\right)\zeta(s) = \frac{1}{s(s-1)} + \int_0^1 x^{\frac{1}{2}s-\frac{3}{2}}\psi\left(\frac{1}{x}\right)dx + \int_1^\infty x^{\frac{1}{2}s-1}\psi(x)dx. \qquad (9.2.8)$$

Next, in the first integral on the right in (9.2.8) make the change of variable $u = \frac{1}{x}$ (and so $dx = -\frac{du}{u^2}$). Then,

$$\int_0^1 x^{\frac{1}{2}s-\frac{3}{2}}\psi\left(\frac{1}{x}\right)dx = \int_\infty^1 \left(\frac{1}{u}\right)^{\frac{1}{2}s-\frac{3}{2}}\psi(u)\left\{-\frac{du}{u^2}\right\} = \int_1^\infty \frac{1}{u^{\frac{1}{2}s+\frac{1}{2}}}\psi(u)du = \int_1^\infty x^{-\frac{1}{2}s-\frac{1}{2}}\psi(x)dx$$

and therefore (9.2.8) becomes

$$\pi^{-\frac{1}{2}s}\Gamma\left(\frac{1}{2}s\right)\zeta(s) = \frac{1}{s(s-1)} + \int_1^\infty \left\{x^{-\frac{1}{2}s-\frac{1}{2}}+x^{\frac{1}{2}s-1}\right\}\psi(x)dx. \qquad (9.2.9)$$

Well!—you might exclaim at this point, as you look at the integral in (9.2.9)—*what do we do now?* You might even feel like repeating the words that Oliver Hardy often directed towards Stan Laurel, after the two old-time movie comedians had stumbled into one of their idiotic jams: "Look at what a fine mess you've gotten us into *this* time!"

In fact, however, we are not in a mess, and all we need do is notice, as did Riemann, that the right-hand side of (9.2.9) is *unchanged* if we replace every occurrence of s with $1 - s$. Try it and see. But that means we can do the same thing on the left-hand side of (9.2.9) because, after all, (9.2.9) is an equality. That is, it must be true that

$$\pi^{-\frac{1}{2}s}\Gamma\left(\frac{1}{2}s\right)\zeta(s) = \pi^{-\frac{1}{2}(1-s)}\Gamma\left(\frac{1-s}{2}\right)\zeta(1-s). \tag{9.2.10}$$

We are now almost done, with just a few more 'routine' steps to go to get the form of the functional equation for $\zeta(s)$ that I gave you in (9.1.4).

Solving (9.2.10) for $\zeta(s)$, we have

$$\zeta(s) = \pi^{s-\frac{1}{2}}\frac{\Gamma\left(\frac{1-s}{2}\right)}{\Gamma\left(\frac{1}{2}s\right)}\zeta(1-s). \tag{9.2.11}$$

Now, recall (4.2.18), one of the forms of Legendre's duplication formula:

$$z!\left(z-\frac{1}{2}\right)! = 2^{-2z}\pi^{\frac{1}{2}}(2z)!$$

or, expressed in gamma notation,

$$\Gamma(z+1)\Gamma\left(z+\frac{1}{2}\right) = 2^{-2z}\pi^{\frac{1}{2}}\Gamma(2z+1). \tag{9.2.12}$$

If we write $2z + 1 = 1 - s$ then $z = -\frac{s}{2}$ and (9.2.12) becomes

$$\Gamma\left(-\frac{s}{2}+1\right)\Gamma\left(-\frac{s}{2}+\frac{1}{2}\right) = 2^s\pi^{\frac{1}{2}}\Gamma(1-s) = \Gamma\left(1-\frac{s}{2}\right)\Gamma\left(\frac{1-s}{2}\right)$$

or,

$$\Gamma\left(\frac{1-s}{2}\right) = \frac{2^s\pi^{\frac{1}{2}}\Gamma(1-s)}{\Gamma\left(1-\frac{s}{2}\right)}. \tag{9.2.13}$$

From (4.2.16), the reflection formula for the gamma function,

$$\Gamma(m)\Gamma(1-m) = \frac{\pi}{\sin(m\pi)}$$

or, with $m = \frac{s}{2}$,

$$\Gamma\left(\frac{s}{2}\right)\Gamma\left(1-\frac{s}{2}\right) = \frac{\pi}{\sin\left(\frac{\pi s}{2}\right)}$$

and this says

$$\Gamma\left(1 - \frac{s}{2}\right) = \frac{\pi}{\Gamma\left(\frac{s}{2}\right) \sin\left(\frac{\pi s}{2}\right)}. \tag{9.2.14}$$

So, putting (9.2.14) into (9.2.13),

$$\Gamma\left(\frac{1-s}{2}\right) = \frac{2^s \pi^{\frac{1}{2}} \Gamma(1-s)}{\frac{\pi}{\Gamma\left(\frac{s}{2}\right) \sin\left(\frac{\pi s}{2}\right)}} = 2^s \pi^{-\frac{1}{2}} \Gamma\left(\frac{s}{2}\right) \sin\left(\frac{\pi s}{2}\right) \Gamma(1-s)$$

or,

$$\frac{\Gamma\left(\frac{1-s}{2}\right)}{\Gamma\left(\frac{s}{2}\right)} = 2^s \pi^{-\frac{1}{2}} \sin\left(\frac{\pi s}{2}\right) \Gamma(1-s). \tag{9.2.15}$$

Inserting (9.2.15) into (9.2.11), we arrive at

$$\zeta(s) = \pi^{s-\frac{1}{2}} 2^s \pi^{-\frac{1}{2}} \sin\left(\frac{\pi s}{2}\right) \Gamma(1-s) \zeta(1-s)$$

or, at last,

$$\zeta(s) = 2(2\pi)^{s-1} \sin\left(\frac{\pi s}{2}\right) \Gamma(1-s) \zeta(1-s), \tag{9.2.16}$$

the functional equation I gave you earlier in (9.1.4).

As a simple 'test' of (9.2.16), suppose $s = \frac{1}{2}$. Then,

$$\zeta\left(\frac{1}{2}\right) = 2(2\pi)^{-\frac{1}{2}} \sin\left(\frac{\pi}{4}\right) \Gamma\left(\frac{1}{2}\right) \zeta\left(\frac{1}{2}\right)$$

which says, once we cancel the $\zeta\left(\frac{1}{2}\right)$ on each side,[8] that

$$1 = \frac{2}{\sqrt{2\pi}} \sin\left(\frac{\pi}{4}\right) \Gamma\left(\frac{1}{2}\right).$$

Is this correct? Yes, because the right-hand side is

$$\left(\sqrt{\frac{2}{\pi}}\right)\left(\frac{1}{\sqrt{2}}\right)(\sqrt{\pi}) = 1.$$

So, (9.2.16) is consistent for $s = \frac{1}{2}$.

As an example of how (9.2.16) works, let's use it to calculate $\zeta(-1)$. Thus, with $s = -1$,

---

[8]We know we can do this because, as shown in the previous section, $\zeta\left(\frac{1}{2}\right) = -1.4\ldots \neq 0$.

$$\zeta(-1) = 2(2\pi)^{-2} \sin\left(-\frac{\pi}{2}\right)\Gamma(2)\zeta(2).$$

Since $\Gamma(2) = 1$, $\sin\left(-\frac{\pi}{2}\right) = -1$, and $\zeta(2) = \frac{\pi^2}{6}$, then

$$\zeta(-1) = \frac{2}{4\pi^2}(-1)\frac{\pi^2}{6} = -\frac{1}{12}.$$

This result is the basis for an interesting story in the history of mathematics co-starring, once again, G. H. Hardy.

In late January 1913 Hardy received the first of several mysterious letters from India. Written by the then unknown, self-taught mathematician Srinivasa Ramanujan (recall his 'master theorem' from Chap. 6), who was employed as a lowly clerk in Madras, the letters were pleas for the world-famous Hardy to look at some of his results. Many of those results were perplexing, but none more than this one:

$$1 + 2 + 3 + 4 + \ldots = -\frac{1}{12}.$$

Most professional mathematicians would have simply chucked that into the trash, and dismissed the clerk as a pathetic lunatic. It was Hardy's genius that he didn't do that, but instead soon made sense of Ramanujan's sum. What the clerk meant (but expressed badly) is understood by writing the sum as

$$1 + 2 + 3 + 4 + \ldots = \frac{1}{\frac{1}{1}} + \frac{1}{\frac{1}{2}} + \frac{1}{\frac{1}{3}} + \frac{1}{\frac{1}{4}} + \ldots = \frac{1}{1^{-1}} + \frac{1}{2^{-1}} + \frac{1}{3^{-1}} + \frac{1}{4^{-1}} + \ldots$$

which is *formally* $\zeta(-1)$. Where Ramanujan's $-\frac{1}{12}$ came from I don't know but, indeed, as we've just shown using (9.2.16), $\zeta(-1) = -\frac{1}{12}$. Ramanujan had discovered a special case of the extended zeta function before he had ever heard of Riemann!

After Ramanujan's death, Hardy remarked (in his 1921 obituary notice that appeared in the *Proceedings of The London Mathematical Society*, as well as in the *Proceedings of the Royal Society*) that his friend's mathematical knowledge had some remarkable gaps: "[He] had found for himself the functional equation of the Zeta-function," but "he had never heard of . . . Cauchy's theorem," "had indeed but the vaguest idea of what a function of a complex variable was," and that "it was impossible to allow him to go through life supposing that all the zeros of the Zeta-function were real." Hardy later said, with perhaps little exaggeration, that Ramanujan was his greatest discovery.[9]

---

[9]You can read more about Ramanujan's amazing life in the biography by Robert Kanigel, *The Man Who Knew Infinity: a life of the genius Ramanujan*, Charles Scribner's Sons 1991.

As a more substantial application of (9.2.16), I'll use it next to calculate the value of $\zeta(0)$. If we do something as crude as just shove $s = 0$ into (9.2.16) we quickly see that we get nowhere:

$$\zeta(0) = 2(2\pi)^{-1} \sin(0)\Gamma(1)\zeta(1) = ????$$

because the zero of $\sin(0)$ and the infinity of $\zeta(1)$ are at war with each other. Which one wins? To find out, we'll have to be a lot more subtle in our calculations. Strange as it may seem to you, we'll get our answer by studying the case of $s = 1$ (not $s = 0$), which I'll simply ask you to take on faith as we start.

Looking back at (9.2.9), we have

$$\zeta(s) = \frac{1}{\pi^{-\frac{1}{2}s}\Gamma(\frac{1}{2}s)s(s-1)} + \frac{1}{\pi^{-\frac{1}{2}s}\Gamma(\frac{1}{2}s)} \int_1^\infty \left\{ x^{-\frac{1}{2}s-\frac{1}{2}} + x^{\frac{1}{2}s-1} \right\} \psi(x)dx.$$

If we let $s \to 1$ then we see that the right-hand side does indeed blow-up (as it should, because $\zeta(1) = \infty$), strictly because of the first term on the right, *alone*, since the integral term is obviously convergent.[10] In fact, since $\lim_{s \to 1} \pi^{-\frac{1}{2}s}\Gamma(\frac{1}{2}s)s = \frac{\Gamma(\frac{1}{2})}{\sqrt{\pi}} = \frac{\sqrt{\pi}}{\sqrt{\pi}} = 1$, then $\zeta(s)$ blows-up like $\frac{1}{s-1}$ as $s \to 1$. *Remember this point — it will prove to be the key to our solution.*

Now, from (9.2.16) we have

$$\zeta(1-s) = \frac{\zeta(s)}{2(2\pi)^{s-1}\sin\left(\frac{\pi s}{2}\right)\Gamma(1-s)}.$$

From the reflection formula for the gamma function we have

$$\Gamma(s)\Gamma(1-s) = \frac{\pi}{\sin(\pi s)}$$

and so

$$\Gamma(1-s) = \frac{\pi}{\Gamma(s)\sin(\pi s)}$$

which says

---

[10] I use the word *obviously* because, over the entire interval of integration, the integrand is finite and goes to zero *very* fast as $x \to \infty$. Indeed, the integrand vanishes even faster than *exponentially* as $x \to \infty$, which you can show by using (9.2.4) to write $\psi(x) = \sum_{n=1}^\infty e^{-n^2\pi x} = e^{-\pi x} + e^{-4\pi x} + e^{-9\pi x} + \dots < e^{-\pi x} + e^{-2\pi x} + e^{-3\pi x} + \dots$, a geometric series easily summed to give $\psi(x) < \frac{1}{e^{\pi x}-1}, x > 0$, which behaves like $e^{-\pi x}$ for x 'large.' With $s = 1$ the integrand behaves (for x 'large') like $\frac{\psi(x)}{x^{3/2}+x^{1/2}} \approx \frac{e^{-\pi x}}{x\sqrt{x}}$ for x 'large.'

$$\zeta(1-s) = \frac{\zeta(s)}{2(2\pi)^{s-1}\sin\left(\frac{\pi s}{2}\right)}\,\frac{\pi}{\Gamma(s)\sin(\pi s)} = \frac{\Gamma(s)\sin(\pi s)\zeta(s)}{2\pi(2\pi)^{s-1}\sin\left(\frac{\pi s}{2}\right)}.$$

Since $\sin(\pi s) = 2\cos\left(\frac{\pi s}{2}\right)\sin\left(\frac{\pi s}{2}\right)$, we arrive at

$$\zeta(1-s) = \frac{\Gamma(s)\cos\left(\frac{\pi s}{2}\right)\zeta(s)}{\pi(2\pi)^{s-1}}, \tag{9.2.17}$$

an alternative form of the functional equation for the zeta function. This is the form we'll use to let $s \to 1$, thus giving $\zeta(0)$ on the left.

So, from (9.2.17) we have

$$\lim_{s\to 1}\zeta(1-s) = \zeta(0) = \lim_{s\to 1}\frac{\Gamma(s)\cos\left(\frac{\pi s}{2}\right)\zeta(s)}{\pi(2\pi)^{s-1}} = \frac{\Gamma(1)}{\pi}\lim_{s\to 1}\cos\left(\frac{\pi s}{2}\right)\zeta(s)$$

$$= \frac{1}{\pi}\lim_{s\to 1}\cos\left(\frac{\pi s}{2}\right)\zeta(s)$$

or,

$$\zeta(0) = \frac{1}{\pi}\lim_{s\to 1}\frac{\cos\left(\frac{\pi s}{2}\right)}{s-1} \tag{9.2.18}$$

where I've used our earlier conclusion (that I told you to remember, remember?): $\zeta(s)$ behaves like $\frac{1}{s-1}$ as $s \to 1$. The limit in (9.2.18) gives the indeterminate result $\frac{0}{0}$, and so we use L'Hospital's rule to compute

$$\zeta(0) = \frac{1}{\pi}\lim_{s\to 1}\frac{\frac{d}{ds}\left\{\cos\left(\frac{\pi s}{2}\right)\right\}}{\frac{d}{ds}\{s-1\}} = \frac{1}{\pi}\lim_{s\to 1}\frac{-\frac{\pi}{2}\sin\left(\frac{\pi s}{2}\right)}{1}$$

or, at last, we have our answer:

$$\zeta(0) = -\frac{1}{2}.$$

To finish our calculation of particular values of $\zeta(s)$, let me now show you a beautiful way to calculate $\zeta(2)$—which we've already done back in Chap. 7—*that also gives us all the other values of* $\zeta(2n)$ *for n > 1 at the same time*. What makes this calculation doubly interesting is that *all* of the analysis is at the level of just freshman calculus. We start by finding the power series expansion for tan(x) around $x \approx 0$, that is, the so-called *Taylor series*.

Why we start with this is, of course, not at all obvious, but go along with me for a bit and you'll see how it will all make sense in the end. So, what we'll do is write

$$\tan(x) = \sum_{k=0}^{\infty} c_k x^k = c_0 + c_1 x + c_2 x^2 + c_3 x^3 + \dots$$

and then find the c's. This is a standard part of freshman calculus, and I'm going to assume that most readers have seen it before (an assumption I made in the last chapter, too, when Taylor series came-up in the discussion in Sect. 8.8 on Laurent series) and, if so, simply skip ahead. If it's been a while for you, however, here's a quick run-through. If you insert $x = 0$ into the series you immediately get $\tan(0) = 0 = c_0$. To find c_1, first differentiate the series with respect to x and *then* set $x = 0$. That is,

$$\frac{d}{dx}\{\tan(x)\} = \frac{d}{dx}\left\{\frac{\sin(x)}{\cos(x)}\right\} = \frac{1}{\cos^2(x)} = c_1 + 2c_2 x + 3c_3 x^2 + \dots$$

and so, setting $x = 0$,

$$\frac{1}{\cos^2(0)} = 1 = c_1.$$

To find c_2, differentiate again and then set $x = 0$ (you should find that $c_2 = 0$), and so on for all the other c's. If you are careful with your arithmetic, you should get

$$\tan(x) = x + \frac{1}{3}x^3 + \frac{2}{15}x^5 + \frac{17}{315}x^7 + \frac{62}{2,835}x^9 + \frac{1,382}{155,925}x^{11} + \dots,$$

and where you stop calculating terms is a function only of your endurance!

Next, we'll use this result to get the power series for cot(x), by writing

$$\cot(x) = \frac{1}{\tan(x)} = \frac{1}{x + \frac{1}{3}x^3 + \frac{2}{15}x^5 + \frac{17}{315}x^7 + \frac{62}{2,835}x^9 + \frac{1,382}{155,925}x^{11} + \dots}$$

$$= \frac{1}{x}\left\{\frac{1}{1 + \frac{1}{3}x^2 + \frac{2}{15}x^4 + \frac{17}{315}x^6 + \dots}\right\}.$$

Then, if you carefully perform the long-division indicated by the fraction inside the curly brackets, you should get

$$\cot(x) = \frac{1}{x}\left\{1 - \frac{1}{3}x^2 - \frac{1}{45}x^4 - \frac{2}{945}x^6 - \dots\right\}$$

or,

$$\cot(x) = \frac{1}{x} - \frac{1}{3}x - \frac{1}{45}x^3 - \frac{2}{945}x^5 - \dots. \qquad (9.2.19)$$

Okay, put (9.2.19) aside for now.

Next, consider this amazing identity: for α any real, *non*-integer value,

$$\cos(\alpha t) = \frac{\sin(\alpha\pi)}{\pi}\left[\frac{1}{\alpha} + 2\alpha\sum_{n=1}^{\infty}\frac{(-1)^n}{\alpha^2 - n^2}\cos(nt)\right]. \tag{9.2.20}$$

This looks pretty spectacular, but in fact (9.2.20) is quickly derived via a simple, routine Fourier series expansion of the periodic function that is the repetition of a single period given by $\cos(\alpha t)$, $-\pi < t < \pi$, extended infinitely far in both directions along the t-axis.[11] (You can now see why α cannot be an integer—if it was, then the $n = \alpha$ term in the sum would blow-up.)

If we set $t = \pi$ in (9.2.20) we get

$$\cos(\alpha\pi) = \frac{\sin(\alpha\pi)}{\pi}\left[\frac{1}{\alpha} + 2\alpha\sum_{n=1}^{\infty}\frac{(-1)^n}{\alpha^2 - n^2}\cos(n\pi)\right]$$

or, since $\cos(n\pi) = (-1)^n$ and since $(-1)^n(-1)^n = \{(-1)^2\}^n = 1$, then

$$\frac{\cos(\alpha\pi)}{\sin(\alpha\pi)} = \cot(\alpha\pi) = \frac{1}{\pi}\left[\frac{1}{\alpha} + 2\alpha\sum_{n=1}^{\infty}\frac{1}{\alpha^2 - n^2}\right].$$

Thus,

$$\cot(\alpha\pi) = \frac{1}{\alpha\pi} + \frac{\alpha\pi}{\pi^2}\sum_{n=1}^{\infty}\frac{2}{\alpha^2 - n^2} = \frac{1}{\alpha\pi} + \sum_{n=1}^{\infty}\frac{2(\alpha\pi)}{(\alpha\pi)^2 - n^2\pi^2}.$$

If we define $x = \alpha\pi$ then we get another series for $\cot(x)$:

$$\cot(x) = \frac{1}{x} + \sum_{n=1}^{\infty}\frac{2x}{x^2 - n^2\pi^2}. \tag{9.2.21}$$

From (9.2.21) it follows that

$$1 - x\cot(x) = -\sum_{n=1}^{\infty}\frac{2x^2}{x^2 - n^2\pi^2} = \sum_{n=1}^{\infty}\frac{2x^2}{n^2\pi^2 - x^2} = \frac{2x^2}{\pi^2}\sum_{n=1}^{\infty}\frac{1}{n^2}\left\{\frac{1}{1 - \frac{x^2}{n^2\pi^2}}\right\}$$

$$= \frac{2x^2}{\pi^2}\sum_{n=1}^{\infty}\frac{1}{n^2}\left\{1 + \frac{x^2}{n^2\pi^2} + \frac{x^4}{n^4\pi^4} + \frac{x^6}{n^6\pi^6} + \cdots\right\}$$

or,

$$1 - x\cot(x) = \frac{2x^2}{\pi^2}\sum_{n=1}^{\infty}\frac{1}{n^2} + \frac{2x^4}{\pi^4}\sum_{n=1}^{\infty}\frac{1}{n^4} + \frac{2x^6}{\pi^6}\sum_{n=1}^{\infty}\frac{1}{n^6} + \cdots. \tag{9.2.22}$$

And from (9.2.19) it follows that

---

[11]All the details in the derivation of (9.2.20) can be found on pp. 154–155 of *Dr. Euler*.

$$1 - x \cot(x) = \frac{1}{3}x^2 + \frac{1}{45}x^4 + \frac{2}{945}x^6 + \ldots \qquad (9.2.23)$$

Equating (9.2.22) and (9.2.23), and then equating the coefficients of equal powers of x, we have

$$\frac{2}{\pi^2} \sum_{n=1}^{\infty} \frac{1}{n^2} = \frac{1}{3},$$

$$\frac{2}{\pi^4} \sum_{n=1}^{\infty} \frac{1}{n^4} = \frac{1}{45},$$

$$\frac{2}{\pi^6} \sum_{n=1}^{\infty} \frac{1}{n^6} = \frac{2}{945},$$

and so on. That is,

$$\zeta(2) = \frac{1}{3}\left(\frac{\pi^2}{2}\right) = \frac{\pi^2}{6},$$

$$\zeta(4) = \frac{1}{45}\left(\frac{\pi^4}{2}\right) = \frac{\pi^4}{90},$$

$$\zeta(6) = \frac{2}{945}\left(\frac{\pi^6}{2}\right) = \frac{\pi^6}{945},$$

and so on. This isn't the way Euler found $\zeta(2n)$ but, believe me, he would have loved this approach!

One of the open questions about the zeta function concerns the values of $\zeta(2n + 1)$, $n \geq 1$. Not one of these values, not even just the very first one for $\zeta(3)$, is known other than through the direct numerical evaluation of its defining infinite series. There are certainly no known formulas involving powers of pi, as there are for $\zeta(2n)$. Why there is this sharp distinction between even and odd arguments of the zeta function is one of the deep mysteries of mathematics, one that has puzzled the world's greatest mathematicians from Euler to the present day.

9.3   Our Final Calculation

As the last calculation of this book, let me end on a dramatic note by showing you how to derive a classic result for the zeta function, one that every modern textbook writer mentions, but one for which nobody (or so it seems) bothers to provide any details! As a glance at Fig. 9.1.1 shows, $\zeta'(0) < 0$. That is, direct computation shows that the first derivative of $\zeta(s)$ at $s = 0$ is clearly negative. But what is its *exact* value? Finding the values of *all* the derivatives of $\zeta(s)$ at $s = 0$ is the basis, of course, for developing the Taylor series expansion of $\zeta(s)$, which has important theoretical applications (which we'll not pursue here).

To find $\zeta'(0)$, we start with a rewrite of (5.3.6):

$$\eta(s) = [1 - 2^{1-s}]\zeta(s). \tag{9.3.1}$$

In (5.3.6) I wrote the left-hand-side as v_s, but here I'm using $\eta(s)$ because mathematicians commonly refer to (9.3.1) as *Dirichlet's eta function*. Taking the logarithm of (9.3.1),

$$\ln\{\eta(s)\} = \ln\{[1 - 2^{1-s}]\} + \ln\{\zeta(s)\},$$

and then differentiating with respect to s, we have

$$\frac{\eta'(s)}{\eta(s)} = \frac{(1 - 2^{1-s})'}{1 - 2^{1-s}} + \frac{\zeta'(s)}{\zeta(s)}. \tag{9.3.2}$$

We know from our earlier work in this chapter that $\zeta(0) = -\frac{1}{2}$, and so (9.3.1) says

$$\eta(0) = [1 - 2^1]\zeta(0) = -\zeta(0) = \frac{1}{2}.$$

With this in hand, 'all' we have left to do is the calculation of $(1 - 2^{1-s})'$ at $s = 0$ and then we'll have everything that appears in (9.3.2), with the exceptions, of course, of $\eta'(0)$ and $\zeta'(0)$. So, if we can calculate $\eta'(0)$, and $(1 - 2^{1-s})'$ at $s = 0$, we can then calculate $\zeta'(0)$.

Since

$$2^{1-s} = e^{\ln(2^{1-s})} = e^{(1-s)\ln(2)} = e^{\ln(2)}e^{-s\ln(2)} = 2e^{-s\ln(2)}$$

then

$$1 - 2^{1-s} = 1 - 2e^{-s\ln(2)}$$

and so

$$(1 - 2^{1-s})' = 2\ln(2)e^{-s\ln(2)} = \ln(2)2^{1-s}.$$

Plugging what we have so far into (9.3.2), for $s = 0$ we arrive at

$$\frac{\eta'(0)}{\frac{1}{2}} = \frac{2\ln(2)}{1 - 2} + \frac{\zeta'(0)}{-\frac{1}{2}}$$

or,

$$\eta'(0) = -\ln(2) - \zeta'(0)$$

and so

$$\zeta'(0) = -\ln(2) - \eta'(0). \tag{9.3.3}$$

At this point we can no longer avoid the question of $\eta'(0) = ?$
We have, from (5.3.7),

$$\eta(s) = \sum_{k=1}^{\infty} \frac{(-1)^{k-1}}{k^s}$$

and so

$$\eta'(s) = \sum_{k=1}^{\infty} (-1)^{k-1} \left(\frac{1}{k^s}\right)'.$$

Since

$$\frac{1}{k^s} = k^{-s} = e^{\ln(k^{-s})} = e^{-s \ln(k)}$$

then

$$\left(\frac{1}{k^s}\right)' = -\ln(k)e^{-s \ln(k)} = -\frac{\ln(k)}{k^s}$$

and so

$$\eta'(s) = -\sum_{k=1}^{\infty} (-1)^{k-1} \frac{\ln(k)}{k^s}.$$

Thus,

$$\eta'(0) = -\sum_{k=1}^{\infty} (-1)^{k-1} \ln(k) = \sum_{k=1}^{\infty} (-1)^k \ln(k)$$

or, since $\ln(1) = 0$, we can drop the $k = 1$ term and write

$$\eta'(0) = \sum_{k=1}^{\infty} (-1)^k \ln(k) = \ln(2) - \ln(3) + \ln(4) - \ln(5) + \ln(6) - \dots.$$

That is,

$$\eta'(0) = \ln\left(\frac{2 \cdot 4 \cdot 6 \cdot 8 \cdots}{3 \cdot 5 \cdot 7 \cdots}\right). \tag{9.3.4}$$

In 1655 Wallis (remember him from Chap. 4?) discovered the wonderful formula[12]

$$\frac{\pi}{2} = \frac{2 \cdot 2 \cdot 4 \cdot 4 \cdot 6 \cdot 6 \cdots}{3 \cdot 3 \cdot 5 \cdot 5 \cdot 7 \cdot 7 \cdots}$$

and so we can write (9.3.4) as

$$\eta'(0) = \ln\left(\sqrt{\frac{2 \cdot 2 \cdot 4 \cdot 4 \cdot 6 \cdot 6 \cdots}{3 \cdot 3 \cdot 5 \cdot 5 \cdot 7 \cdot 7 \cdots}}\right) = \ln\left(\sqrt{\frac{\pi}{2}}\right).$$

Putting this result into (9.3.3), we have

$$\zeta'(0) = -\ln(2) - \ln\left(\sqrt{\frac{\pi}{2}}\right) = -\ln\left(2\sqrt{\frac{\pi}{2}}\right) = -\ln\left(\sqrt{2\pi}\right)$$

or, finally, we at last have our answer:

$$\zeta'(0) - -\frac{1}{2}\ln(2\pi) = -0.9189385\ldots . \tag{9.3.5}$$

9.4 Adieu

All that I've told you here about the zeta function and the RH is but a *very tiny* fraction of what is known. And yet, the RH remains today a vast, dark mystery, with its resolution hidden somewhere in a seemingly very long (perhaps endless), twisting tunnel with not even a glimmer of light visible ahead to hint that there is an end to it. Riemann is rightfully famous for many things in mathematical physics,[13] but it is

---

[12]See, for example, my *An Imaginary Tale: the story of* $\sqrt{-1}$, Princeton 2016, pp. 155–156.

[13]When Riemann submitted his doctoral dissertation (*Foundations for a General Theory of Functions of a Complex Variable*) to Gauss, the great man pronounced it to be "penetrating," "creative," "beautiful," and to be a work that "far exceeds" the standards for such works. To actually be able to teach, however, Riemann had to give a trial lecture to an audience of senior professors (including Gauss), and Gauss asked that it be on the foundations of geometry. That Riemann did in June 1854 and—in the words of Caltech mathematician Eric Temple Bell in his famous book *Men of Mathematics*—it "revolutionized differential geometry and prepared the way for the geometrized physics of [today]." What Bell was referring to is Einstein's later description of gravity as a manifestation of curved spacetime, an idea Riemann might have come to himself had he lived. After Riemann's death the British mathematician William Kingdon Clifford (1845–1879) translated Riemann's lecture into English and was perhaps the only man then alive who truly appreciated what Riemann had done. Clifford, himself, came within a whisker of the spirit of Einstein's theory of gravity in a brief, enigmatic note written in 1876 ("On the Space Theory of Matter"), 3 years before Einstein's birth. Sadly, Clifford too died young of the same disease that had killed Riemann.

no small irony that the most famous of all is his creation of a puzzle that has (so far) resisted all attempts by the world's greatest mathematicians to solve. All that mighty effort has not been in vain, however.

One of Isaac Newton's contemporaries was Roger Cotes (1682–1716), who was a professor at Cambridge by age 26, and editor of the second edition of Newton's masterpiece *Principia*, a work that revolutionized physics. His death from a violent fever one month before his thirty-fourth birthday—even younger than were Riemann and Clifford at their deaths—cut short what was a truly promising scientific life. It was reported that, after Cotes' death, Newton himself said of him "If he had lived we might have known something." No one ever said that of Riemann, however, because the world had learned a *lot* from him even before he departed this life.

This book started with stories involving G. H. Hardy, and for symmetry alone let me conclude with yet one more. Hardy was fascinated by the RH all his working life, and never missed a chance to puzzle over it. One famous story, that amusingly reflects just how deep was the hook the RH had in him, relates how, after concluding a visit with a mathematician friend in Copenhagen, Denmark, Hardy was about to return to England. His journey was to be by boat, over a wild and stormy North Sea. Hoping to avoid potential catastrophe, Hardy quickly, just before boarding, wrote and mailed a postcard to his friend declaring "I have a proof of the Riemann hypothesis!" He was confident, Hardy said later, after reaching England safely, that God would not let him die with the false glory of being remembered for doing what he really had not achieved. Hardy, it should be noted, was a devout atheist in all other aspects of his life.

By some incredible coincidence, I wrote those final words for the first edition of this book on the British cruise ship *Queen Elizabeth*, while returning from holiday in Norway. Cunard's QE is home-based in Southampton, England, and my Baltic journey had taken me through the North Sea and right past Copenhagen. So, there I sat, a hundred years after Hardy, with my feet propped-up on the railing of my stateroom balcony, writing of his little joke perhaps right where he had sailed a century before. Somehow, I think Hardy's ghost was not at all surprised, as it peered over my shoulder, that the mystery of the RH was still unresolved.

By yet another coincidence (of a different sort) I write these new, final words for the second edition of this book while sitting on yet another cruise ship (the Norwegian cruise liner *Gem*, out of Boston, sailing to Bermuda). My feet are again propped-up on the railing of my stateroom balcony, but I doubt Hardy's ghost is with me this time. Hardy sailed across the Atlantic to spend a 1928–1929 sabbatical leave from Oxford in America (at Princeton and Caltech), and again in 1936 to visit Harvard, but he never sailed into waters further south than New York. So, alas, I'm afraid I'm mathematically all alone on this trip.

But one thing hasn't changed; the mystery of the RH is as deep as it was when I first wrote, and, as far as anyone can see into the future, it may well *still* be a mystery a hundred years from now. For mathematicians, however, this is as great a lure as an ancient, magical artifact nailed shut in a wooden crate, lost in a vast, secret military warehouse would be to Indiana Jones, and so the hunt goes gloriously on!

9.5 Challenge Questions

Since this chapter deals mostly with unresolved issue of the RH, I've decided to be more philosophical than analytical with the challenge problems. So, rather than *problems*, here are four *questions* for you to ponder.

(C9.1) As in the text, let p_n denote the n-*th* prime: $p_1 = 2$, $p_2 = 3$, and so on. Suppose we define q_n to be the first prime greater than $p_1 p_2 \ldots p_n + 1$, and then calculate $q_n - (p_1 p_2 \ldots p_n)$. If we do this for the first seven values of n we get the following table:

n	$p_1 p_2 \ldots p_n$	$p_1 p_2 \ldots p_n + 1$	q_n	$q_n - (p_1 p_2 \ldots p_n)$
1	2	3	5	3
2	6	7	11	5
3	30	31	37	7
4	210	211	223	13
5	2310	2311	2333	23
6	30,030	30,031	30,047	17
7	510,510	510,511	510,529	19

Looking at the right-most column of the table, does a conjecture *strongly* suggest itself to you? Can you *prove* your conjecture? It's questions like this about the primes, so easy to cook-up and oh-so-hard to prove, that fascinated Riemann in 1859 and continue to fascinate today. For example, the twin prime conjecture that I mentioned in Section 9.1 has been the basis for recent (2014) excitement in the world of mathematics, with the proof that there is an infinity of prime pairs separated by gaps of no more than 246. A long way from the original conjecture with a gap of 2, yes, but still ....

(C9.2) While numerical computations generally don't prove theorems (although they might well *disprove* a conjecture by finding a counterexample), they can be quite useful in suggesting a possibility worthy of further examination. For example, since Fig. 9.1.1 indicates that, for $\varepsilon > 0$, $\lim_{\varepsilon \to 0} \zeta(1 - \varepsilon) = -\infty$ and $\lim_{\varepsilon \to 0} \zeta(1 + \varepsilon) = +\infty$, then it is at least conceivable that $\lim_{\varepsilon \to 0} \{\zeta(1 - \varepsilon) + \zeta(1 + \varepsilon)\}$ could be finite. In fact, it is easy to use MATLAB to compute the following table:

ε	$\{\zeta(1 - \varepsilon) + \zeta(1 + \varepsilon)\}/2$
0.1	0.577167222774279
0.01	0.577215180384336
0.001	0.577215660056368
0.0001	0.577215664853611

From these calculations it *appears* that $\lim_{\varepsilon \to 0} \{\zeta(1 - \varepsilon) + \zeta(1 + \varepsilon)\}/2 = 0.57721566\ldots$. Does this suggest to you that the limit is a well-known constant? Can you *prove* your conjecture?

(C9.3) Define $h(q) = \sum_{k=1}^{q} \frac{1}{k} = 1 + \frac{1}{2} + \frac{1}{3} + \ldots + \frac{1}{q}, q \geq 1$. In 1737 Euler showed that $\sum_{q=1}^{\infty} \frac{h(q)}{q^2} = 2\zeta(3)$ and $\sum_{q=1}^{\infty} \frac{h(q)}{q^3} = \frac{5}{4}\zeta(4)$. These two results are famous in mathematics as *Euler's sums*, and they can be derived using nothing but high school algebra (but, of course, it's *very clever* algebra). You might wonder, after looking at these two sums, what Euler had to say about the obvious sum he apparently skipped-over: $\sum_{q=1}^{\infty} \frac{h(q)}{q} = ?$ Euler didn't say much about this sum because he 'instantly' saw that the sum diverges. In the spirit of this book, the proof of that divergence is a one-liner *if you make a crucial observation*. Do you see it?

(C9.4) Two undergraduate math majors, Sally and Sam, are discussing the RH in general and, in particular, the view held by some mathematicians that the reason nobody has solved the problem, even after almost two centuries of intense effort, is simply because it's unsolvable. Sally says while that may well be true, it's also true that nobody will ever be able to *prove* the RH is unsolvable. When Sam asks why not, Sally replies "If the RH could be shown to be unsolvable, that would mean nobody could ever find a complex zero off the critical line, even by accident, no matter how long you looked. Not even if you could check zeros at the rate of $10^{10^{10}}$ each nanosecond. In fact, you would fail no matter *how high up* that exponential stack of tens you went. That's because if you did find such a zero then you would have proven the RH to be false. That's a contradiction with the initial premise that there exists a proof that the problem is unsolvable. But the impossibility of finding a zero off the critical line means there *are* no zeros off the critical line, and *that* would mean the RH *had* been solved by showing it is true. That's a contradiction, too. The only way out of this logical quagmire is to conclude that *no such proof exists*"

Sam thinks that over and, finally, replies with "Well, I'm not so sure about all that. There are infinitely many complex zeros, and so no matter how many of them you could check each nanosecond, it would still take infinite time to check them all. So, you'd *never* be done, and a contradiction doesn't occur *until* you are done."

What do you think of each of these arguments? Is one of them on to something, or have Sam and Sally both sadly morphed, in different ways, from being hard-nosed mathematicians into being windbag metaphysicians?

Solutions to the Challenge Problems

Original Preface

Since $x + \frac{1}{x} = 1$, then multiplying through by x gives $x^2 + 1 = x$ or, rearranging, $x^2 = x - 1$. Multiplying through by x again, $x^3 = x^2 - x$ or, substituting our expression for x^2, $x^3 = x - 1 - x = -1$. Squaring this gives $x^6 = 1$, and then multiplying through by x we have $x^7 = x$. So, substituting x^7 for x in the original $x + \frac{1}{x} = 1$ we have $x^7 + \frac{1}{x^7} = 1$ and we are done.

Chapter 1

(C1.1) In the first integral, change variable to $u = x - 2$ (and so $du = dx$) which says

$$\int_0^8 \frac{dx}{x-2} = \int_{-2}^6 \frac{du}{u} = \lim_{\varepsilon \to 0} \left[\int_{-2}^{-\varepsilon} \frac{du}{u} + \int_\varepsilon^6 \frac{du}{u} \right] = \lim_{\varepsilon \to 0} \left[\ln(u) \big|_{-2}^{-\varepsilon} + \ln(u) \big|_\varepsilon^6 \right]$$

$$= \lim_{\varepsilon \to 0} [\ln(-\varepsilon) - \ln(-2) + \ln(6) - \ln(\varepsilon)] = \lim_{\varepsilon \to 0} \left[\ln\left(\frac{-\varepsilon}{-2}\right) + \ln\left(\frac{6}{\varepsilon}\right) \right]$$

$$= \lim_{\varepsilon \to 0} \left[\ln \left\{ \left(\frac{\varepsilon}{2}\right) \left(\frac{6}{\varepsilon}\right) \right\} \right] = \ln(3).$$

That is, $\int_0^8 \frac{dx}{x-2} = \ln(3)$. For the second integral, $\int_0^3 \frac{dx}{(x-1)^{2/3}} = \lim_{\varepsilon \to 0} \int_0^{1-\varepsilon} \frac{dx}{(x-1)^{2/3}} + \lim_{\varepsilon \to 0} \int_{1+\varepsilon}^3 \frac{dx}{(x-1)^{2/3}}$. In both integrals on the right, change variable to $u = x - 1$ and so $du = dx$. Then

© Springer Nature Switzerland AG 2020
P. J. Nahin, *Inside Interesting Integrals*, Undergraduate Lecture Notes in Physics,
https://doi.org/10.1007/978-3-030-43788-6

$$\int_0^3 \frac{dx}{(x-1)^{2/3}} = \lim_{\varepsilon \to 0} \left\{ \int_{-1}^{-\varepsilon} \frac{du}{u^{2/3}} + \int_{\varepsilon}^{2} \frac{du}{u^{2/3}} \right\}$$

$$= \lim_{\varepsilon \to 0} \left\{ \left(3u^{1/3}\right) \Big|_{-1}^{-\varepsilon} + \left(3u^{1/3}\right) \Big|_{\varepsilon}^{2} \right\}$$

$$= 3 \lim_{\varepsilon \to 0} \left\{ (-\varepsilon)^{1/3} - (-1)^{1/3} + (2)^{1/3} - (\varepsilon)^{1/3} \right\} = 3 \left\{ 1 + 2^{1/3} \right\}.$$

(C1.2) Write $\int_1^\infty \frac{dx}{\sqrt{x^3-1}} = \int_1^\infty \frac{dx}{\sqrt{x-1}\sqrt{x^2+x+1}} < \int_1^\infty \frac{dx}{\sqrt{x-1}\sqrt{x^2}} = \int_1^\infty \frac{dx}{x\sqrt{x-1}}$, where the inequality follows because in the integral just before the one at the far-right I've replaced a denominator factor in the integrand with a quantity that is *smaller*. Now, change variable to $u = x - 1$ in the last integral. So, $\int_1^\infty \frac{dx}{\sqrt{x^3-1}} < \int_0^\infty \frac{du}{(u+1)\sqrt{u}} = \int_0^\infty \frac{du}{u^{3/2}+u^{1/2}}$ or, writing the last integral as the sum of two integrals, we have $\int_1^\infty \frac{dx}{\sqrt{x^3-1}} < \int_0^1 \frac{du}{u^{3/2}+u^{1/2}} + \int_1^\infty \frac{du}{u^{3/2}+u^{1/2}}$. We make the inequality even stronger by replacing the denominators of the integrands on the right-hand-side with smaller quantities, and so $\int_1^\infty \frac{dx}{\sqrt{x^3-1}} < \int_0^1 \frac{du}{u^{1/2}} + \int_1^\infty \frac{du}{u^{3/2}} = \left(2u^{1/2}\right)\Big|_0^1 - \left(2u^{-\frac{1}{2}}\right)\Big|_1^\infty = 2+2 = 4.$

(C1.3) In Feynman's integral $\int_0^1 \frac{dx}{[ax+b(1-x)]^2}$ change variable to $u = ax + b(1 - x)$ and so $dx = \frac{du}{a-b}$ and the integral becomes $\int_b^a \frac{1}{u^2} \left(\frac{du}{a-b}\right) = \frac{1}{a-b} \left\{ -\frac{1}{u} \right\} \Big|_b^a = \frac{1}{a-b} \times \left\{ \frac{1}{b} - \frac{1}{a} \right\} = \frac{a-b}{(a-b)ab} = \frac{1}{ab}$, just as Feynman claimed. This all makes sense as long as the integrand doesn't blow-up somewhere inside the integration interval (that is, as long as there is no x such that $ax + b(1 - x) = 0$ in the interval $0 \leq x \leq 1$. Now, the quantity $ax + b(1 - x)$ *does* equal zero when $x = \frac{b}{b-a}$ and so we therefore require either $\frac{b}{b-a} < 0$ (call this Condition 1) *or* that $\frac{b}{b-a} > 1$ (call this Condition 2). For Condition 1 we can write, with k some positive number, $\frac{b}{b-a} = -k$. Or, $b = -bk + ak$ and so, multiplying through by a, $ab = -abk + a^2k$. Solving for ab, we have $ab = \frac{a^2k}{1+k}$. Since k is positive (and, obviously, so is a^2) we then conclude that $ab > 0$. For Condition 2 we can write, with k some number greater than 1, $\frac{b}{b-a} = k$. Or, $b = bk - ak$ and so, multiplying through by a, we have $ab = abk - a^2k$. Solving for ab, we have $ab = \frac{a^2k}{k-1}$. Since $k > 1$ we conclude that $ab > 0$. That is, for both Conditions (conditions that insure the integrand doesn't blow-up somewhere inside the integration interval) we conclude that $ab > 0$. That is, a and b must have the same algebraic sign and that eliminates the puzzle of Feynman's integral.

(C1.4) Transforming $\int_0^\infty \frac{e^{-cx}}{x} dx$ into $\int_0^\infty \frac{e^{-y}}{y} dy$ with the change of variable $y = cx$ (assuming $c > 0$) is straightforward. With $x = \frac{y}{c}$ we have $dx = \frac{1}{c} dy$ and so $\int_0^\infty \frac{e^{-cx}}{x} dx = \int_0^\infty \frac{e^{-c\frac{y}{c}}}{\frac{y}{c}} \left(\frac{1}{c} dy\right) = \int_0^\infty \frac{e^{-y}}{y} dy$. This is *formally* okay, but it is really just an exercise in symbol-pushing. That's because the key to understanding the puzzle is to realize that the integral $\int_0^\infty \frac{e^{-y}}{y} dy$ *doesn't exist*! Here's how to see that. For any given finite value of the upper limit greater than zero (call it δ), the value of $\int_0^\delta \frac{e^{-y}}{y} dy = \infty$. That's because for $0 \le y \le \delta$ we have $e^{-y} \ge e^{-\delta}$ and so $\int_0^\delta \frac{e^{-y}}{y} dy \ge \int_0^\delta \frac{e^{-\delta}}{y} dy = e^{-\delta} \{ \ln(y) \} \Big|_0^\delta = \infty$. Thus, when we write $\int_0^\infty \frac{e^{-ax} - e^{-bx}}{x} dx = \int_0^\infty \frac{e^{-ax}}{x} dx - \int_0^\infty \frac{e^{-bx}}{x} dx = \int_0^\infty \frac{e^{-y}}{y} dy - \int_0^\infty \frac{e^{-y}}{y} dy$ what we are actually doing is subtracting an *infinite* quantity from another *infinite* quantity and the result is undefined. A deeper analysis is required to determine the result and that, as stated in the challenge statement, is done in Chap. 3. The problem of subtracting one infinity from another was given an interesting treatment in the short science fiction story "The Brink of Infinity" by Stanley Wienbaum (1902–1935), which appeared posthumously in the December 1936 issue of *Thrilling Wonder Stories* magazine (it's reprinted in the Wienbaum collection *A Martian Odyssey and Other Science Fiction Tales*, Hyperion Press 1974). A man is terribly injured when an experiment goes wrong, and he blames the mathematician who did the preliminary analysis incorrectly. In revenge on mathematicians in general, he lures one to his home. Holding him at gun point, he tells him he'll be shot unless he answers the following question correctly: the gunman is thinking of a mathematical expression and the mathematician must figure out what that expression is by asking no more than ten questions. One of the mathematician's questions is "What is the expression equal to?" and the answer is "anything." And that is just what the end of the story reveals to the reader—the gunman is thinking of "$\infty - \infty$."

(C1.6) Following the suggestion, using the by-parts formula $\int_0^1 u \, dv = (uv) \Big|_0^1 - \int_0^1 v \, du$ we write $u = \ln(x)$ and $dv = x^r$ and so $du = \frac{1}{x} dx$ and $v = \frac{x^{r+1}}{r+1}$. Thus,

$$\int_0^1 x^r \ln(x) dx = \left(\frac{x^{r+1}}{r+1} \right) \ln(x) \Big|_0^1 - \int_0^1 \frac{x^r \, 1}{r+1} \left(\frac{1}{x} \right) dx$$

$$= -\lim_{x \to 0} \frac{x^{r+1} \ln(x)}{r+1} - \frac{1}{r+1} \int_0^1 x^r dx$$

$$= -\lim_{x \to 0} \frac{x^{r+1} \ln(x)}{r+1} - \frac{1}{(r+1)^2} (x^{r+1}) \Big|_0^1 = -\lim_{x \to 0} \frac{x^{r+1} \ln(x)}{r+1} - \frac{1}{(r+1)^2},$$

which is finite for $r > -1$ if the limit is finite. This because any positive power of x goes to zero faster (as $x \to 0$) than $\ln(x)$ blows-up. We can see this with the aid of L'Hôpital's rule:

$$\lim_{x \to 0} x^{r+1} \ln(x) = \lim_{x \to 0} \frac{\ln(x)}{x^{-(r+1)}} = \lim_{x \to 0} \frac{\frac{1}{x}}{-(r+1)x^{-(r+1)-1}}$$

$$= -\frac{1}{(r+1)} \lim_{x \to 0} \frac{1}{x} x^{(r+1)+1}$$

$$= -\frac{1}{(r+1)} \lim_{x \to 0} x^{(r+1)} = 0 \text{ if } r + 1 > 0,$$

that is, if $r > -1$.

(C1.7) Multiplying top and bottom of the integrand by e^x, we have $\int_1^\infty \frac{dx}{e^{x+1}+e^{3-x}} =$
$\int_1^\infty \frac{e^x}{e^{2x+1}+e^3}dx = \frac{1}{e^3}\int_1^\infty \frac{e^x}{e^{2x-2}+1}dx = \frac{1}{e^2}\int_1^\infty \frac{e^{x-1}}{e^{2(x-1)}+1}dx$. Let $u = e^{x-1} = \frac{e^x}{e}$ and so $\frac{du}{dx} =$
$\frac{e^x}{e} = \frac{ue}{e} = u$ or, $dx = \frac{du}{u}$. Thus, $\int_1^\infty \frac{dx}{e^{x+1}+e^{3-x}} = \frac{1}{e^2}\int_1^\infty \frac{u}{u} \cdot \frac{du}{u} = \frac{1}{e^2}\int_1^\infty \frac{du}{u^2+1} = \frac{1}{e^2} \times$
$\{\tan^{-1}(u)\}\big|_1^\infty = \frac{1}{e^2}\{\tan^{-1}(\infty) - \tan^{-1}(1)\} = \frac{1}{e^2}\{\frac{\pi}{2} - \frac{\pi}{4}\} = \frac{\pi}{4e^2}$.

(C1.8) Following the hint, in the integral $I(t) = \int_0^{2\pi}\{\cos(\varphi) +$
$\sin\{\varphi\}\sqrt{\frac{1-\sqrt{t}\sin(\varphi)}{1-\sqrt{t}\cos(\varphi)}}d\varphi$ make the change of variable $\theta = \varphi + \frac{\pi}{4}$ (and so $d\theta = d\varphi$).
We have $\cos(\varphi) = \cos\left(\theta - \frac{\pi}{4}\right) = \cos(\theta)\cos\left(\frac{\pi}{4}\right) + \sin(\theta)\sin\left(\frac{\pi}{4}\right) = \frac{1}{\sqrt{2}} \times$
$[\cos(\theta) + \sin(\theta)]$ and $\sin(\varphi) = \sin\left(\theta - \frac{\pi}{4}\right) = \sin(\theta)\cos\left(\frac{\pi}{4}\right) - \cos(\theta)\sin\left(\frac{\pi}{4}\right) =$
$\frac{1}{\sqrt{2}}[\sin(\theta) - \cos(\theta)]$. So, $I(t) = \int_{\frac{\pi}{4}}^{2\pi+\frac{\pi}{4}} \frac{2}{\sqrt{2}}\sin(\theta)\sqrt{\frac{1-\sqrt{t}\frac{1}{\sqrt{2}}[\sin(\theta)-\cos(\theta)]}{1-\sqrt{t}\frac{1}{\sqrt{2}}[\cos(\theta)+\sin(\theta)]}}d\theta$ or, writing
$u = \sqrt{\frac{t}{2}}$, $I(u) = \sqrt{2}\int_{\frac{\pi}{4}}^{2\pi+\frac{\pi}{4}}\sin(\theta)\sqrt{\frac{1+u[\cos(\theta)-\sin(\theta)]}{1-u[\cos(\theta)+\sin(\theta)]}}d\theta = \sqrt{2}\left[\int_0^{2\pi} + \int_{2\pi}^{2\pi+\frac{\pi}{4}} - \int_0^{\frac{\pi}{4}}\right]$. By
the periodicity of the sine and cosine functions, the integrand value over the
middle integration interval is, for every value of θ, identical to the integrand
value over the third integration interval (make a sketch of the two functions to
see this). So the last two integrals in the square brackets cancel, and we have
$I(u) = \sqrt{2}\int_0^{2\pi}\sin(\theta)\sqrt{\frac{1+u[\cos(\theta)-\sin(\theta)]}{1-u[\cos(\theta)+\sin(\theta)]}}d\theta$, where u varies from 0 to $\frac{1}{\sqrt{2}}$ as t varies
from 0 to 1. Now, since $\int_0^{2\pi} = \int_0^\pi + \int_\pi^{2\pi}$, and since $\cos(\theta)$ is symmetrical around π
while $\sin(\theta)$ is anti-symmetrical around π (again, make sketches), we can write the π
to 2π integral as an integral from 0 to π if we replace every $\cos(\theta)$ with $\cos(\theta)$ and
every $\sin(\theta)$ with $-\sin(\theta)$. That is,

$$I(u) = \sqrt{2}\left[\int_0^\pi \sin(\theta)\sqrt{\frac{1+u[\cos(\theta)-\sin(\theta)]}{1-u[\cos(\theta)+\sin(\theta)]}}d\theta - \int_0^\pi \sin(\theta)\sqrt{\frac{1+u[\cos(\theta)+\sin(\theta)]}{1-u[\cos(\theta)-\sin(\theta)]}}d\theta\right]$$

or,

$$I(u) = \sqrt{2}\left[\int_0^\pi \sin(\theta)\left\{\sqrt{\frac{1+u[\cos(\theta)-\sin(\theta)]}{1-u[\cos(\theta)+\sin(\theta)]}} - \sqrt{\frac{1+u[\cos(\theta)+\sin(\theta)]}{1+u[\sin(\theta)-\cos(\theta)]}}\right\}d\theta\right].$$

Since $\int_0^\pi = \int_0^{\pi/2} + \int_{\pi/2}^\pi$ and since $\sin(\theta)$ is symmetrical around $\frac{\pi}{2}$ and $\cos(\theta)$ is
anti-symmetrical around $\pi/2$, we can write the $\pi/2$ to π integral as an integral
from 0 to $\pi/2$ if we replace every $\cos(\theta)$ with $-\cos(\theta)$ and every $\sin(\theta)$ with
$\sin(\theta)$. That is,

$$I(u) = \sqrt{2}\int_0^{\pi/2} \sin(\theta) \left\{ \begin{bmatrix} \sqrt{\dfrac{1+u[\cos(\theta)-\sin(\theta)]}{1-u[\cos(\theta)+\sin(\theta)]}} + \sqrt{\dfrac{1-u[\cos(\theta)+\sin(\theta)]}{1+u[\cos(\theta)-\sin(\theta)]}} - \sqrt{\dfrac{1+u[\cos(\theta)+\sin(\theta)]}{1+u[\sin(\theta)-\cos(\theta)]}} \\ - \sqrt{\dfrac{1-u[\cos(\theta)-\sin(\theta)]}{1+u[\sin(\theta)+\cos(\theta)]}} \end{bmatrix} \right\} d\theta .$$

This may look as though we've made a bad situation even more awful, but that's not so. Here's why. If we call the first radical x, then notice that the second radical is 1/x. Also, if we call the fourth radical y, then the third radical is 1/y. So, what we'll do next is show that, over the entire interval of integration, the integrand is non-negative (and it is obviously not identically zero), and so $I(u) > 0$. That is, $I(t) > 0$. (The factor of $\sin(\theta)$ in front of all the radicals is never negative over the integration interval, and so we can ignore it.) Mathematically, what we'll do is demonstrate that $x + \frac{1}{x} \geq y + \frac{1}{y}$, that is, that $x^2 + 1 \geq xy + \frac{x}{y}$, that is, that $x^2 y + y \geq xy^2 + x$, that is, that $x^2 y - xy^2 > x - y$, that is, that $xy(x-y) \geq x - y$. Now, if $x \geq y$ then this last inequality says $xy \geq 1$. What we'll do next is show that both relations, $x \geq y$ and $xy \geq 1$ are true, and so our starting inequality (that is, $x + \frac{1}{x} \geq y + \frac{1}{y}$) is true and that will establish $I(t) \geq 0$ for $0 \leq t \leq 1$. So, first, is $x \geq y$?

Yes, because that says $\sqrt{\dfrac{1+u[\cos(\theta)-\sin(\theta)]}{1-u[\cos(\theta)+\sin(\theta)]}} \geq \sqrt{\dfrac{1-u[\cos(\theta)-\sin(\theta)]}{1+u[\sin(\theta)+\cos(\theta)]}}$ which is the assertion that

$$\{1 + u[\cos(\theta) - \sin(\theta)]\}\{1 + u[\sin(\theta) + \cos(\theta)]\}$$
$$\geq \{1 - u[\cos(\theta) - \sin(\theta)]\}\{1 - u[\cos(\theta) + \sin(\theta)]\} \text{ or}$$
$$1 + 2u\cos(\theta) - u^2[\sin^2(\theta) - \cos^2(\theta)] \geq 1$$
$$- 2u\cos(\theta) + u^2[\cos^2(\theta) - \sin^2(\theta)] \text{ or}$$
$$4u\cos(\theta) \geq 0,$$

which is obviously true for all θ in the integration interval 0 to $\pi/2$. Now, is $xy \geq 1$? Again, the answer is *yes* because that is the claim $\left(\sqrt{\dfrac{1+u[\cos(0)-\sin(\theta)]}{1-u[\cos(\theta)+\sin(\theta)]}}\right) \times$ $\left(\sqrt{\dfrac{1-u[\cos(\theta)-\sin(\theta)]}{1+u[\sin(\theta)+\cos(\theta)]}}\right) \geq 1$, that is, that $1 - u^2[\cos(\theta) - \sin(\theta)]^2 \geq 1 - u^2[\cos(\theta) + \sin(\theta)]^2$, that is, that $[\cos(\theta) + \sin(\theta)]^2 \geq [\cos(\theta) - \sin(\theta)]^2$ which is obviously true for all θ in the integration interval 0 to $\pi/2$ because, in that interval, both $\cos(\theta)$ and $\sin(\theta)$ are never negative and the sum of two non-negative numbers is always at least as large as their difference. This is long and grubby, yes, but it's nothing but algebra and trig, as claimed.

(C1.9) Following the hint, let $2y = 1 - \cos(2x)$. Then $2\frac{dy}{dx} = 2\sin(2x)$ and so $dx = \frac{dy}{\sin(2x)} = \frac{dy}{\sqrt{1-\cos^2(2x)}}$. Also, $\cos(2x) = 1 - 2\sin^2(x)$ and so $2\sin^2(x) = 1 - \cos(2x) = 2y$ or, $\sin^2(x) = y$. Thus, our integral is

$$\int_0^{\pi/2} \frac{\sin^2(x)}{\sqrt{\{2 - \cos(2x)\}^2 - 1}} dx$$

$$= \int_0^1 \frac{y}{\sqrt{(1 + 2y)^2 - 1}} \cdot \frac{dy}{\sqrt{1 - \cos^2(2x)}} = \int_0^1 \frac{y}{\sqrt{4y + 4y^2}} \cdot \frac{dy}{\sqrt{1 - (1 - 2y)^2}}$$

$$= \int_0^1 \frac{y}{2\sqrt{y + y^2}} \cdot \frac{dy}{\sqrt{1 - (1 - 4y + 4y^2)}} = \frac{1}{2}\int_0^1 \frac{y}{\sqrt{y + y^2}} \cdot \frac{dy}{\sqrt{4y - 4y^2}}$$

$$= \frac{1}{4}\int_0^1 \frac{y}{\sqrt{y + y^2}} \cdot \frac{dy}{\sqrt{y - y^2}} = \frac{1}{4}\int_0^1 \frac{y}{\sqrt{y^2 - y^4}} dy = \frac{1}{4}\int_0^1 \frac{y}{y\sqrt{1 - y^2}} dy = \frac{1}{4}\int_0^1 \frac{dy}{\sqrt{1 - y^2}}$$

$$= \frac{1}{4}\{\sin^{-1}(y)\}\Big|_0^1 = \frac{1}{4}\sin^{-1}(1) = \frac{1}{4}\left(\frac{\pi}{2}\right) = \frac{\pi}{8}.$$

(C1.10) In $\int_0^1 \frac{dx}{\sqrt{x(1-x)}}$ let $x = \sin^2(t)$ and so $\frac{dx}{dt} = 2\sin(t)\cos(t)$ or $dx = 2\sin(t)\cos(t)dt$. Thus, our integral becomes $\int_0^{\pi/2} \frac{2\sin(t)\cos(t)dt}{\sqrt{\sin^2(t)\{1 - \sin^2(t)\}}} = 2\int_0^{\pi/2} \frac{\sin(t)\cos(t)dt}{\sin(t)\cos(t)} = \pi$, as the MATLAB calculation in the problem statement strongly suggested.

(C1.11) Suppose that $\tau > 0$. Then and

$$R_f(\tau) = \int_{-\infty}^{\infty} f(t)f(t - \tau)dt = \int_{-\infty}^0 + \int_0^{\tau} + \int_{\tau}^{\infty}$$

$$= \int_{-\infty}^0 e^t e^{t-\tau} dt + \int_0^{\tau} e^{-t} e^{t-\tau} dt + \int_{\tau}^{\infty} e^{-t} e^{-(t-\tau)} dt$$

$$= \int_{-\infty}^0 e^{2t-\tau} dt + \int_0^{\tau} e^{-\tau} dt + \int_{\tau}^{\infty} e^{-2t+\tau} dt$$

Fig. C1.11a Autocorrelation calculation ($\tau > 0$)

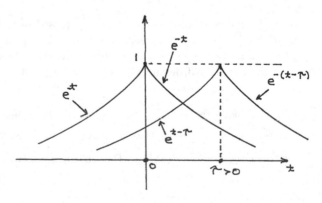

Fig. C1.11b Autocorrelation
calculation ($\tau < 0$)

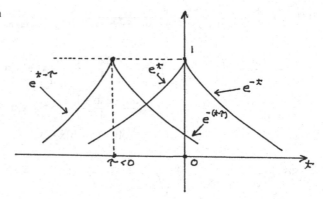

$$= e^{-\tau} \int_{-\infty}^{0} e^{2t}dt + \tau e^{-\tau} + e^{\tau} \int_{\tau}^{\infty} e^{-2t}dt = e^{-\tau}\left(\frac{e^{2t}}{2}\right)\Big|_{-\infty}^{0} + \tau e^{-\tau} + e^{\tau}\left(\frac{e^{-2t}}{-2}\right)\Big|_{\tau}^{\infty}$$

$$= \frac{1}{2}e^{-\tau} + \tau e^{-\tau} + \frac{1}{2}e^{\tau}e^{-2\tau} = \frac{1}{2}e^{-\tau} + \tau e^{-\tau} + \frac{1}{2}e^{-\tau} = \tau e^{-\tau} + e^{-\tau} = e^{-\tau}(\tau + 1).$$

Next, suppose that $\tau < 0$. Then and

$$R_f(\tau) = \int_{-\infty}^{\infty} f(t)f(t-\tau)dt = \int_{-\infty}^{\tau} + \int_{\tau}^{0} + \int_{0}^{\infty} = \int_{-\infty}^{\tau} e^t e^{t-\tau}dt + \int_{\tau}^{0} e^t e^{-(t-\tau)}dt$$

$$+ \int_{0}^{\infty} e^{-t}e^{-(t-\tau)}dt = e^{-\tau}\int_{-\infty}^{\tau} e^{2t}dt + \int_{\tau}^{0} e^{\tau}dt + e^{\tau}\int_{0}^{\infty} e^{-2t}dt$$

$$= e^{-\tau}\left(\frac{e^{2t}}{2}\right)\Big|_{-\infty}^{\tau} - \tau e^{\tau} + e^{\tau}\left(\frac{e^{-2t}}{-2}\right)\Big|_{0}^{\infty} = \frac{1}{2}e^{\tau} - \tau e^{\tau} + \frac{1}{2}e^{\tau} = e^{\tau} - \tau e^{\tau} = (1-\tau)e^{\tau}.$$

So, in summary, $R_f(\tau) = \begin{cases} (1-\tau)e^{\tau}, & \tau < 0 \\ (1+\tau)e^{-\tau}, & \tau > 0 \end{cases}$ or, more compactly,
$R_f(\tau) = (1 + |\tau|)e^{-|\tau|}$ which is obviously even. Equally obvious is that $R_f(0) = 1$
(Figs. C1.11a and C1.11b).

(C1.12) Following the hint, we make the change of variable $y = \frac{1}{x}$ in the second
integral of $S = \int_0^1 e^{\frac{1}{2}\left(x - \frac{1}{x}\right)}dx + \int_1^{\infty} e^{-\frac{1}{2}\left(x - \frac{1}{x}\right)}dx$. So, $\int_1^{\infty} e^{-\frac{1}{2}\left(x - \frac{1}{x}\right)}dx = \int_1^0 e^{-\frac{1}{2}\left(\frac{1}{y} - y\right)}$
$\left(-\frac{dy}{y^2}\right) = \int_0^1 e^{\frac{1}{2}\left(y - \frac{1}{y}\right)}\frac{dy}{y^2} = \int_0^1 e^{\frac{1}{2}\left(x - \frac{1}{x}\right)}\frac{dx}{x^2}$. Thus, $S = \int_0^1 e^{\frac{1}{2}\left(x - \frac{1}{x}\right)}dx + \int_0^1 e^{\frac{1}{2}\left(x - \frac{1}{x}\right)}\frac{dx}{x^2} =$
$\int_0^1 e^{\frac{1}{2}\left(x - \frac{1}{x}\right)}\left(1 + \frac{1}{x^2}\right)dx$. Now, let $u = x - \frac{1}{x}$. Then, $\frac{du}{dx} = 1 + \frac{1}{x^2}$ or, $dx = \frac{du}{1 + \frac{1}{x^2}}$ and $S =$
$\int_{-\infty}^{0} e^{\frac{1}{2}u}\left(1 + \frac{1}{x^2}\right)\frac{du}{1 + \frac{1}{x^2}} = \int_{-\infty}^{0} e^{\frac{1}{2}u}du = \left(2e^{\frac{1}{2}u}\right)\Big|_{-\infty}^{0}$ or, $S = 2$.

(C1.13) $\int_a^b \frac{dx}{\sqrt{(x-a)(b-x)}} = \int_a^b \frac{dx}{\sqrt{xb-x^2-ab+ax}} = \int_a^b \frac{dx}{\sqrt{-ab+(a+b)x-x^2}}$. Notice that

$$\frac{(a-b)^2}{4} - \left(x - \frac{a+b}{2}\right)^2 = \frac{a^2-2ab+b^2}{4} - \left[x^2 - x(a+b) + \frac{(a+b)^2}{4}\right]$$

$$= \frac{a^2-2ab+b^2}{4} - x^2 + x(a+b) - \frac{a^2+2ab+b^2}{4}$$

$$= -ab + (a+b)x - x^2.$$

So, $\int_a^b \frac{dx}{\sqrt{(x-a)(b-x)}} = \int_a^b \frac{dx}{\sqrt{\frac{(a-b)^2}{4} - \left(x - \frac{a+b}{2}\right)^2}}$. Let $u = x - \frac{a+b}{2}$ (and so $du = dx$). Then

$$\int_a^b \frac{dx}{\sqrt{(x-a)(b-x)}} = \int_{a-\frac{b}{2}}^{\frac{b-a}{2}} \frac{du}{\sqrt{\frac{(a-b)^2}{4} - u^2}} = \int_{-\frac{b-a}{2}}^{\frac{b-a}{2}} \frac{du}{\sqrt{\frac{(b-a)^2}{4} - u^2}}$$

$$= \int_{-\frac{b-a}{2}}^{\frac{b-a}{2}} \frac{du}{\sqrt{\left\{\frac{b-a}{2}\right\}^2 - u^2}} = \sin^{-1}\left(\frac{u}{\frac{b-a}{2}}\right)\Big|_{-\frac{b-a}{2}}^{\frac{b-a}{2}}$$

$$= \sin^{-1}(1) - \sin^{-1}(-1) = \frac{\pi}{2} - \left(-\frac{\pi}{2}\right) = \pi,$$

independent of a and b as long as $b > a \geq 0$.

Chapter 2

(C2.1) In $\int_0^4 \frac{\ln(x)}{\sqrt{4x-x^2}} dx$ change variable to $y = \frac{x}{4}$. Then $x = 4y$ and so $dx = 4\,dy$. The integral then becomes $\int_0^1 \frac{\ln(4y)}{\sqrt{16y-16y^2}} 4dy = \int_0^1 \frac{\ln(4)+\ln(y)}{\sqrt{y-y^2}} dy = \ln(4)\int_0^1 \frac{dy}{\sqrt{y}\sqrt{1-y}} + \int_0^1 \frac{\ln(y)}{\sqrt{y}\sqrt{1-y}} dy$. In both of these last two integrals make the change of variable to $y = \sin^2(\theta)$, which means that when $y = 0$ we have $\theta = 0$, and when $y = 1$ we have $\theta = \frac{\pi}{2}$. Also, $\frac{dy}{d\theta} = 2\sin(\theta)\cos(\theta)$ and so $\int_0^1 \frac{dy}{\sqrt{y}\sqrt{1-y}} = \int_0^{\pi/2} \frac{2\sin(\theta)\cos(\theta)}{\sqrt{\sin^2(\theta)}\sqrt{1-\sin^2(\theta)}} d\theta = 2\int_0^{\pi/2} \frac{\sin(\theta)\cos(\theta)}{\sin(\theta)\cos(\theta)} d\theta = 2\int_0^{\pi/2} d\theta = \pi$, and $\int_0^1 \frac{\ln(y)}{\sqrt{y}\sqrt{1-y}} dy = \int_0^{\pi/2} \frac{\ln\left\{\sin^2(\theta)\right\}}{\sin(\theta)}\cos(\theta)2\sin(\theta)\cos(\theta)d\theta = 4\int_0^{\pi/2} \ln\{\sin(\theta)\}d\theta = 4\left[-\frac{\pi}{2}\ln(2)\right] = -2\pi\ln(2)$ where I've used the Euler log-sine integral we derived in (2.4.1). So, what we have is $\int_0^4 \frac{\ln(x)}{\sqrt{4x-x^2}} dx = \ln(4)\int_0^1 \frac{dy}{\sqrt{y}\sqrt{1-y}} + \int_0^1 \frac{\ln(y)}{\sqrt{y}\sqrt{1-y}} dy = \ln(4)\pi - 2\pi\ln(2) = \ln(2^2)\pi - 2\pi\ln(2) = 2\ln(2)\pi - 2\pi\ln(2) = 0$ and we are done.

(C2.2) We start with $\int_0^1 \frac{dx}{x^3+1} = \frac{1}{3}\int_0^1 \frac{dx}{x+1} - \frac{1}{3}\int_0^1 \frac{x-2}{x^2-x+1}\,dx$. In the first integral on the right let $u = x + 1$ ($dx = du$). Then, $\frac{1}{3}\int_0^1 \frac{dx}{x+1} = \frac{1}{3}\int_1^2 \frac{du}{u} = \frac{1}{3}\{\ln(u)\}\Big|_1^2 = \frac{1}{3}\ln(2)$. In the second integral write (following the hint) $\frac{1}{3}\int_0^1 \frac{x-2}{x^2-x+1}\,dx = \frac{1}{3}\int_0^1 \frac{x-2}{x^2-x+\frac{1}{4}+\frac{3}{4}}\,dx = \frac{1}{3}\int_0^1 \frac{x-2}{\left(x-\frac{1}{2}\right)^2+\left(\frac{\sqrt{3}}{2}\right)^2}\,dx$. Next, let $u = x - \frac{1}{2}$ (and so $dx = du$) and we get $\frac{1}{3}\int_0^1 \frac{x-2}{x^2-x+1}\,dx = \frac{1}{3}\int_{-\frac{1}{2}}^{\frac{1}{2}} \frac{u-\frac{3}{2}}{u^2+\left(\frac{\sqrt{3}}{2}\right)^2}\,du = \frac{1}{3}\int_{-\frac{1}{2}}^{\frac{1}{2}} \frac{u}{u^2+\left(\frac{\sqrt{3}}{2}\right)^2}\,du - \frac{1}{2}\int_{-\frac{1}{2}}^{\frac{1}{2}} \frac{du}{u^2+\left(\frac{\sqrt{3}}{2}\right)^2}$. The first integral on the right is zero as the integrand is odd and the limits are symmetrical around zero, and so

$$\frac{1}{3}\int_0^1 \frac{x-2}{x^2-x+1}\,dx = -\frac{1}{2}\int_{-\frac{1}{2}}^{\frac{1}{2}} \frac{du}{u^2+\left(\frac{\sqrt{3}}{2}\right)^2} = -\frac{1}{2}\left\{\frac{2}{\sqrt{3}}\tan^{-1}\left(\frac{2u}{\sqrt{3}}\right)\right\}\Big|_{-\frac{1}{2}}^{\frac{1}{2}}$$

$$= -\frac{1}{\sqrt{3}}\left\{\tan^{-1}\left(\frac{1}{\sqrt{3}}\right) - \tan^{-1}\left(-\frac{1}{\sqrt{3}}\right)\right\}$$

$$= -\frac{1}{\sqrt{3}}\left\{2\tan^{-1}\left(\frac{1}{\sqrt{3}}\right)\right\} = -\frac{2}{\sqrt{3}}\left(\frac{\pi}{6}\right) = -\frac{\pi}{3\sqrt{3}}.$$

So, $\int_0^1 \frac{dx}{x^3+1} = \frac{1}{3}\ln(2) + \frac{\pi}{3\sqrt{3}}$ or,

$$\int_0^1 \frac{dx}{x^3+1} = \frac{1}{3}\left\{\ln(2) + \frac{\pi}{\sqrt{3}}\right\}.$$

This equals 0.8356488 and MATLAB agrees, as $integral(@(x)1./(x.$
$\land 3+1),0,1) = 0.8356488\ldots$.

(C2.3) Define the integral $I(m-1) = \int_0^\infty \frac{dx}{(x^4+1)^m}$ which we'll then integrate by parts. That is, in the standard notation of $\int u\,dv = uv - \int v\,du$ let $dv = dx$ and $u = \frac{1}{(x^4+1)^m}$. Then, $v = x$ and $\frac{du}{dx} = -4m\frac{x^3}{(x^4+1)^{m+1}}$ and so

$$\int_0^\infty \frac{dx}{(x^4+1)^m} = \frac{x}{(x^4+1)^m}\Big|_0^\infty + 4m\int_0^\infty \frac{x^4}{(x^4+1)^{m+1}}\,dx = 4m\int_0^\infty \frac{x^4}{(x^4+1)^{m+1}}\,dx.$$

Now, write

$$\int_0^\infty \frac{x^4}{(x^4+1)^{m+1}}\,dx = \int_0^\infty \frac{x^4+1}{(x^4+1)^{m+1}}\,dx - \int_0^\infty \frac{1}{(x^4+1)^{m+1}}\,dx$$

$$= \int_0^\infty \frac{1}{(x^4+1)^m}\,dx - \int_0^\infty \frac{1}{(x^4+1)^{m+1}}\,dx.$$

So, $\int_0^\infty \frac{dx}{(x^4+1)^m} = 4m\left[\int_0^\infty \frac{1}{(x^4+1)^m} dx - \int_0^\infty \frac{1}{(x^4+1)^{m+1}} dx\right]$ or, rearranging, $4m\int_0^\infty$ $\frac{dx}{(x^4+1)^{m+1}} = 4m\int_0^\infty \frac{dx}{(x^4+1)^m} - \int_0^\infty \frac{dx}{(x^4+1)^m} = (4m-1)\int_0^\infty \frac{dx}{(x^4+1)^m}$. Thus, $\int_0^\infty \frac{dx}{(x^4+1)^{m+1}} = \frac{4m-1}{4m}\int_0^\infty \frac{dx}{(x^4+1)^m}$ and we are done.

(C2.4) The denominator of the integrand is zero at the values given by $x = \frac{b\pm\sqrt{b^2-4}}{2}$. If $-2 < b < 2$ then both of these values are complex and so not on the real axis, which means that the denominator never vanishes in the interval of integration. And if $b \le -2$ then both values of $x < 0$, which again means that the denominator never vanishes in the interval of integration. So, the integral in (2.4.5) exists only if $b < 2$. MATLAB agrees, as experimenting with *integral* shows that random choices for b that are less than 2 always give values for the integral that are either zero or pretty darn close to zero, while any $b \ge 2$ results in MATLAB responding with 'NaN' (not a number), MATLAB's way of saying the integral doesn't exist.

(C2.5) In $\int_0^\infty \frac{\ln(1+x)}{x\sqrt{x}} dx$ let $u = \ln(1+x)$ and $dv = \frac{dx}{x^{3/2}}$. Then, $du = \frac{dx}{1+x}$ and $v = -\frac{2}{\sqrt{x}}$, and so $\int_0^\infty \frac{\ln(1+x)}{x\sqrt{x}} dx = -2\frac{\ln(1+x)}{\sqrt{x}}\Big|_0^\infty + 2\int_0^\infty \frac{1}{\sqrt{x}}\frac{dx}{1+x} = 2\int_0^\infty \frac{dx}{(1+x)\sqrt{x}}$. Next, change variable to $t = \sqrt{x}$ and so $x = t^2$ and $dx = 2t\,dt$. Thus, $\int_0^\infty \frac{\ln(1+x)}{x\sqrt{x}} dx = 2\int_0^\infty \frac{2t\,dt}{(1+t^2)t} = 4\int_0^\infty \frac{dt}{(1+t^2)} = 4\tan^{-1}(t)\Big|_0^\infty = 4\left(\frac{\pi}{2}\right) = 2\pi.$

(C2.6) We start with (2.4.1): $\int_0^{\pi/2} \ln\{a\sin(x)\}dx = \frac{\pi}{2}\ln\left(\frac{a}{2}\right)$. From integration-by-parts, we have $\int u\,dv = uv - \int v\,du$. Let $u = \ln\{a\sin(x)\}$ and $dv = dx$. Then, $v = x$ and $du/dx = (a\cos(x))/(a\sin(x)) = \cot(x)$ or, $du = \cot(x)dx$. So, $\int_0^{\pi/2} \ln\{a\sin(x)\}dx = x\ln\{a\sin(x)\}\Big|_0^{\pi/2} - \int_0^{\frac{\pi}{2}}x\cot(x)dx = \frac{\pi}{2}\ln(a) - \int_0^{\frac{\pi}{2}}x\cot(x)dx = \frac{\pi}{2}\ln\left(\frac{a}{2}\right)$. Thus, $\frac{\pi}{2}\ln(a) - \frac{\pi}{2}\ln\left(\frac{a}{2}\right) = \int_0^{\frac{\pi}{2}}x\cot(x)dx = \frac{\pi}{2}\ln\left(\frac{a}{a/2}\right) = \frac{\pi}{2}\ln(2)$. MATLAB agrees, as $\frac{\pi}{2}\ln(2) = 1.0887930451518\ldots$ while *integral*($@(x)x.*cot(x),0,pi/2$)= *1.0887930451518....*

(C2.7) In (2.5.3) let $m = 0$ and $n = 1$, and so $R(1,1) + R(-1,1) + R(0,2) + R(0,0) - 4R(0,1) = 0$. By symmetry $R(1,1) = R(-1,1)$ and, as $R(0,0) = 0$, we therefore have $2R(1,1) + R(0,2) = 4R(0,1)$ or $R(0,2) = 4R(0,1) - 2R(1,1) = 4\left(\frac{1}{2}\right) - 2\left(\frac{2}{\pi}\right) = 2 - \frac{4}{\pi} = 0.72676\ldots$ Next, to find $R(3,1)$ let $m = 2$ and $n = 1$ in (2.5.3), and so $R(3,1) + R(1,1) + R(2,2) + R(2,0) - 4R(2,1) = 0$. We know from (2.5.8) that $R(1,1) = \frac{2}{\pi}$ and $R(2,2) = \frac{2}{\pi}\left(1+\frac{1}{3}\right) = \frac{8}{3\pi}$. From the end of Sect. 2.5 we have $R(1,2) = R(2,1) = \frac{4}{\pi} - \frac{1}{2}$, and we just calculated $R(2,0) = R(0,2) = 2 - \frac{4}{\pi}$. Thus, $R(3,1) + \frac{2}{\pi} + \frac{8}{3\pi} + 2 - \frac{4}{\pi} - 4\left(\frac{4}{\pi} - \frac{1}{2}\right) = 0$. That is, $R(3,1) + \frac{2}{\pi} + \frac{8}{3\pi} + 2 - \frac{4}{\pi} - \frac{16}{\pi} + 2 = 0$, or $R(3,1) + \frac{6}{3\pi} + \frac{8}{3\pi} + 4 - \frac{12}{3\pi} - \frac{48}{3\pi} + 2 = 0$. So, $R(3,1) = \frac{46}{3\pi} - 4 = 0.88075\ldots$

(C2.8) $\int \frac{x^2}{x^4+1} dx = \frac{1}{2} \int \frac{2x^2}{x^4+1} dx = \frac{1}{2} \int \frac{x^2-1}{x^4+1} dx + \frac{1}{2} \int \frac{x^2+1}{x^4+1} dx = \frac{1}{2} \int \frac{1-\frac{1}{x^2}}{x^2+\frac{1}{x^2}} dx + \frac{1}{2} \int \frac{1+\frac{1}{x^2}}{x^2+\frac{1}{x^2}} dx =$

$\frac{1}{2} \int \frac{\left(1-\frac{1}{x^2}\right)}{\left(x+\frac{1}{x}\right)^2-2} dx + \frac{1}{2} \int \frac{\left(1+\frac{1}{x^2}\right)}{\left(x-\frac{1}{x}\right)^2+2} dx.$ Now, since $\frac{d}{dx}\left(x+\frac{1}{x}\right) = 1-\frac{1}{x^2}$ then $d\left(x+\frac{1}{x}\right) =$

$\left(1-\frac{1}{x^2}\right) dx$ and, similarly, $d\left(x-\frac{1}{x}\right) = \left(1+\frac{1}{x^2}\right) dx$. Thus, $\int \frac{x^2}{x^4+1} dx = \frac{1}{2} \int \frac{d\left(x+\frac{1}{x}\right)}{\left(x+\frac{1}{x}\right)^2-2} +$

$\frac{1}{2} \int \frac{d\left(x-\frac{1}{x}\right)}{\left(x-\frac{1}{x}\right)^2+2}$. In the first integral on the right let $u = x+\frac{1}{x}$ and so

$$\frac{1}{2} \int \frac{d\left(x+\frac{1}{x}\right)}{\left(x+\frac{1}{x}\right)^2-2} = \frac{1}{2} \int \frac{du}{u^2-2} = \frac{1}{2} \int \frac{du}{\left(u-\sqrt{2}\right)\left(u+\sqrt{2}\right)}$$

$$= \frac{1}{2} \int \frac{1}{2\sqrt{2}} \left(\frac{1}{u-\sqrt{2}} - \frac{1}{u+\sqrt{2}}\right) du$$

$$= \frac{1}{4\sqrt{2}} \left[\ln\left(u-\sqrt{2}\right) - \ln\left(u+\sqrt{2}\right)\right]$$

$$= \frac{1}{4\sqrt{2}} \ln\left(\frac{u-\sqrt{2}}{u+\sqrt{2}}\right) = \frac{1}{4\sqrt{2}} \ln\left(\frac{x+\frac{1}{x}-\sqrt{2}}{x+\frac{1}{x}+\sqrt{2}}\right)$$

$$= \frac{1}{4\sqrt{2}} \ln\left(\frac{x^2-x\sqrt{2}+1}{x^2+x\sqrt{2}+1}\right).$$

In the second integral on the right let $u = x - \frac{1}{x}$ and so $\frac{1}{2} \int \frac{d\left(x-\frac{1}{x}\right)}{\left(x-\frac{1}{x}\right)^2+2} = \frac{1}{2} \times$

$\int \frac{du}{u^2+2} = \frac{1}{2} \left[\frac{1}{\sqrt{2}} \tan^{-1}\left(\frac{u}{\sqrt{2}}\right)\right] = \frac{1}{2\sqrt{2}} \tan^{-1}\left(\frac{x-\frac{1}{x}}{\sqrt{2}}\right) = \frac{1}{2\sqrt{2}} \tan^{-1}\left(\frac{x^2-1}{x\sqrt{2}}\right)$. Thus, to within

an arbitrary constant, $\int \frac{x^2}{x^4+1} dx = \frac{1}{4\sqrt{2}} \ln\left(\frac{x^2-x\sqrt{2}+1}{x^2+x\sqrt{2}+1}\right) + \frac{1}{2\sqrt{2}} \tan^{-1}\left(\frac{x^2-1}{x\sqrt{2}}\right)$, a pretty
complicated looking formula for such a 'simple looking' integrand.

(C2.9) $\int_0^\infty \frac{x^2}{1-x^4} dx = \int_0^1 \frac{x^2}{1-x^4} dx + \int_1^\infty \frac{x^2}{1-x^4} dx$. In the second integral on the right,

let $y = \frac{1}{x}$ and so $dx = -\frac{dy}{y^2}$. So, $\int_1^\infty \frac{x^2}{1-x^4} dx = \int_1^0 \frac{\frac{1}{y^2}}{1-\frac{1}{y^4}} \left(-\frac{dy}{y^2}\right) = \int_0^1 \frac{1}{1-\frac{1}{y^4}} dy =$

$\int_0^1 \frac{y^4}{y^4-1} dy = \int_0^1 \frac{1}{x^4-1} dx = -\int_0^1 \frac{1}{1-x^4} dx$. So, $\int_0^\infty \frac{x^2}{1-x^4} dx = \int_0^1 \frac{x^2}{1-x^4} dx - \int_0^1 \frac{1}{1-x^4} dx =$

$\int_0^1 \frac{x^2-1}{1-x^4} dx = -\int_0^1 \frac{1-x^2}{(1-x^2)(1+x^2)} dx = -\int_0^1 \frac{dx}{1+x^2} = -\tan^{-1}(x)\big|_0^1 = -\tan^{-1}(1) = -\frac{\pi}{4}$.
As a MATLAB check,
integral(@(x)(x.^2)./(1-(x.^4)),0,inf) = -0.785277....

(C2.10) $\int_0^\infty \frac{dx}{bx^4+2ax^2+c} = \frac{1}{c} \int_0^\infty \frac{dx}{\frac{b}{c}x^4+2\frac{a}{c}x^2+1} = \frac{1}{c} \left(\frac{\pi}{2\sqrt{2}}\right) \frac{1}{\sqrt{\frac{a}{c}}} + \sqrt{\frac{b}{c}} = \left(\frac{\pi}{2\sqrt{2}}\right) \frac{1}{\sqrt{c^2\left(\frac{a}{c}+\sqrt{\frac{b}{c}}\right)}} =$

$\left(\frac{\pi}{2\sqrt{2}}\right) \frac{1}{\sqrt{c(a+\sqrt{bc})}}$. To check this, suppose $b = 4$, $a = 3$, and $c = 2$. Then our integral

equals $\left(\frac{\pi}{2\sqrt{2}}\right) \frac{1}{\sqrt{6+2\sqrt{8}}} = 0.325322571142143$, and MATLAB agrees, as
*integral(@(x)1./((4*x.^4)+(2*3*x.^2)+2),0,inf)= 0.325322571142143*.

Chapter 3

(C3.1) Start with $I(a) = \int_0^\infty \frac{\ln(1+a^2x^2)}{b^2+x^2} dx$ and differentiate with respect to a to get

$\frac{dI}{da} = \int_0^\infty \frac{2ax^2}{(1+a^2x^2)(b^2+x^2)} dx = \int_0^\infty \left\{ \frac{A}{1+a^2x^2} + \frac{B}{b^2+x^2} \right\} dx$ and so $Ab^2 + Ax^2 + B + Ba^2x^2 = x^2(A + Ba^2) + Ab^2 + B = 2ax^2$. Thus, $A + Ba^2 = 2a$ and $Ab^2 + B = 0$. Solving for A and B, we have $A = \frac{2a}{1-a^2b^2}$ and $B = -\frac{2ab^2}{1-a^2b^2}$. So,

$$\frac{dI}{da} = \frac{2a}{1-a^2b^2} \int_0^\infty \frac{dx}{1+a^2x^2} - \frac{2ab^2}{1-a^2b^2} \int_0^\infty \frac{dx}{b^2+x^2}$$

$$= \frac{\frac{2}{a}}{1-a^2b^2} \int_0^\infty \frac{dx}{\frac{1}{a^2}+x^2} - \frac{2ab^2}{1-a^2b^2} \int_0^\infty \frac{dx}{b^2+x^2}$$

$$= \frac{\frac{2}{a}}{1-a^2b^2} (a) \tan^{-1}(xa)\Big|_0^\infty - \frac{2ab^2}{1-a^2b^2} \left(\frac{1}{b}\right) \tan^{-1}\left(\frac{x}{b}\right)\Big|_0^\infty$$

$$= \frac{\pi}{2} \left[\frac{2}{1-a^2b^2} - \frac{2ab}{1-a^2b^2}\right] = \frac{\pi}{2} 2 \left[\frac{1-ab}{1-a^2b^2}\right] = \pi \frac{1}{1+ab}.$$

So, $\frac{dI}{da} = \pi\frac{1}{1+ab}$ and thus $I = \int \pi\frac{1}{1+ab} da$. Let $u = 1 + ab$ and so $\frac{du}{da} = b$ and therefore $da = \frac{1}{b}du$. Thus, $I = \pi \int \frac{\frac{1}{b} du}{u} = \frac{\pi}{b} \ln(u) + C$ where C is a constant of integration. We know from the very definition of $I(a)$ that $I(0) = 0$, which says $C = 0$. So, at last,

$$\int_0^\infty \frac{\ln(1 + a^2x^2)}{b^2 + x^2} dx = \pi \frac{\ln(1 + ab)}{b}.$$

(C3.2) Write $\int_{-\infty}^\infty \frac{\cos(ax)}{b^2-x^2} dx = \int_{-\infty}^\infty \cos(ax) \left\{ \frac{1}{(b-x)(b+x)} \right\} dx = \frac{1}{2b} \int_{-\infty}^\infty \cos(ax) \times \left\{ \frac{1}{(b+x)} + \frac{1}{(b-x)} \right\} dx = \frac{1}{2b} \left[\int_{-\infty}^\infty \frac{\cos(ax)}{(b+x)} dx + \int_{-\infty}^\infty \frac{\cos(ax)}{(b-x)} dx \right]$. In the first integral let $u = b + x$ (and so $du = dx$), and in the second integral let $u = b - x$ (and so $du = -dx$). Then, $\int_{-\infty}^\infty \frac{\cos(ax)}{b^2-x^2} dx = \frac{1}{2b} \left[\int_{-\infty}^\infty \frac{\cos(au-ab)}{u} du + \int_\infty^{-\infty} \frac{\cos(ab-au)}{u} (-du) \right] = \frac{1}{2b} \left[\int_{-\infty}^\infty \frac{\cos(au-ab)}{u} du + \int_{-\infty}^\infty \frac{\cos(ab-au)}{u} du \right]$ or, as $\cos(-\theta) = \cos(\theta)$, we have $\int_{-\infty}^\infty \frac{\cos(ax)}{b^2-x^2} dx = \frac{1}{b} \int_{-\infty}^\infty \frac{\cos(au-ab)}{u} du$. Now, since $\cos(x - y) = \cos(x)\cos(y) + \sin(x)\sin(y)$, then $\int_{-\infty}^\infty \frac{\cos(ax)}{b^2-x^2} dx = \frac{1}{b} \int_{-\infty}^\infty \frac{\cos(au)\cos(ab)}{u} du + \frac{1}{b} \int_{-\infty}^\infty \frac{\sin(au)\sin(ab)}{u} du = \frac{\cos(ab)}{b} \int_{-\infty}^\infty \frac{\cos(au)}{u} du + \frac{\sin(ab)}{b} \int_{-\infty}^\infty \frac{\sin(au)}{u} du$. The first integral has a Cauchy Principal Value of zero as $\frac{\cos(au)}{u}$ is an odd function. For $a > 0$ the second integral is π from Dirichlet's integral. So,

$$\int_{-\infty}^{\infty} \frac{\cos(ax)}{b^2 - x^2} dx = \pi \frac{\sin(ab)}{b}.$$

(C3.3) Since $\int_0^\infty \frac{\cos(ax)}{b^2+x^2} dx = \frac{\pi}{2b} e^{-ab}$ then as the integrand is an even function we can write $\int_{-\infty}^{\infty} \frac{\cos(ax)}{b^2+x^2} dx = \frac{\pi}{b} e^{-ab}$. Thus, $\int_{-\infty}^{\infty} \frac{\cos(ax)}{b^2+x^2} dx + \int_{-\infty}^{\infty} \frac{\cos(ax)}{b^2-x^2} dx = \frac{\pi}{b} e^{-ab} + \pi \frac{\sin(ab)}{b} = \int_{-\infty}^{\infty} \cos(ax) \left\{ \frac{1}{b^2+x^2} + \frac{1}{b^2-x^2} \right\} dx = 2b^2 \int_{-\infty}^{\infty} \frac{\cos(ax)}{b^4-x^4} dx$ and so we immediately have our result:

$$\int_{-\infty}^{\infty} \frac{\cos(ax)}{b^4 - x^4} dx = \frac{\pi\{e^{-ab} + \sin(ab)\}}{2b^3}.$$

(C3.4) Notice that since the integrand is an even function we can write

$$\int_0^\infty \frac{x \sin(ax)}{x^2 - b^2} dx = \frac{1}{2} \int_{-\infty}^{\infty} x \sin(ax) \left\{ \frac{1}{(x-b)(x+b)} \right\} dx$$

$$= -\frac{1}{2} \int_{-\infty}^{\infty} x \sin(ax) \left\{ \frac{1}{(b-x)(b+x)} \right\} dx$$

$$= -\frac{1}{4b} \int_{-\infty}^{\infty} x \sin(ax) \left\{ \frac{1}{b+x} + \frac{1}{b-x} \right\} dx$$

$$= -\frac{1}{4b} \left[\int_{-\infty}^{\infty} \frac{x \sin(ax)}{b+x} dx + \int_{-\infty}^{\infty} \frac{x \sin(ax)}{b-x} dx \right].$$

In the first integral change variable to $u = b + x$ and in the second integral to $u = b - x$. Then,

$$\int_0^\infty \frac{x \sin(ax)}{x^2 - b^2} dx = -\frac{1}{4b} \left[\int_{-\infty}^{\infty} \frac{(u-b) \sin(au - ab)}{u} du + \int_{\infty}^{-\infty} \frac{(b-u) \sin(ab - au)}{u} (-du) \right]$$

$$= -\frac{1}{4b} \left[\int_{-\infty}^{\infty} \frac{(u-b) \sin(au - ab)}{u} du + \int_{-\infty}^{\infty} \frac{(u-b) \sin(au - ab)}{u} du \right]$$

$$= -\frac{1}{4b} \int_{-\infty}^{\infty} \frac{2(u-b) \sin(au - ab)}{u} du$$

$$= -\frac{1}{2b} \left[\int_{-\infty}^{\infty} \sin(au - ab) du - b \int_{-\infty}^{\infty} \frac{\sin(au - ab)}{u} du \right].$$

In the first integral change variable to $s = au - ab$. Then, $\int_{-\infty}^{\infty} \sin(au - ab) du = \frac{1}{a} \int_{-\infty}^{\infty} \sin(s) ds$ which has a Cauchy Principal Value of zero. Thus, $\int_0^\infty \frac{x \sin(ax)}{x^2 - b^2} dx = \frac{1}{2} \int_{-\infty}^{\infty} \frac{\sin(au - ab)}{u} du$ or, as $\sin(\alpha - \beta) = \sin(\alpha) \cos(\beta) -$

$\cos{(\alpha)}\sin{(\beta)}$, $\int_0^\infty \frac{x\sin(ax)}{x^2-b^2}dx = \frac{1}{2}\int_{-\infty}^\infty \frac{\sin{(au)}\cos{(ab)}-\cos{(au)}\sin{(ab)}}{u}du = \frac{1}{2}\left[\cos{(ab)}\int_{-\infty}^\infty\right.$
$\frac{\sin{(au)}}{u}du - \sin{(ab)}\int_{-\infty}^\infty \frac{\cos{(au)}}{u}du]$. The first integral is, from Dirichlet's integral, π, and the second integral is zero because the integrand is an odd function. Thus,

$$\int_0^\infty \frac{x\sin(ax)}{x^2-b^2}dx = \frac{\pi}{2}\cos{(ab)}.$$

(C3.5) Start by recalling the trigonometric identity $\cos{(ax)}\sin{(bx)} = \frac{1}{2}\times$ $[\sin{\{(b-a)x\}}+\sin{\{(b+a)x\}}]$. Since the cosine is an even function we have $\cos(ax) = \cos{(|a|x)}$ and so $\cos{(ax)}\sin{(bx)} = \frac{1}{2}[\sin{\{(b-|a|)x\}}+\sin{\{(b+|a|)x\}}]$ and so

$$\int_0^\infty \cos{(ax)}\frac{\sin{(bx)}}{x}dx = \frac{1}{2}\int_0^\infty \frac{\sin{\{(b-|a|)x\}}}{x}dx + \frac{1}{2}\int_0^\infty \frac{\sin{\{(b+|a|)x\}}}{x}dx.$$

From (3.2.1) we have

$$\int_0^\infty \frac{\sin{(cx)}}{x}dx = \begin{cases} \frac{\pi}{2}, & c > 0 \\ 0, & c = 0 \\ -\frac{\pi}{2}, & c < 0 \end{cases}$$

and so,

(1) if $|a| < b$ then $b - |a| > 0$ and $b + |a| > 0$ and so

$$\int_0^\infty \cos{(ax)}\frac{\sin{(bx)}}{x}dx = \frac{1}{2}\left(\frac{\pi}{2}\right)+\frac{1}{2}\left(\frac{\pi}{2}\right) = \frac{\pi}{2};$$

(2) if $|a| > b$ then $b - |a| < 0$ and $b + |a| > 0$ and so

$$\int_0^\infty \cos{(ax)}\frac{\sin{(bx)}}{x}dx = \frac{1}{2}\left(-\frac{\pi}{2}\right)+\frac{1}{2}\left(\frac{\pi}{2}\right) = 0;$$

(3) if $|a| = b$ then $b - |a| = 0$ and $b + |a| > 0$ and so

$$\int_0^\infty \cos{(ax)}\frac{\sin{(bx)}}{x}dx = \frac{1}{2}(0)+\frac{1}{2}\left(\frac{\pi}{2}\right) = \frac{\pi}{4}.$$

(C3.6) We start by following the hint and let $x = \cos{(2u)}$. Then $\frac{du}{dx} = -2\sin{(2u)}$ and so (remembering the double-angle identity) $dx = -2\sin{(2u)}du = -4\sin{(u)}\cos{(u)}du$. Thus, $\int_{-1}^1 \sqrt{\frac{1+x}{1-x}}dx = \int_{\pi/2}^0 \sqrt{\frac{1+\cos{(2u)}}{1-\cos{(2u)}}}\{-4\sin{(u)}\cos{(u)}du\} = 4\int_0^{\frac{\pi}{2}}\sin{(u)}\cos{(u)}\sqrt{\frac{1+\cos{(2u)}}{1-\cos{(2u)}}}du$. Next, remembering the identities $1 + \cos{(2u)} = 2\cos^2(u)$ and $1 - \cos{(2u)} = 2\sin^2(u)$, we have

$$\int_{-1}^{1}\sqrt{\frac{1+x}{1-x}}dx = 4\int_{0}^{\frac{\pi}{2}}\sin(u)\cos(u)\sqrt{\frac{2\cos^2(u)}{2\sin^2(u)}}du = 4\int_{0}^{\frac{\pi}{2}}\cos^2(u)du$$

$$= 4\int_{0}^{\pi/2}\left\{\frac{1}{2}+\frac{1}{2}\cos(2u)\right\}du = 4\left\{\frac{1}{2}u+\frac{1}{4}\sin(2u)\right\}\Big|_{0}^{\pi/2} = 4\left\{\frac{\pi}{4}\right\}$$

and so

$$\int_{-1}^{1}\sqrt{\frac{1+x}{1-x}}dx = \pi.$$

(C3.7) In $\int_{0}^{1}\left\{\int_{0}^{1}\frac{x-y}{(x+y)^3}dx\right\}dy$ let $t = x + y$ in the inner integral. Then $dx = dt$ and $\int_{0}^{1}\frac{x-y}{(x+y)^3}dx = \int_{y}^{1+y}\frac{t-2y}{t^3}dt = \int_{y}^{1+y}\frac{dt}{t^2} - 2y\int_{y}^{1+y}\frac{dt}{t^3} = \left(-\frac{1}{t}\right)\Big|_{y}^{1+y} - 2y\left(-\frac{1}{2t^2}\right)\Big|_{y}^{1+y} =$ $\left(\frac{1}{y}-\frac{1}{1+y}\right) + y\left[\frac{1}{(1+y)^2}-\frac{1}{y^2}\right]$ or, after a little simple algebra, this reduces to $-\frac{1}{(1+y)^2}$. So, $\int_{0}^{1}\left\{\int_{0}^{1}\frac{x-y}{(x+y)^3}dx\right\}dy = -\int_{0}^{1}\frac{dy}{(1+y)^2}$. Let $t = 1 + y$, and this integral becomes $-\int_{1}^{2}\frac{dt}{t^2} = -\left(-\frac{1}{t}\right)\Big|_{1}^{2} = -\left(-\frac{1}{2}+1\right) = -\frac{1}{2}$. That is, $\int_{0}^{1}\left\{\int_{0}^{1}\frac{x-y}{(x+y)^3}dx\right\}dy = -\frac{1}{2}$. If you repeat this business for $\int_{0}^{1}\left\{\int_{0}^{1}\frac{x-y}{(x+y)^3}dy\right\}dx$ you'll get $+\frac{1}{2}$. The lack of equality is caused by the integrand blowing-up as we approach the lower left corner ($x = y = 0$) of the region of integration. Don't forget the lesson of Chap. 1—*beware of exploding integrands*!

(C3.8) We know from (3.7.1) that (with $b = 0$) $\int_{-\infty}^{\infty}e^{-ax^2}dx = \sqrt{\frac{\pi}{a}}$. Now, following the hint, consider

$$\int_{-\infty}^{\infty}e^{-ax^2+bx}dx = \int_{-\infty}^{\infty}e^{-(ax^2-bx)}dx = \int_{-\infty}^{\infty}e^{-a\left(x^2-\frac{b}{a}x\right)}dx$$

$$= \int_{-\infty}^{\infty}e^{-a\left(x^2-\frac{b}{a}x+\frac{b^2}{4a^2}-\frac{b^2}{4a^2}\right)}dx = \int_{-\infty}^{\infty}e^{\frac{b^2}{4a}}e^{-a\left(x-\frac{b}{2a}\right)^2}dx = e^{\frac{b^2}{4a}}\int_{-\infty}^{\infty}e^{-a\left(x-\frac{b}{2a}\right)^2}dx.$$

Next, change variable to $y = x - \frac{b}{2a}$. Then, $e^{\frac{b^2}{4a}}\int_{-\infty}^{\infty}e^{-a\left(x-\frac{b}{2a}\right)^2}dx = e^{\frac{b^2}{4a}}\int_{-\infty}^{\infty}e^{-ay^2}dy = e^{\frac{b^2}{4a}}\sqrt{\frac{\pi}{a}}$ and so $I(a,b) = \int_{-\infty}^{\infty}e^{-ax^2+bx}dx = e^{\frac{b^2}{4a}}\sqrt{\frac{\pi}{a}}$. Differentiating with respect to b gives us $\frac{\partial I}{\partial b} = \int_{-\infty}^{\infty}xe^{-ax^2+bx}dx = \frac{2b}{4a}e^{\frac{b^2}{4a}}\sqrt{\frac{\pi}{a}} = \frac{b}{2a}e^{\frac{b^2}{4a}}\sqrt{\frac{\pi}{a}}$. Setting $a = 1$ and $b = -1$ says $\int_{-\infty}^{\infty}xe^{-x^2-x}dx = -\frac{1}{2}\sqrt{\pi}e^{\frac{1}{4}} = -\frac{1}{2}\sqrt{\pi}\sqrt[4]{e} = -1.137937\ldots$. MATLAB agrees, as *integral(@(x)x.*exp(-x.^2-x),-inf,inf)* $= -1.137938\ldots$. If we differentiate $I(a,b)$ with respect to a, we get $\frac{\partial I}{\partial a} = \int_{-\infty}^{\infty}-x^2e^{-ax^2+bx}dx = \sqrt{\frac{\pi}{a}}\times$ $\left[-\frac{4b^2}{16a^2}e^{\frac{b^2}{4a}}\right] + e^{\frac{b^2}{4a}}\left[\frac{-\frac{1}{2}a^{-1/2}\sqrt{\pi}}{a}\right] = -\sqrt{\frac{\pi}{a}}\frac{b^2}{4a^2}e^{\frac{b^2}{4a}} - \frac{\sqrt{\pi}}{2a^{3/2}}e^{\frac{b^2}{4a}}$. So, with $a = 1$ and $b = -1$

we have $\int_{-\infty}^{\infty} x^2 e^{-x^2-x} dx = \sqrt{\pi} \frac{1}{4} e^{\frac{1}{4}} + \frac{\sqrt{\pi}}{2} e^{\frac{1}{4}} = \frac{3}{4} \sqrt{\pi} \sqrt{e} = 1.7069068 \ldots$. MATLAB agrees, as $integral(@(x)x.\char`\^2.*exp(-x.\char`\^2-x),-inf,inf) = 1.7069068 \ldots$.

(C3.9) Following the hint, differentiation gives $\int_0^\infty \frac{\sin(mx)}{x} \left\{ \frac{-2a}{(x^2+a^2)^2} \right\} dx = \frac{\pi}{2} \times \left\{ \frac{a^2(me^{-am})-(1-e^{-am})2a}{a^4} \right\}$. So, $\int_0^\infty \frac{\sin(mx)}{x(x^2+a^2)^2} dx = \frac{\pi}{2} \left\{ \frac{a^2(me^{-am})-(1-e^{-am})2a}{-2a^5} \right\}$ which, after a bit of simple algebra becomes $\int_0^\infty \frac{\sin(mx)}{x(x^2+a^2)^2} dx = \frac{\pi}{2a^4} \left[1 - \frac{2+ma}{2} e^{-am} \right]$.

(C3.10) As declared in the problem statement, while important this is *so* easy to establish that I leave the solution entirely for you.

(C3.11) Let $x = \tan(\theta)$. Then $\theta = \tan^{-1}(x)$ and $\frac{d\theta}{dx} = \frac{1}{1+x^2}$ or, $d\theta = \frac{dx}{1+x^2}$. Thus, $\int_0^{\frac{\pi}{2}} \cos\{m \tan(\theta)\} d\theta = \int_0^\infty \frac{\cos(mx)}{1+x^2} dx$. From (3.1.7) we have $\int_0^\infty \frac{\cos(ax)}{1+b^2} dx = \frac{\pi}{2b} e^{-ab}$ and so, with $a = m$ and $b = 1$, $\int_0^{\frac{\pi}{2}} \cos\{m \tan(\theta)\} d\theta = \frac{\pi}{2} e^{-m}$. Actually, since it's clear that the sign of m doesn't matter, we more generally can write $\int_0^{\frac{\pi}{2}} \cos\{m \tan(\theta)\} d\theta = \frac{\pi}{2} e^{-|m|}$.

(C3.12) $I = \int_1^\infty \int_1^\infty \frac{1}{x^2+y^2} dxdy = \int_1^\infty \int_1^\infty \frac{1}{y^2\left(1+\frac{x^2}{y^2}\right)} dxdy = \int_1^\infty \frac{1}{y^2} \left\{ \int_1^\infty \frac{dx}{1+\left(\frac{x}{y}\right)^2} \right\} dy$. Let $u = \frac{x}{y}$ in the inner integral and so $dx = y \, du$. Thus,

$$I = \int_1^\infty \frac{1}{y^2} \left\{ \int_{\frac{1}{y}}^\infty \frac{y \, du}{1+u^2} \right\} dy = \int_1^\infty \frac{1}{y} \left\{ \int_{\frac{1}{y}}^\infty \frac{du}{1+u^2} \right\} dy = \int_1^\infty \frac{1}{y} \left\{ \tan^{-1}(u) \Big|_{\frac{1}{y}}^\infty \right\} dy$$

$$= \int_1^\infty \frac{\tan^{-1}(\infty) - \tan^{-1}\left(\frac{1}{y}\right)}{y} dy = \int_1^\infty \frac{\frac{\pi}{2} - \tan^{-1}\left(\frac{1}{y}\right)}{y} dy.$$

Since $\frac{\pi}{2} - \tan^{-1}\left(\frac{1}{y}\right) = \tan^{-1}(y)$—which is geometrically obvious if you draw a right triangle with perpendicular sides of lengths 1 and y—and since we have $y \geq 1$ over the entire interval of integration, and since $\tan^{-1}(y)$ increases with increasing y, then $I = \int_1^\infty \frac{\tan^{-1}(y)}{y} dy > \int_1^\infty \frac{\tan^{-1}(1)}{y} dy = \frac{\pi}{4} \int_1^\infty \frac{dy}{y}$ which blows-up logarithmically. Thus, the double integral does *not* exist.

(C3.13) Following the hint, setting $a = 0$ and $b = 1$ in (3.4.5), then writing $t = a$, and then $t = b$, gives $\int_0^\infty e^{-ax} \left\{ \frac{1-\cos(x)}{x} \right\} dx = \ln\left(\frac{\sqrt{a^2+1}}{a} \right)$ and $\int_0^\infty e^{-bx} \left\{ \frac{1-\cos(x)}{x} \right\} dx = \ln\left(\frac{\sqrt{b^2+1}}{a} \right)$. Thus, $\int_0^\infty \left[e^{-ax} \left\{ \frac{1-\cos(x)}{x} \right\} - e^{-bx} \left\{ \frac{1-\cos(x)}{x} \right\} \right] dx = \int_0^\infty \left(e^{-ax} - e^{-bx} \right) \frac{1-\cos(x)}{x} dx = \ln\left(\frac{\sqrt{a^2+1}}{a} \right) - \ln\left(\frac{\sqrt{b^2+1}}{a} \right)$. Therefore, $\int_0^\infty \frac{e^{-ax}-e^{-bx}}{x} dx - \int_0^\infty \left(e^{-ax} - e^{-bx} \right) \frac{\cos(x)}{x} dx = \ln\left(\frac{\sqrt{a^2+1}}{a} \right) - \ln\left(\frac{\sqrt{b^2+1}}{a} \right)$. But (3.3.3) says the first integral on the left is $\ln\left(\frac{b}{a} \right)$, and so

$$\int_0^\infty \left(e^{-ax} - e^{-bx}\right) \frac{\cos(x)}{x} dx = \ln\left(\frac{b}{a}\right) - \ln\left(\frac{\sqrt{a^2+1}}{a}\right) + \ln\left(\frac{\sqrt{b^2+1}}{a}\right)$$

$$= \ln(b) - \ln(a) - \ln\left(\sqrt{a^2+1}\right) + \ln(a) + \ln\left(\sqrt{b^2+1}\right) - \ln(b)$$

$$= \ln\left(\sqrt{b^2+1}\right) - \ln\left(\sqrt{a^2+1}\right).$$

(C3.14)

$$\int_0^\infty \left(e^{-ax} - e^{-bx}\right) \frac{\sin(x)}{x} dx = \int_0^\infty \left(\int_a^b e^{-xy} dy\right) \sin(x) dx$$

$$\int_a^b \left\{\int_0^\infty e^{-xy} \sin(x) dx\right\} dy = \int_a^b \frac{1}{1+y^2} dy = \tan^{-1}(y)\Big|_a^b = \tan^{-1}(b) - \tan^{-1}(a).$$

Compare this result with that of the previous problem.

(C3.15) Following the hint,

$$I = \int_0^\infty \frac{\left(1 - e^{-ax}\right)\left(1 - e^{-bx}\right)}{x^2} dx = \int_0^a \left\{\int_0^b \left\{\int_0^\infty e^{-x(y+z)} dx\right\} dz\right\} dy$$

$$= \int_0^a \left\{\int_0^b \left(-\frac{e^{-x(y+z)}}{y+z}\right)\Big|_0^\infty dz\right\} dy = \int_0^a \left\{\int_0^b \left(\frac{1}{y+z}\right) dz\right\} dy = \int_0^a \left\{\ln(y+z)\Big|_0^b\right\} dy$$

$$= \int_0^a \left\{\ln(y+b) - \ln(y)\right\} dy = \int_0^a \ln(y+b) dy - \int_0^a \ln(y) dy.$$

In the first integral on the right, let $u = y + b$. Then, $dy = du$ and

$$I = \int_b^{a+b} \ln(u) du - \int_0^a \ln(y) dy$$

$$= \{u \ln(u) - u\}\Big|_b^{a+b} - \{y \ln(y) - y\}\Big|_0^a$$

$$= (a+b) \ln(a+b) - (a+b) - b \ln(b) + b - a \ln(a) + a$$

$$= (a+b) \ln(a+b) - a \ln(a) - b \ln(b).$$

(C3.16) Since $x^2 - ax - 1 = x\left(x - a - \frac{1}{x}\right) = x\left(x - \frac{1}{x} - a\right)$ then $\frac{\left(x^2-ax-1\right)^2}{bx^2} =$
$\frac{x^2\left(x-\frac{1}{x}-a\right)^2}{bx^2} = \frac{\left(x-\frac{1}{x}-a\right)^2}{b}$ and so $\int_{-\infty}^\infty e^{-\frac{\left(x^2-ax-1\right)^2}{bx^2}} dx = \int_{-\infty}^\infty e^{-\frac{\left(x-\frac{1}{x}-a\right)^2}{b}} dx$. Using (3.7.2), this
last integral becomes $\int_{-\infty}^\infty e^{-\frac{(x-a)^2}{b}} dx$. Next, change variable to $u = \frac{x-a}{\sqrt{b}}$ (and so $dx =$
$\sqrt{b}\, du$, which says $\int_{-\infty}^\infty e^{-\frac{\left(x^2-ax-1\right)^2}{bx^2}} dx = \int_{-\infty}^\infty e^{-u^2} \sqrt{b}\, du = \sqrt{b}\int_{-\infty}^\infty e^{-u^2} du = \sqrt{\pi b}$

because we know $\int_{-\infty}^{\infty} e^{-x^2}\,dx = \sqrt{\pi}$ from (3.7.1). Notice that there is *no* dependency on a, but only on b. We can experimentally explore this (perhaps) surprising result with MATLAB. Holding $b = 1$, first for $a = 2$ and then for $a = 3$, for example, MATLAB computes

$$integral\left(@(x)\,exp\left((-(x.^2 - 2*x - 1).^2)./(x.^2)\right), -\inf, \inf\right) = 1.77245385\ldots$$

and

$$integral\left(@(x)\,exp\left((-(x.^2 - 3*x - 1).^2)./(x.^2)\right), -\inf, \inf\right) = 1.77245385\ldots$$

compared to the theoretical value of $\sqrt{\pi} = 1.77245385\ldots$. The claim for the independence of the value of the integral on the value of a appears to have merit.

(C3.17) Since $x^b + 1 = e^{\ln(x^b)} + 1 = e^{b\,\ln(x)} + 1$, then $\frac{d}{db}(x^b + 1) = \ln(x)e^{b\ln(x)}$ $= x^b \ln(x)$ and so, for $I(b) = \int_0^{\infty} \frac{f(x)}{x^b+1}\,dx$, we can write $\frac{dI(b)}{db} = \int_0^{\infty} \frac{-f(x)x^b\ln(x)}{(x^b+1)^2}\,dx$, which I'll call Result A. Now, in Result A change variable to $u = \frac{1}{x}$ (and so $dx =$ $-\frac{du}{u^2}$). Thus, $\frac{dI(b)}{db} = \int_{\infty}^{0} \frac{-f\left(\frac{1}{u}\right)\frac{1}{u^b}\ln\left(\frac{1}{u}\right)}{\left[\frac{1}{u^b}+1\right]^2}\left(-\frac{du}{u^2}\right) = \int_0^{\infty} \frac{f\left(\frac{1}{u}\right)\ln(u)}{u^b\left[\frac{1+u^b}{u^b}\right]^2}\frac{du}{u^2}$. Since we are given that $f\left(\frac{1}{x}\right) = x^2 f(x)$, we can write $\frac{dI(b)}{db} = \int_0^{\infty} \frac{u^2 f(u)\ln(u)}{u^b\frac{(1+u^b)^2}{u^{2b}}}\frac{du}{u^2} = \int_0^{\infty} \frac{u^b f(u)\ln(u)}{(1+u^b)^2}\,du$, which I'll call

Result B. Comparing Result A with Result B, we see that $\frac{dI(b)}{db} = -\frac{dI(b)}{db}$. Thus, $\frac{dI(b)}{db} = 0$.

(C3.18) $I = \int_0^{\infty} f\left\{\left(x - \frac{1}{x}\right)^2\right\}dx$ or, with $y = \frac{1}{x}$ (and so $dx = -\frac{dy}{y^2}$) we have $I =$

$\int_{\infty}^{0} f\left\{\left(\frac{1}{y} - y\right)^2\right\}\left(-\frac{dy}{y^2}\right) = \int_0^{\infty} f\left\{\left(y - \frac{1}{y}\right)^2\right\}\frac{dy}{y^2}$. That is, $I = \int_0^{\infty} f\left\{\left(x - \frac{1}{x}\right)^2\right\}\frac{dx}{x^2}$. So,

adding our two expressions for I, we have $2I = \int_0^{\infty} f\left\{\left(x - \frac{1}{x}\right)^2\right\}\left(1 + \frac{1}{x^2}\right)dx$. Now, let

$u = x - \frac{1}{x}$. Then $\frac{du}{dx} = 1 + \frac{1}{x^2}$ or, $dx = \frac{du}{1+\frac{1}{x^2}}$. So, noticing that $x = 0 \to u = -\infty$ and

$x = \infty \to u = \infty$, then $2I = \int_{-\infty}^{\infty} f(u^2)\left(1 + \frac{1}{x^2}\right)\frac{du}{1+\frac{1}{x^2}} = \int_{-\infty}^{\infty} f(u^2)du = 2\int_0^{\infty} f(u^2)du$

(since $f(u^2)$ is even) and we have $I = \int_0^{\infty} f(u^2)du = \int_0^{\infty} f(x^2)dx$ and we are done.

Chapter 4

(C4.1) Change variable to $u = \sqrt{x} = x^{\frac{1}{2}}$ and so $\frac{du}{dx} = \frac{1}{2\sqrt{x}}$ which means that $dx = 2\sqrt{x} = 2u\,du$. Thus, $I(n) = \int_0^1 (1 - \sqrt{x})^n dx = \int_0^1 (1 - u)^n 2u\,du = 2\int_0^1 u(1 - u)^n du = 2B(2, n+1) = 2\frac{\Gamma(2)\Gamma(n+1)}{\Gamma(n+3)} = 2\frac{1!n!}{(n+2)!}$ or, finally,

$$I(n) = \int_0^1 \left(1 - \sqrt{x}\right)^n dx = \frac{2}{(n+1)(n+2)}.$$

In particular, $I(9) = \frac{2}{(10)(11)} = \frac{1}{55}$.

(C4.2) Make the change of variable $u = -\ln(x)$. Since $\ln(x) = -u$ then $x = e^{-u}$ and so $dx = -e^{-u}du$. So, $\int_0^1 x^m \ln^n(x)\, dx = \int_\infty^0 e^{-um}(-u)^n(-e^{-u}du) = (-1)^n \int_0^\infty u^n e^{-(m+1)u}\, du$. Change variable again, now to $t = (m+1)u$ and so $du = \frac{dt}{m+1}$. Thus, $\int_0^1 x^m \ln^n(x)\, dx = (-1)^n \int_0^\infty \left(\frac{t}{m+1}\right)^n e^{-t} \frac{dt}{m+1} = \frac{(-1)^n}{(m+1)^{n+1}} \int_0^\infty t^n e^{-t}\, dt$. From (4.1.1) we have $\Gamma(q) = \int_0^\infty t^{q-1} e^{-t}\, dt$ and so our integral is the $q - 1 = n$ case, that is, $q = n + 1$ and so $\int_0^1 x^m \ln^n(x)\, dx = \frac{(-1)^n}{(m+1)^{n+1}} \Gamma(n+1)$. But by (4.1.3), $\Gamma(n+1) = n!$ and so

$$\int_0^1 x^m \ln^n(x)\, dx = \frac{(-1)^n n!}{(m+1)^{n+1}}.$$

(C4.3) Write the double integral over the triangular region R as $\iint_R x^a y^b\, dx\, dy =$

$\int_0^1 x^a \left\{ \int_0^{1-x} y^b dy \right\} dx = \int_0^1 x^a \left\{ \left(\frac{y^{b+1}}{b+1}\right)\Big|_0^{1-x} \right\} dx = \frac{1}{b+1} \int_0^1 x^a (1-x)^{b+1} dx$. From the defining integral for the beta function, $B(m,n) = \int_0^1 x^{m-1}(1-x)^{n-1} dx$, we have $m - 1 = a$ and $n - 1 = b + 1$. Thus, $m = 1 + a$ and $n = b + 2$. So, $\iint_R x^a y^b\, dx\, dy = \frac{B(1+a, b+2)}{b+1} = \frac{\Gamma(1+a)\Gamma(b+2)}{(b+1)!\,(a+b+3)} = \frac{a!(b+1)!}{(b+1)(a+b+2)!} = \frac{a!b!}{(a+b+2)!}$.

(C4.4) Simply set $b = 1$ and $p = \frac{1}{2}$ in (4.3.2) to get $\int_0^\infty \frac{\sin(x)}{\sqrt{x}}\, dx = \frac{\pi}{2\Gamma(\frac{1}{2})\sin(\frac{\pi}{4})} = \frac{\pi}{2\sqrt{\pi}\frac{1}{\sqrt{2}}} = \sqrt{\frac{\pi}{2}}$. Do the same in (4.3.9) to get $\int_0^\infty \frac{\cos(x)}{\sqrt{x}}\, dx = \frac{\pi}{2\Gamma(\frac{1}{2})\cos(\frac{\pi}{4})} = \frac{\pi}{2\sqrt{\pi}\frac{1}{\sqrt{2}}} = \sqrt{\frac{\pi}{2}}$.

(C4.5) $\int_c^\infty \left\{ \int_0^\infty e^{-xy} \sin(bx) dx \right\} dy = \int_c^\infty \frac{b}{b^2+y^2}\, dy$ and so, with the integration order reversed on the left, $\int_0^\infty \sin(bx)\left\{ \int_c^\infty e^{-xy} dy \right\} dx = b\int_c^\infty \frac{dy}{b^2+y^2} = b\left[\frac{1}{b} \tan^{-1}\left(\frac{y}{b}\right)\right]\Big|_c^\infty$ or, $\int_0^\infty \sin(bx)\left\{-\frac{e^{-xy}}{x}\right\}\Big|_c^\infty dx = \tan^{-1}(\infty) - \tan^{-1}\left(\frac{c}{b}\right) = \frac{\pi}{2} - \tan^{-1}\left(\frac{c}{b}\right)$. But, since $\tan^{-1}\left(\frac{c}{b}\right) = \frac{\pi}{2} - \tan^{-1}\left(\frac{b}{c}\right)$—if you don't see this, draw the obvious right triangle— we have

$$\int_0^\infty \sin(bx)\frac{e^{-cx}}{x}\, dx = \tan^{-1}\left(\frac{b}{c}\right),$$

a generalization of Dirichlet's integral, (3.2.1), to which it reduces as $c \to 0$. If $b = c = 1$ we have $\int_0^\infty \frac{\sin(x)}{x} e^{-x} dx = \tan^{-1}(1) = \frac{\pi}{4} = 0.78536\ldots$ MATLAB agrees, as *integral(@(x)(sin(x)./x).*exp(-x),0,inf) = 0.78539...*.

(C4.6) Following the hint, make the substitution $x^b = y$ in $\int_0^\infty \frac{x^a}{(1+x^b)^c} dx$. Then,

$\frac{dy}{dx} = bx^{b-1} = b\frac{x^b}{x} = b\frac{y}{y^{1/b}} = by^{1-\frac{1}{b}}$ or, $dx = \frac{dy}{by^{1-\frac{1}{b}}}$. So, $\int_0^\infty \frac{x^a}{(1+x^b)^c} dx = \int_0^\infty \frac{y^{a/b}}{(1+y)^c} \frac{dy}{by^{1-\frac{1}{b}}}$

$= \frac{1}{b} \int_0^\infty \frac{y^{\frac{a}{b}+\frac{1}{b}-1}}{(1+y)^c} dy = \frac{1}{b} \int_0^\infty \frac{y^{\frac{1+a}{b}-1}}{(1+y)^c} dy$. This is in the form of the beta function $B(m,n) =$

$\int_0^\infty \frac{y^{m-1}}{(1+y)^{m+n}} dy$ where $m = \frac{1+a}{b}$ and $m + n = c$ $(n = c - m = c - \frac{1+a}{b})$.

Thus, $\int_0^\infty \frac{x^a}{(1+x^b)^c} dx = \frac{1}{b} B\left(\frac{1+a}{b}, c - \frac{1+a}{b}\right) = \frac{\Gamma\left(\frac{1+a}{b}\right)\Gamma\left(c-\frac{1+a}{b}\right)}{b\Gamma(c)}$. For example, if $a = 3$,

$b = 2$, and $c = 4$, then $\int_0^\infty \frac{x^3}{(1+x^2)^4} dx = \frac{\Gamma\left(\frac{1+3}{2}\right)\Gamma\left(4-\frac{1+3}{2}\right)}{2\Gamma(4)} = \frac{\Gamma(2)\Gamma(2)}{2\Gamma(4)} = \frac{1!1!}{2(3!)} = \frac{1}{12} =$

$0.0833333\ldots$. MATLAB agrees, as $integral(@(x)(x.^3)./((1+x.^2).^4),0,$
$inf) = 0.083333\ldots$.

(C4.7) Writing (4.4.2) for an inverse first power force law, $-\frac{k}{y} = mv\frac{dv}{dy}$, which

separates into $-k\frac{dy}{y} = mv\,dv$ and this integrates indefinitely to $-k\ln(y) +$

$\ln(C) = \frac{1}{2}mv^2$ where C is an arbitrary (positive) constant. Since $v = 0$ at $y = a$

we have $-k\ln(a) + \ln(C) = 0$ and so $\ln(C) = k\ln(a)$. Thus, $-k\ln(y) +$

$k\ln(a) = \frac{1}{2}mv^2 = \frac{1}{2}m\left(\frac{dy}{dt}\right)^2 = k\ln\left(\frac{a}{y}\right)$. That is, $\left(\frac{dy}{dt}\right)^2 = \frac{2k}{m}\ln\left(\frac{a}{y}\right)$. Solving for dt,

$dt = \pm\sqrt{\frac{m}{2k}}\frac{dy}{\sqrt{\ln\left(\frac{a}{y}\right)}}$ and so $\int_0^T dt = T = \pm\sqrt{\frac{m}{2k}}\int_a^0 \frac{dy}{\sqrt{\ln\left(\frac{a}{y}\right)}}$. Let $u = \frac{y}{a}$ and so $dy = a\,du$.

Then, $T = \pm\sqrt{\frac{m}{2k}}\int_1^0 \frac{a\,du}{\sqrt{\ln\left(\frac{1}{u}\right)}} = \pm a\sqrt{\frac{m}{2k}}\int_1^0 \frac{du}{\sqrt{-\ln(u)}}$ or, using the minus sign to give

$T > 0$, $T = a\sqrt{\frac{m}{2k}}\int_0^1 \frac{du}{\sqrt{-\ln(u)}}$. From (3.1.8) the integral is $\sqrt{\pi}$ and so $T = a\sqrt{\frac{m\pi}{2k}}$.

(C4.8) The reflection formula for the gamma function says $\Gamma(1-m) = \frac{\pi}{\Gamma(m)\sin(m\pi)}$.
For $m > 1$ we have $1 - m < 0$, and so on the left of the reflection formula we have the
gamma function with a negative argument. On the right, both $\Gamma(m)$ and π are positive
quantities, while $\sin(m\pi)$ goes through zero for every integer value of m. So, as m
increases from 2, $1 - m$ decreases from -1. As each new negative integer is
encountered the gamma function blows-up, *in both directions*, as the sign of sin
$(m\pi)$ goes positive and negative as it passes periodically through zero as m increases.

(C4.9) Follow the hint.

(C4.10) Follow the hint.

(C4.11) In $\int_0^\infty e^{-ax}x^{n-1}dx$ let $y = ax$ $(dx = \frac{1}{a}dy)$. Then $\int_0^\infty e^{-ax}x^{n-1}dx =$

$\int_0^\infty e^{-y}\left(\frac{y}{a}\right)^{n-1}\frac{1}{a}dy = \frac{1}{a^n}\int_0^\infty e^{-y}y^{n-1}dy = \frac{1}{a^n}\Gamma(n)$. From all this we can then immediately write $\int_0^\infty e^{-bx}x^{n-1}dx = \frac{1}{b^n}\Gamma(n)$ and so we instantly have

$$\int_0^\infty \left(e^{-ax} - e^{-bx}\right)x^{n-1}dx = \left(\frac{1}{a^n} - \frac{1}{b^n}\right)\Gamma(n), n > 0.$$

(C4.12) In $I_{2k} = \int_{-\infty}^{\infty} x^{2k} e^{-x^2} dx = 2\int_0^{\infty} x^{2k} e^{-x^2} dx$ let $y = x^2$ and so $\frac{dy}{dx} = 2x$ and thus $dx = \frac{dy}{2x} = \frac{dy}{2\sqrt{y}}$. So, $I_{2k} = 2\int_0^{\infty} y^k e^{-y} \frac{dy}{2\sqrt{y}} = \int_0^{\infty} y^{k-\frac{1}{2}} e^{-y} dy$. Now, $\Gamma(n) = \int_0^{\infty} e^{-x} x^{n-1} dx$. So, if we let $n - 1 = k - \frac{1}{2}$ then $n = k + \frac{1}{2}$ and we have $\Gamma(k + \frac{1}{2}) = \int_0^{\infty} e^{-x} x^{k-\frac{1}{2}} dx = I_{2k} = \int_{-\infty}^{\infty} x^{2k} e^{-x^2} dx$. As a check, suppose $k = 1$. Then MATLAB computes *gamma(1.5)=0.88622692545275...* and *integral(@(x)(x.^2).*exp(-x. ^2),-inf,inf)= 0.88622692545275....*

Chapter 5

(C5.1) The integrand is $\frac{1-x^m}{1-x^n} = (1 - x^m)(1 + x^n + x^{2n} + x^{3n} + \ldots)$ and so

$$
I(m,n) = \int_0^1 (1 - x^m)\left\{\sum_{k=0}^{\infty} x^{kn}\right\} dx = \sum_{k=0}^{\infty} \int_0^1 \{x^{kn} - x^{kn+m}\} dx
$$

$$
= \sum_{k=0}^{\infty} \left\{\frac{x^{kn+1}}{kn+1} - \frac{x^{kn+m+1}}{kn+m+1}\right\}\bigg|_0^1
$$

$$
= \sum_{k=0}^{\infty} \left\{\frac{1}{kn+1} - \frac{1}{kn+m+1}\right\} = \sum_{k=0}^{\infty} \frac{kn+m+1-kn-1}{(kn+1)(kn+m+1)}
$$

$$
= m\sum_{k=0}^{\infty} \frac{1}{(kn+1)(kn+m+1)}.
$$

For given values of m and n this last summation is easy to code as an accumulating sum inside a loop, which is a near-trivial coding situation for *any* modern language. The MATLAB code named **cp5.m** does the case of $m = 9$, $n = 11$:

cp5.m
```
m=9;n=11;s=0;k=0;
while 1>0
    f1=k*n+1;f2=f1+m;k=k+1;
    s=s+1/(f1*f2);m*s
end
```

Since 1 is *always* greater than 0, **cp5.m** is stuck in an endless *while* loop, and I just let the code run while I watched the successive values of the summation stream upward on my computer screen until I didn't see the sixth decimal digit change anymore (and then I let the code run for 15 or 20 seconds more). I finally stopped the code with a Control-C. This is less than elegant, sure, but totally in keeping with the pretty casual philosophical approach I've taken all through this book. The results: $I(9, 11) = 0.972662$ and $I(11, 9) = 1.030656$.

(C5.2) Start by writing $\int_1^k \frac{dx}{x^2} = \left(-\frac{1}{x}\right)\big|_1^k = 1 - \frac{1}{k}$. Since $\int_1^k \frac{dx}{x^2} = \sum_{j=1}^{k-1} \int_j^{j+1} \frac{dx}{x^2}$ we can write $\sum_{k=1}^n \int_1^k \frac{dx}{x^2} = \sum_{k=1}^n \sum_{j=1}^{k-1} \int_j^{j+1} \frac{dx}{x^2} = \sum_{k=1}^n \left(1 - \frac{1}{k}\right) = n - \sum_{k=1}^n \frac{1}{k}$. Next, you'll recall from the opening of Sect. 5.4 that $\gamma(n) = \sum_{k=1}^{n} \frac{1}{k} - \ln(n)$ where Euler's constant $\gamma = \lim_{n \to \infty} \gamma(n)$, and so $\sum_{k=1}^{n} \frac{1}{k} = \gamma(n) + \ln(n)$ which says $\sum_{k=1}^{n} \sum_{j=1}^{k-1} \int_{j}^{j+1} \frac{dx}{x^2} = n - \gamma(n) - \ln(n)$. Writing out the double summation, term-by-term, we have

$k = 1$: nothing

$k = 2$: $\int_{1}^{2} \frac{dx}{x^2}$

$k = 3$: $\int_{1}^{2} \frac{dx}{x^2} + \int_{2}^{3} \frac{dx}{x^2}$

$k = 4$: $\int_{1}^{2} \frac{dx}{x^2} + \int_{2}^{3} \frac{dx}{x^2} + \int_{3}^{4} \frac{dx}{x^2}$

. . . .

$k = n$: $\int_{1}^{2} \frac{dx}{x^2} + \int_{2}^{3} \frac{dx}{x^2} + \int_{3}^{4} \frac{dx}{x^2} + \ldots + \int_{n-1}^{n} \frac{dx}{x^2}$.

Adding these terms *vertically* to recover the double-sum we have

$$(n-1)\int_{1}^{2} \frac{dx}{x^2} + (n-2)\int_{2}^{3} \frac{dx}{x^2} + (n-3)\int_{3}^{4} \frac{dx}{x^2} + \ldots + (1)\int_{n-1}^{n} \frac{dx}{x^2}$$

$$= \int_{1}^{2} \frac{(n-1)dx}{x^2} + \int_{2}^{3} \frac{(n-2)dx}{x^2} + \int_{3}^{4} \frac{(n-3)dx}{x^2} + \int_{n-1}^{n} \frac{dx}{x^2} = n - \gamma(n) - \ln(n).$$

The general form of the integrals is $\int_{j}^{j+1} \frac{(n-j)}{x^2} dx$. Since the interval of integration is $j \le x \le j+1$ we have, *by definition*, $\lfloor x \rfloor = j$. So, $n - \gamma(n) - \ln(n) = \int_{1}^{n} \frac{n - \lfloor x \rfloor}{x^2} dx$. Or, as $\{x\} = x - \lfloor x \rfloor$ and so $\lfloor x \rfloor = x - \{x\}$, then $n - \lfloor x \rfloor = n - [x - \{x\}] = n - x + \{x\}$. Thus,

$$n - \gamma(n) - \ln(n) = \int_{1}^{n} \frac{n - x + \{x\}}{x^2} dx = n\int_{1}^{n} \frac{dx}{x^2} - \int_{1}^{n} \frac{dx}{x} + \int_{1}^{n} \frac{\{x\}}{x^2} dx$$

$$= n\left(-\frac{1}{x}\right)\Big|_{1}^{n} - [\ln(x)]\Big|_{1}^{n} + \int_{1}^{n} \frac{\{x\}}{x^2} dx = n\left(1 - \frac{1}{n}\right) - \ln(n) + \int_{1}^{n} \frac{\{x\}}{x^2} dx$$

$$= n - 1 - \ln(n) + \int_{1}^{n} \frac{\{x\}}{x^2} dx \text{ or, } \int_{1}^{n} \frac{\{x\}}{x^2} dx = 1 - \gamma(n).$$

Finally, letting $n \to \infty$, we have

$$\int_{1}^{\infty} \frac{\{x\}}{x^2} dx = 1 - \gamma.$$

(C5.3) For the $\zeta(3)$ calculation, things go in much the same way. Starting with $\int_{1}^{k} \frac{dx}{x^4} = \sum_{j=1}^{k} \int_{j}^{j+1} \frac{dx}{x^4} = \left(-\frac{1}{3x^3}\right)\Big|_{1}^{k} = \frac{1}{3}\left(1 - \frac{1}{k^3}\right)$, we have $\sum_{k=1}^{n} \int_{1}^{k} \frac{dx}{x^4} = \sum_{k=1}^{n} \frac{1}{3} \times \left(1 - \frac{1}{k^3}\right) = \frac{1}{3}n - \frac{1}{3}\zeta_n(3)$, where $\lim_{n \to \infty} \zeta_n(3) = \zeta(3)$. Writing out the double summation, term-by-term just as we did in (C5.2), and then doing the 'adding vertically' trick, you should now be able to show that

$$
\frac{1}{3}n - \frac{1}{3}\zeta_n(3) = \int_1^n \frac{n - x + \{x\}}{x^4}\,dx = n\int_1^n \frac{dx}{x^4} - \int_1^n \frac{dx}{x^3} + \int_1^n \frac{\{x\}}{x^4}\,dx
$$

$$
= n\left(-\frac{1}{3x^3}\right)\Big|_1^k - \left(-\frac{1}{2x^2}\right)\Big|_1^n + \int_1^n \frac{\{x\}}{x^4}\,dx
$$

$$
= n\left(\frac{1}{3} - \frac{1}{3n^3}\right) - \left(\frac{1}{2} - \frac{1}{2n^2}\right) + \int_1^n \frac{\{x\}}{x^4}\,dx
$$

$$
= \frac{1}{3}n - \frac{1}{3n^2} - \frac{1}{2} + \frac{1}{2n^2} + \int_1^n \frac{\{x\}}{x^4}\,dx = \frac{1}{3}n - \frac{1}{3}\zeta_n(3).
$$

So, $\int_1^n \frac{\{x\}}{x^4}\,dx = \frac{1}{2} + \frac{1}{3n^2} - \frac{1}{2n^2} - \frac{1}{3}\zeta_n(3)$ or, as $n \to \infty$, $\int_1^\infty \frac{\{x\}}{x^4}\,dx = \frac{1}{2} - \frac{1}{3}\zeta(3)$. Thus, $\zeta(3) = \frac{3}{2} - 3\int_1^\infty \frac{\{x\}}{x^4}\,dx$.

(C5.4) Let $a = 1 + x$, with $x \approx 0$. Then, $\frac{1}{1-a} + \frac{1}{\ln(a)} = \frac{1}{-x} + \frac{1}{\ln(1+x)}$ or, using the power-series expansion for $\ln(1 + x)$ given at the beginning of Sect. 5.2, for $-1 < x < 1$, we have

$$
\frac{1}{1-a} + \frac{1}{\ln(a)} = \lim_{x \to 0}\left\{-\frac{1}{x} + \frac{1}{x - \frac{x^2}{2} + \frac{x^3}{3} - \frac{x^4}{4} + \cdots}\right\}
$$

$$
= \lim_{x \to 0}\left\{-\frac{1}{x} + \frac{1}{x\left(1 - \frac{x}{2} + \frac{x^2}{3} - \frac{x^3}{4} + \cdots\right)}\right\}
$$

$$
= \lim_{x \to 0}\left\{\frac{1}{x}\left[-1 + \frac{1}{1 - \frac{x}{2} + \frac{x^2}{3} - \frac{x^3}{4} + \cdots}\right]\right\}
$$

$$
= \lim_{x \to 0}\left\{\frac{1}{x}\left[-1 + 1 + \frac{x}{2}\right]\right\} = \frac{1}{2}.
$$

(C5.5) We start with $2\left[\frac{1}{1^2} - \frac{1}{2^2} + \frac{1}{3^2} - \frac{1}{4^2} + \cdots\right] = 2\left[\left\{\frac{1}{1^2} + \frac{1}{3^2} + \frac{1}{5^2} + \cdots\right\} - \left\{\frac{1}{2^2} + \frac{1}{4^2} + \frac{1}{6^2} + \cdots\right\}\right]$. The two sums in the curly brackets are the sum of the odd terms of $\zeta(2)$, and of the even terms of $\zeta(2)$, respectively. Now, consider the sum of the even terms: $\frac{1}{2^2} + \frac{1}{4^2} + \frac{1}{6^2} + \cdots = \frac{1}{(2 \cdot 1)^2} + \frac{1}{(2 \cdot 2)^2} + \frac{1}{(2 \cdot 3)^2} + \cdots = \frac{1}{4} \times \left[\frac{1}{1^2} + \frac{1}{2^2} + \frac{1}{3^2} + \frac{1}{4^2} + \cdots\right] = \frac{1}{4}\zeta(2)$. This means the sum of the odd terms is $\frac{3}{4}\zeta(2)$. So, our initial expression is $2\left[\frac{3}{4}\zeta(2) - \frac{1}{4}\zeta(2)\right] = \zeta(2)$, as was to be shown.

(C5.6) In $\int_0^1 \frac{\ln^2(1-x)}{x}\,dx$, follow the hint and change variable to $1 - x = e^{-t}$. Thus, $-\frac{dx}{dt} = -e^{-t}$ or, $dx = e^{-t}\,dt$. Now, when $x = 0$ we have $t = 0$, and when $x = 1$ then

$t = \infty$. So, $\int_0^1 \frac{\ln^2(1-x)}{x} dx = \int_0^\infty \frac{\ln^2(e^{-t})}{1-e^{-t}} e^{-t} dt = \int_0^\infty \frac{t^2}{1-e^{-t}} e^{-t} dt = \int_0^\infty \frac{t^2}{e^t-1} dt$. From (5.3.4) we see that this is $\Gamma(s)\zeta(s)$ for $s = 3$. That is, $\int_0^1 \frac{\ln^2(1-x)}{x} dx = \Gamma(3)\zeta(3) = 2!\zeta(3) = 2\zeta(3)$.

(C5.7) In (5.3.4) write $p = s - 1$, and so we have $\int_0^\infty \frac{x^p}{e^x-1} dx = \Gamma(p+1)\zeta(p+1)$. Continuing, $\int_0^\infty \frac{x^p}{e^x-1} dx = \int_0^\infty \frac{x^p}{e^x(1-e^{-x})} dx$. Let $u = e^{-x}$ and so $e^x = \frac{1}{u}$. Also, $\ln(u) = -x$ or, $x = -\ln(u)$. Finally, $\frac{du}{dx} = -e^{-x} = -u$, and so $dx = -\frac{du}{u}$. So, $\int_0^\infty \frac{x^p}{e^x(1-e^{-x})} dx = \int_1^0 \frac{\{-\ln(u)\}^p}{\frac{1}{u}(1-u)} \left(-\frac{du}{u}\right) = \int_0^1 \frac{\{-\ln(u)\}^p}{1-u} du$. Thus, $\int_0^1 \frac{\{-\ln(x)\}^p}{1-x} dx = \Gamma(p+1)\zeta(p+1)$.

(C5.8) Since $\frac{1}{1-p} = 1 + p + p^2 + \ldots = \sum_{k=0}^\infty p^k$, then $\frac{1}{1-x_1x_2x_3\ldots x_n} = \sum_{k=0}^\infty (x_1x_2x_3\ldots x_n)^k$ and so

$$\int_0^1 \int_0^1 \ldots \int_0^1 \frac{1}{1-x_1x_2x_3\ldots x_n} dx_1 dx_2 dx_3 \ldots dx_n$$

$$= \int_0^1 \int_0^1 \ldots \int_0^1 \sum_{k=0}^\infty (x_1x_2x_3\ldots x_n)^k dx_1 dx_2 dx_3 \ldots dx_n$$

$$= \sum_{k=0}^\infty \int_0^1 \int_0^1 \ldots \int_0^1 x_1^k x_2^k \ldots x_n^k dx_1 dx_2 \ldots dx_n.$$

Now,

$$\int_0^1 \int_0^1 \ldots \int_0^1 x_1^k x_2^k \ldots x_n^k dx_1 dx_2 \ldots dx_n = \int_0^1 x_1^k dx_1 \int_0^1 x_2^k dx_2 \ldots \int_0^1 x_n^k dx_n$$

$$= \left(\frac{x_1^{k+1}}{k+1}\right)\Big|_0^1 \left(\frac{x_2^{k+1}}{k+1}\right)\Big|_0^1 \ldots \left(\frac{x_n^{k+1}}{k+1}\right)\Big|_0^1 = \frac{1}{(k+1)^n}.$$

So,

$$\int_0^1 \int_0^1 \ldots \int_0^1 \frac{1}{1-x_1x_2x_3\ldots x_n} dx_1 dx_2 dx_3 \ldots dx_n = \sum_{k=0}^\infty \frac{1}{(k+1)^n}$$

$$= \frac{1}{1^n} + \frac{1}{2^n} + \frac{1}{3^n} + \frac{1}{4^n} + \ldots = \zeta(n).$$

(C5.9) In $\int_0^\infty \ln\left(\frac{1+e^{-x}}{1-e^{-x}}\right) dx$, let $u = e^{-x}$ and so $\frac{du}{dx} = -u$ which means $dx = -\frac{du}{u}$. Thus,

$$\int_0^\infty \ln\left(\frac{1+e^{-x}}{1-e^{-x}}\right) dx = \int_1^0 \ln\left(\frac{1+u}{1-u}\right)\left(-\frac{du}{u}\right)$$

$$= \int_0^1 \frac{1}{u}\ln\left(\frac{1+u}{1-u}\right) du = \int_0^1 \frac{1}{u}\{\ln(1+u) - \ln(1-u)\} du$$

$$= \int_0^1 \frac{1}{u}\left\{\left(u - \frac{u^2}{2} + \frac{u^3}{3} - \frac{u^4}{4} + \cdots\right) - \left(-u - \frac{u^2}{2} - \frac{u^3}{3} - \frac{u^4}{4} - \cdots\right)\right\} du$$

$$= \int_0^1 \frac{1}{u}\left(2u + 2\frac{u^3}{3} + 2\frac{u^5}{5} + \cdots\right) du = 2\int_0^1 \left(1 + \frac{u^2}{3} + \frac{u^4}{5} + \cdots\right) du$$

$$= 2\left(1 + \frac{u^3}{9} + \frac{u^5}{25} + \cdots\right)\Big|_0^\infty = 2\left(\frac{1}{1^2} + \frac{1}{3^2} + \frac{1}{5^2} + \cdots\right) = 2\left(\frac{\pi^2}{8}\right) = \frac{\pi^2}{4}.$$

(C5.10) Following the hint, write $I(m) = \int_0^\infty x^m e^{-x} dx = \int_0^\infty e^{m \ln(x)} e^{-x} dx$. Then, $\frac{dI}{dm} = \int_0^\infty \ln(x) e^{m \ln(x)} e^{-x} dx$, and so $\frac{d^2 I}{dm^2} = \int_0^\infty \ln^2(x) e^{m \ln(x)} e^{-x} dx$. Thus, $\int_0^\infty e^{-x} \ln^2(x) dx = \frac{d^2 I}{dm^2}\big|_{m=0}$. Notice that $I(m)$ is the gamma function $\Gamma(n)$ for $n - 1 = m$, that is, $n = m + 1$. So, $I(m) = \Gamma(m + 1)$ and $\int_0^\infty e^{-x} \ln^2(x) dx = \left\{\frac{d^2}{dm^2}\Gamma(m+1)\right\}\big|_{m=0}$. Now, the digamma function says $\frac{d\Gamma(z)}{dz} = \Gamma(z) \times \left[-\frac{1}{z} - \gamma + \sum_{r=1}^\infty \left(\frac{1}{r} - \frac{1}{r+z}\right)\right]$ and so, for $z = m + 1$, $\frac{d\Gamma(m+1)}{d(m+1)} = \frac{d\Gamma(m+1)}{dm} = \Gamma(m+1)\left[-\frac{1}{m+1} - \gamma + \sum_{r=1}^\infty \left(\frac{1}{r} - \frac{1}{r+m+1}\right)\right]$. Differentiating again, $\frac{d^2\Gamma(m+1)}{dm^2} = \frac{d\Gamma(m+1)}{dm}\left[-\frac{1}{m+1} - \gamma + \sum_{r=1}^\infty \left(\frac{1}{r} - \frac{1}{r+m+1}\right)\right] + \Gamma(m+1)\left[\frac{1}{(m+1)^2} + \sum_{r=1}^\infty \frac{1}{(r+m+1)^2}\right]$. So, since $\Gamma(1) = 1$,

$$\frac{d\Gamma(m+1)}{dm}\Big|_{m=0} = \Gamma(1)\left[-1 - \gamma + \sum_{r=1}^\infty \left(\frac{1}{r} - \frac{1}{r+1}\right)\right]$$

$$= -1 - \gamma + \left(1 - \frac{1}{2}\right) + \left(\frac{1}{2} - \frac{1}{3}\right) + \cdots$$

$$= -1 - \gamma + 1 = -\gamma \text{ and } \frac{d^2\Gamma(m+1)}{dm^2}\Big|_{m=0}$$

$$= -\gamma\left[-1 - \gamma + \sum_{r=1}^\infty \left(\frac{1}{r} - \frac{1}{r+1}\right)\right] + \Gamma(1)\left[1 + \sum_{r=1}^\infty \frac{1}{(r+1)^2}\right]$$

$$= \gamma + \gamma^2 - \gamma\left[\left(1 - \frac{1}{2}\right) + \left(\frac{1}{2} - \frac{1}{3}\right) + \cdots\right] + 1 + \frac{1}{2^2} + \frac{1}{3^2} + \cdots$$

$$= \gamma + \gamma^2 - \gamma + 1 + \frac{1}{2^2} + \frac{1}{3^2} + \cdots$$

$$= \gamma^2 + \zeta(2) = \gamma^2 + \frac{\pi^2}{6} = \int_0^\infty e^{-x} \ln^2(x) dx.$$

(C5.11) From (5.4.1), $\gamma = \int_0^1 \frac{1-e^{-u}}{u} du - \int_1^\infty \frac{e^{-u}}{u} du$. In the second integral, let $t = \frac{1}{u}$ and so $du = -\frac{1}{t^2} dt$. Then, $\int_1^\infty \frac{e^{-u}}{u} du = \int_1^0 \frac{e^{-\frac{1}{t}}}{\frac{1}{t}}(-\frac{1}{t^2} dt) = \int_0^1 \frac{e^{-\frac{1}{t}}}{t} dt$. So, $\gamma = \int_0^1 \frac{1-e^{-u}}{u} du - \int_0^1 \frac{e^{-\frac{1}{u}}}{u} du$ or, as was to be shown, $\gamma = \int_0^1 \frac{1-e^{-u}-e^{-\frac{1}{u}}}{u} du$.

(C5.12) Following the hint,

$$
\int_0^\infty \frac{x^{s-1}}{\sinh(x)} dx = 2\int_0^\infty x^{s-1} \frac{e^{-x}}{1-e^{-2x}} dx
$$

$$
= 2\int_0^\infty x^{s-1} e^{-x}\left(1 + e^{-2x} + e^{-4x} + e^{-6x} + \ldots\right) dx
$$

$$
= 2\int_0^\infty x^{s-1}\left(\sum_{k=0}^\infty e^{-(2k+1)x}\right) dx = 2\sum_{k=0}^\infty \int_0^\infty x^{s-1} e^{-(2k+1)x} dx.
$$

Let $u = (2k+1)x$ and so $dx = \frac{du}{2k+1}$. Then, $\int_0^\infty \frac{x^{s-1}}{\sinh(x)} dx = 2\sum_{k=0}^\infty \int_0^\infty \frac{u^{s-1}}{(2k+1)^{s-1}} e^{-u} \frac{du}{2k+1} = 2\sum_{k=0}^\infty \frac{1}{(2k+1)^s} \int_0^\infty u^{s-1} e^{-u} du$. But, by (4.1.1), $\int_0^\infty u^{s-1} e^{-u} du = \Gamma(s)$ and so $\int_0^\infty \frac{x^{s-1}}{\sinh(x)} dx = 2\Gamma(s)\sum_{k=0}^\infty \frac{1}{(2k+1)^s}$. Now, $\sum_{k=0}^\infty \frac{1}{(2k+1)^s} = \frac{1}{1^s} + \frac{1}{3^s} + \frac{1}{5^s} + \ldots$ which by (5.3.5) is $\zeta(s)\left[1 - \frac{1}{2^s}\right] = \zeta(s)(1 - 2^{-s})$. So, $\int_0^\infty \frac{x^{s-1}}{\sinh(x)} dx = 2(1 - 2^{-s})\Gamma(s)\zeta(s)$.

(C5.13)

$$
\int_0^1\int_0^1 \frac{dxdy}{1+x^2y^2} = \int_0^1 dx\left\{\int_0^1 (1 - x^2y^2 + x^4y^4 - x^6y^6 + \ldots) dy\right\}
$$

$$
= \int_0^1 \left(y - x^2\frac{y^3}{3} + x^4\frac{y^5}{5} - x^6\frac{y^7}{7} + \ldots\right)\Big|_0^1 dx = \int_0^1 \left(1 - \frac{x^2}{3} + \frac{x^4}{5} - \frac{x^6}{7} + \ldots\right) dx
$$

$$
= \left(x - \frac{x^3}{3^2} + \frac{x^5}{5^2} - \frac{x^7}{7^2} + \ldots\right)\Big|_0^1 = \left(1 - \frac{1}{3^2} + \frac{1}{5^2} - \frac{1}{7^2} + \ldots\right) = G = 0.9159655\ldots.
$$

That is, Catalan's constant (see just before (5.1.1)). We can numerically check this result by typing the MATLAB command $integral2(@(x,y)1./(1 + (x.*y).\hat{}\ 2), 0, 1, 0, 1)$, which returns the value $0.9159655\ldots$.

(C5.14) $\int_0^\infty \frac{\ln^2(x)}{1+x^2} dx = \int_0^1 \frac{\ln^2(x)}{1+x^2} dx + \int_1^\infty \frac{\ln^2(x)}{1+x^2} dx$. In the last integral, let $u = \frac{1}{x}$ (and so $dx = -\frac{du}{u^2}$). So, $\int_1^\infty \frac{\ln^2(x)}{1+x^2} dx = \int_1^0 \frac{\ln^2(\frac{1}{u})}{1+\frac{1}{u^2}}(-\frac{du}{u^2}) = \int_0^1 \frac{u^2 \ln^2(u)}{1+u^2}\frac{du}{u^2} = \int_0^1 \frac{\ln^2(u)}{1+u^2} du = \int_0^1 \frac{\ln^2(x)}{1+x^2} dx$. Thus, $\int_0^\infty \frac{\ln^2(x)}{1+x^2} dx = 2\int_0^1 \frac{\ln^2(x)}{1+x^2} dx = 2\int_0^1 (1 - x^2 + x^4 - x^6 + \ldots)\ln^2(x) dx$. Now, from (C4.2) $\int_0^1 x^m \ln^n(x) dx = (-1)^n \frac{n!}{(m+1)^{n+1}}$ and so with $n = 2$ we have $\int_0^1 x^m \ln^2(x) dx = \frac{2}{(m+1)^3}$. Therefore,

$$\int_0^\infty \frac{\ln^2(x)}{1+x^2}\,dx = 2\left[\frac{2}{1^3} - \frac{2}{3^3} + \frac{2}{5^3} - \frac{2}{7^3} + \dots\right]$$

$$= 4\left[\frac{1}{1^3} - \frac{1}{3^3} + \frac{1}{5^3} - \frac{1}{7^3} + \dots\right] = 4\left(\frac{\pi^3}{32}\right) = \frac{\pi^3}{8}.$$

(C5.15) Following the hint, since

$$\{\ln(1-x) + \ln(1+x)\}^2 = \ln^2(1-x) + 2\ln(1-x)\ln(1+x) + \ln^2(1+x)$$
$$= \ln^2(1-x^2)$$

then

$$\int_0^1 \frac{\ln(1-x)\ln(1+x)}{x}\,dx = \frac{1}{2}\int_0^1 \frac{\ln^2(1-x^2)}{x}\,dx - \frac{1}{2}\int_0^1 \frac{\ln^2(1-x)}{x}\,dx$$
$$- \frac{1}{2}\int_0^1 \frac{\ln^2(1+x)}{x}\,dx$$

$$= \frac{1}{2}\zeta(3) - \frac{1}{2}(2\zeta(3)) - \frac{1}{2}\left(\frac{1}{4}\zeta(3)\right) = \zeta(3)\left[\frac{1}{2} - 1 - \frac{1}{8}\right] = -\frac{5}{8}\zeta(3).$$

(C5.16) Letting $y = 1 - x$ ($dx = -dy$) gives

$$\int_0^1 \frac{\ln^3(1-x)}{x}\,dx = \int_1^0 \frac{\ln^3(y)}{1-y}\,(-dy)$$

$$= \int_0^1 \frac{\ln^3(y)}{1-y}\,dy = \int_0^1 (1 + y + y^2 + y^3 + \dots)\ln^3(y)\,dy$$

$$= \int_0^1 \sum_{k=1}^\infty y^{k-1}\ln^3(y)\,dy = \sum_{k=1}^\infty \int_0^1 y^{k-1}\ln^3(y)\,dy.$$

Now, from (C4.2) $\int_0^1 x^m \ln^n(x)\,dx = (-1)^n \frac{n!}{(m+1)^{n+1}}$ and so in $\int_0^1 y^{k-1}\ln^3(y)\,dy$ we have $m = k-1$ and $n = 3$. Thus, $\int_0^1 y^{k-1}\ln^3(y)\,dy = (-1)^3\frac{3!}{k^4} = -\frac{6}{k^4}$. Thus,

$$\int_0^1 \frac{\ln^3(1-x)}{x}\,dx = \sum_{k=1}^\infty -\frac{6}{k^4} = -6\sum_{k=1}^\infty \frac{1}{k^4} = -6\zeta(4) = -6\left(\frac{\pi^4}{90}\right) = -\frac{\pi^4}{15}.$$

(C5.17) From (5.3.2) $\int_0^1 \int_0^1 \frac{(xy)^a \ln^{s-2}(xy)}{1-xy}\,dxdy = (-1)^s(s-1)!\sum_{n=1}^\infty \frac{1}{(n+a)^s}$. If $a = 2$ and $s = 3$ then

$$\int_0^1 \int_0^1 \frac{(xy)^2 \ln(xy)}{1-xy} \, dx dy = -(2!) \sum_{n=1}^{\infty} \frac{1}{(n+2)^3} = -2\left(\frac{1}{3^3} + \frac{1}{4^3} + \cdots\right)$$

$$= -2\left(\frac{1}{1^3} + \frac{1}{2^3} + \frac{1}{3^3} + \frac{1}{4^3} + \cdots - \frac{1}{1^3} - \frac{1}{2^3}\right) = -2\zeta(3) - 2\left(-1 - \frac{1}{8}\right)$$

$$= 2\left(\frac{9}{8}\right) - 2\zeta(3) = \frac{9}{4} - 2\zeta(3),$$

which MATLAB calculates to be $-0.154113806\ldots$, in nice agreement with the numerical evaluation of the double integral I gave you in the problem statement.

(C5.18) With $u = x^2$ ($du = 2x dx$) and $v = y^2$ ($dv = 2y dy$) we have $D = \int_0^1 \int_0^1 \frac{2xy dx dy}{1-x^2y^2} = \int_0^1 \int_0^1 \frac{2xy \frac{du}{2x} \frac{dv}{2y}}{1-uv} = \frac{1}{2} \int_0^1 \int_0^1 \frac{du dv}{1-uv}$. That is, $= \frac{1}{2} \int_0^1 \int_0^1 \frac{dx dy}{1-xy}$. Now, adding the two boxed expressions in the problem statement, we get $D + S = 2\int_0^1 \int_0^1 \frac{dx dy}{1-xy}$. But $D + S$ is also given by $D + S = \frac{1}{2} \int_0^1 \int_0^1 \frac{dx dy}{1-xy} + 2\int_0^1 \int_0^1 \frac{dx dy}{1-x^2y^2}$ and so $2\int_0^1 \int_0^1 \frac{dx dy}{1-x^2y^2} = \frac{3}{2} \int_0^1 \int_0^1 \frac{dx dy}{1-xy}$ or, $\int_0^1 \int_0^1 \frac{dx dy}{1-x^2y^2} = \frac{3}{4} \int_0^1 \int_0^1 \frac{dx dy}{1-xy} = \frac{3}{4}\zeta(2) = \frac{3}{4}\left(\frac{\pi^2}{6}\right) = \frac{\pi^2}{8}$ which equals $1.2337\ldots$, while $integral2(@(x,y)1./(1-(x.*y).\hat{}2),0,1,0,1) = 1.2337\ldots$. Finally, in the development of the boxed expression for S notice that we have the statement

$$\int_0^1 \int_0^1 \frac{dx dy}{1+xy} = 2\int_0^1 \int_0^1 \frac{dx dy}{1-x^2y^2} - \int_0^1 \int_0^1 \frac{dx dy}{1-xy} = 2\left(\frac{\pi^2}{8}\right) - \zeta(2) = \frac{\pi^2}{4} - \frac{\pi^2}{6} = \frac{\pi^2}{12}$$

which equals $0.822467\ldots$ while $integral2(@(x,y)1./(1+x.*y),0,1,0,1) = 0.822467\ldots$.

(C5.19) Starting with $\int_0^{\infty} e^{-(n+a)x} x^{s-1} dx$, let $u = (n + a)x$. Then $\int_0^{\infty} e^{-(n+a)x} x^{s-1} dx = \int_0^{\infty} e^{-u} \left(\frac{u}{n+a}\right)^{s-1} \frac{du}{n+a} = \frac{1}{(n+a)^s} \int_0^{\infty} e^{-u} u^{s-1} du = \frac{\Gamma(s)}{(n+a)^s}$. Thus, $\sum_{n=0}^{\infty} \int_0^{\infty} e^{-(n+a)x} x^{s-1} dx = \sum_{n=0}^{\infty} \frac{\Gamma(s)}{(n+a)^s} = \Gamma(s)\zeta_h(s,a)$. Rearranging a bit, $\Gamma(s)\zeta_h(s,a) = \int_0^{\infty} x^{s-1} \left\{\sum_{n=0}^{\infty} e^{-(n+a)x}\right\} dx$. But

$$\sum_{n=0}^{\infty} e^{-(n+a)x} = e^{-ax} + e^{-(1+a)x} + e^{-(2+a)x} + e^{-(3+a)x} + \cdots$$

$$= e^{-ax}\left(1 + e^{-x} + e^{-2x} + e^{-3x} + \cdots\right) = e^{-ax} \frac{1}{1-e^{-x}}.$$

Thus, $(s)\zeta_h(s,a) = \int_0^{\infty} \frac{e^{-ax}}{1-e^{-x}} x^{s-1} dx = \int_0^{\infty} \frac{e^x e^{-ax}}{e^x-1} x^{s-1} dx = \int_0^{\infty} \frac{e^{-(1-a)x}}{e^x-1} x^{s-1} dx.$

Chapter 6

(C6.1) Follow the same path described in Sect. 6.3.

(C6.2) Following the hint, consider $f(x) = \frac{x}{x^n+1} - \frac{1}{x^{n-1}+x^{n-2}+\ldots+x+1}$. By direct multiplication, you can confirm that $x^n - 1 = (x-1)(x^{n-1} + x^{n-2} + \ldots + x + 1)$, and so $f(x) = \frac{x}{x^n+1} - \frac{1}{\frac{x^n-1}{x-1}} = \frac{x}{x^n+1} - \frac{x-1}{x^n-1} = \left[\frac{x}{x^n+1} - \frac{x}{x^n-1}\right] + \frac{1}{x^n-1} = -\frac{2x}{x^{2n}-1} + \frac{1}{x^n-1}$. Next, pick a value for a such that $0 \leq a \leq 1$. Then

$$\int_0^a f(x)dx = \int_0^a \frac{dx}{x^n-1} - \int_0^a \frac{2x}{x^{2n}-1}dx.$$

In the second integral on the right in the box let $y = x^2$ and so $\frac{dy}{dx} = 2x$ or, $dx = \frac{dy}{2x}$ and so $\int_0^a \frac{2x}{x^{2n}-1}dx = \int_0^{a^2} \frac{2x}{y^n-1}\frac{dy}{2x} = \int_0^{a^2} \frac{dy}{y^n-1}$. Thus, $\int_0^a f(x)dx = \int_0^a \frac{dx}{x^n-1} - \int_0^{a^2} \frac{dy}{y^n-1}$ or, as $a^2 < a$ (since $a < 1$), we can write

$$\int_0^a f(x)dx = \int_{a^2}^a \frac{dx}{x^n-1}.$$

Next, return to the first box and replace the integration limits with b to infinity, where $b > 1$. That is, $\int_b^\infty f(x)dx = \int_b^\infty \frac{dx}{x^n-1} - \int_b^\infty \frac{2x}{x^{2n}-1}dx$. In the second integral on the right let $y = x^2$ (just as before) and so $\int_b^\infty f(x)dx = \int_b^\infty \frac{dx}{x^n-1} - \int_{b^2}^\infty \frac{dy}{y^n-1} = \int_b^{b^2} \frac{dx}{x^n-1}$ because $b^2 > b$ (because $b > 1$). Now, in the integral on the right, let $y = \frac{1}{x}$ and so $\frac{dy}{dx} = -\frac{1}{x^2} = -y^2$ or, $dx = -\frac{dy}{y^2}$. So, $\int_b^{b^2} \frac{dx}{x^n-1} = \int_{1/b}^{1/b^2} \frac{-\frac{dy}{y^2}}{\frac{1}{y^n}-1} = -\int_{\frac{1}{b}}^{\frac{1}{b^2}} \frac{y^{n-2}}{1-y^n}dy = \int_{\frac{1}{b^2}}^{\frac{1}{b}} \frac{y^{n-2}}{1-y^n}dy$ and we have

$$\int_b^\infty f(x)dx = \int_{\frac{1}{b^2}}^{\frac{1}{b}} \frac{x^{n-2}}{1-x^n}dx.$$

Thus, adding the two results in the second and third boxes, we get $\int_0^a f(x)dx + \int_b^\infty f(x)dx = \int_{a^2}^a \frac{dx}{x^n-1} + \int_{\frac{1}{b^2}}^{\frac{1}{b}} \frac{x^{n-2}}{1-x^n}dx$ or, $\int_0^a f(x)dx + \int_b^\infty f(x)dx = \int_{a^2}^a \frac{dx}{x^n-1} - \int_{\frac{1}{b^2}}^{\frac{1}{b}} \frac{x^{n-2}}{x^n-1}dx$. Let $b = \frac{1}{a}$ (since $a < 1$ then $b > 1$, just as we supposed), and so $\int_0^a f(x)dx + \int_{\frac{1}{a}}^\infty f(x)dx = \int_{a^2}^a \frac{dx}{x^n-1} - \int_{a^2}^a \frac{x^{n-2}}{x^n-1}dx = \int_{a^2}^a \frac{1-x^{n-2}}{x^n-1}dx$ or, $\int_0^a f(x)dx + \int_{\frac{1}{a}}^\infty f(x)dx = -\int_{a^2}^a \frac{1-x^{n-2}}{1-x^n}dx$. Our last step is to now let $a \to 1$. Notice that the integrand in the integral on the right is a continuous function that always has a finite value, even when $x = 1$, because of L'Hospital's rule: $\lim_{x \to 1} \frac{1-x^{n-2}}{1-x^n} = \lim_{x \to 1} \frac{-(n-2)x^{n-3}}{-nx^{n-1}} = \frac{n-2}{n}$. Thus, the integral always exists. So, as $a \to 1$, we have $\int_0^1 f(x)dx + \int_1^\infty f(x)dx = -\int_1^1 \frac{1-x^{n-2}}{1-x^n}dx = 0 = \int_0^\infty f(x)dx$ and we are done.

(C6.3) Following the hint, write $\int_0^1 c^x \, dx = \int_0^1 \left(e^{-\lambda}\right)^x \, dx = \int_0^1 e^{-\lambda x} \, dx =$ $\left(\frac{e^{-\lambda x}}{-\lambda}\right)\big|_0^1 = \frac{e^{-\lambda}-1}{-\lambda} = \frac{1-e^{-\lambda}}{\lambda}$. Also, $\sum_{k=1}^{\infty} c^k = \sum_{k=1}^{\infty} e^{-\lambda k}$ which is, of course, simply a geometric series easily summed using the standard trick (which I'm assuming you know—if you need a reminder, look in any high school algebra text) to give $\frac{e^{-\lambda}}{1-e^{-\lambda}}$. So, $\frac{1-e^{-\lambda}}{\lambda} = \frac{e^{-\lambda}}{1-e^{-\lambda}}$ which, with just a bit of elementary algebra becomes $e^{2\lambda} - (2 + \lambda)$ $e^{\lambda} + 1 = f(\lambda) = 0$. Note, *carefully*, that while $f(0) = 0$, it is *not* true that $\lambda = 0$ is a solution to the problem. That's because then $c = e^{-\lambda} = 1$ and this c obviously fails to satisfy $\int_0^1 c^x \, dx = \sum_{k=1}^{\infty} c^k$. Now, observe that $f(1) = e^2 - 3e + 1 = 0.23 > 0$ and that $f\left(\frac{1}{2}\right) = e - 2.5\sqrt{e} + 1 = -0.4 < 0$. So, $f(\lambda) = 0$ for some λ in the interval $\frac{1}{2} <$ $\lambda < 1$. We can get better and better (that is, narrower and narrower) intervals in which this λ lies using the simple, easy-to-program 'binary chop' method. That is, we start by defining two variables, *lower* and *upper*, and set them to the initial bounds on λ of 0.5 and 1, respectively. We then set the variable *lambda* to the value $\frac{1}{2}(lower + upper)$. If $f(lambda) > 0$ then set $upper = lambda$, and if $f(lambda) < 0$ then set $lower = lambda$. Each time we do this cycle of operations we reduce the interval in which the solution λ lies by one-half (if, at some point, $f(lambda) = 0$ then, of course, we are immediately done). So, running through 100 such cycles (done in a flash on a modern computer) we reduce the initial interval, of width $\frac{1}{2}$, by a factor of 2^{100}, and so both *lower* and *upper* will have converged toward each other to squeeze *lambda* closer to the solution λ by many more than the 13 decimal digits requested. The result (see the following code **cp6.m**) is $\lambda = 0.9308211936517\ldots$ and so $c = e^{-\lambda} = 0.3942298383683\ldots$. The common value of the integral and the sum is $0.65079\ldots$.

```
cp6.m
lower=0.5;upper=1;
for loop=1:100
    lambda=(lower+upper)/2;
    term=exp(lambda);
    f=term^2-(2+lambda)*term+1;
    if f<0
        lower=lambda;
    else
        upper=lambda;
    end
end
c=exp(-lambda)
```

(C6.4)

(a) Taking advantage of the suggestive nature of Leibniz's differential notation that I hinted at in the problem statement, write the *differentiation operator* $\frac{d}{dt} =$ $\frac{dx}{dt}\frac{d}{dx} = \dot{x}\frac{d}{dx}$. Thus, $\frac{d}{dx}\left(\dot{x}^2\right) = 2\dot{x}\frac{d\dot{x}}{dx} = 2\left[\{\dot{x}\frac{d}{dx}\}\dot{x}\right]$. So, replacing the *operator* $\dot{x}\frac{d}{dx}$ in the curly brackets with the equivalent operator $\frac{d}{dt}$, we have $\frac{d}{dx}\left(\dot{x}^2\right) = 2\frac{d\dot{x}}{dt} = 2\ddot{x}$.

Thus, $\frac{x}{2}\frac{d}{dx}\left(\dot{x}^2\right) = \frac{x}{2}2\ddot{x} = x\ddot{x}$ and so, starting with Sommerfeld's equation $x\ddot{x} + \dot{x}^2 = gx$, we can rewrite it as $\frac{x}{2}\frac{d}{dx}\left(\dot{x}^2\right) + \dot{x}^2 = gx$ or, $\frac{d}{dx}\left(\dot{x}^2\right) + \frac{2}{x}\dot{x}^2 = g$ or, as $\dot{x}^2 = v^2$, we arrive at $\frac{d}{dx}\left(v^2\right) = -\frac{2}{x}v^2 + 2g$, which is Keiffer's equation.

(b) Following the hint, multiply through Keiffer's equation by x^2 to get $x^2\frac{d}{dx} \times \left(v^2\right) + 2xv^2 = 2gx^2$. Then, $\frac{d}{dx}\left(x^2v^2\right) = 2gx^2$ or, integrating, $\int_0^x \frac{d}{dx'}\left(x'^2v'^2\right)$

$dx' = \int_0^x 2gx'dx'$ and so $\int_0^x d\left(x'^2v'^2\right) = 2g\frac{1}{3}x^3 = x^2v^2$. Thus, $v^2 = \frac{2}{3}gx = \left(\frac{dx}{dt}\right)^2$

and so $\frac{dx}{dt} = \sqrt{\frac{2}{3}g}\sqrt{x}$.

(c) From (b), $\frac{d^2x}{dt^2} = \sqrt{\frac{2}{3}g}\left(\frac{1}{2\sqrt{x}}\right)\frac{dx}{dt} = \sqrt{\frac{2}{3}g}\left(\frac{1}{2\sqrt{x}}\right)\sqrt{\frac{2}{3}g}\sqrt{x} = \frac{2}{3}g\frac{1}{2} = \frac{1}{3}g$.

(d) From (b), $\frac{dx}{\sqrt{x}} = \sqrt{\frac{2}{3}g}\,dt$ and so $\int_0^L\frac{dx}{\sqrt{x}} = \sqrt{\frac{2}{3}g}\int_0^T dt = T\sqrt{\frac{2}{3}g} = \left(2x^{1/2}\right)\Big|_0^L = 2\sqrt{L}$.

Thus, $T = 2\sqrt{L}\sqrt{\frac{3}{2g}} = \sqrt{\frac{6L}{g}}$.

(e) When a length x of the chain (with mass μx) has slid over the edge, it is moving at speed $v = \frac{dx}{dt}$ and so its K.E. is $\frac{1}{2}\mu x\left(\frac{dx}{dt}\right)^2$. The center of mass is $\frac{1}{2}x$ below the table top and so the P.E. is $-\mu xg\frac{1}{2}x = -\frac{1}{2}\mu gx^2$. Assuming conservation of energy says K. E. + P. E. = 0, and so $\frac{1}{2}\mu x\left(\frac{dx}{dt}\right)^2 - \frac{1}{2}\mu gx^2 = 0$ or, $\left(\frac{dx}{dt}\right)^2 = \frac{gx^2}{x} = gx$. Thus, $\frac{dx}{dt} = \sqrt{g}\sqrt{x}$ and the acceleration of the chain is $\frac{d^2x}{dt^2} = \sqrt{g}\left(\frac{1}{2\sqrt{x}}\right) \times \frac{dx}{dt} = \sqrt{g}\left(\frac{1}{2\sqrt{x}}\right)\sqrt{g}\sqrt{x} = \frac{1}{2}g$.

(C6.5) As the problem statement says; *You* are to supply the details!

(C6.6) $\frac{d^2x}{dt^2} = \frac{dv}{dt} = v^3$. Thus, $\frac{dv}{v^3} = dt$ or, $t = -\frac{1}{2v^2} + C$. Since $v(0) = 1$ we therefore have $0 = -\frac{1}{2} + C$ and so $C = \frac{1}{2}$. Thus, $t = -\frac{1}{2v^2} + \frac{1}{2}$ or, $v = \frac{1}{\sqrt{1-2t}}$. So, $v = \frac{dx}{dt} = \frac{1}{\sqrt{1-2t}}$ and we see that $v \to \infty$ as $t \to \frac{1}{2}$ second. Integrating again, $\int_0^x du = \int_0^t \frac{du}{\sqrt{1-2u}} = \left\{-(1-2u)^{1/2}\right\}\Big|_0^t$ or, $x = 1 - \sqrt{1-2t}$. So, $x\left(\frac{1}{2}\right) = 1$ meter. If $t > \frac{1}{2}$ then x is complex (has the mass, like Marty McFly in *Back to the Future*, perhaps left the real axis and traveled back through time?—probably not, but who *really* knows???)!

(C6.7) The improved inequality is $\int_0^{2\pi}\sqrt{a^2\sin^2(t) + b^2\cos^2(t)}\,dt \geq$ $\sqrt{4\pi\left\{\pi ab + \frac{3\pi}{8}(a-b)^2\right\}}$ which looks much like the original inequality except for the factor of $\frac{3\pi}{8}$ which replaces the original factor of 1. Since $\frac{3\pi}{8} > 1$ (to see this, simply notice that $\pi > 3$ and so $\frac{3\pi}{8} > \frac{9}{8} > 1$) then this new lower bound is greater than the original lower bound. For the details of the derivation, see my book *Number—Crunching*, Princeton 2011, pp. 345–346.

(C6.8) Putting Murphy's solution for $f(t)$ into the integral we have $\int_0^1 t^x \frac{t^{m-1}}{(n-1)!} \ln^{n-1}\left(\frac{1}{t}\right) dt = \frac{1}{(n-1)!} \int_0^1 t^{x+m-1}(-1)^{n-1} \ln^{n-1}(t)dt$. Now, from C4.2 we have $\int_0^1 x^m \ln^n(x)dx = (-1)^n \frac{n!}{(m+1)^{n+1}}$ or, $\frac{1}{n!} \int_0^1 x^m (-1)^n \ln^n(x)dx = \frac{1}{(m+1)^{n+1}}$ and this integral becomes Murphy's integral if we replace m with $x + m - 1$ and n with $n - 1$. So, doing that on the right, $\frac{1}{(m+1)^{n+1}}$ becomes $\frac{1}{(x+m-1+1)^{n-1+1}} = \frac{1}{(x+m)^n}$ which is, indeed, Murphy's assumed $F(x)$.

(C6.9) Putting Murphy's solution for $f(t)$ into the integral we have $\int_0^1 t^x t^{m-1} \frac{t^{a}-1}{\ln(t)} dt = \int_0^1 \frac{t^{x+m+a-1}-t^{x+m-1}}{\ln(t)} dt$. Now, from (3.4.4) we have $\int_0^1 \frac{x^a - x^b}{\ln(x)} dx = \ln\left(\frac{a+1}{b+1}\right)$. So, in Murphy's integral we have a in (3.4.4) replaced with $x + m + a - 1$ and b replaced with $x + m - 1$, and thus Murphy's integral becomes the claim $\int_0^1 \frac{t^{x+m+a-1}-t^{x+m-1}}{\ln(t)} dt = \ln\left(\frac{x+m+a-1+1}{x+m-1+1}\right) = \ln\left(\frac{x+m+a}{x+m}\right) = \ln\left(1 + \frac{a}{x+m}\right)$ which is, indeed, Murphy's assumed $F(x)$.

Chapter 7

(C7.1) For the $n = 5$ case we have p odd and so use (7.7.2) to write (with $p = q = 5$)

$$\int_0^\infty \left\{\frac{\sin(x)}{x}\right\}^5 dx = \frac{5!}{4!} \int_0^\infty \frac{u^4}{(u^2+1^2)(u^2+3^2)(u^2+5^2)} du = 5\left[\int_0^\infty \frac{A}{u^2+1^2} du + \int_0^\infty \frac{B}{u^2+3^2} du + \int_0^\infty \frac{C}{u^2+5^2} du\right].$$

So, we have $A[u^2 + 9][u^2 + 25] + B[u^2 + 1][u^2 + 25] + C[u^2 + 1][u^2 + 9] = u^4$. Multiplying out, we arrive at $A[u^4 + 34u^2 + 225] + B[u^4 + 26u^2 + 25] + C[u^4 + 10u^2 + 9] = u^4$. Then, equating coefficients of equal powers of u on each side of the equality gives us three simultaneous algebraic equations for A, B, and C:

$$A + B + C = 1$$

$$34A + 26B + 10C = 0$$

$$225A + 25B + 9C = 0$$

This system is easily solved, as follows, using determinants (*Cramer's rule*, which you can find in any good book on algebra). First, the system determinant is

$$D = \begin{vmatrix} 1 & 1 & 1 \\ 34 & 26 & 10 \\ 225 & 25 & 9 \end{vmatrix} \quad \text{which expands as} \quad D = \begin{vmatrix} 26 & 10 \\ 25 & 9 \end{vmatrix} - \begin{vmatrix} 34 & 10 \\ 225 & 9 \end{vmatrix} + \begin{vmatrix} 34 & 26 \\ 225 & 25 \end{vmatrix} =$$

$(234 - 250) - (306 - 2,250) + (850 - 5,850) = -16 + 1,944 - 5,000 = -3,072$. Thus, the values of A, B, and C are:

$$A = \dfrac{\begin{vmatrix} 1 & 1 & 1 \\ 0 & 26 & 10 \\ 0 & 25 & 9 \end{vmatrix}}{D} = \dfrac{\begin{vmatrix} 26 & 10 \\ 25 & 9 \end{vmatrix}}{-3{,}072} = \dfrac{-16}{-3{,}072} = \dfrac{16}{3{,}072},$$

$$B = \dfrac{\begin{vmatrix} 1 & 1 & 1 \\ 34 & 0 & 10 \\ 225 & 0 & 9 \end{vmatrix}}{D} = \dfrac{-\begin{vmatrix} 34 & 10 \\ 225 & 9 \end{vmatrix}}{-3{,}072} = \dfrac{1{,}944}{-3{,}072} = -\dfrac{1{,}944}{3{,}072},$$

and so

$$C = 1 - A - B = 1 - \frac{16}{3{,}072} + \frac{1{,}944}{3{,}072} = \frac{3{,}072 - 16 + 1{,}944}{3{,}072} = \frac{5{,}000}{3{,}072}.$$

So,

$$\int_0^\infty \left\{ \frac{\sin(x)}{x} \right\}^5 dx = 5 \left[\frac{16}{3{,}072}\left(\frac{1}{1}\right)\tan^{-1}(\infty) - \frac{1{,}944}{3{,}072}\left(\frac{1}{3}\right)\tan^{-1}(\infty) + \frac{5{,}000}{3{,}072}\left(\frac{1}{5}\right)\tan^{-1}(\infty) \right]$$

$$= \frac{\pi}{2}\left[\frac{80}{3{,}072} - \frac{9{,}720}{3{,}072}\left(\frac{1}{3}\right) + \frac{25{,}000}{3{,}072}\left(\frac{1}{5}\right) \right] = \frac{\pi}{2}\left[\frac{80 - 3{,}240 + 5{,}000}{3{,}072} \right]$$

$$= \frac{\pi}{2}\left[\frac{1{,}840}{3{,}072} \right] = \frac{920}{3{,}072}\pi$$

or, at last,

$$\int_0^\infty \left\{ \frac{\sin(x)}{x} \right\}^5 dx = \frac{115}{384}\pi.$$

For the $n = 6$ case we have p even and so use (7.7.1) to write (with $p = q = 6$)
$\int_0^\infty \left\{ \frac{\sin(x)}{x} \right\}^6 dx = \frac{6!}{5!}\int_0^\infty \frac{u^4}{(u^2+2^2)(u^2+4^2)}\,(u^2 + 6^2)\,du = 6\left[\int_0^\infty \frac{A}{u^2+4}\,du + \int_0^\infty \frac{B}{u^2+16}\,du + \int_0^\infty \frac{C}{u^2+36}\,du \right]$ and so $A[u^2 + 16][u^2 + 36] + B[u^2 + 4][u^2 + 36] + C[u^2 + 4][u^2 + 16] = u^4$ and therefore $A[u^4 + 52u^2 + 576] + B[u^4 + 40u^2 + 144] + C[u^4 + 20u^2 + 64] = u^4$ and so

$$A + B + C = 1$$

$$52A + 40B + 20C = 0$$

$$576A + 144B + 64C = 0.$$

The system determinant is $D = \begin{vmatrix} 1 & 1 & 1 \\ 52 & 40 & 20 \\ 576 & 144 & 64 \end{vmatrix} = -7{,}680$, and so $A =$

$$\frac{\begin{vmatrix} 1 & 1 & 1 \\ 0 & 40 & 20 \\ 0 & 144 & 64 \end{vmatrix}}{D} = \frac{320}{7{,}680}, \quad C = \frac{\begin{vmatrix} 1 & 1 & 1 \\ 52 & 40 & 0 \\ 576 & 144 & 0 \end{vmatrix}}{D} = \frac{15{,}552}{7{,}680},$$ and then $B = 1 - A - C =$

$-\frac{8{,}192}{7{,}680}$. Thus, $\int_0^\infty \left\{ \frac{\sin(x)}{x} \right\}^6 dx = 6\left[\frac{320}{7{,}680} \left(\frac{1}{2}\right) \tan^{-1}(\infty) - \frac{8{,}192}{7{,}680} \left(\frac{1}{4}\right) \tan^{-1}(\infty) + \right.$

$\left. \frac{15{,}552}{7{,}680} \left(\frac{1}{6}\right) \tan^{-1}(\infty) \right]$ and so, after just a bit of trivial (but tedious) arithmetic,

$$\int_0^\infty \left\{ \frac{\sin(x)}{x} \right\}^6 dx = \frac{11}{40}\pi.$$

For the $n = 7$ case we are back to p odd and so back to (7.7.2) with $p = q = 7$, which says

$$\int_0^\infty \left\{ \frac{\sin(x)}{x} \right\}^7 dx = \frac{7!}{6!} \int_0^\infty \frac{u^6}{(u^2 + 1^2)(u^2 + 3^2)(u^2 + 5^2)(u^2 + 7^2)} du$$

$$= 7\left[\int_0^\infty \frac{A}{u^2 + 1^2} du + \int_0^\infty \frac{B}{u^2 + 3^2} du + \int_0^\infty \frac{C}{u^2 + 5^2} du + \int_0^\infty \frac{D}{u^2 + 7^2} du \right]$$

and so now you see we are going to be working with $4{\times}4$ determinants which is starting to be *really* tedious and so I'll let *you* grind through the routine arithmetic to arrive at

$$\int_0^\infty \left\{ \frac{\sin(x)}{x} \right\}^7 dx = \frac{5{,}887}{23{,}040}\pi.$$

This is equal to $0.8027151\ldots$, and MATLAB agrees as *integral*(@ $(x)(\sin(x)./x)$.
^7,0,inf$) = 0.8027151\ldots$.

(C7.2) To start, write $\int_x^\infty e^{it^2} dt = \int_x^\infty \cos(t^2)dt + i\int_x^\infty \sin(t^2)dt = C(x) + iS(x)$.
Then, $\int_x^\infty e^{-it^2} dt = \int_x^\infty \cos(t^2)dt - i\int_x^\infty \sin(t^2)dt = C(x) - iS(x) = \int_x^\infty e^{-iu^2} du$.
(Remember, t and u are *dummy* variables.) So, $\{C(x) + iS(x)\}\{C(x) - iS(x)\} =$
$\int_x^\infty e^{it^2} \int_x^\infty e^{-iu^2} du$ and so $C^2(x) + S^2(x) = \int_x^\infty \int_x^\infty e^{i(t^2-u^2)} dt\, du = \int_x^\infty \int_x^\infty \cos(t^2 - u^2)$
$dt\, du + i\int_x^\infty \int_x^\infty \sin(t^2 - u^2)dt\, du$. Since $C^2(x) + S^2(x)$ is *purely real* then its imaginary
part must vanish, that is, $\int_x^\infty \int_x^\infty \sin(t^2 - u^2)dt\, du = 0$ for any x and we are done.

(C7.3) To start, write $1 - ix^3 = \sqrt{1 + x^6}e^{i\theta}$ where θ puts this complex vector in the fourth quadrant if $x > 0$ and in the second quadrant if $x < 0$. Since $|e^{i\theta}| = 1$ for *any* θ then the specific value of θ doesn't actually matter when we calculate the absolute value of $1 - ix^3$ as $\sqrt{1 + x^6}$. Thus, $\left|\frac{1}{1-ix^3}\right| = \frac{1}{|1-ix^3|} = \frac{1}{\sqrt{1+x^6}} = \frac{1}{(1+x^6)^{1/2}}$ and so $\left|\frac{1}{\sqrt{1-ix^3}}\right| = \frac{1}{(1+x^6)^{1/4}}$. Thus, $\left|\int_{-\infty}^{\infty} \frac{dx}{\sqrt{1-ix^3}}\right| \leq \int_{-\infty}^{\infty} \frac{dx}{(1+x^6)^{\frac{1}{4}}} = 2\int_0^{\infty} \frac{dx}{(1+x^6)^{\frac{1}{4}}}$ because the integrand in the middle integral is even. Now, $\int_0^{\infty} \frac{dx}{(1+x^6)^{\frac{1}{4}}} = \int_0^1 \frac{dx}{(1+x^6)^{\frac{1}{4}}} + \int_1^{\infty} \frac{dx}{(1+x^6)^{\frac{1}{4}}}$. Since $\int_0^1 \frac{dx}{(1+x^6)^{\frac{1}{4}}} = M$, where M is some *finite* number because the integrand is finite over the entire interval of integration, and noticing that $1 + x^6 > x^6$ we can write $\int_0^{\infty} \frac{dx}{(1+x^6)^{\frac{1}{4}}} \leq M + \int_1^{\infty} \frac{dx}{x^{3/2}}$. And since $\int_1^{\infty} \frac{dx}{x^{3/2}} = \left. (-2x^{-1/2})\right|_1^{\infty} = 2$. then $\left|\int_{-\infty}^{\infty} \frac{dx}{\sqrt{1-ix^3}}\right| \leq M + 2$ and so $\left|\int_{-\infty}^{\infty} \frac{dx}{\sqrt{1-ix^3}}\right|$ exists.

(C7.4) Start with $\int_1^k \frac{dx}{x^3} = \sum_{j=1}^{k-1} \int_j^{j+1} \frac{dx}{x^3} = \left. \left(-\frac{1}{2x^2}\right)\right|_1^k = \frac{1}{2}\left(1 - \frac{1}{k^2}\right)$. Then,

$$\sum_{k=1}^n \int_1^k \frac{dx}{x^3} = \sum_{k=1}^n \sum_{j=1}^{k-1} \int_j^{j+1} \frac{dx}{x^3} = \sum_{k=1}^n \frac{1}{2}\left(1 - \frac{1}{k^2}\right) = \frac{1}{2}(1 - S(n))$$

where $S(n) = \sum_{k=1}^n \frac{1}{k^2}$. (Remember, $\lim_{n \to \infty} S(n) = \frac{\pi^2}{6}$.) Writing out the double summation, term-by-term just as we did in (C5.2), and then doing the 'adding vertically' trick, you should now be able to show that $\frac{n}{2} - \frac{1}{2}S(n) = \int_1^n \frac{n-x+\{x\}}{x^3} dx = \frac{n}{2} - \frac{1}{2n} - 1 + \frac{1}{n} + \int_1^n \frac{\{x\}}{x^3} dx$. That is, $\int_1^n \frac{\{x\}}{x^3} dx = 1 - \frac{1}{2}S(n) - \frac{1}{n}$ and so, letting $n \to \infty$ we have $\int_1^{\infty} \frac{\{x\}}{x^3} dx = 1 - \frac{1}{2}S(\infty)$ or, at last,

$$\int_1^{\infty} \frac{\{x\}}{x^3} dx = 1 - \frac{\pi^2}{12}.$$

(C7.5) Starting with $I(a) = \int_0^{\infty} \frac{\sin^2(ax)}{x^2} dx$, differentiate with respect to the parameter a. Then, $\frac{dI}{da} = \int_0^{\infty} \frac{2x \sin(ax) \cos(ax)}{x^2} dx = \int_0^{\infty} \frac{2 \sin(ax) \cos(ax)}{x} dx$. Recalling the identity $\sin(ax)\cos(ax) = \frac{1}{2}\sin(2ax)$, we have $\frac{dI}{da} = \int_0^{\infty} \frac{\sin(2ax)}{x} dx = \pm\frac{\pi}{2}$ from (3.2.1), where the sign on the right depends on the sign of a (+ if $a > 0$ and − if $a < 0$). Then doing the *indefinite* integral, $\int dI = \pm\int \frac{\pi}{2} da + C$ or, $I(a) = \pm\frac{\pi}{2}a + C$ where C is an arbitrary constant. Since $I(0) = 0$ we know that $C = 0$ and so

$$I(a) = \int_0^{\infty} \frac{\sin^2(ax)}{x^2} dx = \frac{\pi}{2}|a|.$$

(C7.6) If you set $b = 1$ and $k = 2$ in (4.3.10) and (4.3.11) then the Fresnel integrals follow immediately.

(C7.7) Following the hint, the Fourier transform of f(t) is (I'll let you fill-in the easy integration details) $F(\omega) = \frac{1-e^{-ma}\cos(m\omega)+ie^{-ma}\sin(m\omega)}{a+i\omega}$ and so $|F(\omega)|^2 = \frac{1+e^{-2ma}-2e^{-ma}\cos(m\omega)}{\omega^2+a^2}$. The time integral for the energy of f(t) is $\int_{-\infty}^{\infty}f^2(t)dt = \int_0^m e^{-2at}dt = \frac{1-e^{-2ma}}{2a}$ while the frequency integral for the energy is $\int_{-\infty}^{\infty}\frac{|F(\omega)|^2}{2\pi}d\omega = \frac{1}{2\pi}\int_{-\infty}^{\infty}\frac{1+e^{-2ma}-2e^{-ma}\cos(m\omega)}{\omega^2+a^2}d\omega$. Equating these two integrals, and then doing some easy algebra (where I've also changed the dummy variable of integration from ω to x), gives $\int_{-\infty}^{\infty}\frac{\cos(mx)}{x^2+a^2}dx = \frac{\pi}{a}e^{-ma}$. You'll recognize this as (3.1.7), where in that result I wrote b in place of m, and the integration interval (of an even integrand) is 0 to ∞ rather than $-\infty$ to ∞ (thus accounting for the 2 on the right in (3.1.7)).

(C7.8)

(a) $\int_{-\infty}^{\infty}\frac{1}{t^2+1}e^{-i\omega t}dt = \int_{-\infty}^{\infty}\frac{\cos(\omega t)}{t^2+1}dt - i\int_{-\infty}^{\infty}\frac{\sin(\omega t)}{t^2+1}dt$. The last integral is zero because the integrand is odd. Now, $\int_{-\infty}^{\infty}\frac{1}{t^2+1}e^{-i\omega t}dt = 2\int_0^{\infty}\frac{\cos(\omega t)}{t^2+1}$ and the last integral is, with b = 1 and a = ω in (3.1.7), equal to $\frac{\pi}{2}e^{-\omega}$. So, for $\omega > 0$ the transform is $\pi e^{-\omega}$. Since the time signal is even, we have a purely real transform and we know that is even. So, for $\omega < 0$ the transform is πe^{ω}. Thus, for all ω, $\frac{1}{t^2+1} \leftrightarrow \pi e^{-|\omega|}$.

(b) Since the Fourier transform of f(t) is $F(\omega) = \int_{-\infty}^{\infty}f(t)e^{-i\omega t}dt$, then $\frac{dF}{d\omega} = \int_{-\infty}^{\infty}-itf(t)e^{-i\omega t}dt$. Thus, $\int_{-\infty}^{\infty}tf(t)e^{-i\omega t}dt = -\frac{1}{i}\frac{dF}{d\omega} = i\frac{dF}{d\omega}$. So, if f(t) $\leftrightarrow$ F(ω) then tf(t) $\leftrightarrow$ $i\frac{dF}{d\omega}$. Now, as shown in (a), for $\omega > 0$ the transform of $\frac{1}{t^2+1}$ is $\pi e^{-\omega}$, and so the transform of $\frac{t}{t^2+1}$ is $i\frac{d}{d\omega}(\pi e^{-\omega}) = -i\pi e^{-\omega}$. Since the time function is odd we have a purely imaginary transform which we know is odd. That is, for $\omega < 0$ the transform must be $i\pi e^{\omega}$. So, for all ω we write $\frac{t}{t^2+1} \leftrightarrow -i\pi e^{-|\omega|}\text{sgn}(\omega)$.

(c) $\int_{-\infty}^{\infty}\{\frac{1}{2}\delta(t) + i\frac{1}{2\pi t}\}e^{-i\omega t}dt = \frac{1}{2}\int_{-\infty}^{\infty}\delta(t)e^{-i\omega t}dt + \frac{i}{2\pi}\int_{-\infty}^{\infty}\frac{e^{-i\omega t}}{t}dt$. From (7.9.14) the first integral on the right is 1. And from (7.9.3) we have $\int_{-\infty}^{\infty}\frac{e^{-i\omega t}}{t}dt = i\pi\text{sgn}(-\omega)$. So, the transform is $\frac{1}{2} + \frac{i}{2\pi}[i\pi\text{sgn}(-\omega)] = \frac{1}{2} - \frac{1}{2}\text{sgn}(-\omega)$. Since sgn(x) = +1 if x > 0 and −1 if x < 0, then the transform is $\frac{1}{2} - \frac{1}{2} \times (-1) = \frac{1}{2} + \frac{1}{2} = 1$ if $\omega > 0$ and $\frac{1}{2} - \frac{1}{2}(+1) = \frac{1}{2} - \frac{1}{2} = 0$ if $\omega < 0$. But this is just u(ω), the step function in the ω-domain. So, $\frac{1}{2}\delta(t) + i\frac{1}{2\pi t} \leftrightarrow u(\omega)$.

(d) Following the hint, let $x = \frac{u}{t}$ and so xt = u and du = t dx. Thus, $E_i(t) =$

$$\int_t^{\infty}\frac{e^{-u}}{u}du = \int_1^{\infty}\frac{e^{-xt}}{xt}t\,dx = \begin{cases}\int_1^{\infty}\frac{e^{-xt}}{x}dx, t \geq 0 \\ 0, t < 0\end{cases}.$$ So, the Fourier transform is

$$\int_0^{\infty}\left\{\int_1^{\infty}\frac{e^{-xt}}{x}dx\right\}e^{-i\omega t}dt = \int_1^{\infty}\frac{1}{x}\left\{\int_0^{\infty}e^{-(x+i\omega)t}dt\right\}dx = \int_1^{\infty}\frac{1}{x}\left\{\frac{e^{-(x+i\omega)t}}{-(x+i\omega)}\right\}\Big|_0^{\infty}dx$$

$$= \int_1^{\infty}\frac{1}{x(x+i\omega)}dx = \frac{1}{i\omega}\int_1^{\infty}\left(\frac{1}{x} - \frac{1}{x+i\omega}\right)dx = \frac{1}{i\omega}[\ln(x) - \ln(x+i\omega)]\Big|_1^{\infty}$$

$$= \frac{1}{i\omega}\ln\left(\frac{x}{x+i\omega}\right)\Big|_1^{\infty} = -\frac{1}{i\omega}\ln\left(\frac{1}{1+i\omega}\right) = \frac{\ln(1+i\omega)}{i\omega}.$$

That is, $\int_t^\infty \frac{e^{-u}}{u} du \leftrightarrow \frac{\ln(1+i\omega)}{i\omega}$.

(C7.9) The second Hilbert transform integral in (7.10.11) is $\pi X(\omega) = -\int_{-\infty}^\infty \frac{R(u)}{\omega-u} du = -\int_{-\infty}^0 \frac{R(u)}{\omega-u} du - \int_0^\infty \frac{R(u)}{\omega-u} du$. If, in the first integral on the right of the equals sign, we make the change of variable $s = -u$, we have $\pi X(\omega) = -\int_\infty^0 \frac{R(-s)}{\omega+s}(-ds) - \int_0^\infty \frac{R(u)}{\omega-u} du =$ (when x(t) is real) $-\int_0^\infty \frac{R(s)}{\omega+s} ds - \int_0^\infty \frac{R(u)}{\omega-u} du = -\int_0^\infty R(u)\left[\frac{1}{\omega-u} + \frac{1}{\omega+u}\right] du = -\int_0^\infty R(u)\frac{2\omega}{\omega^2-u^2} du$ and so, *when x(t) is real,* $X(\omega) = -\frac{2\omega}{\pi}\int_0^\infty \frac{R(u)}{\omega^2-u^2} du$.

(C7.10) By Rayleigh's theorem the energy of x(t) is $\frac{1}{2\pi}\int_{-\infty}^\infty |X(\omega)|^2 d\omega$ and, since the energy is given as finite, we have (A) $\int_{-\infty}^\infty |X(\omega)|^2 d\omega < \infty$. Also, since $y(t) = x(t) * h(t)$ we have $Y(\omega) = X(\omega)H(\omega)$. (See note 14 again in Chap. 7.) Thus, $|Y(\omega)| = |X(\omega)H(\omega)| = |X(\omega)||H(\omega)|$ and so $|Y(\omega)|^2 = |X(\omega)|^2|H(\omega)|^2$. From this we conclude (B), $\frac{1}{2\pi}\int_{-\infty}^\infty |Y(\omega)|^2 d\omega = \frac{1}{2\pi}\int_{-\infty}^\infty |X(\omega)|^2|H(\omega)|^2 d\omega$. Finally, $H(\omega) = \int_{-\infty}^\infty h(t)e^{-i\omega t} dt$ and so $|H(\omega)| = |\int_{-\infty}^\infty h(t)e^{-i\omega t} dt| \leq \int_{-\infty}^\infty |h(t)e^{-i\omega t}| dt$. (See the hint in Challenge Problem C7.3.) Continuing, $|H(\omega)| \leq \int_{-\infty}^\infty |h(t)||e^{-i\omega t}| dt = \int_{-\infty}^\infty |h(t)| dt$ because $|e^{-i\omega t}| = 1$. This last integral is given to us as finite, and so $|H(\omega)| < \infty$, that is $|H(\omega)|^2 < \infty$. This means $|H(\omega)|^2$ has a maximum value, which we'll call M. Putting that into (B) we have $\frac{1}{2\pi}\int_{-\infty}^\infty |Y(\omega)|^2 d\omega \leq \frac{1}{2\pi}\int_{-\infty}^\infty |X(\omega)|^2 M \, d\omega = \frac{M}{2\pi}\int_{-\infty}^\infty |X(\omega)|^2 \, d\omega$. From (A) we know that this last integral is finite, and so $\frac{1}{2\pi}\int_{-\infty}^\infty |Y(\omega)|^2 d\omega < \infty$. But this integral *is* the energy of y(t), and so the energy of y(t) is finite.

(C7.11) From (7.10.12), the Hilbert transform of $x(t) = \cos(\omega_0 t)$, with $\omega_0 > 0$, is $\overline{x(t)} = \frac{1}{\pi}\int_{-\infty}^\infty \frac{\cos(\omega_0 u)}{t-u} du = \frac{1}{2\pi}\int_{-\infty}^\infty \frac{e^{i\omega_0 u}+e^{-i\omega_0 u}}{t-u} du$. Let $s = t - u$ (du $= -$ ds). Then $\overline{x(t)} = \frac{1}{2\pi}\int_\infty^{-\infty} \frac{e^{i\omega_0(t-s)}}{s}(-ds) + \frac{1}{2\pi}\int_\infty^{-\infty} \frac{e^{-i\omega_0(t-s)}}{s}(-ds) = \frac{e^{i\omega_0 t}}{2\pi}\int_{-\infty}^\infty \frac{e^{-i\omega_0 s}}{s} ds + \frac{e^{-i\omega_0 t}}{2\pi}\int_{-\infty}^\infty \frac{e^{i\omega_0 s}}{s} ds$. From (7.9.3) we have $\int_{-\infty}^\infty \frac{e^{i\omega t}}{\omega} d\omega = i\pi \text{sgn}(t)$, and so $\int_{-\infty}^\infty \frac{e^{i\omega_0 s}}{s} ds = i\pi \text{sgn}(\omega_0) = i\pi$ because $\omega_0 > 0$ and sgn(positive argument) $= +1$. Also, $\int_{-\infty}^\infty \frac{e^{-i\omega_0 s}}{s} ds = i\pi \text{sgn}(-\omega_0) = -i\pi$ because $\omega_0 > 0$ and sgn(negative argument) $= -1$. So, $\overline{x(t)} = \frac{1}{2\pi}[-i\pi e^{i\omega_0 t} + i\pi e^{-i\omega_0 t}] = \frac{-i\pi}{2\pi}[e^{i\omega_0 t} - e^{-i\omega_0 t}] = \frac{-i\pi}{2\pi}2i\sin(\omega_0 t) = \sin(\omega_0 t)$, the Hilbert transform of $\cos(\omega_0 t)$. If you re-do all this for the Hilbert transform of $\sin(\omega_0 t)$, you should find that it is $-\cos(\omega_0 t)$.

(C7.12) From (7.5.2), $\int_0^\infty \frac{\cos(x)}{\sqrt{x}} dx = \int_0^\infty \frac{\sin(x)}{\sqrt{x}} dx = \sqrt{\frac{\pi}{2}}$. In the first integral, change variable to $x = u^2$ (and so $dx = 2u\,du$). Thus, $\int_0^\infty \frac{\cos(u^2)}{u}2u\,du = 2\int_0^\infty \cos(u^2) du = \sqrt{\frac{\pi}{2}}$ or, $\int_0^\infty \cos(x^2) dx = \frac{1}{2}\sqrt{\frac{\pi}{2}}$. Similarly for $\int_0^\infty \sin(x^2) dx$.

(C7.13) $\pounds\{\int_0^t f(x)dx\} = \int_0^\infty \{\int_0^t f(x)dx\}e^{-st}dt$. Integrating by-parts, $\int u\,dv = uv - \int v\,du$ with $u = \int_0^t f(x)dx$ (and so $du = f(t)dt$), and $dv = e^{-st}$, and $v = -\frac{e^{-st}}{s}$. Thus, $\pounds\{\int_0^t f(x)dx\} = \{-\frac{e^{-st}}{s}\int_0^t f(x)dx\}|_0^\infty + \int_0^\infty \frac{e^{-st}}{s} f(t)dt = \frac{F(s)}{s}$, which is (7.11.13). The

Laplace transform changes integration in the time domain into division in the s-domain. Also, $F(s) = \int_0^\infty f(t)e^{-st}dt$, and so $\frac{dF(s)}{ds} = \int_0^\infty f(t)(-t)e^{-st}dt$ or, $-\frac{dF(s)}{ds} = \int_0^\infty t f(t)e^{-st}dt = \mathcal{L}\{tf(t)\}$ which is (7.11.14). And finally,

$$\int_s^\infty F(x)dx = \int_s^\infty \left\{\int_0^\infty f(t)e^{-xt}dt\right\}dx = \int_0^\infty f(t)\left\{\int_s^\infty e^{-xt}dx\right\}dt$$

$$= \int_0^\infty f(t)\left(-\frac{e^{-xt}}{t}\right)\Big|_s^\infty dt$$

$$= \int_0^\infty \frac{f(t)}{t}e^{-st}dt = \mathcal{L}\left\{\frac{f(t)}{t}\right\},$$

which establishes (7.11.15).

(C7.14) In $F(s) = \int_0^\infty \frac{e^{-k^2/4t}}{\sqrt{\pi t^3}}e^{-st}dt$, change variable to $\frac{k^2}{4t} = u^2$ and so, differentiating, $\frac{k^2}{4}\left(-\frac{1}{t^2}\right) = 2u\frac{du}{dt}$. Thus, $dt = -\frac{8ut^2}{k^2}du = -\frac{8t^2}{k^2}\left(\frac{k}{2t^2}\right)du = -\frac{4t^{3/2}}{k}du$ or $\frac{dt}{\sqrt{t^3}} = -\frac{4}{k}du$ and so $F(s) = -\frac{4}{k\sqrt{\pi}}\int_\infty^0 e^{-u^2}e^{-s\left(\frac{k^2}{4u^2}\right)}du = \frac{4}{k\sqrt{\pi}}\int_0^\infty e^{-\left(u^2 + s\frac{k^2}{4u^2}\right)}du$. Now, from (3.7.1), with $a = 1$ and $b = \frac{sk^2}{4}$, we have $\int_0^\infty e^{-\left(u^2 + s\frac{k^2}{4u^2}\right)}du = \frac{1}{2}\sqrt{\pi}e^{-2\frac{k}{2}\sqrt{s}} = \frac{1}{2}\sqrt{\pi}e^{-k\sqrt{s}}$ and so $F(s) = \left(\frac{4}{k\sqrt{\pi}}\right)\left(\frac{1}{2}\sqrt{\pi}e^{-k\sqrt{s}}\right) = \frac{2}{k}e^{-k\sqrt{s}}$.

(C7.15)

$$\mathcal{L}\left\{\int_0^\infty \frac{\cos(tx)}{x^2+b^2}dx\right\} = \int_0^\infty \left\{\int_0^\infty \frac{\cos(tx)}{x^2+b^2}dx\right\}e^{-st}dt$$

$$= \int_0^\infty \left(\frac{1}{x^2+b^2}\right)\left\{\int_0^\infty \cos(tx)e^{-st}dt\right\}dx$$

$$= \int_0^\infty \left(\frac{1}{x^2+b^2}\right)\left(\frac{s}{x^2+s^2}\right)dx$$

where I've used (7.11.4), the Laplace transform of $\cos(tx)$. Continuing,

$$\int_0^\infty \left(\frac{1}{x^2+b^2}\right)\left(\frac{s}{x^2+s^2}\right)dx = \frac{s}{s^2-b^2}\int_0^\infty \left(\frac{1}{x^2+b^2} - \frac{1}{x^2+s^2}\right)dx$$

$$= \frac{s}{s^2-b^2}\left\{\frac{1}{b}\tan^{-1}\left(\frac{x}{b}\right) - \frac{1}{s}\tan^{-1}\left(\frac{x}{s}\right)\right\}\Big|_0^\infty = \frac{s}{s^2-b^2}\left\{\frac{\pi/2}{b} - \frac{\pi/2}{s}\right\}$$

$$= \frac{\frac{\pi}{2}s}{s^2-b^2}\left(\frac{s-b}{bs}\right) = \frac{\frac{\pi}{2}}{b(s+b)} = \frac{\pi}{2b}\left(\frac{1}{s+b}\right).$$

This is the Laplace transform of the time function $\int_0^\infty \frac{\cos(tx)}{x^2+b^2}dx$. But, using (7.11.3), $\frac{\pi}{2b}\left(\frac{1}{s+b}\right)$ is the Laplace transform of the time function $\frac{\pi}{2b}e^{-bt}$, and so

$$\int_0^\infty \frac{\cos(tx)}{x^2+b^2}dx = \frac{\pi}{2b}e^{-bt}.$$

(C7.16) The Fourier transform of $R_f(t) = \int_{-\infty}^\infty f(\tau)f(\tau-t)d\tau$ is $\int_{-\infty}^\infty R_f(t)e^{-i\omega t}dt = \int_{-\infty}^\infty \{\int_{-\infty}^\infty f(\tau)f(\tau-t)d\tau\}e^{-i\omega t}dt = \int_{-\infty}^\infty f(\tau)\{\int_{-\infty}^\infty f(\tau-t)e^{-i\omega t}dt\}d\tau$. In the inner integral change variable to $s = \tau - t$ (and so $ds = -dt$). Then, the transform becomes $\int_{-\infty}^\infty f(\tau)\{\int_{\infty}^{-\infty} f(s)e^{-i\omega(\tau-s)}(-ds)\}d\tau = \int_{-\infty}^\infty f(\tau)e^{-i\omega\tau}\{\int_{-\infty}^\infty f(s)e^{i\omega s}ds\}d\tau$.

The inner integral is $F^*(\omega)$, and so the transform of $R_f(t)$ is $\int_{-\infty}^\infty f(\tau)e^{-i\omega\tau}F^*(\omega)d\tau = F^*(\omega)\int_{-\infty}^\infty f(\tau)e^{-i\omega\tau}d\tau = F^*(\omega)F(\omega) = |F(\omega)|^2$. The energy spectral density of $f(t)$ is $\frac{|F(\omega)|^2}{2\pi}$, which (as just shown) is the Fourier transform of $R_f(t)$ (to within a factor of 2π).

(C7.17) From (3.7.4) we have $\int_0^\infty e^{-ax^2-\frac{b}{x^2}}dx = \frac{1}{2}\sqrt{\frac{\pi}{a}}e^{-2\sqrt{ab}}$. Letting $a = i$ and $b = -i$ (and so $ab = 1$), we have $\int_0^\infty e^{-ix^2+\frac{i}{x^2}}dx = \int_0^\infty e^{-i\left(x^2-\frac{1}{x^2}\right)}dx = \frac{1}{2} \times \sqrt{\frac{\pi}{i}}e^{-2} = \frac{\sqrt{\pi}}{2e^2}\sqrt{-i}$. So, applying Euler's identity to the integrand, $\int_0^\infty \cos\left(x^2 - \frac{1}{x^2}\right)dx = \frac{\sqrt{\pi}}{2e^2}Re\{\sqrt{-i}\}$ and $-\int_0^\infty \sin\left(x^2 - \frac{1}{x^2}\right)dx = \frac{\sqrt{\pi}}{2e^2}Im\{\sqrt{-i}\}$.

Now, $-i = e^{-i\frac{\pi}{2}}$ so $\sqrt{-i} = \sqrt{e^{-i\frac{\pi}{2}}}$ where (because of Hint 2) we use the $+$ root and so $\sqrt{-i} = +e^{-i\frac{\pi}{4}} = \cos\left(\frac{\pi}{4}\right) - i\sin\left(\frac{\pi}{4}\right)$. Thus, $Re\{\sqrt{-i}\} = \cos\left(\frac{\pi}{4}\right) = \frac{1}{\sqrt 2}$ and $\{\sqrt{-i}\} = -\frac{1}{\sqrt 2}$. Therefore, $\int_0^\infty \cos\left(x^2 - \frac{1}{x^2}\right)dx = \int_0^\infty \sin\left(x^2 - \frac{1}{x^2}\right)dx = \frac{\sqrt\pi}{2e^2} \times \frac{1}{\sqrt 2} = \frac{1}{2e^2}\sqrt{\frac{\pi}{2}} = 0.084\ldots$, in agreement with the preliminary MATLAB calculation in Hint 2.

(C7.18) With $a = b = i$ in (3.7.4),

$$\int_0^\infty e^{-ix^2-\frac{i}{x^2}}dx = \int_0^\infty e^{-i\left(x^2+\frac{1}{x^2}\right)}dx = \frac{1}{2}\sqrt{\frac{\pi}{i}}e^{-2\sqrt{-1}}$$

$$= \frac{\sqrt\pi}{2}\frac{e^{-i2}}{\sqrt{e^{i\frac{\pi}{2}}}} = \frac{\sqrt\pi}{2}\frac{e^{-i2}}{e^{i\frac{\pi}{4}}} = \frac{\sqrt\pi}{2}e^{-i2}e^{-i\frac{\pi}{4}} = \frac{\sqrt\pi}{2}e^{-i\left(2+\frac{\pi}{4}\right)}$$

or,

$$\int_0^\infty \cos\left(x^2 + \frac{1}{x^2}\right)dx - i\int_0^\infty \sin\left(x^2 + \frac{1}{x^2}\right)dx = \frac{\sqrt\pi}{2}\left[\cos\left(2 + \frac{\pi}{4}\right) - i\sin\left(2 + \frac{\pi}{4}\right)\right].$$

Thus, $\int_0^\infty \cos\left(x^2 + \frac{1}{x^2}\right)dx = \frac{\sqrt\pi}{2}\cos\left(2 + \frac{\pi}{4}\right)$ and $\int_0^\infty \sin\left(x^2 + \frac{1}{x^2}\right)dx = \frac{\sqrt\pi}{2}\sin\left(2 + \frac{\pi}{4}\right)$.

(C7.19) Following the hint,

$$\int_0^\infty x^s \cos(xt)dx = \int_0^\infty x^s \frac{e^{ixt}+e^{-ixt}}{2}dx$$

$$= \frac{1}{2}\int_0^\infty x^s e^{ixt}\,dx + \frac{1}{2}\int_0^\infty x^s e^{-ixt}\,dx.$$

Now, from the solution in C4.11 we have $\int_0^\infty x^{s-1}e^{-px}dx = \frac{\Gamma(s)}{p^s}$ and so if $p = \pm it$
then

$$\int_0^\infty x^s \cos(xt)dx = \frac{1}{2}\frac{\Gamma(s+1)}{(it)^{s+1}} + \frac{1}{2}\frac{\Gamma(s+1)}{(-it)^{s+1}}$$

$$= \frac{1}{2}\frac{\Gamma(s+1)}{t^{s+1}}\left[\frac{1}{\left(e^{i\frac{\pi}{2}}\right)^{s+1}} + \frac{1}{\left(e^{-i\frac{\pi}{2}}\right)^{s+1}}\right] = \frac{\Gamma(s+1)}{t^{s+1}}\left[\frac{e^{-i(s+1)\frac{\pi}{2}}+e^{i(s+1)\frac{\pi}{2}}}{2}\right]$$

$$= \frac{\Gamma(s+1)}{t^{s+1}}\left[\frac{e^{-i\frac{\pi s}{2}}e^{-i\frac{\pi}{2}} + e^{i\frac{\pi s}{2}}e^{i\frac{\pi}{2}}}{2}\right] = \frac{\Gamma(s+1)}{t^{s+1}}\left[\frac{-ie^{-i\frac{\pi s}{2}} + ie^{i\frac{\pi s}{2}}}{2}\right]$$

$$= \frac{\Gamma(s+1)}{t^{s+1}}\left[\frac{-i\left\{\cos\left(\frac{\pi s}{2}\right) - i\sin\left(\frac{\pi s}{2}\right)\right\} + i\left\{\cos\left(\frac{\pi s}{2}\right) + i\sin\left(\frac{\pi s}{2}\right)\right\}}{2}\right]$$

$$= \frac{\Gamma(s+1)}{t^{s+1}}\left[\frac{-2\sin\left(\frac{\pi s}{2}\right)}{2}\right] = -\frac{\Gamma(s+1)}{t^{s+1}}\sin\left(\frac{\pi s}{2}\right).$$

Chapter 8

(C8.1) Since $f(z) = g(z)(z - z_0)^m$ then $\oint_C \frac{f'(z)}{f(z)}dz = \oint_C \frac{g(z)m(z-z_0)^{m-1}+g'(z)(z-z_0)^m}{g(z)(z-z_0)^m}$
$dz = \oint_C \frac{m}{z-z_0}dz + \oint_C \frac{g'(z)}{g(z)}dz = m\oint_C \frac{dz}{z-z_0} + \oint_C \frac{g'(z)}{g(z)}dz$. Now, the first integral is $2\pi i$
by (8.7.1), and the second integral is zero by (8.6.1) because $g(z)$ is analytic which
means $g'(z)$ is analytic and so $\frac{g'(z)}{g(z)}$ is analytic *with no zeros inside* C (by the given
statement of the problem). Thus, $\oint_C \frac{f'(z)}{f(z)}dz = 2\pi i\, m$ and we are done.

(C8.2) Since we are going to work with $f(z) = \frac{e^{imz}}{z(z^2+a^2)}$ we see that we have three
first-order singularities to consider: $z = 0$, $z = -ia$, and $z = ia$. This suggests the
contour C shown in Fig. C8, where eventually we'll let $\varepsilon \to 0$ and $R \to \infty$. The first
two singularities will always be outside of C, while the third one will, as $R \to \infty$, be
inside C (remember, $a > 0$).

Fig. C8 The contour for
Challenge Problem **C8.2**

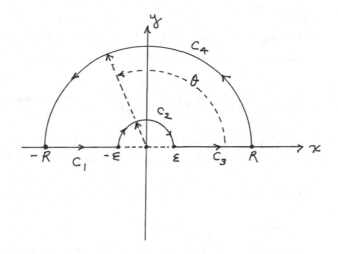

Now, $\oint_C f(z)\,dz = \int_{C_1} + \int_{C_2} + \int_{C_3} + \int_{C_4}$. On C_1 and C_3 we have $z = x$ and $dz = dx$. Thus, $\int_{C_1} + \int_{C_3} = \int_{-R}^{-\varepsilon} \frac{e^{imx}}{x(x^2+a^2)}\,dx + \int_{\varepsilon}^{R} \frac{e^{imx}}{x(x^2+a^2)}\,dx$ or, as $\varepsilon \to 0$ and $R \to \infty$, $\int_{C_1} + \int_{C_3} = \int_{-\infty}^{\infty} \frac{e^{imx}}{x(x^2+a^2)}\,dx$. On C_2 we have $z = \varepsilon e^{i\theta}$, $-\pi < \theta < 0$, $dz = i\varepsilon e^{i\theta}\,d\theta$. Thus, $\int_{C_2} = \int_{\pi}^{0} \frac{e^{im\varepsilon e^{i\theta}}}{\varepsilon e^{i\theta}(\varepsilon^2 e^{i2\theta}+a^2)} i\varepsilon e^{i\theta}\,d\theta$ or, as $\varepsilon \to 0$, $\int_{C_2} = -i\int_{0}^{\pi} \frac{d\theta}{a^2} = -i\frac{\pi}{a^2}$. On C_4 we have $z = Re^{i\theta}$, $0 < \theta < \pi$, and $dz = iRe^{i\theta}\,d\theta$. Thus, $\int_{C_4} = \int_{0}^{\pi} \frac{e^{imRe^{i\theta}}}{Re^{i\theta}(R^2 e^{i2\theta}+a^2)} iRe^{i\theta}\,d\theta = i\int_{0}^{\pi} \frac{e^{imRe^{i\theta}}}{R^2 e^{i2\theta}+a^2}\,d\theta$ The absolute value of the integrand behaves, as $R \to \infty$, like $\frac{1}{R^2}$ because the absolute value of the numerator is 1 for all R and θ. Thus, as $R \to \infty$ the integral behaves as $\frac{\pi}{R^2}$ which goes to zero as $R \to \infty$. Thus, $\oint_C = \int_{-\infty}^{\infty} \frac{e^{imx}}{x(x^2+a^2)}\,dx - i\frac{\pi}{a^2}$. But we know that $\oint_C = 2\pi i$ (residue of singularity at $z = ia$). That residue is $\lim_{z \to ia} (z - ia) \frac{e^{imz}}{z(z-ia)(z+ia)} = \frac{e^{imia}}{ia(2ia)} = -\frac{e^{-ma}}{2a^2}$. So, $\int_{-\infty}^{\infty} \frac{e^{imx}}{x(x^2+a^2)}\,dx - i\frac{\pi}{a^2} = 2\pi i\left(-\frac{e^{-ma}}{2a^2}\right) = -i\frac{\pi e^{-ma}}{a^2}$. And thus, $\int_{-\infty}^{\infty} \frac{e^{imx}}{x(x^2+a^2)}\,dx = i\frac{\pi}{a^2} - i\frac{\pi e^{-ma}}{a^2} = i\frac{\pi}{a^2}(1 - e^{-ma})$. Using Euler's identity on the numerator of the integrand, $\int_{-\infty}^{\infty} \frac{\cos(mx)}{x(x^2+a^2)}\,dx + i\int_{-\infty}^{\infty} \frac{\sin(mx)}{x(x^2+a^2)}\,dx = i\frac{\pi}{a^2}(1 - e^{-ma})$. Equating real parts of this last equation we get $\int_{-\infty}^{\infty} \frac{\cos(mx)}{x(x^2+a^2)}\,dx = 0$, which is no surprise since the integrand is odd. More interesting is the result from equating imaginary parts: $\int_{-\infty}^{\infty} \frac{\sin(mx)}{x(x^2+a^2)}\,dx = \frac{\pi}{a^2}(1 - e^{-ma})$ or, since $\int_{0}^{\infty} = \frac{1}{2}\int_{-\infty}^{\infty}$, we have

$$\int_{0}^{\infty} \frac{\sin(mx)}{x(x^2+a^2)}\,dx = \frac{\pi}{2}\left(\frac{1 - e^{-ma}}{a^2}\right),$$

which earlier we simply assumed.

(C8.3)

(a) Writing $z = e^{i\theta}$ (and so $d\theta = \frac{dz}{iz}$) on the unit circle C, we have $\cos(\theta) = \frac{z+z^{-1}}{2}$. Now, consider $\oint_C f(z)dz$ with $f(z) = \frac{1}{1-2a\frac{z+z^{-1}}{2}+a^2}$. Then $\int_0^{2\pi} \frac{d\theta}{1-2a\cos(\theta)+a^2} =$

$\oint_C \frac{dz}{1-2a\frac{z+z^{-1}}{2}+a^2}\left(\frac{dz}{iz}\right) = \frac{i}{a}\oint_C \frac{dz}{z^2-\frac{a^2+1}{a}z+1}$. There are two first-order singularities at $z =$

$\frac{\frac{a^2+1}{a}\pm\sqrt{\left(\frac{a^2+1}{a}\right)^2-4}}{2}$ which, after a bit of algebra, reduces to $z = a$ and $z = \frac{1}{a}$. Since $0 < a < 1$, the first is inside C and the second is outside C. So, $\oint_C \frac{dz}{z^2-\frac{a^2+1}{a}z+1} = 2\pi i$

(residue at $z = a$). From (8.8.9) that residue is $\frac{1}{\frac{d}{dz}\left(z^2-\frac{a^2+1}{a}z+1\right)\Big|_{z=}}$ $a =$

$\frac{1}{\left(2z-\frac{a^2+1}{a}\right)\Big|_{z=a}} = \frac{1}{\left(2a-\frac{a^2+1}{a}\right)} = \frac{a}{a^2-1}$. So, $\int_0^{2\pi} \frac{d\theta}{1-2a\cos(\theta)+a^2} = \frac{i}{a}2\pi i\frac{a}{a^2-1} = \frac{2\pi}{1-a^2}, 0 <$

$a < 1$. If $a = \frac{1}{2}$, for example, we have $\int_0^{2\pi} \frac{d\theta}{\frac{5}{4}-\cos(\theta)} = \frac{2\pi}{1-\frac{1}{4}} = \frac{8\pi}{3} = 8.37758\ldots$

and MATLAB agrees because *integral(@(x)1./(1.25-cos(x)),0,2*pi)* = 8.37758....

(b) Let $f(z) = \frac{e^{iz}}{(z+a)^2+b^2}$, that is consider the contour integral $\oint_C \frac{e^{iz}}{(z+a)^2+b^2}dz$ where C is shown in Fig. 8.7.1. The integrand has two singularities (are each first-order), at $z = -a \pm ib$. Since a and b are both positive, only the $z = -a + ib$ singularity is inside C. Thus, $\oint_C \frac{e^{iz}}{(z+a)^2+b^2}dz = 2\pi i$ (residue at $z = -a + ib$). On the real axis $z = x$, $dz = dx$, and on the semi-circular arc $z = Te^{i\theta}$, $dz = iTe^{i\theta}d\theta$, $0 < \theta < \pi$. So, on the semi-circular arc we have the integral $\int_0^\pi \frac{e^{iTe^{i\theta}}}{(Te^{i\theta}+a)^2+b^2}iTe^{i\theta}d\theta$. The absolute value of the numerator is T for all θ and, as $T \to \infty$, the absolute value of the denominator behaves like $\frac{1}{T^2}$. So, the integral behaves as $\pi\frac{T}{T^2} = \frac{\pi}{T} \to 0$ as $T \to \infty$. Thus, $\oint_C \frac{e^{iz}}{(z+a)^2+b^2}dz = \lim_{T\to\infty}\int_{-T}^T \frac{e^{ix}}{(x+a)^2+b^2}dx = 2\pi i$ (residue at $z = -a + ib) = \int_{-\infty}^\infty \frac{e^{ix}}{(x+a)^2+b^2}dx$. Now, the residue at $z = -a + ib$ is given by

$\lim_{z\to-a+ib}[z-(-a+ib)]\frac{e^{iz}}{[z-(-a+ib)][z-(-a-ib)]} = \frac{e^{i(-a+ib)}}{-a+ib+a+ib} = \frac{e^{-ia}e^{-b}}{i2b} = \frac{e^{-b}\cos(a)-ie^{-b}\sin(a)}{i2b}$

and so $\int_{-\infty}^\infty \frac{e^{ix}}{(x+a)^2+b^2}dx = 2\pi i\left[\frac{e^{-b}\cos(a)-ie^{-b}\sin(a)}{i2b}\right]$. Expanding the numerator of the integral with Euler's identity and equating real parts, we get $\int_{-\infty}^\infty \frac{\cos(x)}{(x+a)^2+b^2}dx = \frac{\pi}{b}e^{-b}\cos(a)$, while equating imaginary parts gives $\int_{-\infty}^\infty \frac{\sin(x)}{(x+a)^2+b^2}dx = -\frac{\pi}{b}e^{-b}\sin(a)$. If $a = 1$ and $b = 1$, for example these two integrals are equal to 0.6244... and $-0.97251...$, respectively, and MATLAB agrees since *integral(@(x)cos(x)./((x+1).^2+1),-inf,inf)* = 0.6244... and *integral(@(x)sin(x)./((x+1).^2+1),-inf,inf)* = $-0.9725...$.

(c) Let $f(z) = \frac{e^{iz}}{(z^2+a^2)(z^2+b^2)}$, that is consider the contour integral $\oint_C \frac{e^{iz}}{(z^2+a^2)(z^2+b^2)}dz$ where C is shown in Fig. 8.7.1. The integrand has four first-order singularities at

$z = \pm ia$ and at $z = \pm ib$, but since a and b are positive only $z = +ia$ and $z = +ib$ are inside C. So, $\oint_C \frac{e^{iz}}{(z^2+a^2)(z^2+b^2)} dz = 2\pi i$ (residue at $z = +ia$ plus residue at $z = +ib$). Those residues are:

at $z = +ia$, $\lim_{z \to ia}(z - ia) \frac{e^{iz}}{(z - ia)(z + ia)(z - ib)(z + ib)} = \frac{e^{i(ia)}}{i2a(ia - ib)(ia + ib)}$

$$= \frac{e^{-a}}{-i2a(a - b)(a + b)} = i\frac{e^{-a}}{2a(a^2 - b^2)},$$

and at

$z = +ib$, $\lim_{z \to ib}(z - ib) \frac{e^{iz}}{(z - ia)(z + ia)(z - ib)(z + ib)} = \frac{e^{i(ib)}}{(ib - ia)(ib + ia)i2b}$

$$= \frac{e^{-b}}{-i2b(b - a)(b + a)} - i\frac{e^{-b}}{2b(b^2 - a^2)}.$$

On the semi-circular arc $z = Te^{i\theta}$, $dz = iTe^{i\theta}d\theta$, $0 < \theta < \pi$, and so that integral is $\int_0^\pi \frac{e^{iTe^{i\theta}}}{(T^2 e^{i2\theta}+a^2)(T^2 e^{i2\theta}+b^2)} iTe^{i\theta}d\theta$ which, as $T \to \infty$, behaves like $\pi\frac{T}{T^4} \to 0$. So, all that we have left is the portion of C that lies along the real axis, which says that

$$\int_{-\infty}^\infty \frac{e^{ix}}{(x^2 + a^2)(x^2 + b^2)} dx = 2\pi i\left[i\frac{e^{-a}}{2a(a^2 - b^2)} + i\frac{e^{-b}}{2b(b^2 - a^2)}\right]$$

$$- 2\pi\left[\frac{e^{-a}}{2a(a^2 - b^2)} + \frac{e^{-b}}{2b(b^2 - a^2)}\right]$$

$$= \pi\left[\frac{e^{-b}}{b(a^2 - b^2)} - \frac{e^{-a}}{a(a^2 - b^2)}\right] = \int_{-\infty}^\infty \frac{\cos(x)}{(x^2 + a^2)(x^2 + b^2)} dx$$

$$+ i\int_{-\infty}^\infty \frac{\sin(x)}{(x^2 + a^2)(x^2 + b^2)} dx.$$

Equating real parts gives us our answer: $\int_{-\infty}^\infty \frac{\cos(x)}{(x^2+a^2)(x^2+b^2)} dx = \frac{\pi}{(a^2-b^2)} \times \left[\frac{e^{-b}}{b} - \frac{e^{-a}}{a}\right]$. If $a = 2$ and $b = 1$ this reduces to $\frac{\pi(2e-1)}{6e^2} = 0.31438\ldots$, and MATLAB agrees:

integral(@(x)cos(x)./((x.^2+1).(x.^2+4)),-inf,inf) = 0.31438....*

(d) Let $f(z) = \frac{e^{iaz}}{(z^2+b^2)^2}$, that is consider the contour integral $\oint_C \frac{e^{iaz}}{(z^2+b^2)^2} dz$ where C is shown in Fig. 8.7.1. The integrand has two *second*-order singularities at $z = \pm ib$

and, since b is positive, only $z = +ib$ is inside C. So, $\oint_C \frac{e^{iaz}}{(z^2+b^2)^2} dz = 2\pi i$

(residue at $z = ib$). We calculate that residue as follows, from (8.8.8), using $m = 2$:

$$\lim_{z \to ib} \frac{d}{dz} \left\{ (z - ib)^2 \frac{e^{iaz}}{(z^2+b^2)^2} \right\}$$

$$= \frac{d}{dz} \left\{ \frac{e^{iaz}}{(z+ib)^2} \right\} \bigg|_{z=ib} = \frac{(z+ib)^2 iae^{iaz} - e^{iaz} 2(z+ib)}{(z+ib)^4} \bigg|_{z=ib}$$

which reduces (after just a bit of algebra) to $-i\frac{1+ab}{4b^3} e^{-ab}$. Thus, $\oint_C \frac{e^{iaz}}{(z^2+b^2)^2} dz =$

$2\pi i \left(-i\frac{1+ab}{4b^3} e^{-ab} \right) = \frac{\pi}{2b^3}(1 + ab)e^{-ab}$. On the semi-circular arc $z = Te^{i\theta}$, $dz = iTe^{i\theta}d\theta$,

$0 < \theta < \pi$, and so the integrand is $\frac{e^{iaTe^{i\theta}}}{(T^2 e^{i2\theta}+b^2)^2} Te^{i\theta}$, with an absolute value that behaves

as $\frac{T}{T^4} = \frac{1}{T^3}$ as $T \to \infty$. The integral therefore behaves as $\frac{\pi}{T^3}$ which vanishes as $T \to \infty$.

Thus, since on the real axis $z = x$ we have in the limit of $T \to \infty$ that

$\int_{-\infty}^{\infty} \frac{e^{iax}}{(x^2+b^2)^2} dx = \frac{\pi}{2b^3}(1 + ab)e^{-ab}$. Equating real parts (after using Euler's identity

in the numerator of the integrand) gives us the result: $\int_{-\infty}^{\infty} \frac{\cos(ax)}{(x^2+b^2)^2} dx = \frac{\pi}{2b^3} \times$

$(1 + ab)e^{-ab}$ or, since $\int_0^\infty = \frac{1}{2}\int_{-\infty}^\infty$ because the integrand is even, we have our result:

$\int_0^\infty \frac{\cos(ax)}{(x^2+b^2)^2} dx = \frac{\pi}{4b^3}(1 + ab)e^{-ab}$. If $a = b = 1$ this equals $\frac{\pi}{2e} = 0.57786\ldots$ and

MATLAB agrees: *integral(@(x)cos(x)./((x.^2+1).^2),0,inf) = 0.57786....*

(C8.4) In the integral $\int_0^\infty \frac{x^k}{(x^2+1)^2} dx$, the integrand, for large x, behaves like $\frac{x^k}{x^4} = \frac{1}{x^{4-k}}$.

For the integral not to blow-up as $x \to \infty$ we must have the exponent $4 - k > 1$ or

$3 > k$ or, $k < 3$. For small x the integrand behaves like $x^k = \frac{1}{x^{-k}}$. For the integral not to

blow-up as $x \to 0$ we must have the exponent $-k < 1$ or $k > -1$ or, $-1 < k$. Thus,

$-1 < k < 3$. Now, following the hint, let's consider the integral $\oint_C f(z)dz$ where C is

the contour in Fig. 8.9.1 and $f(z) = \frac{e^{k \ln(z)}}{(z^2+1)^2}$. This integrand has two second-order

singularities, at $z = \pm i$, and both will be inside C as we let $R \to \infty$ and both ε and ρ

$\to 0$. On C_2, $z = Re^{i\theta}$ and $dz = iRe^{i\theta}d\theta$ and we see that the absolute value of the

integrand behaves as $\frac{1}{R^3}$ as $R \to \infty$. That is, the integral on C_2 will vanish as $R \to \infty$.

On C_4, $z = \rho e^{i\theta}$ and $dz = i\rho e^{i\theta}d\theta$ and we see that the absolute value of the integrand

behaves as ρ as $\rho \to 0$. That is, the integral on C_4 will vanish as $\rho \to 0$. So, all we have

left to calculate is $\int_{C_1} + \int_{C_3}$. On C_1, $z = re^{i\varepsilon}$ and $dz = e^{i\varepsilon}dr$, and on C_3, $z = re^{i(2\pi - \varepsilon)}$

and $dz = e^{i(2\pi - \varepsilon)}dr$. So, $\int_{C_1} + \int_{C_3} = \int_\rho^R \frac{e^{k \ln(re^{i\varepsilon})}}{(r^2 e^{i2\varepsilon}+1)^2} e^{i\varepsilon}dr + \int_R^\rho \frac{e^{k \ln\{re^{i(2\pi-\varepsilon)}\}}}{(r^2 e^{i2(2\pi-\varepsilon)}+1)^2} e^{i(2\pi-\varepsilon)}dr$ or,

as we let $\varepsilon \to 0$, $\int_{C_1} + \int_{C_3} = \int_\rho^R \frac{e^{k \ln(r)}}{(r^2+1)^2} dr + \int_R^\rho \frac{e^{k\{\ln(r)+i2\pi\}}}{(r^2+1)^2} dr$. If we now let $\rho \to 0$ and

$R \to \infty$, we have

$$\int_{C_1} + \int_{C_3} = \int_0^\infty \frac{e^{\ln(r^k)}}{(r^2+1)^2} dr - \int_0^\infty \frac{e^{\ln(r^k)+i2\pi k}}{(r^2+1)^2} dr = \int_0^\infty \frac{r^k - r^k e^{i2\pi k}}{(r^2+1)^2} dr$$

$$= \int_0^\infty \frac{r^k[1 - e^{i2\pi k}]}{(r^2+1)^2} dr = \int_0^\infty \frac{r^k e^{i\pi k}[e^{-i\pi k} - e^{i\pi k}]}{(r^2+1)^2} dr$$

$$= \int_0^\infty \frac{r^k e^{i\pi k}[-2i \sin(\pi k)]}{(r^2+1)^2} dr.$$

Now, $\oint_C f(z)dz = \int_0^\infty \frac{r^k e^{i\pi k}[-2i \sin(\pi k)]}{(r^2+1)^2} dr = 2\pi i$(residue at $z = -i$ plus residue at $z = +i$). Since $f(z) = \frac{z^k}{(z+i)^2(z-i)^2}$, and since for a second-order singularity at $z = z_0$ we have from (8.8.8) that the residue is $\lim_{z \to z_0} \frac{d}{dz}\{(z - z_0)^2 f(z)\}$, then for $z_0 = -i$, the residue is R_1 where $R_1 = \lim_{z \to -i} \frac{d}{dz}\{(z + i)^2 \frac{z^k}{(z+i)^2} (z - i)^2\} = \lim_{z \to -i} \frac{d}{dz} \times$

$\left\{\frac{z^k}{(z-i)^2}\right\} = \lim_{z \to -i} \frac{(z-i)^2 k z^{k-1} - z^k 2(z-i)}{(z-i)^4} = \frac{(-2i)^2 k(-i)^{k-1} - (-i)^k 2(-2i)}{(-2i)^4} = \frac{-4k(-i)^{k-1} + (-i)^k 4i}{16}$ or,

$R_1 = \frac{-k(-i)^{k-1} + i(-i)^k}{4}$. If you repeat this for the other residue, R_2, you'll find that

$R_2 = \lim_{z \to i} \frac{d}{dz}\left\{(z - i)^2 \frac{z^k}{(z+i)^2(z-i)^2}\right\} = \lim_{z \to i} \frac{d}{dz}\left\{\frac{z^k}{(z+i)^2}\right\} = \frac{-k(i)^{k-1} - i(i)^k}{4}$. So,

$$R_1 + R_2 = -\frac{k}{4}\left[(-i)^{k-1} + (i)^{k-1}\right] + \frac{i}{4}\left[(-i)^k - (i)^k\right]$$

$$= -\frac{k}{4}\left[\frac{(-i)^k}{-i} + \frac{(i)^k}{i}\right] + \frac{i}{4}\left[(-i)^k - (i)^k\right]$$

$$= \frac{k}{4i}\left[(-i)^k - (i)^k\right] + \frac{i}{4}\left[(-i)^k - (i)^k\right] = \left[-i\frac{k}{4} + \frac{i}{4}\right]\left[(-i)^k - (i)^k\right]$$

$$= \frac{i}{4}(1 - k)\left[(-i)^k - (i)^k\right].$$

Thus, $2\pi i(R_1 + R_2) = -\frac{\pi}{2}(1 - k)\left[(-i)^k - (i)^k\right]$ and so $\int_0^\infty \frac{r^k e^{i\pi k}[-2i \sin(\pi k)]}{(r^2+1)^2} dr = -\frac{\pi}{2}(1 - k)\left[(-i)^k - (i)^k\right]$ or, $\int_0^\infty \frac{r^k}{(r^2+1)^2} dr = \frac{\pi(1-k)[(-i)^k - (i)^k]}{4i \sin(\pi k) e^{i\pi k}}$. Since $-i = e^{i\frac{3\pi}{2}}$ and $= e^{i\frac{\pi}{2}}$, we have

$$(-i)^k - (i)^k = e^{ik\frac{3\pi}{2}} - e^{ik\frac{\pi}{2}} = e^{i\pi k}\left\{e^{i(k\frac{3\pi}{2} - \pi k)} - e^{i(k\frac{\pi}{2} - \pi k)}\right\} = e^{i\pi k}\left\{e^{i(\frac{k\pi}{2})} - e^{-i(\frac{k\pi}{2})}\right\}$$

$$= e^{i\pi k} 2i \sin\left(\frac{k\pi}{2}\right).$$

Thus, $\int_0^\infty \frac{r^k}{(r^2+1)^2}\,dr = \frac{\pi(1-k)e^{i\pi k}2i\,\sin\left(\frac{k\pi}{2}\right)}{4i\,\sin(\pi k)e^{i\pi k}} = \frac{\pi(1-k)}{2}\left\{\frac{\sin\left(\frac{k\pi}{2}\right)}{\sin(\pi k)}\right\}$. Or, since $\sin(\pi k) =$
$2\sin\left(\frac{k\pi}{2}\right)\cos\left(\frac{k\pi}{2}\right)$, and changing the dummy variable of integration from r to x, we
arrive at $\int_0^\infty \frac{x^k}{(x^2+1)^2}\,dx = \frac{\pi(1-k)}{4\cos\left(\frac{k\pi}{2}\right)}$, $-1 < k < 3$. MATLAB agrees, as if $k = \frac{1}{2}$ then
$\int_0^\infty \frac{\sqrt{x}}{(x^2+1)^2}\,dx = \frac{\pi}{8\cos\left(\frac{\pi}{4}\right)} = \frac{\pi}{8\frac{1}{\sqrt{2}}} = \frac{\pi\sqrt{2}}{8} = 0.55536\ldots$, and *integral(@(x)sqrt(x)./(x.*
^2+1).^2,0,inf) = 0.55536...*, while if $k = \frac{1}{3}$ then $\int_0^\infty \frac{x^{1/3}}{(x^2+1)^2}\,dx = \frac{\frac{\pi 2}{3}}{4\cos\left(\frac{\pi}{6}\right)} = \frac{\pi}{6\frac{\sqrt{3}}{2}} =$
$\frac{\pi}{3\sqrt{3}} = 0.60459\ldots$, and *integral(@(x)(x.^(1/3))./(x.^2+1).^2,0,inf) = 0.60459....*

(C8.5) Consider $\oint_C f(z)dz$, where $f(z) = \frac{e^{imz}}{az^2+bz+c}$ where $m \geq 0$, $b^2 > 4ac$, and C is the
contour in Fig. 8.6.2. As in the derivation of (8.6.5), there are two singularities on the
real axis, at x_1 and x_2, as shown in Fig. 8.6.2. The values of x_1 and x_2 are as given in
the text. The analysis here goes through just as in the text, taking into account the
change in f(z). That is, the three integrals along C_1, C_3, and C_5 will combine (as we
let $\varepsilon \to 0$ and $T \to \infty$) to give us the integral we are after, and its value will be
$-\left\{\int_{C_2} f(z)dz + \int_{C_4} f(z)dz + \int_{C_6} f(z)dz\right\}$. So, let's calculate each of these three line
integrals. For C_2,

$$\int_{C_2} f(z)dz = \int_\pi^0 \frac{e^{im\left(x_2+\varepsilon e^{i\theta}\right)}i\varepsilon e^{i\theta}}{a\left(x_2+\varepsilon e^{i\theta}\right)^2 + b\left(x_2+\varepsilon e^{i\theta}\right) + c}\,d\theta$$

$$= \int_\pi^0 \frac{e^{imx_2}e^{im\varepsilon e^{i\theta}}i\varepsilon e^{i\theta}}{a\left(x_2^2+2x_2\varepsilon e^{i\theta}+\varepsilon^2 e^{i2\theta}\right) + b\left(x_2+\varepsilon e^{i\theta}\right) + c}\,d\theta$$

$$= \int_\pi^0 \frac{e^{imx_2}e^{im\varepsilon e^{i\theta}}i\varepsilon e^{i\theta}}{\left(ax_2^2+bx_2+c\right) + \left(2ax_2\varepsilon e^{i\theta}+a\varepsilon^2 e^{i2\theta}+b\varepsilon e^{i\theta}\right)}\,d\theta.$$

Since $ax_2^2 + bx_2 + c = 0$ because x_2 is a zero of the denominator, and as $\varepsilon^2 \to 0$
faster than $\varepsilon \to 0$, then for 'small' ε we have $\lim_{\varepsilon\to 0}\int_{C_2} f(z)dz =$
$\lim_{\varepsilon\to 0}\int_\pi^0 \frac{e^{imx_2}i\varepsilon e^{i\theta}}{2ax_2\varepsilon e^{i\theta}+b\varepsilon e^{i\theta}}\,d\theta = i\int_\pi^0 \frac{e^{imx_2}}{2ax_2+b}\,d\theta = -\pi i\frac{e^{imx_2}}{2ax_2+b}$. In the same way,
$\lim_{\varepsilon\to 0}\int_{C_4} f(z)dz = -\pi i\frac{e^{imx_1}}{2ax_1+b}$. Also, as before, $\lim_{\varepsilon\to 0}\int_{C_6} f(z)dz = 0$. Thus,
$\int_{-\infty}^\infty \frac{\cos(mx)+i\sin(mx)}{ax^2+bx+c}\,dx = \pi i\left[\frac{e^{imx_1}}{2ax_1+b} + \frac{e^{imx_2}}{2ax_2+b}\right]$. Now, $2ax_1+b = \sqrt{b^2-4ac}$ and
$2ax_2+b = -\sqrt{b^2-4ac}$ and so $\int_{-\infty}^\infty \frac{\cos(mx)+i\sin(mx)}{ax^2+bx+c}\,dx = \pi i\left[\frac{e^{imx_1}}{\sqrt{b^2-4ac}} - \frac{e^{imx_2}}{\sqrt{b^2-4ac}}\right] =$
$i\frac{\pi}{\sqrt{b^2-4ac}}\left(e^{imx_1}-e^{imx_2}\right) = i\frac{\pi}{\sqrt{b^2-4ac}}[\cos(mx_1)+i\sin(mx_1)-\cos(mx_2)-i\sin(mx_2)]$
or, equating real parts on each side of the equality, $\int_{-\infty}^\infty \frac{\cos(mx)}{ax^2+bx+c}\,dx = \frac{\pi}{\sqrt{b^2-4ac}}\times$
$[\sin(mx_2)-\sin(mx_1)]$. Using the trigonometric identity $\sin(A)-\sin(B) =$
$2\cos\left(\frac{A+B}{2}\right)\sin\left(\frac{A-B}{2}\right)$, we have

$$\int_{-\infty}^{\infty} \frac{\cos{(mx)}}{ax^2 + bx + c}\,dx = \frac{\pi}{\sqrt{b^2 - 4ac}}\left[2\cos\left(\frac{mx_2 + mx_1}{2}\right)\sin\left(\frac{mx_2 - mx_1}{2}\right)\right]$$

$$= \frac{2\pi}{\sqrt{b^2 - 4ac}}\left[\cos\left(-\frac{mb}{2a}\right)\sin\left(\frac{m}{2}\left\{-\frac{1}{a}\sqrt{b^2 - 4ac}\right\}\right)\right]$$

$$= -\frac{2\pi}{\sqrt{b^2 - 4ac}}\left[\cos\left(\frac{mb}{2a}\right)\sin\left(\frac{m\sqrt{b^2 - 4ac}}{2a}\right)\right].$$

(C8.6) Following the hint and using Fig. 8.9.1, there are two first-order singularities inside C (as we let $\varepsilon \to 0$, $\rho \to 0$, and $R \to \infty$), one at $z = -1 = e^{i\pi}$ and one at $z = -2 = 2e^{i\pi}$. The residue of a first-order singularity at $z = z_0$ in the integrand function $f(z) = \frac{g(z)}{h(z)}$ is $\frac{g(z_0)}{h'(z_0)}$. For $f(z) = \frac{z^p}{(z+1)(z+2)}$ we have $g(z) = z^p$ and $h(z) = (z+1)(z+2)$ and so $h'(z) = (z+1) + (z+2) = 2z + 3$. Thus, the residue at -1 is $\frac{(e^{i\pi})^p}{2(-1)+3} = \frac{e^{ip\pi}}{-2+3} = e^{ip\pi}$, and the residue at -2 is $\frac{(2e^{i\pi})^p}{2(-2)+3} = \frac{2^p e^{ip\pi}}{-4+3} = -2^p e^{ip\pi}$. Thus, $2\pi i$ times the sum of the residues is $2\pi i(e^{ip\pi} - 2^p e^{ip\pi}) = 2\pi i e^{ip\pi}(1 - 2^p)$. So, $\int_{C_1} + \int_{C_2} + \int_{C_3} + \int_{C_4} = 2\pi i e^{ip\pi}(1 - 2^p)$. Now, $\int_{C_1} = \int_{\rho}^{R} \frac{(re^{i\varepsilon})^p}{(re^{i\varepsilon}+1)(re^{i\varepsilon}+2)}e^{i\varepsilon}\,dr$ and $\int_{C_3} = \int_{R}^{\rho} \frac{(re^{i(2\pi-\varepsilon)})^p}{(re^{i(2\pi-\varepsilon)}+1)(re^{i(2\pi-\varepsilon)}+2)}e^{i(2\pi-\varepsilon)}\,dr$ and so, as $\varepsilon \to 0$, $\rho \to 0$, and $R \to \infty$, $\int_{C_1} + \int_{C_3} = \int_{0}^{\infty} \frac{r^p}{(r+1)(r+2)}\,dr - \int_{0}^{\infty} \frac{r^p e^{i2p\pi}}{(r+1)(r+2)}\,dr = \int_{0}^{\infty} \frac{r^p(1-e^{i2p\pi})}{(r+1)(r+2)}\,dr$. Next, $\int_{C_2} = \int_{\varepsilon}^{2\pi-\varepsilon} \frac{(Re^{i\theta})^p}{(Re^{i\theta}+1)(Re^{i\theta}+2)}iRe^{i\theta}\,d\theta$. Since as $R \to \infty$ the numerator of the integrand blows-up like R^{p+1} for any θ in the integration interval, while the denominator blows-up like R^2, then the integrand behaves like $\frac{R^{p+1}}{R^2} = R^{p-1} = \frac{1}{R^{1-p}}$. The C_2 integral thus behaves like $\frac{2\pi}{R^{1-p}}$. For $p < 1$, $\int_{C_2} \to 0$ as $R \to \infty$. Also, $\int_{C_4} = \int_{2\pi-\varepsilon}^{\varepsilon} \frac{(\rho e^{i\theta})^p}{(\rho e^{i\theta}+1)(\rho e^{i\theta}+2)}i\rho e^{i\theta}\,d\theta$. Now, as $\rho \to 0$ the numerator behaves like ρ^{p+1} while the denominator behaves like 2. So, the integral behaves like $\pi\rho^{p+1}$ which clearly $\to 0$ as $\rho \to 0$ as long as $p > -1$. So, we have $\int_{C_2} + \int_{C_4} = 0$ which means $\int_{C_1} + \int_{C_3} = \int_{0}^{\infty} \frac{r^p(1-e^{i2p\pi})}{(r+1)(r+2)}\,dr = 2\pi i e^{ip\pi}(1 - 2^p)$ or, $\int_{0}^{\infty} \frac{x^p}{(x+1)(x+2)}\,dx = 2\pi i \frac{e^{ip\pi}(1-2^p)}{1-e^{i2p\pi}} = 2\pi i \frac{e^{ip\pi}(1-2^p)}{e^{ip\pi}(e^{-ip\pi}-e^{ip\pi})} = 2\pi i \frac{(1-2^p)}{-2i\sin{(p\pi)}} = \pi\frac{(2^p-1)}{\sin{(p\pi)}}, -1 < p < 1$.

(C8.7) Following the hint, we'll use the contour in Fig. 8.6.1 to compute $\oint_C \frac{e^{iz}}{z}\,dz = \int_{C_1} + \int_{C_2} + \int_{C_3} + \int_{C_4} = 0$ because C keeps the lone singularity of the integrand (at the origin) in its exterior. So, $\int_{\varepsilon}^{T} \frac{e^{ix}}{x}\,dx + \int_{0}^{\frac{\pi}{2}} \frac{e^{iTe^{i\theta}}}{Te^{i\theta}}iTe^{i\theta}\,d\theta + \int_{T}^{\varepsilon} \frac{e^{i(iy)}}{iy}i\,dy + \int_{\frac{\pi}{2}}^{0} \frac{e^{i\varepsilon e^{i\theta}}}{\varepsilon e^{i\theta}}i\varepsilon e^{i\theta}\,d\theta = 0$. That is, $\int_{\varepsilon}^{T} \frac{e^{ix}}{x}\,dx + i\int_{0}^{\frac{\pi}{2}}e^{iTe^{i\theta}}\,d\theta - \int_{\varepsilon}^{T} \frac{e^{-y}}{y}\,dy - i\int_{0}^{\frac{\pi}{2}}e^{i\varepsilon e^{i\theta}}\,d\theta = 0$. In the last integral, as we let $\varepsilon \to 0$, $e^{i\varepsilon e^{i\theta}} \to e$ and so the integral is $\frac{\pi}{2}e$. In the second integral, $e^{iTe^{i\theta}} = e^{iT[\cos{(\theta)}+i\sin{(\theta)}]} = e^{iT\cos{(\theta)} - T\sin{(\theta)}} = e^{\left\{\frac{e^{iT\cos{(\theta)}}}{e^{T\sin{(\theta)}}}\right\}}$ and so as $T \to \infty$ the integrand goes to $e^0 = 1$ and so the integral is $\frac{\pi}{2}$. Thus, as $\varepsilon \to 0$ and

$T \to \infty$, $\int_0^\infty \frac{e^{ix}}{x} dx + i\frac{\pi}{2} - \int_0^\infty \frac{e^{e^{-y}}}{y} dy - i\frac{\pi}{2}e = 0$. So, $\int_0^\infty \frac{e^{\cos(x)+i\,\sin(x)}}{x} dx -$

$\int_0^\infty \frac{e^{e^{-y}}}{y} dy = i\frac{\pi}{2}e - i\frac{\pi}{2} = i\frac{\pi}{2}(e-1)$ or, $\int_0^\infty \frac{e^{\cos(x)}e^{i\,\sin(x)}}{x} dx - \int_0^\infty \frac{e^{e^{-y}}}{y} dy = i\frac{\pi}{2}(e-1)$

or, $\int_0^\infty \frac{e^{\cos(x)}[\cos\{\sin(x)\}+i\,\sin\{\sin(x)\}]}{x} dx - \int_0^\infty \frac{e^{e^{-y}}}{y} dy = i\frac{\pi}{2}(e-1)$ or, equating imagi-

nary parts, we have Cauchy's result: $\int_0^\infty \frac{e^{\cos(x)}\sin\{\sin(x)\}}{x} dx = \frac{\pi}{2}(e-1)$.

(C8.8) Following the hint, write $\int_{-\infty}^\infty \frac{x^2}{(x^2+a^2)(x^2+b^2)} dx = \int_{-\infty}^\infty \frac{A}{x^2+a^2} dx + \int_{-\infty}^\infty \frac{B}{x^2+b^2} dx$

and so $\int_{-\infty}^\infty \frac{x^2}{(x^2+a^2)(x^2+b^2)} dx = \int_{-\infty}^\infty \frac{Ax^2+Ab^2+Bx^2+Ba^2}{(x^2+a^2)(x^2+b^2)} dx$. This means that $A + B = 1$

and $Ab^2 + Ba^2 = 0$. These two equations are easily solved to give $A = \frac{a^2}{a^2-b^2}$, $B = -\frac{b^2}{a^2-b^2}$. So,

$$\int_{-\infty}^\infty \frac{x^2}{(x^2+a^2)(x^2+b^2)} dx = \frac{a^2}{a^2-b^2}\int_{-\infty}^\infty \frac{1}{x^2+a^2} dx - \frac{b^2}{a^2-b^2}\int_{-\infty}^\infty \frac{1}{x^2+b^2} dx$$

$$= \frac{a^2}{a^2-b^2}\left\{\frac{1}{a}\tan^{-1}\left(\frac{x}{a}\right)\right\}\Big|_{-\infty}^\infty - \frac{b^2}{a^2-b^2}\left\{\frac{1}{b}\tan^{-1}\left(\frac{x}{b}\right)\right\}\Big|_{-\infty}^\infty$$

$$= \frac{a}{a^2-b^2}\pi - \frac{b}{a^2-b^2}\pi = \frac{\pi}{a+b}.$$

Now, let $b \to a$. Then, $\int_{-\infty}^\infty \frac{x^2}{(x^2+a^2)^2} dx = \frac{\pi}{2a}$. Finally, differentiating with respect to

a, $\int_{-\infty}^\infty \frac{-2x^2(x^2+a^2)2a}{(x^2+a^2)^4} dx = -\frac{2\pi}{4a^2} = -\frac{\pi}{2a^2}$. Thus, $\int_{-\infty}^\infty \frac{x^2}{(x^2+a^2)^3} dx = \frac{\pi}{8a^3}$. For $a = 1$ this is

$0.392699\ldots$, and $integral(@(x)(x.\wedge 2)./((x.\wedge 2+1).\wedge 3),-inf,inf) = 0.392699\ldots$.

(C8.9) We start with $\psi(1) = \lim_{\varepsilon \to 0}\left\{\int_\varepsilon^\infty \frac{e^{-z}}{z} dz - \int_\varepsilon^\infty \frac{dz}{(1+z)z}\right\}$, as suggested in the

hint. In the first integral, use integration by-parts. That is, $\int_\varepsilon^\infty u\, dv = (uv)\big|_\varepsilon^\infty -$

$\int_\varepsilon^\infty v\, du$, with $u = e^{-z}$ ($du = -e^{-z}\, dz$) and $dv = \frac{dz}{z}$ ($v = \ln(z)$). Thus, $\int_\varepsilon^\infty \frac{e^{-z}}{z} dz =$

$(e^{-z}\ln(z))\big|_\varepsilon^\infty + \int_\varepsilon^\infty e^{-z}\ln(z)dz$ and so $\int_\varepsilon^\infty \frac{e^{-z}}{z} dz = -\ln(\varepsilon) + \int_\varepsilon^\infty e^{-z}\ln(z)dz$. In the

second integral of our opening line, write $\int_\varepsilon^\infty \frac{dz}{(1+z)z} = \int_\varepsilon^\infty \left(\frac{1}{z} - \frac{1}{1+z}dz\right) =$

$(\ln(z) - \ln(1+z))\big|_\varepsilon^\infty = \ln\left(\frac{z}{1+z}\right)\big|_\varepsilon^\infty = -\ln(\varepsilon)$, where I've written $-\ln\left(\frac{\varepsilon}{1+\varepsilon}\right) =$

$-\ln(\varepsilon)$ as $\varepsilon \to 0$. Thus, $\psi(1) = \lim_{\varepsilon \to 0}\left\{-\ln(\varepsilon) + \int_\varepsilon^\infty e^{-z}\ln(z)dz + \ln(\varepsilon)\right\} =$

$\int_0^\infty e^{-z}\ln(z)dz = -\gamma$ by (5.4.3).

(C8.10) Let $e^y = x^n$ or, $e^{y/n} = x$ and so $x^m = e^{y\frac{m}{n}}$. Now, $y = \ln(x^n) = n\ln(x)$ and so

$\frac{dy}{dx} = \frac{n}{x}$ or, $dx = \frac{x}{n} dy = \frac{e^{y/n}}{n} dy$. Thus, $\int_0^\infty \frac{x^m}{1-x^n} dx = \int_{-\infty}^\infty \frac{e^{y\frac{m}{n}}}{1-e^y}\frac{e^{y/n}}{n} dy = \frac{1}{n}\int_{-\infty}^\infty \frac{e^{y\frac{(m+1)}{n}}}{1-e^y} dy$.

So, $\frac{1}{n}(m+1)$ plays the role of a in (8.6.9), as $\frac{1}{n}(m+1) < 1$, and we have

$\int_0^\infty \frac{x^m}{1-x^n} dx = \frac{1}{n}\frac{\pi}{\tan\left(\frac{m+1}{n}\pi\right)}$.

(C8.11) From (8.7.8) we have $\int_0^\infty \frac{x^m}{1+x^n} dx = \frac{\pi/n}{\sin\left\{\frac{(m+1)}{n}\pi\right\}}$. With $x = ya^{\frac{1}{n}}$ $(dx = a^{\frac{1}{n}} dy)$

we then have $\int_0^\infty \frac{y^m a^{\frac{m}{n}}}{1+y^n a} a^{\frac{1}{n}} dy = \frac{\pi/n}{\sin\left\{\frac{(m+1)}{n}\pi\right\}} = a^{\frac{m+1}{n}} \int_0^\infty \frac{y^m}{1+y^n a} dy$ or, $\int_0^\infty \frac{y^m}{1+y^n a} dy =$

$\frac{\pi}{na^{\frac{m+1}{n}}\sin\left\{\frac{(m+1)}{n}\pi\right\}} = \int_0^\infty \frac{y^m}{a\left(\frac{1}{a}+y^n\right)} dy$. Let $b = \frac{1}{a}$. Then this last integral is $\int_0^\infty \frac{y^m}{\frac{1}{b}(b+y^n)} dy =$

$b\int_0^\infty \frac{y^m}{b+y^n} dy$ or,

$$\int_0^\infty \frac{y^m}{b+y^n} dy = \frac{\pi}{nba^{\frac{m+1}{n}}\sin\left\{\frac{(m+1)}{n}\pi\right\}} = \frac{\pi}{nb\left(\frac{1}{b}\right)^{\frac{m+1}{n}}\sin\left\{\frac{(m+1)}{n}\pi\right\}}$$

$$= \frac{\pi}{nbb^{\frac{-m-1}{n}}\sin\left\{\frac{(m+1)}{n}\pi\right\}} = \frac{\pi}{nb^{1+\frac{-m-1}{n}}\sin\left\{\frac{(m+1)}{n}\pi\right\}}$$

or, changing our dummy variable from y back to x, $\int_0^\infty \frac{x^m}{x^n+b} dx = \frac{\pi}{nb^{\frac{n-m-1}{n}}\sin\left\{\frac{(m+1)}{n}\pi\right\}}$.
Notice that this generalizes (8.7.8) to values of b other than 1.

(C8.12) From our result in C8.11, with $m = a$ and $n = 1$, we have $\int_0^\infty \frac{x^a}{x+b} dx =$

$\frac{\pi}{b^{-a}\sin\{(a+1)\pi\}} = \frac{\pi b^a}{\sin\{(a+1)\pi\}}$. Thus, differentiating with respect to b, $\int_0^\infty \frac{-x^a}{(x+b)^2} dx =$

$\frac{\pi \frac{d}{db}b^a}{\sin\{(a+1)\pi\}} = \frac{\pi ab^{a-1}}{\sin\{(a+1)\pi\}} = -\frac{\pi ab^{a-1}}{\sin\{a\pi\}}$ as $\sin(a\pi + \pi) = -\sin(a\pi)$. Or, $\int_0^\infty \frac{x^a}{(x+b)^2} dx =$

$b^{a-1}\frac{\pi a}{\sin\{a\pi\}}$.

(C8.13) Following the hint, we have from the previous solution $\int_0^\infty \frac{x^a}{(x+b)^2} dx =$

$b^{a-1}\frac{\pi a}{\sin\{a\pi\}}$ and so, differentiating with respect to a, we start on the left by writing $x^a = e^{\ln(x^a)} = e^{a\ln(x)}$ and so $\frac{d}{da}x^a = \ln(x)e^{a\ln(x)} = x^a\ln(x)$. Next, we move to the right to write

$$\frac{d}{da}\left\{b^{a-1}\frac{\pi a}{\sin\{a\pi\}}\right\} = \frac{d}{da}\left\{e^{(a-1)\ln(b)}\frac{\pi a}{\sin\{a\pi\}}\right\} = \frac{d}{da}\left\{e^{a\ln(b)}e^{-\ln(b)}\frac{\pi a}{\sin\{a\pi\}}\right\}$$

$$= e^{-\ln(b)}\frac{d}{da}\left\{e^{a\ln(b)}\frac{\pi a}{\sin\{a\pi\}}\right\}$$

$$= e^{-\ln(b)}\left\{e^{a\ln(b)}\left[\frac{\pi\sin\{a\pi\} - \pi^2 a\cos(a\pi)}{\sin^2(a\pi)}\right] + \frac{\pi a}{\sin\{a\pi\}}\ln(b)e^{a\ln(b)}\right\}.$$

Setting $a = \frac{1}{2}$ in our two differentiations, on the left we get $\sqrt{x}\ln(x)$ and on the right we have $e^{\ln(b^{-1})}\{e^{1/2\ln(b)}\pi + \frac{1}{2}\pi\ln(b)e^{1/2\ln(b)}\} = \frac{1}{b} \times \{\pi\sqrt{b} + \frac{1}{2}\pi\ln(b)\sqrt{b}\} = \frac{\pi\sqrt{b}[1+\frac{1}{2}\ln(b)]}{b}$. Thus, $\int_0^\infty \frac{\sqrt{x}\ln(x)}{(x+b)^2} dx = \frac{\pi[1+\frac{1}{2}\ln(b)]}{\sqrt{b}}$. As a numerical check, suppose $b = 3$. Then $(1+0.5*log(3))*pi/sqrt(3) = 2.81013...$ while $integral(@(x)(sqrt(x).*log(x))./((x+3).^2),0,inf) = 2.81013....$

Index

A
Abel, N.H., 260
Aerodynamic integral, 11, 23, 133
Ahmed, Z., 230–233
Alford, J., xxix
Algebraic trick, 199
Analytic function, 361–364, 368, 382, 383,
    385, 386, 399, 401, 403, 407, 408, 419
Analytic signal, 330, 361
a priori plausible, 223
Auto-correlation detector, 47
Autocorrelation function, 45–47, 58, 349
Average value, 33

B
Bartle, R., 10
Bell, E.T., 445
Bernoulli, J., xxxiv, 227–229
Bertrand, J., 426
Bessel, F.W., 368
Binary chop algorithm, 270
Binomial coefficients, 307
Binomial theorem, 260, 304, 357, 404
Bohr, H., 430
Boras, G., xxxiii, xxxiv, xxxvii
Brachistochrone, 273
Branch
    cut, 408, 409, 432
    point, 408, 409
Brown, E., 273
Brownian motion, 31, 32
Brown, R., 31

C
Cadwell, J., 419
Cantor, G., 8
Catalan, Eugène, xxxiii
Catalan's constant, xxxiii, 181–183, 185,
    218–222, 225
Catalan's function, 221
Cauchy, A.-L., xxxii, 368, 420
Cauchy Principal Value, 20, 144, 145
Cauchy-Riemann equations, 361–364
Cauchy-Schlömilch transformation, 136, 148
Cauchy-Schwarz inequality, 44–46
Cauchy's integral theorems
    first, 368–372, 374–381, 385, 387, 390
    second, 381–399
Causality, 326–334
Cayley, A., 256, 270, 271
Chain-rule, 173
Chebyshev, P.L., 426
Clifford, W.K., 445, 446
Conjugate (complex), 314, 325, 346
Conrey, J.B., 430
Contour integration, 262
Convolution, 325, 328, 329
Copson, E., 420, 421
Cornu's spiral, 277, 280
Cotes, R., 446
Coxeter, H.S.M., xxxvi, 234, 269
Critical line, 429, 448

Brun, V., 426
Buchanan, H., xxxv

© Springer Nature Switzerland AG 2020

P. J. Nahin, *Inside Interesting Integrals*, Undergraduate Lecture Notes in Physics,
https://doi.org/10.1007/978-3-030-43788-6

Critical strip, 427, 429
Cross-cut, 384, 385, 393, 401
Current injection, 87, 93

D

Dalzell, D.P., 25
Davis, B., 234
dblquad (MATLAB command), xxxvi
de Moivre, A., 36
Diagonal resistances, 89
Diffusion constant, 33
Dini, U., 140
Dirac, P., 318, 319, 322
Dirichlet, L., 7, 111
Dirichlet's eta function, 443
Double-angle formulas (trigonometry), 66, 95,
    158, 234, 249, 291, 350, 356
Double-root' solution, 97
Duality theorem, 323, 331
Dummy variable, (of integration), 3, 15, 17, 64,
    66, 68, 69, 73, 77, 97, 203, 300, 344,
    416, 417

E

Edwards, H.M., 429, 431
Edwards, J., xxxvii, 28, 98
Einstein, A., 31, 32, 34, 176
Equation
    difference, 75, 76
    functional, 149, 164, 178
    integral, 143
Erdös, P., 426
Error function, 340, 343, 344
Euclid, 425, 426
Euler, L., xxxiii, 9, 15, 77, 79, 111, 149, 150,
    160, 186, 193, 196, 223, 269, 272, 277,
    280, 288, 299, 424, 428, 442
Euler's constant, 204, 206, 346
Euler's identity, 276, 277, 280, 286–289, 298,
    300, 303, 312, 313, 315, 317, 345, 348,
    350, 356, 370, 375, 380, 384, 388, 405
Exponential integral, 336

F

Fagnano, G., 15, 16
Fermat, P. de, 41, 42
Fessenden, R., 326
Feynman, R., 19, 43, 44, 52, 53, 55, 101, 117,
    131, 194, 216, 230, 347, 421
Flanders, H., 85

floor (MATLAB command), 15, 35, 223, 347
Flux capacitor, 273
Forward shift operator, 258
Fourier, J., 313
Fowler, R., 246
Frequency convolution, 341
Fresnel, A.J., 347
Fresnel Integrals, 277
Frullani, G., 113
Function
    beta, 151, 152, 154–158, 160, 161, 175
    complex, 355–361
    digamma, 210, 217, 224
    Dirichlet, 8
    even, 14, 19, 34, 73, 74, 103, 161, 211, 313,
        328
    factorial, 150
    gamma, 149, 150, 177, 178, 202, 216, 224,
        246, 309
    gate, 331
    impulse, 318–323
    odd, 13, 20, 33, 74, 106, 313, 328, 384
    probability density, 32, 103
    Riemann's, 1–8
    signum, 317
    unit step, 318, 320, 322

G

Gabor, D., 330
Galilei, G., 246
Gamma
    function, 149, 150, 158, 177, 178, 180, 202,
        228, 246, 262, 300, 415, 418
    number, 205, 262
Gamma (MATLAB command), 150, 151, 158,
    159, 166, 180
Gauss, C.F., 176, 368, 425, 431, 445
Gibbs phenomenon, 112
Glaisher, J.W.L., 260, 340
Goldscheider, F., 294–296
Green, G., 364
Green's integral theorem, 365–367

H

Hadamard, J., 425
Hamming, R., 10
Hardy, G.H., xxx, xxxvi, xxxvii, 127, 241, 246,
    260, 269, 329
Hardy, O., 434
Heterodyne theorem, 326
Hilbert, D., 423

Hilbert transform, 326–334
Hurwitz, A., 226

I
Indent, 372, 374, 376
Indiana Jones, 446
Ingham, A.E., 426
int (MATLAB command), 28, 132
Integral
    Ahmed's, 230–233, 240
    area interpretation, 154, 155, 204
    convolution, 325
    Coxeter's, xxxvi, 234–240, 269
    Dalzell's, 23–25
    derivative of, 101
    differentiation trick, 203
    Dini's, 140–142
    Dirichlet's, 111, 112, 116, 144, 145, 163,
       180, 302, 307, 317, 339
    elliptic, 246, 252, 254–257
    equations, 143
    Fresnel's, xxxiii, 20, 241, 277–280, 345,
       348
    Frullani's, 112–115, 212
    Hardy–Schuster optical, 132, 241, 242, 244,
       245, 345
    improper, 18, 55, 371
    Lebesgue's, 7–10
    lifting theory, 11
    line, 352–355, 367, 372, 384, 387
    logarithmic, 424
    log-sine, 79–84, 284–288
    probability, 103, 134, 136–139
    Ramanujan's, 211, 212
    Riemann, 1–5, 18, 19, 346
    Serret's, 69
Integrating factor, 143, 271
Integration by-parts, 61, 105, 107, 109, 111,
    126, 128, 144, 149, 162, 166, 182, 184,
    188, 191, 209, 228, 240, 283, 292, 308,
    310
Irresistible Integrals, xxxiii–xxxv, 25, 35,
    37–39
Isoperimetric theorem, 273
Ivić, A., 429

J
Jordan curve theorem, 365

K
Keiffer, D., 271
Kepler, J., 361
Khinchin, A., 349
Kirchhoff's law, 88
Kramers, H.A., 246
Kronig, R., 329
Krusty the Clown, 238

L
Lagrange, J.L., 272
Landau, E., 430
Laplace, P.-S., 110, 137
Laplace transform, 334–345, 348, 485–487
Laurel, S., 434
Laurent, P.A., 400
Laurent series, 400, 403
Lebesgue, H., 7
Legendre, A.-M., 149, 246
Legendre duplication formula, 161, 435
Leibniz, G., 54, 270
Levinson, N., 430
L' Hospital's rule, 53, 222, 390, 397, 413, 439
Littlewood, J.E., 431
Lodge, O., 41
Lord Rayleigh, 314

M
Mascheroni, L., 204
Mathematica, 23
MATLAB, xxxi, xxxii, xxxiv, xxxvi, 21, 31,
    35, 48, 57, 85, 147, 349, 350, 447
    See also dblquad (MATLAB command);
       floor (MATLAB command); int
       (MATLAB command); quad
       (MATLAB command); rand (MATLAB
       command); syms (MATLAB
       command); triplequad (MATLAB
       command)
Mellin, R.H., 260
Mellin transform, 260
Mercator, G., 56
Mirsky, L., 419
Moll, V., xxxiii–xxxv, xxxvii
Monte Carlo simulation, 28, 30, 48
Mordell, L.J., 419
Morris, J.C., xxxv
Multiply connected, 365

See also Simply connected
Murphy, R., 274, 480

N
Newton, I., xxx, 173, 270, 358, 359, 446

O
Ohm's law, 312
Operators, 258, 259
Optical integral, xxxvi, 132, 241, 242, 244, 245,
   277, 345

P
Parseval's theorem, 314
Partial fraction expansion, 21, 67, 70–72, 98,
   107, 144, 231, 238, 275, 311, 382, 386,
   389, 394, 421
Pascal, B., 260
Perera, M.W., 273
Poisson summation, 433
Polar coordinates, 27, 103, 152, 272, 355
Pole, *see* Singularity
Prime
   numbers, 424–426, 447
   number theorem, 425
Putnam, 52

Q
quad (MATLAB command), xxxi, xxxvi

R
Radius vector, 355
Ramanujan, S., 211, 437
Ramanujan's master theorem (RMT), 258–269
rand (MATLAB command), 30
Rayleigh's energy theorem, 313, 314, 333, 347,
   348
Recursion, 76, 77, 98, 126, 127
Reflection formula (gamma function), 160, 169,
   435, 438
Rejection method, 48
Residue, 402–404, 406, 410–412
Residue theorem, 399–407, 410
Riemann, B., 1, 8, 150, 160, 196, 423–433,
   445–447

Riemann hypothesis, 423, 424, 431, 446, 447
Riemann-Lebesgue lemma, 279
Riemann surface, 408
Rotation of axes, 90

S
Sampling property, 323, 326
Schuster, A., 241
Schwartz, L., 318, 322
Seitz, E., 31
Selberg, A., 430
Series
   geometric, 196, 198, 215, 281, 438
   harmonic, 193, 198, 426
   "merabili", xxxiv
   power, 181–226, 287, 292, 356, 377
      (*see also* Taylor series)
Series expansions
   ex, 228, 357, 376
   ln(2), xxxiii, 217
   log(1+x), 185, 187, 205, 217, 223, 225, 280,
      288
   tan-1 (x), 185
Serret, J., 69
Simple curve, 385
Simply connected, 365
Singularity, 18–23, 55, 330, 332, 364, 369,
   374–376, 378, 379, 386, 387
   first order, 382, 388, 393, 394, 402, 410
   high order, 382, 404
Skewes, S., 431
Sokhotsky, V., 329
Sommerfeld, A., 270
Sophomore's dream, 229, 270
"Special ingenuity", 224
Spectrum
   baseband, 325
   energy, 312, 314, 325, 326, 384
Spherical coordinates, 249
Spiegel, M., xxxii, xxxiii, xxxvi
Stalker, J., 352
Stirling, J., 36
Stirling's asymptotic formula, 36
Strutt, J.W., *see* Lord Rayleigh
Stäckel, P., 294, 298
Symbolic Math Toolbox (MATLAB), 132, 133,
   180, 195, 242, 245, 346
syms (MATLAB command), 132, 133, 195,
   346

T
Taylor series, 399, 439
Thomas, G.B., 11, 23
Time/frequency scaling theorem, 324, 332
Tisserand, F., 422
Transform
    Fourier, xxxvi, 241, 313, 315, 319–321,
        324, 326, 341, 347
    Hilbert, 326–334, 347, 348
Trier, P.E., 85–94
triplequad (MATLAB command), xxxvi

U
Uhler, H.S., 131, 132
Uniform convergence, xxx, 188

V
Vallée-Poussin, Charles-Joseph de la, 425
van Peype, W.F., 246

Volterra, V., 143

W
Wallis, J., 151, 152, 154
Watson, G.N., 11, 245–251
Weiner-Khinchin theorem, 349
Wiener, N., 31
Wiener random walk, 32, 33
Wolstenholme, J., 284
Woods, F., 131
Woolhouse, W., 31

Z
Zeta function, 150, 160, 193–195, 197, 198,
        200–204, 226, 280, 424, 442
    functional equation, 427, 428, 432–434,
        436–442
    zero, 429–431, 448

Printed in the United States
By Bookmasters